1 MONTH OF
FREE
READING

at

www.ForgottenBooks.com

By purchasing this book you are eligible for one month membership to ForgottenBooks.com, giving you unlimited access to our entire collection of over 1,000,000 titles via our web site and mobile apps.

To claim your free month visit:

www.forgottenbooks.com/free897183

ISBN 978-0-266-83926-2
PIBN 10897183

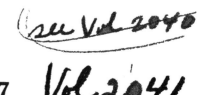

No. 8107

IN THE

United States Circuit Court of Appeals

For the Ninth Circuit

MIKE ERCEG, Guardian of the Estate of
George Gartner (an insane person),

Appellant,

vs.

FAIRBANKS EXPLORATION COMPANY
(a corporation),

Appellee.

BRIEF FOR APPELLEE, FAIRBANKS EXPLORATION COMPANY,
ON THE CROSS-APPEAL.

PILLSBURY, MADISON & SUTRO,
Standard Oil Building, San Francisco,
Attorneys for Appellee,
Fairbanks Exploration Company,
on the Cross-Appeal.

ALFRED SUTRO,
FRANCIS N. MARSHALL,
FRANCIS R. KIRKHAM,
  Standard Oil Building, San Francisco,
  Of Counsel.

Table of Contents

Table of Authorities Cited

Cases

Statutes and Codes

No. 8107

United States Circuit Court of Appeals
For the Ninth Circuit

MIKE ERCEG, Guardian of the Estate of
George Gartner (an insane person),

Appellant,

vs.

FAIRBANKS EXPLORATION COMPANY
(a corporation),

Appellee.

BRIEF FOR APPELLEE, FAIRBANKS EXPLORATION COMPANY, ON THE CROSS-APPEAL.

STATEMENT OF THE CASE.

The facts of this case are stated in defendant's brief on the main appeal, filed October 20, 1937. Those particularly pertinent to the cross-appeal are as follows:

The trial court found that the injury to the Gartner claims was permanent (Conclusion No. 3, R. 39)[1]. It held that the measure of plaintiff's damages was the cost of removing the silt, or the value of the property at the time of the injury—whichever was less (R. 50-52).

The Gartner claims had no market value (R., main appeal, 240), and no value other than as mining property

---

1. References to the record in this brief, unless otherwise indicated, are to the record on the cross-appeal.

(R. 46). In order, therefore, to show the value of the property at the time of the injury, plaintiff introduced evidence (a) as to the value of the gold in the ground, and (b) as to the cost of mining the gold with the property in its original condition prior to the commencement of defendant's mining operations. Evidence addressed to the same ultimate facts was presented by defendant. Upon this evidence the court (a) found the value of the gold— $320,750, (b) found the cost of mining—$290,750, and (c) subtracted the latter from the former to arrive at the value of the property—$30,000 (Finding No. 23, R. 37). In its opinion the court concluded that the cost of removing the silt would exceed this figure (R. 51), and, therefore, took the value of the property as the measure of damages (Findings Nos. 23 and 24, R. 38, 39; Opinion, R. 52).[2]

Among the costs of mining allowed by the court was an item of $8,442.05 based upon the court's "taking into consideration the five-year period that it would take to mine out said claims."[3] Plaintiff took no exception to the action of the court in allowing this item of cost (R. 55-61), nor was it assigned as error (R. 65-67).

At the time of the injury to plaintiff's property, and at the date of the commencement of this action, September

---

2. As pointed out at pp. 35-38, infra, on the court's own basis of computation, the cost of removing the silt is actually much *less* than $30,000.

3. The court found the value of the gold in the ground to be $320,750. It then found that the 1,045,585 cubic yards of gravel containing this gold could be mined for 27 cents per cubic yard—a total of $282,307.95. It then found that, "taking into consideration the five-year period that it would take to mine out said claims," the value of the ground was $30,000 (Finding No. 23, R. 37, 38). Since the difference between $320,750 and $282,307.95 is $38,442.05, the court obviously allowed $8,442.05 as the item in question.

Plaintiff's figure of $8,686.63 for the same item (Opening Brief for Appellant Erceg, p. 20) is based upon its revision of the court's finding that the value of the gold in the Gartner claims was $320,750 (see Opening Brief for Appellant Erceg, p. 6).

8, 1932 (R. 53; R., main appeal, 16), the price of gold was $20.67 per ounce. On January 31, 1934, prior to the trial, the price of gold was increased by Presidential proclamation to $35 per ounce. At the trial, plaintiff requested the court to compute damages on a basis of the $35 price of gold. The court concluded that the value of the land as of the time of the injury was the proper measure of damages, and based its valuation of the property, for the purpose of ascertaining the amount of damages plaintiff was entitled to recover, on the $20.67 price of gold (Conclusion No. 4, R. 39; Opinion, R. 53).

At the close of all the testimony plaintiff requested the court to assess punitive damages against defendant (R. 60), and also requested the court to render a judgment in favor of plaintiff for the gross value of the gold in the property, without any deduction for costs of mining, for the reason that defendant's acts were "wilful and malicious" (R. 61). The court found that defendant, in carrying on the stripping operations by which the channel of Goldstream was filled and the Gartner claims covered with silt, acted wilfully (Finding No. 21, R. 36); but it refused to assess punitive damages against defendant (R. 60), and refused to render judgment for the gross value of the gold content of the claims (R. 61).

Throughout his brief plaintiff repeatedly refers to defendant's "depredations" (pp. 6, 14); to its "atrocious trespass" (p. 25); to its "malicious" (p. 7), "ruthless" (p. 15), "flagitious" (p. 18) acts; to its "wilful and wanton trespass" (p. 25); and to the taking of plaintiff's "land and gold" by defendant—"not innocently, not inadvertently, but * * * knowingly, wantonly, greedily" (p. 14).

We respectfully submit that these charges are without support in the record or in the finding of the court. At the close of all the testimony in the case, the court expressly refused to assess punitive damages against defendant (R. 60). In its opinion[4] it emphatically stated that the evidence showed that defendant had carried on its operations "carefully"; that "the defendant has for the most part conducted its mining operations on Goldstream Creek and its tributaries in a minerlike manner. That there is no serious or actionable negligence in its actual mining operations" (R. 44). And the court held defendant liable—not because of any "malicious", "ruthless" or "flagitious" conduct—but because, under applicable principles of law, defendant's "whole duty to downstream owners on Goldstream [did not end] with merely carrying on its operations carefully" (R. 44), and defendant was liable to owners whose property had been injured, regardless of the care used by it in carrying on its mining operations (R. 44-46).

It is thus apparent that the court's finding that defendant acted wilfully and in disregard and defiance of plaintiff's rights has reference to the fact that defendant, in carrying on its mining operations—"in a minerlike manner"—intentionally discharged into Goldstream the stripping waters which thereafter deposited silt on plaintiff's

---

4. As plaintiff himself urges (Opening Brief for Appellant Erceg, pp. 40-43) this court may consider the opinion of the court below to inform itself of the reasons relied on to support the conclusions of that court, and to aid in understanding its decision. See *Hall v. United States* (5th C.C.A., 1920), 267 Fed. 795, 797 (quoted at p. 23, note 10, Brief for Appellant on the main appeal); *Bankers Trust Co. v. Henwood* (8th C.C.A., 1937), 88 F. (2d) 163. In the latter case the court said (p. 166):

"Where, as here, the orders and findings leave it uncertain as to the basis of the decision of the court and such basis is material in a determination of an appeal, the opinion of the court may be examined to aid in understanding the decision."

property. Nothing in the findings of the court, or in its opinion, or in the record, justifies plaintiff's repeated assertions of wanton and malicious bad faith on the part of the defendant.[5]

At the commencement of the trial, and at the close of plaintiff's case in chief, and again at the close of all the testimony, plaintiff asked leave to amend his amended complaint so as to increase the amount of damages alleged and prayed. Each of these motions was denied (R. 56-58).

STATEMENT OF THE ISSUES.

There are two main issues involved on this cross-appeal: First, whether the trial court erred in its application of the rule of damages; and, second, whether the trial court erred in denying plaintiff's motions to amend the amended complaint.

With respect to the first issue, plaintiff complains that the trial court erred in its application of the rule of damages in three particulars: (1) in refusing to award damages equal to the value of the gold in the claims without deduction of costs of mining, (2) in making an allowance, in computing mining costs, for the five-year period it would require to mine the claims, and (3) in awarding damages based upon the value of the land at the time of the injury, when the price of gold was $20.67 per ounce, instead of

---

5. In this connection we also point out that the references in plaintiff's brief to defendant's "trespass" are incorrect. The amended complaint alleges an indirect and consequential injury; the findings of the trial court establish such an injury. Defendant's alleged wrong was to strip the overburden from its upstream claims into Goldstream Creek, whence said overburden was carried down by the force of the water and, as a result of this intervening agency, was deposited in the channel and on the surface of the Gartner claims (see Brief for Appellant on the main appeal, pp. 60-65).

upon the value of the land at the time of the trial, when the price had been increased to $35 per ounce.

SUMMARY OF ARGUMENT.

With respect to the foregoing issues, defendant contends:

First, the measure of damages applicable to the instant case is the one applied by the court below, namely, the cost of removing the silt, or the value of the property at the time of the injury—whichever is less.

Second, each one of the plaintiff's three criticisms of the trial court's application of the rule of damages is unsound.

Third, the trial court did not abuse its discretion in denying plaintiff's motions to amend the amended complaint.

Fourth, in any event, at plaintiff's own figures, which were accepted by the court in computing the cost of mining, the cost of removing the silt from the claims is $15,590.51, which is the maximum plaintiff is entitled to recover as damages.

ARGUMENT.

FIRST: THE MEASURE OF DAMAGES APPLICABLE TO THE INSTANT CASE IS THE ONE APPLIED BY THE COURT BELOW, NAMELY, THE COST OF REMOVING THE SILT, OR THE VALUE OF THE PROPERTY AT THE TIME OF THE INJURY—WHICHEVER IS LESS.

The court below found that the silting of the Gartner claims occurred between the summer of 1925 and September 8, 1932, the date of the commencement of this action (Findings Nos. 7-13, R. 28-31). The injury for which plaintiff sued was complete and permanent as of the date of the commencement of the action (Finding No. 13, R. 31; Conclusion No. (3), R. 39). The court held, therefore, that plaintiff was entitled to recover all damages, "past, present and future," proximately caused by the injury to the claims (R. 53), and that the measure of such damages was the cost of removing the silt, or the value of the property at the time of the injury—whichever was less (R. 52). Having concluded that the cost of removing the silt would exceed the value of the property at the time of the injury (R. 51), the court awarded plaintiff the full value of the property.

The rule of damages applied by the court is correct and is in accord with well established principles of law.

Since all damages proximately caused by a permanent injury to land are recoverable in one action, the primary measure of damages is the diminution in the value of the land by reason of the injury. Ordinarily, as the court below said (R. 51), this diminution is conveniently and accurately measured by the cost of restoration or repair. If this were the sole test, however, injustice would result where the cost of restoration exceeds the value of the property at the time of the injury, for, in such a case, an award in the former amount obviously would "mulct [defendant] in a far greater sum than his adversary actually lost" (*Pedelty v. Wisconsin Zinc Co.* (1912), 148 Wis. 245, 134 N.W. 356, 358; *Mustang Reservoir Canal & Land Co. v. Hissman* (1911), 49 Colo. 308, 112 Pac. 800, 801). It is, therefore, well settled that the measure of damages for permanent injury to land is the cost of restoration, or the diminution in the value of the land by reason of the injury—whichever is less. The rule and the principles underlying its application are well stated in *Swanson v. Nelson* (1906), 127 Ill. App. 144, 149:

"In our opinion the primary and logical rule is that the cost of restoration or repair is to be considered the measure of damages in cases of injuries to real property. This makes the damages conform to the general theory of the law that they are to be indemnity or compensation. But it frequently happens that an injury or trespass to real property may be in a certain sense irreparable. The condition of things before the trespass was committed or the injury done cannot be restored at all, or can be restored only at a very great and disproportionate expense. In either of these cases another measure of damages is properly adopted by the courts, namely, the difference in market value of the property before and after the act complained of. The result of this reasonable view which has been taken by almost all courts of last resort where they have had occasion to pass on the matter has been to establish a rule that makes the measure of damages, in cases of injury to real estate, the cost of restoration or the difference in market value, as one or the other is the less amount. [Cases]"

In *Seely v. Alden* (1869), 61 Pa. 302, an action to recover for permanent injury to plaintiff's land resulting from the deposit of debris flowing from defendant's upstream tannery, the Supreme Court of Pennsylvania said (pp. 305-306):

> "We think the court erred also in refusing to admit both methods of computing the permanent damages, to wit, that which measures the damages by the different values of the land, with and without the deposit; and that which measures them by the cost of removing the deposit. It is often difficult for a court to determine the true measure until all the evidence is in. It may turn out that the cost of removing the deposit in a certain case would be less than the difference in the value of the land, and then the cost of removal would be the proper measure of the damages; or it may be that the cost of removal would be much greater than the injury by the deposit, when the true measure would be the difference in value merely."

Like principles are stated by the Supreme Court of Wisconsin in *Pedelty v. Wisconsin Zinc Co.* (1912), 148 Wis. 245, 134 N.W. 356, supra, a suit to recover for damage to plaintiff's land caused by the deposition of waste material from defendant's upstream mineral concentrating mill (134 N.W. 358):

> "The actual loss sustained is the just measure of reparation in any case. That may well be arrived at in such a case as this by proving the former value and the cost of accomplishing the restoration, or by proving the value of the land in its former and in its changed condition. Either method would reach the same result in case of the cost of restoration being less than the value of the restored property, while the

cost of restoration, which appellant contends is invariably the proper measure of damages, might lead to a result very unjust to the wrongdoer—it might mulct him in a far greater sum than his adversary actually lost.''

Robinson v. Moark-Nemo Consol. Mining Co. (Mo. App., 1914), 163 S.W. 885 (affirmed by the Supreme Court of Missouri, 196 S.W. 1131), like the case at bar, was a case where the defendant had dumped refuse and tailings from its mine upon plaintiff's property. Speaking of the measure of damages, the court said (163 S.W. 886):

''The usual measure of damages in a case of this kind is the difference in the value of the land with and without the tailings thereon; but, where the cost of removal is less than such difference, such cost of removal is the proper measure of damages.''

Similar decisions are:

City of Globe v. Rabogliatti (1922), 24 Ariz. 392, 210 Pac. 685, 687, 688;

Salstrom v. Orleans Bar Gold Min. Co. (1908), 153 Cal. 551, 96 Pac. 292;

Green v. General Petroleum Corp. (1928), 205 Cal. 328, 336, 270 Pac. 952;

Perkins v. Blauth (1912), 163 Cal. 782, 792, 127 Pac. 50, 54-55;

Mustang Reservoir Canal & Land Co. v. Hissman (1911), 49 Colo. 308, 112 Pac. 800, 801, supra;

Fine v. Beck (1922), 140 Md. 317, 117 Atl. 754, 756;

Harvey v. Sides Silver Mining Co. (1865), 1 Nev. 539;

Hartshorn v. Chaddock (1892), 135 N.Y. 116, 31 N.E. 997, 998;

Ohio Collieries Co. v. Cocke (1923), 107 Ohio St.
238, 140 N.E. 356, 359;

Weaver v. Berwind-White Coal Co. (1907), 216 Pa.
195, 65 Atl. 545, 547;

Rabe v. Shoenberger Coal Co. (1906), 213 Pa. 252,
62 Atl. 854.

Lindley, in his work on Mines (3d ed.), vol. 3, sec. 844,
states the rule as applied to injuries to land by the deposits of mining debris, as follows:

> "*Measure of damages for unlawfully depositing
> debris on another's land.*—The general rules govern-
> ing the question of determining the measure of dam-
> ages where mining debris is deposited on the mining
> or agricultural land of another seem to be that when
> the reasonable cost of repairing the injury, or of
> restoring the land to its former condition, is less than
> what the diminution of the market value of the whole
> property by reason of the injury is shown to be, such
> cost of restoration is the proper measure of damages.
> On the other hand, when the cost of restoring it is
> more than such diminution, the latter is, generally
> speaking, the true measure of damages."

See also:

Sedgwick on Damages (9th ed.), vol. 3, sec. 932, pp.
1916-1920;

Sutherland on Damages (4th ed.), vol. 4, sec. 1017,
pp. 3760, 3761; sec. 1018, pp. 3768, 3771.

It is equally well settled that damages for permanent
injury to land are to be determined as of the time of
injury, and not as of some later date.[6] Thus, the decisions

---

·6. Plaintiff's contention that the court below should have based the award
of damages upon the value of the Gartner claims as of the date of the trial,
when the price of gold had been increased, instead of as of the date of the
injury, is particularly treated at pp. 21-33, infra.

appear to be unanimous that the diminution in the value of
the land by reason of the injury is to be ascertained by the
difference in the value (ordinarily, market value) immediately before and immediately after the injury—a rule
recently approved by this court in *Inland Power & Light
Co. v. Grieger* (9th C.C.A., 1937), 91 F.(2d) 811, 817.
Literally hundreds of cases have stated and applied this
rule.[7] To avoid unnecessary multiplication of authority,
we cite representative cases from a number of jurisdictions
involving damages due to the deposit of debris from mining operations, and similar injuries:

> *Fuller v. Fair* (1919), 202 Ala. 430, 80 So. 814, 816;
>
> *City of Globe v. Rabogliatti* (1922), 24 Ariz. 392,
> 210 Pac. 685, 687, 688, supra;
>
> *St. Louis, I.M. & S. Ry. Co. v. Miller* (1913), 107
> Ark. 276, 154 S.W. 956, 957;
>
> *Sterling Hydraulic Co. v. Galt* (1898), 81 Ill. App.
> 600, 603;
>
> *Jackson Hill Coal & Coke Co. v. Bales* (1915), 183
> Ind. 276, 108 N.E. 962, 964;
>
> *Kopecky v. Benish* (1908), 138 Iowa 362, 116 N.W.
> 118, 119;
>
> *Nisbet v. Lofton* (1925), 211 Ky. 487, 277 S.W. 828,
> 831, 832;
>
> *Butcher v. St. Louis-San Francisco Ry. Co.* (1931),
> 225 Mo. App. 749, 39 S.W. (2d) 1066;
>
> *Pulaski Oil Co. v. Edwards* (1923), 92 Okl. 56, 217
> Pac. 876;

---

7. Many of these cases are collected under the following titles and key
numbers of the American Digest System: Damages, 107, 108, 110-112; Mines
and Mining, 124, 125; Trespass, 50; Waters and Water Courses, 178(1),
178(2).

Avery v. Wallace (1924), 98 Okl. 155, 224 Pac. 515,
516;

Hoffman v. Berwind-White Coal Mining Co. (1920),
265 Pa. 476, 109 Atl. 234, 236;

Morton Salt Co. v. Lybrand (1927, Tex. Civ. App.),
292 S.W. 264;

Norfolk County Water Co. v. Etheridge (1917), 120
Va. 379, 91 S.E. 133, 134.

The instant case involves an injury resulting from the
deposit of mining debris upon land claimed to be valuable
exclusively for placer mining purposes. In such cases, the
rule of damages, of course, is the same, but the *computation* of damages is simpler. In placer mining it is essential to remove the upper soil by ground sluicing or stripping in order to reach the gold scattered in the underlying
gravel. Where silt and debris have been deposited on
placer mining ground, therefore, the value of the property
is diminished—up to a maximum of the full value—in the
exact amount of the cost of removing the surface deposits.
As a result, the applicable rule of damages is held to be:

"As to the land available exclusively for mining purposes, if the cost of repairing the injury by removing
the debris deposited by defendant would amount to
less than the value of the property as it was prior to
the injury, such cost would be the proper measure of
damages. But if such cost of repair or of restoration
would exceed such value, then the value of the property would be the proper measure."

Salstrom v. Orleans Bar Gold Min. Co., 153 Cal. 551,
558, 96 Pac. 292, supra;

Green v. General Petroleum Corp., 205 Cal. 328, 336
270 Pac. 952, supra;

Lindley on Mines (3d ed.), vol. 3, sec. 844, p. 2091
(where the rule just quoted is approved as "the
true rule of damages in such a case");

Sutherland on Damages (4th ed.), vol. 4, sec. 1017.

In the light of the foregoing, we submit that the rule of
damages applied by the court below was unquestionably
correct.

SECOND: EACH ONE OF THE PLAINTIFF'S THREE CRITI-
CISMS OF THE TRIAL COURT'S APPLICATION OF THE
RULE OF DAMAGES IS UNSOUND.

(a) Plaintiff's contention that the trial court should have
awarded damages in an amount equal to the gross value of
the gold, without deduction of costs of mining—as if this
were a case of wilful severance and conversion of minerals—
is without merit. This is an action for permanent injury to
real property, and compensatory damages cannot exceed the
value of the property.

Plaintiff's first complaint of the trial court's ruling with
respect to damages is that it erred in refusing to give
judgment for the gross value of the gold in the Gartner
claims (Opening Brief for Appellant Erceg, pp. 9-20).

A brief resumé of the facts found by the court below
will, we think, assist in weighing the merits of this contention.

Plaintiff's ward owned three claims in Alaska, valuable
only for mining. Except for the effect upon them of
several unprofitable mining ventures (see Brief for Appellant, main appeal, pp. 3-4), the claims, prior to defendant's
operations, lay in their natural condition. In these claims,
buried beneath 24 feet of muck, and scattered throughout
a layer of frozen gravel 25 feet deep, were particles of

gold—30.7 cents worth to a cubic yard. In the aggregate this gold was worth $320,750. It would take five years and cost $290,750 to extract it from the ground. The land, therefore—and so the court found—was worth $30,000.[8] With the land in this condition silt was deposited upon its surface as a result of defendant's mining operations. The court held that defendant was liable for this injury and should compensate plaintiff by paying him $30,000—the full value of the property.

But plaintiff wants more than the full value of the property; he asks for the value of the *gold* in the ground—$350,750—as though that gold, instead of being scattered many feet beneath the surface of the Gartner claims, were mined, refined, and stacked in bars ready for delivery to the mint. Such value of the gold is, concededly, more than ten times greater than the value of the land with the gold in place.

The plaintiff's argument in support of this contention is this: (a) Where, for a public purpose and by virtue of sovereign authority, land is buried by water or other material so as to destroy its usefulness, this constitutes a "taking" of property within the meaning of the constitutional limitation against the taking of private property for public use without just compensation. (b) Defendant, in this case, has flooded plaintiff's *property* and covered it with silt. (c) Therefore, defendant has "taken" plaintiff's *gold*. Arguing further from this conclusion, plaintiff says that since defendant's "taking" was wilful, the trial court should have applied the measure of damages appli-

---

8. Our contention that this finding of the court is without substantial support, and that the land was not shown to have any value as mining property, is presented on the main appeal.

cable to a trespasser who wilfully and in bad faith takes
mineral from another's land and converts it to his own
use, namely, the value of the mineral "when and where it
is finally converted to the use of the trespasser, without
any deduction for expense incurred or for any value he
may have added to the mineral by his labor" (Opening
Brief for Appellant Erceg, pp. 16, 17).

Aside from the fact that the true logical conclusion of
plaintiff's syllogism is that defendant has taken plaintiff's
property, not the *gold* in the property (and should therefore pay the value of the *property*—as the judgment herein
requires it to do—not the value of the *gold* in the property), it is obvious that the argument rests upon a mere
play on the word "take".[9] As appears from the authorities
cited in plaintiff's brief (Opening Brief for Appellant
Erceg, pp. 16-17), the rule of damages invoked by plaintiff
applies to the case where a defendant, by wilful and wanton
trespass, removes mineral from another's land and converts it to his own use, and the owner brings an action in
the nature of trover to recover the value of the removed
mineral. By its own terms this rule is impossible of application to the instant case. Under it, the measure of damages is the value of *severed* mineral, when and where it is
converted to defendant's use. Here defendant has removed
no gold; has converted no gold to its own use.

The absurdity of plaintiff's contention is emphasized
when it is seen that if it were adopted defendant would

---

9. It may further be noted that plaintiff's analogy is directly contrary to
his objectives on this appeal, since it leads only to the rule, settled by the
decisions of the Supreme Court of the United States, that the compensation
to be paid to an owner for property taken for a public use
"* * * is the market value of the property at the time of the taking
contemporaneously paid in money. * * * He must be made whole but is
not entitled to more" (*Olson v. United States* (1934), 292 U.S. 246, 255.
54 S. Ct. 704).

be compelled to pay more than it would be required to pay had it wilfully, maliciously and feloniously run its dredges upon plaintiff's property and extracted and converted to its own use every ounce of gold in the claims. In such case plaintiff could recover $320,750—the value of the gold —*but* defendant could satisfy the judgment *by the return of plaintiff's gold,* and would be out only its costs of mining. In the instant case, defendant, although it has taken no gold with which to satisfy the judgment, would be compelled, under plaintiff's theory, to pay the full $320,750, and would thus pay a penalty in excess of the amount it would have been required to pay had it stolen plaintiff's gold and appropriated it to its own use. Furthermore, plaintiff would be handed a pure gift of $290,750 over and above the value of his entire property at the time of the injury, although he still owns the land with all the gold in it.

It is obvious, we submit, that plaintiff seeks punitive, not compensatory, damages. In this connection we have pointed out that plaintiff's extreme assertions of malice on the part of the defendant are not justified (supra, pp. 3-5). Moreover, the trial court refused to assess punitive damages against defendant (R. 60). This ruling is conclusive, not only because plaintiff's specification of errors fails to present for review the trial court's ruling in this respect (Opening Brief for Appellant Erceg, p. 8), but also because it is settled by the great weight of authority that the question whether punitive damages should be assessed is a matter within the discretion of the trial court. If trial is by jury, the matter lies within the discretion of the jury (*Knight v. Beyers* (1913), 70 Or. 413, 134 Pac.

787; 17 C. J. 971-972); if trial is by the court without a jury, the matter is within the discretion of the court, and no appeal lies from the refusal of the court to award such damages.

See,

> *Carter v. Illinois Cent. R. Co.* (Ky., 1896), 34 S.W. 907;
>
> *Hagerman Irrigation Co. v. McMurry* (1911), 16 N.M. 172, 113 Pac. 823, 826;
>
> *Tilton v. James L. Gates Land Co.* (1909), 140 Wis. 197, 121 N.W. 331, 336.

Finally, plaintiff's contention is entirely inconsistent with the allegations of his amended complaint and the case made at the trial. The amended complaint states a cause of action for injury to the Gartner claims resulting from the deposit of silt and debris (R. 8-11). No allegation therein, and no proof adduced at the trial, would support a judgment for the value of gold "removed" and "converted" by defendant. Indeed, it was merely for want of ordinary proof as to the value of plaintiff's property—such as evidence of market value—that it was necessary to take testimony concerning the amount of gold in the ground and the costs of mining.

As stated in the preceding subdivision of this brief, the settled law is at variance with plaintiff's contention. Numerous decisions, including many cases practically identical with the one at bar, establish that the measure of damages for a permanent injury to real property, such as is here involved, is the cost of restoration, or the value of the property at the time of the injury—whichever is less. Plaintiff has cited no decision to the contrary.

We have examined and made a brief statement in the footnote of each of the cases cited by plaintiff upon this branch of his argument (Opening Brief for Appellant Erceg, pp. 9-20). We submit it clearly appears from the statement of each case, that no one of them is in point.[10]

10. *Dettering v. Nordstrom* (9th C.C.A., 1906), 148 Fed. 81, was a suit for an accounting against a cotenant who had extracted gold from a jointly owned claim. It was held that the cotenant was entitled to a credit for the expense incurred in taking out the gold from the claim, but that the trial court did not err in entering a judgment for a proportion of the gross output in the absence of any proof by defendant of the reasonable expense of extracting the gold.

Koonz v. Hempy (1909), 142 Iowa 337, 120 N.W. 976, was an action to recover for cutting and *removing* trees. The value of the removed trees was held to be the measure of damages. Said the court (p. 977): "The appellant is certainly in no position to complain of this. *He had the timber*, and, if he pays its value, he cannot complain" (italics ours throughout the brief).

Doak v. Mammoth Copper Mining Co. (C.C.,N.D.Cal., 1911), 192 Fed. 748, held that where growing trees are injured by smelter fumes, the measure of damages is the difference in the value of the land before and after the injury.

Galveston, H. & S. A. Ry. Co. v. Warnecke (1906), 43 Tex. Civ. App. 83, 95 S.W. 600, which involved a loss of fruit trees by fire, turned upon a technical point of pleading. It was held that since the pleader had sued to recover the value of the trees, the measure of damages was the value of the trees detached from the soil; if the suit had been for injury to the land, the measure of damages would have been the difference in the value of the land before and after the injury.

Armstrong v. May (1916), 55 Okl. 539, 155 Pac. 238, held that the measure of damages for destruction of growing crops and of certain other crops stacked upon the property is the value of such crops.

Pumpelly v. Green Bay Company (1871), 13 Wall. 166, *United States v. Wabasha-Nelson Bridge Co.* (7th C.C.A., 1936), 83 F.(2d) 852, *United States v. Lynah*, 188 U.S. 445, 23 S.Ct. 349, *United States v. Cress*, 243 U.S. 316, 37 S.Ct. 380, *United States v. Land in Monterey Co.*, 47 Cal. 515, and *Village of St. Johnsville v. Smith*, 184 N.Y. 341, 77 N.E. 617, all involved the taking of property for a public purpose under the power of eminent domain.

Wooden-ware Co. v. United States, 106 U.S. 432, 1 S.Ct. 398, holds that in an action of trover to recover the value of timber from one who has purchased without notice from a wilful trespasser, the measure of damages is the value at the time of such purchase.

Benson Mining Co. v. Alta Mining Co., 145 U.S. 428, 12 S.Ct. 877, and *Pan-American Petroleum Co. v. United States* (9th C.C.A., 1926), 9 F.(2d) 761, hold that where a wilful trespasser extracts and removes minerals from another's property and converts them to his own use, the measure of damages is the value of the minerals without deduction of mining costs.

The quotation from 7 A.L.R. p. 922, supplies its own distinction:
"The American jurisdictions which have had occasion to pass on the question of the measure of damages against a trespasser who *removes* oil or other mineral from the land of another are virtually unanimous in holding that if the taking is reckless, wilful, or intentional, or without claim of right or title, the trespasser is liable for the enhanced value of the product *when and where it is finally converted to the use of the trespasser*, without any deduction for expenses incurred, or for any value he may have added to the mineral by his labor."

(b) Plaintiff's contention that the trial court erred in including within the costs of mining the item of $8,442.05 on account of the five-year period required to mine the claims is unsound. Moreover, the ruling on this question is not before this court.

Without citation of authority, and without argument other than the bare statement "that the deduction was wholly unjustifiable" (Opening Brief of Appellant Erceg, p. 21), plaintiff contends that the trial court erred in including within the costs of mining the item of $8,442.05 (see p. 2, supra), which the court allowed because of "the five-year period that it would take to mine out said claims and extract said gold" (Finding No. 23, R. 37-38).

The item, we submit, was properly allowed.

In computing mining costs, the trial court accepted plaintiff's evidence concerning the cost of equipment and of mining (R., main appeal, 238-239). These costs do not include interest on the $45,170 plaintiff would be required to invest in mining equipment (R., main appeal, 238), which at six per cent over a period of five years, would amount to $13,551. Plaintiff's costs of mining also failed to include any allowances for expenses during a long term operation, such as for a watchman or other labor expense during the long winter shutdown seasons; for conditioning the mining equipment at the beginning of the successive mining seasons; for unpredictable expense which might be caused in the course of a five-year operation by accident, inclement weather, unexpected breakdowns, etc., etc. It is obvious that the court had in mind these and similar costs, and that its allowance of $8,442.05 is an extremely conservative estimate of the difference in cost between a short- and long-term mining operation of the magnitude of the one here involved.

Plaintiff took no exception to the action of the court in allowing this item of cost (R. 55-61) and assigned no error with respect thereto (R. 65-67). The brief for plaintiff states that the "foundation" for urging the point is "assignments Nos. 1 to 6, Tr. p. 65" (Opening Brief for Appellant Erceg, p. 20), but no one of these assignments covers the ruling in question.

(c) **Plaintiff's contention that the trial court, in computing the value of the Gartner claims for the purpose of determining plaintiff's damage, should have valued the land as of the time of the trial, instead of as of the time of the injury, is unsound. The measure of damages for permanent injury to mining land, where the cost of restoration exceeds the value of the land, is the value of the land at the date of the injury, not its value at some later date as affected by subsequent fortuitous events.**

As above shown, the rule of damages for covering land with mining debris is the diminution of the value of the land, or the cost of removing the debris, whichever is less. The reason is that the payment of damages in such amount makes the injured party whole and puts him in exactly as good a position as that in which he was before the injury occurred. In the case of land valuable only for placer mining, where the increase of mining costs—and the consequent diminution of value—is in direct ratio to the cost of removing debris, the rule of damages is taken to be the value of the land if the cost of removing the debris equals or exceeds it. This was the rule applied by the trial court.

The plaintiff contends, however, that the trial court, in determining the value of the land, should have valued the gold in the ground, not at $20.67 per ounce, its value when the property was injured, but at $35 per ounce, its value, by reason of an intervening governmental proclamation, at the time of the trial.

This contention, we submit, is contrary to fundamental principles. An example, used by the House of Lords, is illustrative: If on January 1 the defendant tortiously destroyed three cows of the plaintiff, and if the plaintiff on that date could have gone into the market and bought three similar cows for $150, obviously the plaintiff's damages would be $150, and it would be immaterial on the question of damages that in July similar cows would have cost in the market $300.[11]

This rule is fundamental in the law of damages, and accordingly, the courts have held, in actions for damages for injury to land, that the only material inquiry is as to the value of the land immediately before and immediately after the injury, and that it is reversible error to base an award of damages upon the value at some later time.

McDougald v. Southern Pac. R. Co. (1912) 162 Cal. 1, 120 Pac. 766, for example, was a suit for damages resulting from defendant's construction of a railway and drawbridge which encroached on the plaintiff's land. The

---

11. This, with dollars substituted for pounds sterling, is the exact illustration used by Lord Wrenbury in *The Celia v. The Volturno* [1921] 2 A.C. 544; 20 A.L.R. 884, 895. Lord Wrenbury said:

"The defendant is liable to pay the plaintiff damages—that is to say, money to some amount for the loss of the cows; the only question is, how much? The answer is, such sum as represents the market value *at the date of the tort* of the goods of which the plaintiff was tortiously deprived."

Accordingly, the House of Lords held, in the case cited—an action arising out of a collision between an English ship and an Italian ship, in which both ships were to blame, and cross claims for damages were agreed, subject to a question as to the rate of exchange in respect of the damages calculated in Italian lire for detention of the Italian ship during repairs—that the claim should be converted into English currency at the rate of exchange which obtained at the date of the detention and not that which obtained at the date when the damage was assessed. This leading decision is directly in accord with the trial court's ruling in the instant case. See also:

Moser v. Corn (1931) 249 N.Y.S. 606.

trial took place more than six years after the construction of the railway and drawbridge. The court found that at the time of the trial the plaintiff's land would have been worth $3,000 but for the railway and drawbridge, and that by reason thereof the value at that time was only $1,800. Accordingly, the court gave judgment for $1,200. The Supreme Court reversed the judgment, saying (120 Pac. 766-767):

> "There is no finding as to the value of the land at any other period than the time of the trial, and no evidence was introduced relating to any other period. The evidence all referred to the *values at the time of the trial,* and it was all introduced over the objection of defendant, that the value at that time was irrelevant and immaterial.
>
> These findings are insufficient to support the judgment. * * * The injury to the plaintiff is of a permanent character. * * * For such injuries, the plaintiff may recover in one action, not only the damages occasioned up to the time the action was begun, but also all that she can then show with reasonable certainty that she will suffer in the future. But all this is necessarily given to her if she receives as damages the difference between the value of the entire parcel as it was just before the defendant took permanent possession, and its value immediately after the works of the defendant were completed and put in operation, taking into consideration all the injurious consequences to the part of the land not taken reasonably probable from such works and the operation thereof and from the severance of the land taken. * * * *There is some evidence indicating that the value at the time of the trial had been enhanced by independent causes arising long after the taking by the defendant.* This cause had been previously tried, and from the

judgment the defendant appealed. The appeal was
heard in the District Court of Appeal for the Third
District, 9 Cal. App. 236, 98 Pac. 685. In that case
it was held that the value of the property at the
first trial, four years after the injury, was not sufficient to show its value at the time of the injury. Much
more is this true at a period nearly seven years after
the injury. The evidence should have been confined
to the damage, present and prospective, *based upon
the values existing immediately before and immediately after the construction of the railway and
drawbridge.* Because of this error, the judgment
must be reversed.''

Sinclair Oil & Gas Co. v. Allen (1930) 143 Okl. 290, 288
Pac. 981 was an action for damages alleged to have been
caused by the pollution of a stream running through
plaintiff's farm, by casting in the stream salt water,
crude oil, and base sediment from oil wells. The plaintiff
called a number of witnesses to testify as to the diminution of the market value of the land caused by the pollution of the stream. Of this evidence the court said (288
Pac. 983):

"It will thus be seen that plaintiff made no attempt
whatever to show the reasonable market value of the
premises immediately before the pollution. *All her
evidence was as to the value at the time of the trial:*
First, with the stream unpolluted and the trees not
killed; and, second, with it polluted and the trees
killed as shown by the evidence.

*This was clearly not the proper measure of dam-
ages, as it wholly failed to establish the value of the
land before the injury, and such evidence alone, we
think, has universally been held insufficient to support
a verdict for substantial damages.*''

In *St. Louis, B. & M. Ry. Co. v. West* (Tex.Civ.App.,
1910) 131 S.W. 839, the defendant's railroad construction
had made the plaintiffs' land subject to periodic flooding,
so that the land became "lifeless and unproductive." On
appeal from a judgment for the plaintiffs the defendant
attacked the instruction on the measure of damages for
permanent injury to the land. The court said (p. 842):

> "The correct rule is the difference in the market
> value of the land immediately before the obstruction,
> and its value when the permanent damage was finally
> accomplished due to the particular cause. The peti-
> tion distinctly alleged that the permanent injury was
> completed by the fall of 1908. Accordingly *the dif-*
> *ference in values should have been referred to that*
> *time, and not to the time of trial,* which occurred
> some years later. Trinity & Sabine Ry. Co. v. Scho-
> field, 72 Tex. 496, 10 S.W. 575; Ry. v. Terhune, 94
> S.W. 381. * * * The court in its charge told the
> jury to take the value before and the value after,
> 'under the same conditions, *without reference to the*
> *fact, if any, that the value of said land has been*
> *increased or diminished by any other cause.'* We
> think this was not an improper addition to make in
> connection with the rule."

See also:

> *Schuylkill Navigation Co. v. Thoburn* (1821) 7
> Serg & R. (Pa.) 411, 420-423;
>
> *Twin State Oil Co. v. Long* (1935) 170 Okl. 413, 40
> P.(2d) 650;
>
> *Ward v. Chicago, M. & St. P. Ry. Co.* (1895) 61
> Minn. 449, 63 N.W. 1104, 1105;
>
> *Chicago, B. & Q. R. Co. v. Gelvin* (8th C.C.A., 1916)
> 238 Fed. 14, 20, 22.

In cases of permanent injury to real property, therefore, it is a settled rule that the damages are determined by the value of the land at the date of the injury, not by the value at some later date,

> *Trinity & S. Ry. Co. v. Schofield* (1889) 72 Tex. 496, 10 S. W. 575, 576-577;
>
> *Schuylkill Navigation Company v. Farr* (1842) 4 Watts & S. (Pa.) 362, 375;
>
> *Hidalgo County Water Control, etc. v. Peter* (Tex. Commn. App., 1931) 37 S. W. (2d) 133, 134;
>
> *Texas Cent. R. Co. v. Brown* (1905) 38 Tex. Civ. App. 610, 86 S. W. 659, 660;
>
> *Houston & T. C. R. Co. v. Wright* (Tex. Civ. App., 1917) 195 S. W. 605, 606,[12]

and evidence of value at other times cannot be considered.

> *Zuidema v. Sanitary Dist. of Chicago* (1921) 223 Ill. App. 138, 146-147;
>
> *Missouri K. & T. Ry. Co. of Texas v. Chilton* (1908) 52 Tex. Civ. App. 516, 118 S. W. 779, 780-781;
>
> *Cleveland C. C. & St. L. Ry. Co. v. Vettel* (Ind. App., 1922) 133 N. E. 605, 608, 609;
>
> *Sweeney v. Montana Cent. Ry. Co.* (1897) 19 Mont. 163, 47 Pac. 791, 792.

These decisions and the principles which underlie them show that the damages recoverable in an action for permanent injury to land are not affected by subsequent fortuitous changes of value—value "enhanced by independent causes arising long after" the injury *(McDougald v. Southern Pac. R. Co.,* 120 Pac. 766, 767, supra).

---

12. See, also, cases cited at pp. 12-13, supra.

A case strikingly parallel to the case at bar, in that both
involve a change of value caused by governmental authority after the commission of the injury, is *Stone v. Codman*
(1834) 15 Pick. (32 Mass.) 297. The case was an action for
damage caused by negligence of the defendant's workmen
in digging a drain, which let tide water into the plaintiff's
cellar and caused his merchandise stored therein to become
wet and damaged. Part of the merchandise was a quantity of wool in bags, which had been imported, and on
which duties had been paid. It became necessary to take
the wet wool out of the bags in order to prevent further
injury. A month later, Congress passed an act amending
the duties on wool so as to provide a drawback or rebate
on such wool as was still in the original packages. The
result obviously would have been to increase the value of
the damaged wool if it had not been damaged and hence
had been allowed to remain in the original packages. The
court held that the loss of the rebate was not a loss for
which the plaintiff could recover. The court said (pp.
300-302) :

> "In the case of *Whitwell v. Kennedy*, 4 Pick. 466,
> this [i. e., the rule of compensatory, as distinguished
> from punitive damages], in similar cases, was settled
> to be the value of the property at the time when the
> damages were sustained, where the property is wholly
> lost or converted; and the same applies, in estimating
> the value of property to which any partial damage is
> done, in ascertaining the real damage, which the
> owner has sustained. The property is to be estimated
> in the manner as it would be, if a jury could have
> been called on the spot, at the moment the damage
> was discovered, to estimate it. They would consider
> the market value of the property, on the day, and

the degree, if any, to which that market value was diminished, by the wetting of the wool. If it had sustained no diminution of its market value, then the expense of removing, drying, repacking, and replacing in store, would be the amount of the damage. *The drawback afterwards allowed by the government, upon certain species of imported wool, to which but for the accident in question, this wool would have been entitled, was a casual and accidental advantage, accruing afterwards, and originating in a measure, which took place after the facts fixing the plaintiff's claim to damages had all occurred.* The plaintiff can no more avail himself of this after act of Congress, even if it enhanced the value of the wool, than if the defendant had been under contract to deliver wool of the same quality to the plaintiff on the 21st of June, and had failed to do so. *The measure of damages could not be enhanced by the act which passed on the 14th of July,* although it is quite as probable as in the present case, that the wool, if delivered, would have been kept in the original packages to the 3d of March, deposited in the customhouse for the drawback.

    *   *   *   *   *   *   *

All that can be said is, that if the loss had happened at a later period, and after the right of drawback had attached, the loss might have been greater; but this may be answered by saying, that in the mean time the plaintiff might have sold his goods, at the market price, and so not have realized the benefit arising from the drawback. I cannot distinguish it from a case, where after damage done to goods, and before the time of trial, events happen, by which the value of such goods is greatly enhanced or diminished.

Taking into consideration these subsequent events, and the strong probability that the goods would not have been sold, it may often appear that the value at the time of the damage, may give more or less than a true indemnity; still considering the importance of a fixed and certain rule, the strong probability that in the great majority of cases, *the value at the time* will afford the nearest approximation to an exact indemnity, the rule has been adopted, and we think ought to be adhered to.''

The same principle is illustrated by the case of *The Celia v. The Volturno* [1921] 2 A.C. 544, 20 A.L.R. 884, discussed supra in the footnote, p. 22. In that case a maritime collision had damaged an Italian ship; the owners had suffered a deduction from her hire, covering the period of detention for repairs, of 304,418 lire; between the time of the detention and the time of the trial the value of the lira had changed due to fluctuations of the rate of exchange. The House of Lords held that the damages on this account were measured by the value of the lira according to the rate of exchange prevailing at the time of the detention.

The reasons underlying the principle—that damages for tortious injury to property are determined by the value of the property at the time of the injury—are set forth in the opinions of the Lords in *The Celia* case. One reason appears from Lord Wrenbury's illustration of the cows (supra, p. 22): if the defendant has destroyed three cows of the plaintiff worth £150, the damages are £150 regardless of the fact that the market value of the cows at the time of trial would have been £300, because the plaintiff lost by the defendant's tort only £150 worth, and could have compensated himself immediately by replacing his cows in the market for £150. Another reason is that an alteration in the rate of exchange is an inde-pendent factor, unconnected with the defendant's tort. As Lord Parmoor said (20 A.L.R. p. 893):

> "The probability of the alteration in the rate of ex-change is not an admissible factor in the ascertain-ment of the amount of damage, both on the ground of remoteness, and on the ground that it is a matter which affects generally all financial transactions, and is in no way connected with the tortious act for which the respondents are liable."

And so, in the case at bar, if the Gartner claims at the time of the injury were worth $30,000 with gold at $20.67 per ounce,[13] the damages are $30,000 regardless of later governmental action raising the price of gold, because the plaintiff lost only $30,000 worth, and immediately after the injury he could presumably have purchased for that sum land of equal value for mining purposes. And in the second place, the change in the price of gold was an independent occurrence, wholly unconnected with the de-fendant's tort.

The fallacy of plaintiff's contention strikingly appears, we submit, if it be assumed that after the injury to his land, the price of gold, instead of going up, had gone down. In such case the plaintiff obviously would not concede that he would be compensated by the payment of a sum equal to the value of the land based on the lower price of gold at the time of the trial, instead of the value of the land based on the higher price of gold at the time of the injury. A case in point is *Houston Belt & Terminal Ry.*

---

13. Our contention to the contrary is the subject of the main appeal.

Co. v. Daidone (Tex. Civ. App., 1933) 62 S.W.(2d) 524. There the plaintiff sued for damage to his property caused by the nearby construction of a roadbed for a railroad. Shortly after construction of the roadbed, the buildings on the plaintiff's property were destroyed by fire. The court held (62 S.W.(2d) 528) that the damages were to be ascertained with reference to the plaintiff's property as it was when the injury occurred, and not as it was at the time of the trial. The court said (p. 528):

> "The rule of damage as it respects real estate is the difference between the value of the premises immediately before the injury and such value immediately after the injury" [citing cases].

Similarly, in *Zuidema v. Sanitary Dist. of Chicago* (1921) 223 Ill. App. 138' the court held that the rule of damages, in an action based upon the destruction of plaintiff's growing crop by flooding, was not affected by a subsequent excessive rainfall and flood which in any event would have destroyed the crop. The court said (pp. 146-147):

> "The rule of law in such case is that the measure of damages to growing crops which are not matured is the value of the crop as it was when destroyed with the right of the owner to mature and harvest it at the proper time. * * * The sanitary district offered an instruction, which was refused, to the effect that if the damage to the crops was occasioned by obstructing the drainage facilities, still, if afterwards it appeared from the evidence that the farm was submerged from excessive rainfall which of itself would have destroyed the crops, plaintiff could not recover. It is apparent that this instruction was not in ac-

cordance with the rule which we have stated, for if
the evidence showed what the crop in its immature
condition was worth on May 15, prior to the time it
was damaged, that value would be the measure of
damages for it would be the amount the crop would
then bring, and the fact that afterwards in July or
August there was a flood which submerged the land
would in no way change the amount of damages. The
evidence covering this point should tend to show what
the crop would fairly bring in its immature condition
just prior to the time it was damaged.''

Plaintiff's only answer to the foregoing well-settled
principle is a discussion of the rule in eminent domain
cases, with citation of cases where compensation was determined as of the time of trial (Opening Brief for Appellant Erceg, pp. 29-42). These cases obviously are not
in point. An eminent domain proceeding is a sale compelled by law. It results in a transfer of title. The
court, therefore, may fix the price as of the time of the
sale—i. e., when the award is made and the title is adjudged to be transferred. This is so whether the proceeding is the usual one instituted by the corporation
making the appropriation, as in some of the cases cited
by the plaintiff; or whether it is a statutory proceeding
instituted by the landowner to compel appropriation, as in
others; or whether it is a suit for damages converted
into an eminent domain proceeding by the corporation's
plea, as in still others.

The instant case is not an eminent domain proceeding.
There is no question of transfer of title or possession. The
judgment did not award the defendant title to or any inter-

est in the Gartner claims. The plaintiff still owns his land, with whatever increase in its value the rise in the price of gold has brought about. The plaintiff's contention that the amount of this increase in the value of the land should also be included in the award of damages is based on the idea that *defendant,* somehow, has benefited from the increase and would be unjustly enriched if permitted to retain this benefit (Opening Brief for Appellant Erceg, p. 25). This, clearly, is contrary to fact; the defendant does not have the land or the gold, and the judgment does not award either to it. The plaintiff's contention, we submit, obviously is without merit.[14]

---

14. Although the eminent domain cases are not in point in this action for injury to land, it may be noted that the rule established by the decisions of this court, and of the Supreme Court of the United States, is that just compensation for the taking of property under the power of eminent domain is the value of the property at the time of the taking—not the value at the date of trial, or judgment, or assessment of damages, if that date is different. If payment is deferred, interest must be added to give the owner the present full equivalent of the value of the property at the time of the taking paid contemporaneously with the taking, but compensation is still based upon the value at that time. These rules apply whether a proceeding to condemn is brought prior to the taking,

    Brett v. United States (9th C.C.A., 1936) 86 F. (2d) 305 (certiorari denied, 301 U.S. 682, 57 S.Ct. 782),

or whether such a proceeding is brought after the actual taking,

    United States v. Rogers (1921) 255 U.S. 163, 41 S.Ct. 281;
    Olson v. United States (1934) 292 U.S. 246, 54 S.Ct. 704,

or whether, following a taking without payment of full or any compensation, the owner brings an action to recover just compensation,

    United States v. Creek Nation (1935) 295 U.S. 103, 55 S.Ct. 681;
    Shoshone Tribe v. U. S. (1937) 299 U.S. 476, 57 S.Ct. 244;
    Jacobs v. United States (1933) 290 U.S. 13, 54 S.Ct. 26;
    Brooks-Scanlon Corp. v. U. S. (1924) 265 U.S. 106, 44 S.Ct. 471;
    Phelps v. United States (1927) 274 U.S. 341, 47 S.Ct. 611;
    Liggett & Myers v. U. S. (1927) 274 U.S. 215, 47 S.Ct. 581;
    Seaboard Air Line Ry. v. U. S. (1923) 261 U.S. 299, 43 S.Ct. 354;
    Vogelstein & Co. v. U. S. (1923) 262 U.S. 337, 43 S.Ct. 564;
    United States v. North American Co. (1920) 253 U.S. 330, 40 S.Ct. 518.

Section 1574, Compiled Laws of Alaska, 1933, provides that in condemnation proceedings the compensation shall be the value of the property at the date of the summons—not at the date of the trial or assessment of damages.

THIRD: THE TRIAL COURT DID NOT ABUSE ITS DISCRETION IN DENYING PLAINTIFF'S MOTIONS TO AMEND THE AMENDED COMPLAINT.

Plaintiff's amended complaint alleged and prayed damages in the sum of $100,000 (R. 10). At the opening of the trial, at the close of plaintiff's case in chief, and again at the close of all the testimony, plaintiff asked leave to amend the amended complaint so as to increase the damages to $500,000, to $919,193.72, and to $544,319.48, respectively (R. 56-58). Each of these motions was denied (R. 56-58); plaintiff has assigned these rulings as error (R. 65-66).

The refusal of the court to grant leave to amend the amended complaint is unimportant unless plaintiff's claims for damages in excess of $100,000 are sustained. The matter, moreover, lay in the discretion of the trial court,

> *Vermont Loan & Trust Co. v. Bramel* (1924), 111
> Or. 50, 224 Pac. 1085, 1086;
>
> *Ross v. European Mortg. & Inv. Corporation* (3d
> C.C.A., 1934), 72 F.(2d) 825;
>
> *Tway v. Seneca Motor Car Co.* (5th C.C.A., 1922),
> 284 Fed. 265, 266;
>
> *Pulver v. Union Inv. Co.* (8th C.C.A., 1922), 279
> Fed. 699, 705,

and, for the reasons already stated in this brief, that discretion was not abused.

FOURTH: IN ANY EVENT, AT PLAINTIFF'S OWN FIGURES, WHICH WERE ACCEPTED BY THE COURT BELOW IN COMPUTING THE COSTS OF MINING, THE COST OF REMOVING THE SILT FROM THE CLAIMS IS $15,590.51, WHICH IS THE MAXIMUM PLAINTIFF IS ENTITLED TO RECOVER AS DAMAGES.

Under the unnumbered heading at page 42 of plaintiff's opening brief, "COMPARING POSSIBLE MODES OF RELIEF," plaintiff states "Six alternative plans," the first of which, "Plan A," is the cost of removing the silt. Plaintiff computes this cost at $66,886.19, plus certain suggested items, the amounts of which are not calculated. We are unable to tell from plaintiff's brief whether he contends that damages in this amount should have been awarded. If so, the argument evidently is an afterthought, since the point is not included within either the assigned errors (R. 65-67), or the specification of errors (Opening Brief for Appellant Erceg, p. 8), or the questions which plaintiff says are involved in this appeal (id., p. 7).

In this connection, however, we wish to point out the error, in the computation of the cost of removing the silt, which was made by the court below and which is now repeated in plaintiff's brief.

The only evidence in this record that drainage sufficient to strip the Gartner claims existed on Goldstream was the testimony of plaintiff's expert witness Johnston, who testified that the claims could be stripped by digging an artificial muck drain 9,000 feet long (see Brief for Appellant on the main appeal, pp. 50-53). If—as the court below found[15]—drainage is thus obtainable, the few feet

---

15. Our contention that drainage sufficient to strip the Gartner claims never has existed is presented on the main appeal.

of silt which have been deposited upon the surface of
the ground obviously could be stripped through this drain
at least as readily as the remaining twenty-four feet of
muck which overlie the gravels on the Gartner claims
(id., p. 52).[16]

Two separate estimates of the cost of digging this artificial drain were introduced in evidence at the trial: the
first by plaintiff; the second by defendent. Plaintiff's expert testified that a drain 9,000 feet long would supply
sufficient drainage to transport the muck from the Gartner claims, and that this drain would cost $7,000 if there
were no silt in Goldstream channel (R., main appeal, 230,
234). This same expert testified that the cost of stripping
muck from the Gartner claims would be "just a trifle
under 5 cents per cubic yard" (R., main appeal, 267).

On the other hand, defendant's witness testified that it
would cost approximately $2.50 a lineal foot to build the
drain (R., main appeal, 588). The same witness testified
that the cost of stripping the muck from the Gartner
ground would be more than 9 cents per cubic yard (R.,
main appeal, 593, 602).

---

16. The conclusion of the trial court, "That the acts of the defendant
* * * have rendered it impossible for plaintiff or the said Gartner to mine
the said properties" (Conclusion No. (2), R. 38-39), does not make against
this statement. The conclusion immediately following—Conclusion No. (3),
R. 39—is:

"That the said injuries are permanent in their nature and can not be
· restored or removed except by human labor."

These conclusions, when read together, obviously show that the court concluded that it is impossible to mine the Gartner claims unless the silt is removed first by human labor. That this is the meaning of these conclusions
clearly appears from the opinion of the court (R. 51), in which it holds that
it is possible to remove the silt from the Gartner claims, but at a cost
greater than the value of the property at the time of the injury. As to the
propriety of consulting the opinion of the court on this point, see p. 4,
note 4, supra.

The cost of mining, including the above items, as testified to by defendant's witnesses, exceeded the value of the gold in the ground, and hence showed the claims to be worthless as mining property (R., main appeal, 602; see also opinion below, R. 48). On the other hand, the cost of mining, including the above items, as testified to by plaintiff's witness, was less than the value of the gold (R., main appeal, 238-239, 267).

In computing the value of the Gartner claims, the court accepted the low estimate of mining costs given by plaintiff's witness (R. 48); but in computing the cost of removing the silt, the court took the higher estimate of defendant's witness (R. 51)![17] Obviously, it was erroneous to accept plaintiff's estimate of mining costs in determining the value of the claims; and then to accept defendant's different estimate of the *same costs* for the purpose of determining the cost of removing the silt.

Taking plaintiff's estimate of costs, a simple calculation shows that the silt could be removed from the Gartner claims at a total cost of $15,590.51—a figure arrived at by adding (a) $7,000, the cost of digging the drain if no silt had been deposited in Goldstream channel, (b) $3,375.96, the cost of removing the silt which has been deposited in Goldstream channel over the length of the proposed drain (23,123 cubic yards),[18] at 14.6 cents per

---

17. As to the court's computation, see Brief for Appellant on the main appeal, pp. 25-27.

18. This figure, 23,123, is the difference between 71,000, the total yardage in the proposed drain, and 47,877, the yardage that would have to be removed in constructing the drain were Goldstream in its original condition. See testimony of plaintiff's expert, Johnston, R., main appeal, 230.

cubic yard,[19] and (c) $5,214.55, the cost of removing the silt on the surface of the Gartner claims (104,291 cubic yards)[20] at 5 cents per cubic yard.

We submit that if, on this record, plaintiff is entitled to recover any damages, then the sum of $15,590.51—i.e., the cost of removing the silt at plaintiff's own figures and on the assumption that, as plaintiff contends, sufficient drainage exists—is the maximum amount of damage established by the evidence.

CONCLUSION.

In conclusion, we respectfully submit that the trial court applied the correct measure of damages; that plaintiff's criticisms of the ruling on this point are without merit; that the trial court did not abuse its discretion in denying plaintiff's motions to amend the amended complaint; and that, all other reasons aside, plaintiff is not entitled to recover more than the $30,000 awarded by the court below, because, under plaintiff's own evidence, which was accepted by the court below, the cost of restoring the Gartner claims to their former condition is $15,590.51.

We further submit that the judgment below should be reversed, not for the reasons advanced by plaintiff on the

---

19. This figure, 14.6, is the cost in cents per cubic yard of moving 47,877 cubic yards at a total cost of $7,000 (7,000 ÷ 47,877). See testimony of plaintiff's expert, Johnston, R., main appeal, 230, 234.
20. Finding No. 13, R. 31.

cross-appeal, but for the reasons stated by defendant in its brief on the main appeal.

Dated, San Francisco,
December 27, 1937.

Respectfully submitted,
PILLSBURY, MADISON & SUTRO,
Attorneys for Appellee,
Fairbanks Exploration Company,
on the Cross-Appeal.

ALFRED SUTRO,
FRANCIS N. MARSHALL,
FRANCIS R. KIRKHAM,
Of Counsel.

No. 8107

IN THE

United States Circuit Court of Appeals
For the Ninth Circuit

FAIRBANKS EXPLORATION COMPANY
(a corporation),

Appellant,

vs.

MIKE ERCEG, Guardian of the Estate of
George Gartner, an Insane Person,

Appellee.

REPLY BRIEF FOR APPELLANT.

PILLSBURY, MADISON & SUTRO,
Standard Oil Building, San Francisco,
Attorneys for Appellant.

ALFRED SUTRO,
FRANCIS N. MARSHALL,
FRANCIS R. KIRKHAM,
Standard Oil Building, San Francisco,
Of Counsel.

Table of Contents

Table of Authorities Cited

Court Rules

Statutes and Codes

No. 8107

United States Circuit Court of Appeals
For the Ninth Circuit

FAIRBANKS EXPLORATION COMPANY

(a corporation),

Appellant,

vs.

MIKE ERCEG, Guardian of the Estate of
George Gartner, an Insane Person,

Appellee.

REPLY BRIEF FOR APPELLANT.

REPLY TO PLAINTIFF'S STATEMENT OF FACTS.
(Brief for Appellee, pp. 1-9.)

At pages 2 and 9 of his brief, plaintiff says that defendant has deprived him of the waters of Goldstream
Creek. The court below held that defendant is entitled
to use the waters of Goldstream for mining purposes
until a demand, based upon a bona fide intention to make
a beneficial use of such waters, is made on it by plaintiff;
that no such demand has ever been made; that plaintiff,
therefore, is entitled to no damages for alleged depriva-
tion of waters (Conclusion No. 1, R. 75; Opinion, R. 86[1]).

At page 3, plaintiff refers to the cabins and houses
which were built upon the claims, the shafts sunk, the
ditches constructed, the mining machinery and sluice
boxes installed, etc. The trial court found (Finding No.

---

1. A printer's error deleted the following words from the opinion of the
court in the third line from the bottom of page 86 of the record: "I shall
therefore award no damages" (see p. 50 of the record on the cross-appeal).

2

22, R. 73) that "the buildings of plaintiff located on said mining claims have been unoccupied and subject to natural depreciation for many years, and the damage, if any, to them has been caused by lack of care, exposure to the elements, and other natural causes, and not to any acts of the defendant. That the personal property of plaintiff on said claims has not been injured or destroyed by any acts of the defendant. * * * That old shafts on said claims have been filled with silt, but they were of no real value to plaintiff and had served their purpose * * *. That an old ditch had been cut on the left limit, * * * but said ditch was not on the silted area and was not destroyed by deposition of silt, * * *."

On page 5, plaintiff says that our statement of the mining history of the Gartner claims (Brief for Appellant, pp. 3-5) "is not true." The statement in our brief is correct. It contains numerous and complete record references, and is supported in its entirety by the record. Plaintiff also asserts that defendant has acquired and is now remining at a profit claims which have been "mined out" by earlier methods; that "Because of the cold water process of thawing, the 'mined out' claims of yesterday became the 'bonanza' claims of today" (Brief for Appellee, pp. 4-5). These statements—apparently made to create the impression that the Gartner claims, despite their unsuccessful mining and their partial depletion, are of "bonanza" value—are without support in the record. There is no evidence that defendant has ever acquired "mined out" claims or that it is remining or can remine such claims at a profit.

REPLY TO PLAINTIFF'S ARGUMENT.

(Brief for Appellee, pp. 10-82.)

(1) Plaintiff's contention that this court is without jurisdiction to consider whether there is any evidence to support the findings, conclusions, and judgment of the court, is untenable.

(a) R.S. 649 and 700 (U.S.C. 28:773, 875) are not applicable to the review of the decisions of the District Court for the Territory of Alaska, sitting as a court of general jurisdiction.

This action was brought in the District Court for the Territory of Alaska as a court of general jurisdiction (see *Buckley v. Verhonic* (9th C.C.A., 1936) 82 F.(2d) 730; *Electrical Research Products v. Gross* (9th C.C.A., 1936) 86 F.(2d) 925, 926). Accordingly, the laws of Alaska —not the inconsistent provisions of R.S. 649 and 700— supply the applicable rules of procedure.[2]

In *Buckley v. Verhonic*, 82 F.(2d) 730, supra, this court, referring to the provisions of the Compiled Laws of Alaska relating to the filing of exceptions, said (p. 731, footnote 3):

"These actions were brought in the District Court, not as a court of the United States, but as a court of general jurisdiction. Consequently, they are not federal equity suits. The Federal Equity Rules are, therefore, inapplicable."

In *Shields v. Mongollon Exploration Co.* (9th C.C.A., 1905) 137 Fed. 539, the precise question now raised by plaintiff was adversely disposed of by this court. That case was an action at law, brought in the District Court of Alaska, and tried to the court without a jury. On appeal defendants in error, relying on R.S. 649 and 700, contended that this court was without power

---

2. See,
   Electrical Research Products v. Gross, supra, p. 930;
   Lane v. Jordan (9th C.C.A., 1902) 116 Fed. 623;
   Mackay v. Fox (9th C.C.A., 1903), 121 Fed. 487, 488-489;
   Starklof v. United States (9th C.C.A., 1927) 20 F.(2d) 32, 33;
   Van Dyke v. Midnight Sun Mining & Ditch Co. (9th C.C.A., 1910) 177 Fed. 85, 87-88;
   Northern Mining & Trading Co. v. Alaska Gold Recovery Co. (9th C.C.A., 1927) 20 F.(2d) 5, 7;
   Felder v. Reeth (9th C.C.A., 1929) 34 F.(2d) 744, 745.

"'* * * to consider the assignments of error so far as
they relate to the rulings of the trial court in admitting or excluding evidence, or to the alleged insufficiency of the evidence to support the findings of
fact, on the ground that it does not appear from the
record that written consent of the parties to waive
a jury trial was filed in the court below" (137 Fed.
543).

This court overruled the contention, saying (p. 544):

"If this were the ordinary case of a writ of error to
review the proceedings on a trial in a Circuit Court
of a civil law case without the intervention of a jury,
the objection now interposed would be well taken, for
the record in this case fails to show that there was a
written waiver of a jury trial. But the act of June
6, 1900, adopting a Civil Code for the trial of causes
in Alaska (31 Stat. 363, c. 19), provides that trial
by jury may be waived by the parties to an issue of
fact not only by written consent, but by oral consent
in open court entered in the minutes. It appears both
in the bill of exceptions and in the judgment of the
court below that a jury trial was expressly waived
upon the stipulation of the respective parties. There
can be no doubt, therefore, that such oral consent,
thus evidenced by the record, is as effective to preserve the right of a plaintiff in error to review the
rulings of the court below as would have been a
written consent under the provisions of section 649
of the Revised Statutes."

Chapter LXXXVI of the Compiled Laws of Alaska,
1933, establishes the procedure for "Trial by the Court"
in the courts of Alaska. Section 3619 of that chapter
provides:

"Upon the trial of an issue of fact by the court, its
decision shall be given in writing and filed with the
clerk during the term or within twenty days thereafter. The decisions shall state the facts found and
the conclusion of law separately, without argument or
reason therefor. Such decision shall be entered in the
journal and judgment entered thereon accordingly.
The court may deliver any argument or reason in support of such decision, either orally or in writing,

separate from the decision, and file the same with the clerk.''

Rule 47 of the Rules of the District Court for the Territory of Alaska provides:

"Upon the trial of an issue of fact by the Court and upon the announcement of the ;Court's determination by opinion or otherwise, it shall be the duty of counsel in whose favor the determination is made to prepare findings of fact and conclusions of law, in accordance with the determination of the Court, within five idays and to serve the same upon the opposite party or his counsel and lodge the same with the Clerk of this Court, who shall note thereon the date received. The opposite party may then prepare, serve and lodge his objections and amendments thereto and such findings and conclusions as he may desire within five days thereafter, and the prevailing party may in writing propose amendments or objections to those served on him by serving such proposed amendments or objections upon the opposite party within three days and lodging the same with the Clerk. The settlement thereof before the Court shall be brought on for hearing within the time and in the manner prescribed by rules for hearing motions, or the Court may settle the same summarily upon notice to both parties.''

Since this rule supplements and is entirely consistent with section 3619 of the Compiled Laws of Alaska, 1933, supra, *Century Indemnity Co. v. Nelson* (9th C.C.A., 1937) 90 F.(2d) 644, is not in point (see Brief for Appellee, p. 11).

The procedure provided by section 3619 and Rule 47, supra, rather than the inapplicable procedure provided by R.S. 649 and 700, has long been followed in jury-waived actions at law in the District Court for the Territory of Alaska, and this court has never doubted its power to review.

See,

> *M'Intosh v. Price* (9th C.C.A., 1903) 121 Fed. 716, 717;
>
> *Hemple v. Raymond* (9th C.C.A., 1906) 144 Fed. 796, 800-801;

Cook v. Robinson (9th C.C.A., 1912) 194 Fed. 753,
  759, 761-762;

*Ebner Gold Mining Co. v. Alaska-Juneau Gold Min.
  Co.* (9th C.C.A., 1914), 210 Fed. 599, 601, 603;

Fleischman v. Rahmstorf (9th C.C.A., 1915.) 226
  Fed. 443, 444;

Schoenwald v. Bishop (9th C.C.A., 1917), 244 Fed.
  715, 718, 719;

Felder v. Reeth (9th C.C.A., 1933) 62 F.(2d) 730.

In each of the foregoing cases, the procedure provided
by Rule 47 was followed, and this court, without question,
considered the sufficiency of the evidence to support the
findings and judgment—subject only to the familiar limitation that the findings of the trial court, if supported
by any substantial evidence, are conclusive. We have
examined the records in these cases, and, in the appendix
to this brief, we give a short summary of the procedure
followed in each case, with references to the respective
records in the files of this court.

Plaintiff also contends, supplementary to his argument
under R.S. 700, that certain of defendant's exceptions
were insufficient (Brief for Appellee, pp. 14-19). The
exceptions thus referred to, we submit, are sufficient under
the general federal practice. In any event, the contention
is here foreclosed by the laws of Alaska. Section 3636
of the Compiled Laws of Alaska, 1933, provides in part:

> "No exception need be taken or allowed to any deci-
> sion upon a matter of law when the same is entered
> in the journal or made wholly upon matters in writing
> and on file in the court."

Section 3637 provides:

> "The verdict of the jury, any order or decision, par-
> tially or finally determining the rights of the parties,
> or any of them, or affecting the pleadings, or grant-
> ing or refusing a continuance, or granting or refusing
> a new trial, or admitting or rejecting the evidence,
> provided objection be made to its admission or rejec-
> tion at the time of its offer, or made upon ex parte

application or in the absence of a party, are deemed excepted to without the exception being taken or stated, or entered in the journal.''

The judgment, conclusions, and findings of the court in this case were orders or decisions "partially or finally determining the rights of the parties.'' They were made by the court in the absence of the parties. The "decision,'' consisting of a separate statement of the facts found and the conclusions of law, was entered in the journal (section 3619, supra, p. 4).

(b) In any event, the bill of exceptions shows that proper steps were taken by defendant to preserve the right of review, even under R.S. 649 and 700.

The bill of exceptions clearly shows that defendant preserved the right of review even under R.S. 649 and 700. The bill recites that "defendant duly proposed and lodged with the clerk of the court * * * [its] proposed findings of fact and conclusions of law and * * * filed a written motion for an order of court making, finding and adjudging the findings of fact and each one thereof and conclusions of law and each one thereof as so proposed by the defendant on the ground that each such finding of fact and conclusion of law was established by the evidence at the trial of the case'' (R. 711). Then follows defendant's proposed findings of fact and conclusions of law, with recitals that the court denied such findings and conclusions, defendant duly excepted, and an exception was allowed (R. 711-717). Thereafter the defendant filed its objections to the findings of fact and conclusions of law proposed by plaintiff (R. 717-725). The court overruled these objections, defendant duly excepted and an exception was allowed (R. 725). The bill of exceptions then continues:

> "*Thereafter* the court gave its decision in favor of the plaintiff and made its findings of fact and conclusions of law, and defendant duly excepted to said findings of fact and conclusions of law, and to each thereof, and said exceptions and each of them were allowed.

The trial of the above entitled cause thereupon closed, * * *." (Italics ours throughout the brief.)

These express recitals, we submit, are not overcome by the inferences which the defendant draws from the quotation of Rule 47 of the District Court for the Territory of Alaska and the recital of the date on which *plaintiff* filed its proposed findings of fact and conclusions of law (Brief for Appellee, pp. 10-12).

(c) Defendant in fact filed timely motion and request for special findings of fact, conclusions of law and judgment prior to the decision of the trial court, and if, as defendant contends, the record should be taken to recite a different order of proceedings, it would be contrary to the facts, and may and should be corrected at the direction of this court.

Despite the assertion to the contrary at pages 10-11 of plaintiff's brief, defendant's proposed findings and conclusions were served on plaintiff and were filed on January 18, 1935, and defendant's motion for an order of court adopting these findings and conclusions was served and filed on January 22, 1935, and was submitted to the court at a hearing on January 25, 1935—over three months before the court handed down its opinion on May 3, 1935 (R. 76-90), and more than six months before the court filed its findings of fact and conclusions of law on June 11, 1935 (R. 60-76). As stated in the preceding subdivision of this brief, we think these facts sufficiently appear in the bill of exceptions. If, however, in accordance with plaintiff's contention, the language of the bill should be taken to recite a different order of proceedings, this recital would be contrary to the facts as they appear in the record of this case in the court below; and, pursuant to the settled power of this court,[3] the bill of exceptions should, we submit and respectfully request, be corrected to conform to the facts.

---

3. *Huglin v. H. M. Byllesby & Co.* (8th C.C.A., 1934) 72 F.(2d) 341, 342-343;
Patrick v. United States (9th C.C.A., 1935) 77 F.(2d) 442, 445;
Johnson v. Titanium Pigment Co. (8th C.C.A., 1936) 81 F.(2d) 529;
United States v. F. & F. Lunch Room (2d C.C.A., 1930), 38 F.(2d) 76;
Ohl & Co. v. Smith Iron Works (1933) 288 U.S. 170, 53 S.Ct. 340;
Cyc. Fed. Proc., Vol. 6, §§2832, 2890, 2892, 2893 (and 1937 Cum. Supp.).

(2) Plaintiff's contention that the assignments of error are insufficient is without merit.

Plaintiff's contention (Brief for Appellee, pp. 15-16) that assignments of error X and XII-XV are insufficient is untenable. An assignment that a *finding* is not supported by any evidence is clearly proper and sufficient. It is quite different from the assignments condemned in the cases relied upon by plaintiff, i. e., "That the court erred in entering judgment in favor of the plaintiff" (*Flanagan v. Benson* (8th C.C.A., 1929) 37 F.(2d) 69, 70); "that the judgment * * * is not justified by any evidence" (*Century Indemnity Co. v. Nelson,* 90 F.(2d) 644, 652, supra); that the verdict is contrary to the law and to the evidence (*Washburn v. Douthit* (8th C.C.A., 1934) 73 F.(2d) 23, 24).

Plaintiff also contends that assignments of error V–VII and XXVIII are insufficient because too general (Brief for Appellee, pp. 17-19). Assignment V comprises the substance of these assignments. It is: "The court erred in rendering its judgment * * * for the reason that said judgment is without support in and is contrary to the evidence" (Brief for Appellant, p. 36). This assignment clearly presents the question whether there is any substantial evidence to support the judgment below. It is quite different from the assignments condemned in the cases cited by plaintiff, i. e., "A general assignment that the court erred in rendering judgment or decree for or against a particular party" (*American Surety Co. v. Fischer Warehouse Co.* (9th C. C. A., 1937) 88 F.(2d) 536, 539; *Humphreys Gold Corporation v. Lewis* (9th C.C.A., 1937) 90 F.(2d) 896, 898); "that the trial court erred in rendering judgment against appellant" (*United States v. Shingle* (9th C.C.A., 1937) 91 F.(2d) 85, 90).

Plaintiff also objects to the sufficiency of assignments of error XXIX–XXXIII and XLIII, on the grounds that the bill of exceptions fails to show that the testimony summarized in these assignments of error was the testimony to which objection was made in the court below,

of the evidence" (Brief for Appellee, pp. 20, 46-47, 51).
These objections, as a mere reading of the assignments of
error will disclose, are unsubstantial.[4]

(3) The testimony of the $30,000 offer. (Replying to Plaintiff's Brief, pp. 21-43.)

In our opening brief (Brief for Appellant, pp. 15-20)
we discussed the settled rule that offers to purchase are
inadmissible as evidence of the value of the land; and
we showed how the trial court used inadmissible evidence
of this kind ("the $30,000 offer") as the basis of its
conclusion as to the value of the Gartner claims.

Plaintiff's answers to this showing are without merit.
He first attempts to establish that some of the many
reasons commonly given for the rule are inapplicable to
this case. But the main reason—that extraneous considerations commonly lurk in offers and destroy their
accuracy as measures of value—is strikingly illustrated
by the instant case.

The $30,000 offer at the most was purely speculative.
It could not be an indication of the actual value of the
Gartner claims, because it was made in 1925 before the
defendant had any knowledge of such value. Defendant
did its test drilling, pursuant to permission, in 1928 (R.
297-306, 311-319, 322-324; Brief for Appellee, p. 24); and
after such drilling had disclosed to defendant the actual
gold values in the ground, it offered $5000 for the prop-
erty (R. 161).[5] Obviously this destroyed all probative

---

4. In closing our reply to plaintiff's argument on procedure, we may
point out that under plaintiff's contentions, his own assignments of error on
the cross-appeal are "invalid". See assignments of error Nos. 4, 5, 6, 9, 10
and 11 (R. Cross-Appeal, 66-67), and the corresponding exceptions in the bill
of exceptions (R., Cross-Appeal, 58-61). The recital in the record on the
cross-appeal, as to the time when plaintiff filed his requests for special findings, is substantially the same as the recital in the record on the main
appeal, i.e., "That, thereafter [i.e., after the close of all the evidence] and
before the findings of fact and conclusions of law were made and filed in the
above entitled cause [i.e., before June 11, 1935], and *within the time prescribed by the rules of the Court* [i.e., says plaintiff, at least ten days *after*
the decision of the court was handed down], the plaintiff requested the
Court" etc. (R., Cross-Appeal, 59).

5. Plaintiff refers to page 676 of the record—the testimony of Earl
Young—as showing that at the time of the offer defendant had already
prospected the Gartner ground by drilling. On cross-examination, Young

weight of the evidence of the $30,000 offer. Particularly
pertinent is the reason for the rule given by the Supreme
Court of the United States in the *Sharp* case (*Sharp v.
United States*, 191 U.S. 341, 348):

> "He [the person making the offer] may have so
> slight a knowledge on the subject as to render his
> opinion of no value, and inadmissible for that reason.
> * * * Pure speculation may have induced [the offer],
> a willingness to take chances that some new use of
> the land might in the end prove profitable."

Plaintiff's next contention is that the offer was introduced for the purpose of showing an admission by the
defendant that the Gartner property had some value,
and not for the purpose of showing its real value (Brief
for Appellee, p. 21); that the evidence was thus admissible as a declaration against interest and as an admission
by conduct (Brief for Appellee, p. 32). The record does
not show that the plaintiff so limited the purpose of the
evidence in answer to the defendant's objections to its
admission. On the contrary, it conclusively appears that
the trial court received the testimony as evidence of the

admitted that his drilling was on the Diebold property, and that he did
not know where the boundaries of the Gartner property were; and his
efforts to locate the spot where he "saw the other fellows drilling" (R. 678)
variously hit upon locations on one or another of the Gartner claims and
the ground above those claims (R. 679, and see map, Plaintiff's Exhibit 1).

The defendant offered to prove by its mining engineer that its drilling
records disclosed no such drilling on the Gartner ground (R. 708-709).
The court's ruling excluding this testimony was erroneous (Brief for Appellant, pp. 34-35; infra, p. 14); and the plaintiff's contention now under
consideration emphasizes the prejudicial effect of that ruling.

At the same place in his brief (p. 23) plaintiff refers to the fact that
at the time of the offer defendant was mining on adjacent property, and
evidently considers this fact as giving weight to the offer as an indication
of the value of the property. Defendant's mining method, however, was
large-scale dredging, which makes up for the necessarily immense initial
capital outlay with cheaper operating costs. The cost of mining the Gartner
ground, after stripping and thawing, would not be the same for the defendant as for the small operator using a dragline, and the defendant's
costs would be further and particularly affected by the fact that its heavy
mining machinery was at the time on adjacent ground and would not have
to be moved from a distance. The Gartner ground might thus have a
special value to the defendant which would not accurately reflect its true
value. The special value of lands to a particular person, however, is immaterial on the question of damages (*United States v. Honolulu Plantation Co.*
(9th C.C.A., 1903) 122 Fed. 581, 584, 585). Far from giving force to the
offer as evidence of the value of the land, the fact that defendant was mining
nearby is really another illustration that offers to purchase, by reason of
extraneous factors, are deceptive as measures of value—a consideration which
is particularly pertinent to the $30,000 offer in the case at bar.

value of the property and *used it as the basis of its con-clusion as to how much the land was worth* (Brief for Appellant, pp. 22-23, 25-27). It is well settled that a finding of fact cannot be based upon evidence which was inadmissible to show that fact even though admissible for some other purpose (23 C. J. 56-57; 64 C. J. 1258-1259).

If the $30,000 offer is to be considered as evidence of the value which the defendant at the time of the offer placed upon the land,[6] then it proved nothing; for at that time the defendant had no knowledge of the gold content of the land. Even as an admission, the evidence admitted no more than that the defendant in 1925 was willing to take a chance that the Gartner ground would have some value to defendant, with its expensive and heavy mining equipment already nearby.

Plaintiff's efforts to show that the error in admitting evidence of the $30,000 offer was harmless clearly do not meet the obvious facts, referred to in our brief (Brief for Appellant, pp. 22-23, 25-27), which demonstrate that the court used the evidence as the basis of its finding of value.[7] Plaintiff's intimation that the court's use of the evidence actually benefited the defendant in the sum of $81,631.14 (Brief for Appellee, p. 42), is obviously unwarranted. The trial court's judgment shows that in reckoning mining costs the court did not consider the testimony of plaintiff's witnesses as being credible to show that the Gartner ground was worth $30,000 plus $81,631.14. The only possible conclusion is the one stated in our open-

---

6. Such would be the only purpose of the testimony in accord with the contention in the brief of present counsel for defendant in another case, quoted in plaintiff's brief at page 35. In regard to plaintiff's reference to this former brief, we may say that we know of no principle of judicial estoppel based upon a contention of the same counsel in another case.

7. In this connection we cannot let pass unnoticed plaintiff's statement (Brief for Appellee, p. 41) that "immediately" after the trial court made its remark concerning the prejudicial testimony, the consideration which the lower court would give to "offers to purchase" was concretely put to him in connection with the evidence of the $15,000 offer. The remark of the court quoted by the appellee in connection with the latter offer was subsequent by 14 pages of transcript relating to several different matters (R. 149, 163). Moreover, the court's remark was limited to the $15,000 offer, and stated no more than it purported to state: "I am not going to consider that fifteen thousand offer as having any force in this case" (R. 163).

ing brief, namely, that the court took $30,000 as the value of the land and then attempted a computation which would approximate that result.

(4) The testimony concerning the Diebold land. (Replying to Plaintiff's Brief, pp. 44-55.)

In our opening brief we pointed out that there is no evidence in the record showing the surface elevation of the Diebold land or its other physical characteristics, as compared with the Gartner ground, so as to establish such similarity between the tracts as to make evidence of damage to the Diebold ground material to the issue of damage to the Gartner ground (Brief for Appellant, p. 32). Plaintiff answers that "the record does show surface similarity between the Diebold and the Gartner ground" (Brief for Appellee, p. 53). To support this statement plaintiff refers to a colloquy between court and counsel—which is not "evidence"; and to testimony by the witness Diebold that the "lower end of the Gartner ground and the biggest part of it was all covered [with silt] the same as mine" (Brief for Appellee, pp. 53-54). Obviously, this testimony does not show the physical characteristics of these two tracts of land.

This court, we submit, need read only the substance of the Diebold testimony, set forth in the assignments of error (Brief for Appellant, pp. 28-31[8]), to see that this testimony was outside the issues in the case and prejudicial.

With respect to the court's refusal to allow defendant to show that Diebold's suit against defendant had been settled, plaintiff argues that defendant was attempting to introduce a "collateral matter" (Brief for Appellee, p. 53). This "collateral matter" was raised by *plaintiff*. *Plaintiff* introduced testimony that Diebold had sued defendant for silt damage. Certainly defendant then had

---

8. Through an inadvertence we omitted in our opening brief to give references to the pages of the record containing the testimony summarized in assignments of error XXIX-XXXIII. These references are: assignment XXIX, R. 103; assignment XXX, R. 104; assignment XXXI, R. 105; assignment XXXII, R. 108; assignment XXXIII, R. 108-110.

the right to counteract the resulting prejudice by showing, on cross-examination, that the suit had been settled.

(5) The Nordale testimony. (Replying to Plaintiff's Brief, pp. 55-57.)

Plaintiff's answer to our contention that the court erred in excluding the Nordale testimony concerning drilling records on the Gartner claims is that Nordale merely had "access" to defendant's records and was not the "custodian" of the records. As pointed out in our brief (pp. 34-35), counsel for plaintiff objected to this testimony, and it was excluded by the court, on the ground that defendant's records were the best evidence. As the authorities cited in our brief show, this was patent error. Nordale, defendant's mining engineer, who for years had access to all of its drilling records, was clearly competent to testify. The prejudicial nature of the court's ruling on this question has been emphasized by plaintiff's brief. See pages 10-11, footnote 5, supra.

(6) Plaintiff's argument concerning drainage. (Replying to Plaintiff's Brief, pp. 57-70.)

The crucial factual question on this appeal is whether Goldstream Creek had sufficient grade to permit the stripping of the overburden from the Gartner claims in the manner proposed by plaintiff. If there is no substantial evidence of such drainage, the court's finding that the claims had a value of $30,000, and its consequent award of damages, are without any support.[9]

In our opening brief we pointed out, first, that the silting which actually has occurred on Goldstream demonstrates, as a plain physical fact, that sufficient drainage is wanting; and that, accordingly, even if plaintiff's proposed 9,000-foot drain could carry the muck to the mouth of the drain, the stripping waters would there immediately begin

---

9. Plaintiff begins his argument on "Drainage" with a suggestion that the record does not contain all of the evidence material on this appeal (Brief for Appellee, p. 57). The order settling and allowing the bill of exceptions states that the bill "contains all of the evidence material to the exceptions therein set forth and also all of the evidence relating to the value of the placer mining claims in controversy * * *" R. 731-732).

silting (Brief for Appellant, pp. 42-51). Secondly, we showed by undisputed physical facts that plaintiff's proposed drain would not have sufficient grade to transport the muck to its mouth (id., pp. 51-58).

With respect to the first of these basic contentions:

(a) Plaintiff repeats what the trial court said (Opinion, R. 85; Brief for Appellee, p. 63), that defendant found the drainage on Goldstream sufficient for a time to mine its ground just above the Gartner claims.[10] But, as defendant's experience in mining this ground conclusively demonstrates, this does not show that sufficient drainage existed. As we pointed out in our opening brief (pp. 57-58), even if muck could have been washed away from the Gartner claims for all or part of the time, silting below indisputably would have occurred; plaintiff would have faced exactly the situation with which defendant has had to cope because of lack of drainage; and plaintiff's costs of mining would have included the cost of silt easements, of settling just claims for silt damage, and of defending actions for claims in amounts deemed to be unjust.

In addition, defendant's dredging operations do not require that all the muck be removed from the gravel, as does plaintiff's proposed dragline method; and the record shows that the area just above the Gartner claims was not stripped to gravel, but to "grade," which was "a little above the level of the creek bed" (R. 362; and see R. 507). Drainage to a little above the level of the creek bed on the Gartner claims would leave about a third of the muck on the claims (see pp. 19-20, infra.)

(b) Plaintiff says that defendant's "contention" at the trial that there was a good channel through the Gartner

---

10. To this statement plaintiff adds that defendant up to August, 1930, had stripped 25,000,000 cubic yards of overburden into Goldstream Creek. This assertion has no support in the record and is erroneous on plaintiff's own showing. At page 31 plaintiff says that "a mere calculation" shows that defendant up to November, 1934, had stripped in excess of 25,000,000 cubic yards of overburden. (The record furnishes no basis, either at the references given by plaintiff or in any other place, for this "calculation.") At page 62 of his brief plaintiff says that up to August, 1930 (four years earlier), defendant had stripped 25,000,000 cubic yards, and to support this assertion he refers to his prior "computation" as of November, 1934. The record contains no testimony as to the total yardage of overburden stripped by defendant.

property until the flood of 1930 is "evidence" that the drainage of Goldstream was sufficient to carry off defendant's overburden (Brief for Appellee, pp. 59-63). Obviously defendant's "contentions" at the trial are not "evidence". The court found that by August 28, 1930, the bottom of the channel of Goldstream on the Gartner claims and for three miles below had been raised by silt deposits to an average of 4 or 5 feet from the surface, and that in 1928 silt had been deposited on the lower corner of the Gartner ground (Findings Nos. 11, 12, R. 67).[11]

(c) Plaintiff characterizes certain testimony concerning the 1933 drain as an "admission" by defendant that the drainage on Goldstream was sufficient to strip the Gartner claims (Brief for Appellee, pp. 61-62). The 1933 drain commences at an elevation of 733.3 feet (734.4 on top of ice) just above the Gartner ground (R. 451). To strip the Gartner claims it would be necessary to start a drain at an elevation of 705.4 feet—the low point on the top of the Gartner gravels (R. 298; see Brief for Appellant, p. 54).

(d) Plaintiff refers to the testimony of defendant's witness Fenton as opposed to our statement that there was no evidence that defendant failed to keep the channel clear after 1927 (Brief for Appellee, p. 60). We referred to this same testimony in support of our statement (Brief for Appellant, p. 47). Fenton testified to defendant's activities *in keeping the channel of Goldstream clear.* He testified that, with the exception of certain jams which he and his men found and cleared out, the channel was in good condition (R. 331-332, 342, 344-345). But regardless of how this testimony is read, it does not go to the essential fact, namely, that even if a straight and clear drain, commencing at the bottom of the natural channel of Gold-

---

11. Defendant never did contend that there was sufficient drainage on Goldstream to carry its stripping waters without deposition of silt. It urged at the trial that most of the silt damage, especially to the surface, came from the flood; but it conceded that the channel of Goldstream had been raised by silt deposits to an average of 5 or 6 feet from the surface before the flood in August, 1930 (R. 712).

stream on the Gartner claims, could carry the muck to a
point 9,000 feet below the Gartner ground, where plaintiff's proposed drain would end, or even 16,000 feet below
the claims, where Big Eldorado flows into Goldstream, the
stripping waters would begin silting at the mouth of such
drain. Goldstream itself has demonstrated this and it is
idle for plaintiff to contend otherwise. We do not argue
that it is impossible to strip the Gartner ground. We
simply point out that plaintiff could not have done it without either building a drain to some undisclosed point
beyond the silt deposits shown by this record, or securing
easements for the deposition of silt on lands below the
Gartner claims.

(e) Finally plaintiff says we did not call a single witness to testify in opposition to plaintiff's testimony on the
issue of drainage (Brief for Appellee, p. 61). All of the
testimony of defendant's witnesses concerning the levels
on Goldstream and on the 1933 drain showed lack of drainage. Plaintiff criticizes this testimony as "indirect"
(Brief for Appellee, p. 67). We submit, however, that it
was evidence of much greater probative value than a mere
assertion that Goldstream lacked sufficient drainage. In
addition, defendant's witness Patty "directly" testified
concerning drainage for the Gartner claims (R. 506, 507):

> "I wouldn't attempt to strip the overburden; if I
> were called upon to do it, I think I would pass it and
> say get some one else, because there was not drainage
> enough for it.
> * * * * * * *
> My reasons for testifying that I did not think it
> was practicable to work the Gartner ground by the
> dragline method, are, * * * you have no cut drain-
> age for getting that muck off the ground."

With respect to our second point, i. e., that plaintiff's
proposed drain would not have sufficient grade to transport the muck, plaintiff says that our argument is based
upon "misstatements of the oral testimony and calculations based thereon" (Brief for Appellee, pp. 70, 65).
Plaintiff, we think, is mistaken, as an examination of his

(a) Plaintiff refers to the fact that the grade on Big
Eldorado from the mouth of the 1933 drain to Goldstream
is .15%; says that this grade carried off defendant's stripping water during 1933 and 1934; and concludes that this
shows a grade of .15% is sufficient to transport muck
(Brief for Appellee, pp. 65-66). This argument has the
same vice as the argument that defendant found the grade
on Goldstream sufficient to strip the claims above the Gartner ground. It is true that the grade at the mouth of the
1933 drain is .15% for over a thousand feet; but there is
no testimony that this grade carries stripping water without silting. On the contrary, the evidence shows that
heavy silt deposits have occurred in the channel at and
below the confluence of Big Eldorado and Goldstream.[12]
Indeed, the channel between Big Eldorado and Happy
Station, with a grade of .219%, is filling with silt (see
Brief for Appellant, p. 54).

(b) Plaintiff says that there has been no deposition of
silt in the channel of Goldstream at Happy Station; that
this shows that the grade of .219% between Big Eldorado
and Happy Station is sufficient to carry stripping water
(Brief for Appellee, pp. 65-66). To support his statement
that the channel at Happy Station was the same depth in
1934 as in 1912, plaintiff refers to testimony of defendant's
engineer that the channel *one quarter of a mile* above
Happy Station was 14 or 15 feet deep in 1934 (R. 460, 706),
and then to the testimony of the witness Harbell that the
channel *one mile* above Happy Station was about 15 feet
deep in 1912 (R. 704-705). Plaintiff's comparative points
are three quarters of a mile apart.[13] The record in-

---

12. For example, Saulich, a witness for the plaintiff, testified: "At that
time [September, 1933] I observed the condition of Goldstream where Big
Eldorado emptied into it; it was all muck and sand and mud and that sort
of stuff; to the best of my knowledge, from the top of the bank of Goldstream
down to the water was never deeper than over two feet at that time; that
extended from where Big Eldorado comes into Goldstream between two and
three miles down Goldstream, near to one mile above Happy Station; from
Happy Station on down below the channel is now four and five and as high
as eight feet" (R. 673-674).
13. The witness Harbell added that he did not "remember exactly what
was the comparative depth of the channel of Goldstream from that bridge
down Goldstream to opposite Happy Station; like all these streams it would
vary some; it would be a little deeper in one place and shallower in another
place" (R. 705).

disputably shows heavy silting on lower Goldstream to Happy Station (see Brief for Appellant, p. 54).

(c) Plaintiff calculates the grade between the Gartner property and Big Eldorado as .298% (Brief for Appellee, p. 67), instead of .207%, as computed in our brief (Brief for Appellant, pp. 54-55). But plaintiff's grade runs from the surface of the Gartner claims, not from the top of gravel.[14]

(d) Plaintiff says there is testimony in conflict with our statement that 12 feet of silting in the channel of lower Goldstream is the maximum established by the evidence (Brief for Appellee, p. 68). The testimony plaintiff refers to does not show the difference in the depth of the channel, before and after silting, at any one place. As stated by the witness Harbell (supra, p. 18), the channel of Goldstream, like that of any other stream, varies in depth at different places. Husack was the only witness who testified to the difference in depth of the channel, before and after silting, at one place (R. 667-673). He stated that at one point, where the channel is now 2½ feet deep (to the top of ice), it formerly was "about 15 feet" (R. 671); that he could not "swear to any more than twelve feet"; that "it was better than 10 feet" (R. 672). Twelve feet, obviously, is the maximum of silting shown by the evidence.

(e) Finally, plaintiff points out that if Discovery Queenie Claim is eliminated, the average depth of the muck on the Gartner claims is less than 24 feet. This is true, but does not affect the figures concerning drainage given in our opening brief (Brief for Appellant, pp. 53-56). To insure accuracy, our computations were based upon the actual elevation of the top of the gravel. With Discovery Queenie omitted, however, the average depth of muck on

---

14. In addition, plaintiff's grade of .298% is calculated upon a distance of 16,272 feet between drill hole 26-30 and the mouth of Big Eldorado. The record shows that the measured distance from the junction of Big Eldorado and Goldstream to drill hole 18-26 is 16,500 feet (R. 463), and from that drill hole up to drill hole 26-30, 1,100 feet (R. 463)—making a total of 17,600 feet between drill hole 26-30 and the mouth of Big Eldorado. Plaintiff says our calculation "means nothing"; that "it is incorrect," because "The distance between the two points were not correctly 'calculated'" (Brief for Appellee, p. 67). We took the measured distances shown by the record. We do not know the source of plaintiff's figures.

the remaining three claims is still approximately 21 feet.[15]
The top of gravel on these claims is, therefore, on the
average about 6 feet below the bottom of the natural
15-foot channel of Goldstream. Plaintiff, by referring to
certain points on the Gartner gravels near the old channel
of Goldstream, does not, we feel sure, intend to suggest
that drainage to the level of the bottom of the old channel
would strip the Gartner claims (see Brief for Appellee, p.
69). The average elevation of the top of the gravel at the
four drill holes referred to by plaintiff is 716.2 feet.[16] To
strip the Gartner claims, drainage must exist to a point
nearly 11 feet lower, i. e., to an elevation of 705.4 feet, the
top of gravel at drill hole 16-22 (R. 298).[17]

We submit that plaintiff has failed to point to any substantial evidence which sustains the finding of the court
that the drainage of Goldstream Creek at and below the
Gartner claims was, prior to the deposit of silt by defendant, of sufficient grade to permit the stripping of the muck
overburden on those claims.

**(7) The statute of limitations. (Replying to Plaintiff's Brief,
pp. 70-82.)**

Plaintiff, without citation of authority, contends that
the defense of the statute of limitations was waived. The
record, we submit, does not bear this out. Defendant
raised the defense in its amended answer (R. 56), moved
to strike portions of the amended complaint referring to
time beyond the statutory period (R. 41), moved to have
the amended complaint made more definite and certain
by setting forth the time the alleged damage to the
Gartner claims occurred (R. 42), demanded a bill of

---

15. The depth of the muck at the drill holes on these three claims, as
shown by the drill hole logs, is as follows: 14-23, 22 feet (R. 297); 14-25,
15 feet (R. 311); 16-22, 26 feet (R. 298); 16-25, 17 feet (R. 312); 18-22,
30.5 feet (R. 324); 18-24, 24 feet (R. 313); 18-26, 13.5 feet (R. 314);
22-21, 36.5 feet (R. 299); 22-23, 28 feet (R. 300); 22-25, 23 feet (R. 301);
22-27, 17 feet (R. 302); 22-29, 15 feet (R. 323); 22-31, 19 feet (R. 315);
26-24, 21 feet (R. 317); 26-26, 19 feet (R. 318); 26-28, 12 feet (R. 319);
26-30, 16.5 feet (R. 322).

16. Drill hole 26-30, elevation at top of gravel 717.2 feet (R. 322); 22-27,
715.5 feet (R. 302); 18-26, 717.6 feet (R. 314); 14-25, 714.5 feet (R. 311).

17. Other drill holes show top of gravel elevations practically as low,
e.g., 18-24, 706.5 feet (R. 313); 14-23, 708 feet (R. 297); 18-22, 708.8 feet
(R. 324); 22-23, 709.7 feet (R. 300).

particulars setting forth when the alleged damage occurred (R. 46), and, at the conclusion of the trial, requested a conclusion of law that any damage occurring more than two years before the commencement of the action was barred (R. 716).

With respect to the applicable period of limitations, we agree with plaintiff that if defendant had loaded its overburden into trucks and carried it to and dumped it upon the Gartner ground, or had carried muck and tailings through a flume and discharged them directly upon plaintiff's land (as in *Atkinson v. A. & S. Canal Co.* (1878) 53 Cal. 102) or had forceably entered upon plaintiff's property and built a railroad across it (as in the cases cited by plaintiff at pages 74-75 of his brief), the statute of limitations relating to trespass would apply. The case at bar, obviously, is different. As pointed out in our opening brief (pp. 60-65), it involves an indirect or consequential injury. *Conniff v. San Francisco* (1885) 67 Cal. 45, 7 Pac. 41, upon which plaintiff chiefly relies (Brief for Appellee, p. 75), is not "approved," but, on the contrary, is sharply limited and distinguished in *Hicks v. Drew* (1897) 117 Cal. 305, 49 Pac. 189, and *Law v. Smith* (9th C.C.A., 1923) 288 Fed. 7. The latter cases, as pointed out in our opening brief (p. 62), state the applicable rule.

We also agree that if defendant intentionally had conducted its operations for the purpose of injuring plaintiff's ground, the action would be in trespass. No such purpose or intention was proved or is present in this case (see Brief for Appellee, Fairbanks Exploration Company, on the Cross-Appeal, pp. 3-5). Defendant, as the court below held, carried on its operations "carefully" and "in a minerlike manner" (R. 80). It intentionally stripped its overburden into Goldstream, not for the purpose of injuring plaintiff's property, but in furtherance of its own legitimate mining operations. The distinction between the consequential injury resulting from such acts, for which the remedy is an action on the case, and an inten-

tional injury for which trespass lies, is well illustrated by
Kelly v. Lett (1851) 35 N.C. 53. In that case the declaration alleged that plaintiff owned a mill shortly downstream
from defendant's mill; that defendant, particularly on
Sundays and at night, shut down his gates so as to accumulate as large a head of water as possible in his mill
pond, and then raised the gates, thus discharging an immense volume of water which ran with great force
against plaintiff's dam below and swept it away; that
these acts were done by defendant wilfully and with an
intent to do the injury. The court pointed out that the
damage was the immediate result of a wilful act, and held
that the declaration stated an action in trespass, as distinguished from an action on the case.[18]

With respect to the effect Gartner's insanity may have
upon the period of limitations, nothing, we think, need
be added to the statement in our opening brief save to
point out that plaintiff is clearly in error in suggesting
that defendant is precluded by its amended answer from
asserting that the cause of action sued upon is vested in
plaintiff. The amended answer stated the defense of limitations in the alternative. The *second* further defense
(R. 56-57) contains the language quoted by plaintiff
(Brief for Appellee, p. 78). The *first* further defense
alleged (R. 56):

> "That if the matters set forth in said Amended
> Complaint as an alleged cause of action in favor of
> the plaintiff, as Guardian of the estate of George
> Gartner, an insane person, occurred at all, they oc-

---

18. *Drake v. Lady Ensley Coal, Iron & R. Co.* (1894) 102 Ala. 501, 14
So. 749, also states the distinction. There refuse and debris cast into a
stream by defendant in the course of its mining operations were carried down
and deposited upon plaintiff's land. The court held that the action for this
injury was an action on the case, saying (14 So. 752):
"The boundary line between where trespass ends and case begins is not
always easily determined. Under the law the defendant had the right to
divert the water from its channel and utilize it in washing the ore.
His duty was to return the water to its proper channel. This was done.
The tort to plaintiff was neither in the diversion of the water from its
channel, nor that defendant used it for his own purposes, but that the
use to which it was applied rendered it impure, filled it with clay and
objectionable ore and debris, and in this condition it was carried by the
flow of the water to plaintiff's farm. The damage inflicted was neither
intentional nor direct nor immediate, but was consequential."

curred more than two years before the commencement of this action and after the first day of August, 1927, before which last mentioned date, the plaintiff had taken custody, control, and possession of all real and personal property of said George Gartner.''

CONCLUSION.

We have purposely disregarded, as absolutely foreign to the merits of the case, the abuse and vilification in plaintiff's brief. Thus, plaintiff refers to defendant as a ''brazen racketeer'' and ''predatory corporation'' (Brief for Appellee, p. 30), and speaks of ''its greed to take from an insane man his property'' (pp. 39-40); and plaintiff charges counsel for defendant with ''evasion, equivocation and deception'' (pp. 30, 31), with ''welshing'' (p. 48), with making ''insidious'' contentions (p. 40), ''pathetic pleas'' (p. 30) and ''chimerical and occult assumptions in the face of the record, inspired by greed'' (p. 39).

To consider in detail a typical example of this sort of tirade—not argument—on pages 30-32 of plaintiff's brief: The attack here is based upon the fact that counsel for defendant objected to the introduction in evidence of *all* of defendant's mining costs. Obviously, such evidence would have been wholly irrelevant. In the first place, as the trial court ruled in sustaining our objection (see below), the physical conditions of the Gartner ground were not shown to be the same as those of the areas mined by defendant. In the second place, the mining method used by defendant, at least with respect to *digging* and *sluicing* the gravels, is different from that proposed by plaintiff for the Gartner claims. Plaintiff proposed to dig and sluice the gravels on the Gartner claims with dragline equipment costing some $45,000 (R. 238); defendant, with an investment of several million dollars, uses large and extremely expensive dredges. The record

in this connection concerning *stripping* and *thawing* costs
is as follows: Counsel for plaintiff asked defendant's
general manager to produce statements showing the costs
of stripping on Goldstream Creek for the seasons of
1928-1932 (R. 246). Counsel for defendant objected that
this evidence would refer to the whole of Goldstream,
that no showing had been made that the Gartner claims
were the same as those all over the creek, and that the
evidence should be limited to the ground which is adjacent to the Gartner ground and similar to it. The
court sustained "the objection as to the whole of the
creek (R. 246). It is too general. It is too large" (R.
247). Counsel for plaintiff thereupon called for the production of statements showing the cost of stripping and
thawing on "properties located in the vicinity of the
Gartner ground" (R. 247-248). Thereupon defendant
produced statements showing "the cost of the thawing
and the stripping in the vicinity of the Gartner ground"
(R. 248), but, after counsel for plaintiff had examined
these documents, he stated that "he did not care to examine the witness any further" (R. 256). Counsel for
defendant then questioned the witness concerning these
mining costs, and counsel for *plaintiff* objected. Counsel
for defendant pointed out that it had taken two days
to prepare these statements and he offered to show that
the actual cost to defendant of stripping on Creek Claim
21 Below on Goldstream and on the Morrison Bench
Claim, adjacent to the Gartner ground, was over 17 cents
per cubic yard, and that the cost of thawing in the same
area was a little over 11 cents per cubic yard (R. 256-258).
This would have shown that the *actual* cost of stripping the
ground adjacent to the Gartner claims was more than three
times, and that the *actual* cost of thawing was more than
twice, what plaintiff's expert witness *estimated* it would
cost to strip and thaw the Gartner claims. *Plaintiff's*
objections to this offer of proof were sustained (R. 256-
258). *It was at plaintiff's insistence and over defendant's
protest and offer that defendant's actual costs of stripping*

and thrawing on ground adjacent to the Gartner ground were excluded.[19]

We respectfully submit that the judgment should be reversed.

Dated, San Francisco,
January 17, 1938.

PILLSBURY, MADISON & SUTRO,
Attorneys for Appellant.

ALFRED SUTRO,
FRANCIS N. MARSHALL,
FRANCIS R. KIRKHAM,
Of Counsel.

---

19. Plaintiff also refers to page 610 of the record, which shows that counsel for plaintiff asked defendant's witness, McFarland, how much it cost defendant "to strip per square yard." Counsel for defendant objected on the ground that this was not cross-examination (R. 610). The court *over-ruled* the objection and directed defendant's witness to answer with respect to the costs of "comparatively adjacent ground" (R. 613), whereupon defendant's witness testified that defendant's actual costs of thawing and stripping were, as he recalled, approximately the same as those given in his estimate. McFarland's estimate of stripping and thawing costs was about double that given by plaintiff's witness (R. 602, 267).

(Appendix Follows.)

Appendix

(The following statements disclose the procedure followed in the cases cited at pp. 5-6 of this brief. The record references are to the records in the respective cases on file with this court.)

M'Intosh v. Price (9th C.C.A., 1903) 121 Fed. 716, Docket No. 856, February 2, 1903.

After the close of all of the evidence the "cause was submitted to the court" (R. 134). Thereafter, on November 16, 1901, the court handed down its opinion and decision (R. 33-51), which concluded: "Findings and decision will be in favor of the plaintiff" (R. 51). The record further states: "* * * after said cause had been submitted as aforesaid on the————day of November, A. D. 1901," defendants served and filed requests for findings of fact and conclusions of law (R. 134-137). On December 23, 1901, the application of defendants to make findings, etc., came on for hearing, and was denied and an exception allowed (R. 137). On December 23, 1901, the court made and filed its findings of fact and conclusions of law (R. 138-151). Defendants excepted and, on December 26, 1901, filed written objections and exceptions to said findings (R. 151, 152, 155). On January 4, 1902, judgment for plaintiff was entered (R. 52-54).

Hemple v. Raymond (9th C.C.A., 1906) 144 Fed. 796, Docket No. 1,202, February 5, 1906.

The trial commenced December 16, 1904, "and the evidence being closed, the cause was submitted to the court" (R. 19). Thereafter, on January 4, 1905, the court delivered its opinion and decision, concluding "judgment for plaintiff, as demanded." (The opinion is not in the record. It is reported, 2 Alaska 343.) Thereafter, January 7, 1905, the court made and filed findings of fact and conclusions of law (R. 20-26). Thereafter, on January 16, 1905, judgment was entered. On January 23, 1905, defendant filed a "Bill of Exceptions," excepting to the findings, conclusions and certain other rulings of the court, which exceptions were allowed (R. 13-15).

Cook v. Robinson (9th C.C.A., 1912), 194 Fed. 753, Docket No. 2,012, March 18, 1912.

The opinion was handed down January 9, 1911, stating that garnishee's motion to dismiss should be granted (opinion not in record; reported 4 Alaska 285). Thereafter, January 14, 1911, a copy of the findings and conclusions was served on plaintiff's attorneys (R. 62). Thereafter, on January 19, plaintiff filed exceptions to the findings and conclusions, and the exceptions were allowed (R. 62-65). On January 19, 1911, plaintiff also filed proposed findings (R. 66-70). On January 23, 1911, plaintiff excepted to the court's refusal to sign plaintiff's proposed findings (R. 70-71). On January 23, 1911, the court made and entered its findings and conclusions (which theretofore had been served on plaintiff as aforesaid) (R. 57-61). The findings recite that plaintiff, having closed his case and defendant having moved for a dismissal, and the court having taken the matter under advisement on briefs, etc. and having "thereafter and *heretofore* rendered its written opinion," etc. (R. 57-58). On January 23, 1911, judgment was entered (R. 72-73).

Ebner Gold Mining Co. v. Alaska-Juneau Gold Min. Co. (9th C.C.A., 1914) 210 Fed. 599, Docket No. 2,155, January 5, 1914.

After the evidence was all in and the cause had been submitted to the court (see Brief of Plaintiff in Error, p. 15), the court ordered, pursuant to stipulation, that the files in the cause be transferred from the first to the fourth division, for the court to settle the findings, enter judgment and rule upon the findings requested by either side—the papers then to be returned to the first division for final judgment (R. 94-95). The order and stipulation further provided that each side should have thirty days from the return of the findings and judgment to the first division within which to file written exceptions to such findings and judgment. On June 12, 1911, plaintiff filed proposed findings of fact and conclusions of law (R. 1672-1677). On June 15, 1911, plaintiff filed objections to defendant's proposed findings and conclusions (R. 133-136; 1679-1682). On July 5, 1911, the court denied defendant's requests other than those given in the findings already filed (R. 84), and overruled plaintiff's objections to defendant's proposed findings, etc. (R. 1683). On the same day, the court denied the findings requested by plaintiff other than those given in the findings already on file (R. 1678). On the same day, July 5, 1911, nunc pro tunc, as of June 12, 1911, the court made and entered its findings of fact and conclusions of law (R. 85-91), and handed down its judgment and decree (R. 92-94). On October 17, 1911, plaintiff filed written objections and exceptions to the findings and conclusions made by the court (R. 1687-1692).

Fleischman v. Rahmstorf (9th C.C.A., 1915), 226 Fed. 443, Docket No. 2,574, October 4, 1915.

After the "action was submitted to the court," and within the time allowed by law (R. 56), the defendant presented and moved the court to make findings and conclusions. "Thereafter," the defendant filed objections to plaintiff's findings and conclusions (R. 62). Thereafter, on January 24, 1914, the court approved and entered findings in favor of the plaintiff (R. 64-68). On March 9, 1914, judgment for plaintiff was entered (R. 27).

Schoenwald v. Bishop (9th C.C.A., 1917), 244 Fed.
715, Docket No. 2,817, October 8, 1917.

After the plaintiff's evidence was in, it was agreed that
the jury should be dismissed and the court should render
a decision on the pleadings and the evidence. Both sides
moved for judgment (R. 212-213). On January 3, 1916,
the court rendered its decision in favor of defendants and
directed findings to be presented accordingly (R. 213).
The opinion was filed January 6, 1916 (R. 54). Plaintiff presented proposed findings of fact and conclusions
of law, which the court, on January 26, 1916, refused—
"exception allowed" (R. 211). Defendants' proposed
findings and conclusions, served February 3, 1916 (R.
223) and filed the same day (R. 223), were adopted
(R. 212).

Felder v. Reeth (9th C.C.A., 1933), 62 F.(2d)
730, Docket No. 6,717, January 9, 1933.

The trial began September 3, 1931 (R. 73) and evidence was taken on the 3rd and 4th (R. 73-74). The
case was argued on September 29, 1931, "at the close of
which the court orally announced its decision in favor
of the defendant" and directed defendant to prepare
findings and conclusions (R. 74). Defendant's proposed
findings were served on plaintiff's attorneys on October
3, 1931 (R. 78). Plaintiff's exceptions to said findings
were served on defendant on October 6 (R. 81). These
exceptions were denied (R. 81) and defendant's proposed
findings adopted and filed October 9 (R. 78). Judgment
(reciting the foregoing) was filed October 21 (R. 84-85).

No. 8107

IN THE

United States Circuit Court of Appeals
For the Ninth Circuit

MIKE ERCEG, Guardian of the Estate of
George Gartner (an Insane Person),
Appellant,

vs.

FAIRBANKS EXPLORATION COMPANY
(a corporation),
Appellee.

REPLY BRIEF FOR APPELLANT.

ROBERT L. MCWILLIAMS,
Financial Center Building, San Francisco, California,
JULIAN A. HURLEY,
Fairbanks, Alaska,
JOHN L. MCGINN,
Fairbanks, Alaska,
Attorneys for Appellant.

Subject Index

Table of Authorities Cited

Cases

No. 8107

United States Circuit Court of Appeals
For the Ninth Circuit

MIKE ERCEG, Guardian of the Estate of
George Gartner (an Insane Person),
Appellant,

vs.

FAIRBANKS EXPLORATION COMPANY
(a corporation),
Appellee..

REPLY BRIEF FOR APPELLANT.

Plaintiff and defendant, in the lower Court, proceeded upon the theory that the value of the Gartner
mining claims were to be determined by their productiveness, i. e., the profits which their mining would
bring to the owner. To this end both plaintiff and
defendant introduced evidence as to the gold content
of the Gartner property and the cost of mining the
same. Indeed, the trial Court's finding as to the gold
content of the Gartner claims was based upon defendant's estimate. (Court's Opinion, R. 47.) Neither
plaintiff nor defendant offered any evidence whatsoever as to the cost of removing the silt and debris
deposited on the Gartner claims or the cost of "re-

storing'' the channel of Goldstream Creek and the
water which formerly ran through it and which has
been diverted and appropriated by the defendant.
Neither did plaintiff or defendant offer any other evidence by which the damage to the Gartner property
could be estimated. The value of the gold content of
the property at the time the action was commenced
and at the time of trial was determined by no uncertain or speculative factor. It was determined by the
value of gold as fixed by law; and the element of a
fluctuating market price was non-existent.

It must be conceded that plaintiff is entitled to
recover an amount ''which will compensate him for all
detriment proximately caused'' by the acts of the
defendant. This is the underlying principle in all
awards of damages. The ordinary rule that in torts—
not amounting to a *wilful or wanton wrong*—the
wrongdoer is only liable for such consequences as were
or should have been contemplated, foreseen, or anticipated is not applicable here. The rule which must
govern here is that where the injury is the result of
a *reckless, wanton, and malicious trespass* plaintiff
is entitled to recover all damages proximately caused
by the injury even though they could not have been
contemplated, foreseen or anticipated and no matter
how *unlikely, unusual, and unforeseen* they may have
been. It becomes important therefore to determine the
nature of defendant's wrongful acts.

DEFENDANT'S TRESPASS WANTON, RECKLESS AND MALICIOUS.

The lower Court found that defendant in filling and destroying the channel of Goldstream Creek and in flooding the Gartner claims with water and debris,

> "acted willfully, and in utter *disregard and defiance of the rights* of said plaintiff and the said Gartner";

and that *defendant utterly failed, refused, and neglected to pay any attention to the repeated demands of plaintiff to desist from such acts and conduct.* (Finding 21, R. 36.)

This is a finding that the defendant acted wantonly and maliciously. (38 C. J. 353.)

This finding is plain and unambiguous. It needs no clarification. Yet defendant undertakes to modify and emasculate it by saying that the Court in its opinion stated that the defendant for the most part had conducted its mining operations in a minerlike manner. While reference to the opinion of the trial Court may be had to clarify the meaning of a finding that is otherwise doubtful, the opinion may not be referred to "for the purpose of eking out, controlling, or modifying the scope of the finding".

> *American P. & Mfg. Co. v. United States*, 300 U.S. 475, 479, 81 L. ed. 751.

The lower Court's opinion states that the defendant for the "most part" had conducted its mining operations in a minerlike manner—not that it had conducted *all* of its mining operations in a minerlike manner. The Court found that the defendant not only acted willfully and maliciously (Finding 21, R. 31) but also recklessly. (Findings 8-13, R. 28-31.)

The defendant in knowingly hydraulicking moss and roots into the channel of Goldstream Creek when it was full of trunks of trees, roots, windfalls, etc.; and in failing to straighten out the "horseshoe bends"; and in refusing to heed the demands of plaintiff that it desist from flooding and covering the Gartner claims with silt and debris, acted recklessly and with gross negligence.

DEFENDANT "TOOK" PLAINTIFF'S PROPERTY WITHOUT DUE PROCESS OF LAW.

In our opening brief (pages 12-15) we cited cases to support the Court's conclusion that the acts of the defendant (1) in filling and destroying the channel of Goldstream Creek which deprived the Gartner claims of all water; (2) in taking away all drainage; and (3) in flooding and covering the Gartner claims with water, silt and debris, which made them inaccessible for entry, work or mining, constituted a "taking" within the 5th Amendment for which plaintiff was entitled to recover "just compensation".

The defendant does not question the cases cited by us. It brushes them all aside, however, with the sweeping statement that they are not in point. That the instant case is not an eminent domain proceeding.

It is our contention that the defendant was (1) clothed with the power of eminent domain; and, that (2) it wantonly, recklessly and maliciously "destroyed'' and ''took'' plaintiff's property, for which plaintiff is entitled to recover just compensation; and, that these facts bring the defendant within the rule that:

"Where a corporation entitled to exercise the power of eminent domain wrongfully enters upon property before condemnation, the corporation is not entitled to have the *damage assessed* as of the *date of the wrongful entry,* but as of the time of trial, or of the date when, according to the law of the particular jurisdiction, the property is lawfully appropriated."

20 *C. J.* 831, Sec. 266.

Generally speaking the land may be regarded as taken at the moment when by the terms of the statute the owner is divested of his title and it vests in the condemning party. This rule is conceded by the defendant in its brief on page 32. (20 C. J. 826-828, Sec. 262.)

U. S. v. Cary (D.C. Va.), 2 F. Supp. 870, 878.

In Alaska property is lawfully appropriated in condemnation proceedings when payment has been made and a copy of the order of condemnation has been filed in the Recorder's Office. (C. L. A. 1933, Sec. 1580, p. 350.)

The provision of the Alaska Statute, Sec. 1574, cited by defendant (Brief 33) that in condemnation proceedings the compensation shall be the value of the property as of the date the summons is issued, is for the benefit of those who pursue the lawful and orderly procedure prescribed by the statute, not for malefactors—those guilty of wanton and malicious trespass.

> 20 *C. J.* 1207, Sec. 566;
>
> *Bear Gulch Placer Min. Co. v. Walsh,* 198 Fed. 346, 351;
>
> *Hazard Dean Coal Co. v. McIntosh,* 183 Ky. 316, 209 S. W. 364.

DEFENDANT CLOTHED WITH POWER OF EMINENT DOMAIN.

In our opening brief (page 24) we called the Court's *particular attention* to the fact that the defendant could have taken the Gartner property by condemnation. Defendant does not question this statement or the authorities cited to support it. By its silence it concedes them.

DIVESTMENT OF TITLE AND "ABSOLUTE CONVERSION" TO PUBLIC USE NOT ESSENTIAL TO A TAKING.

The general rule deducible from the authorities cited by us in our opening brief, page 13, is that to bring a case within the term "taking" as used in the Constitutional Amendment it is not necessary that the owner be actually divested of all title and control over the property or that it be unqualifiedly appropriated to the public.

> 20 *C. J.* 666, Sec. 138.

That it is not necessary that there be an "absolute conversion" to the use of the public is shown by the decision of the Supreme Court of the United States in *Pumpelly v. Green Bay etc. Co.,* 80 U.S. 166, 177, 20 L. ed. 557.

And to show that the rule is general and not limited to the government we cited many cases of .wrongful invasion. (Opening Brief 29-42.) Among the cases cited is *Dripps v. Allison's Mines Co.,* 45 Cal. App. 95, 187 Pac. 448, in which case the plaintiff sued to recover damages because his mining claim was invaded by tailings, slickings, and other material from the mining claim of another locator. The Court held that the invasion constituted a "taking" and quoted from *Pumpelly v. Green Bay Co.,* supra, as follows:

> "Where real estate is actually invaded by superinduced additions of water, earth, sand or other material * * * so as to effectually destroy or impair its usefulness it is a taking within the meaning of the constitution."

PAYMENT OF JUDGMENT IN AN ACTION WHERE ALL DAM-AGES PAST, PRESENT AND FUTURE ARE RECOVERED OPERATES TO TRANSFER TITLE.

The defendant says that this is not an eminent domain proceeding. That there is no question of title and possession and that the judgment did not award the defendant title to or any interest in the Gartner claims. (Defendant's Brief 32.)

In the strictest sense this is true. This is not an eminent domain proceeding. Such a proceeding under the Alaska Code (C. L. A. 1933, Sec. 1568) can only be instituted by the party seeking condemnation. But in its broadest sense it is an action founded upon the

5th Amendment for a "taking" without "due process of law".

Jacobs v. U. S., 290 U.S. 13, 16, 78 L. ed. 142.

WAIVER OF STATUTORY REQUIREMENTS.

"Where land is wrongfully taken for public use the owner has several remedies which he may pursue; he may bring an action in ejectment, may proceed by injunction, or may revert to the *common law action of trespass.*"

10 *Cal. Jur.* 443.

"In the latter proceeding if he sues for and recovers judgment for the value of the land taken, he thereby ratifies the wrongful taking."

10 *Cal. Jur.* 387.

10 *R. C. L.* 232, Sec. 195;

Newberry v. Evans, 97 Cal. App. 120, 275 Pac. 465;

Soulard v. St. Louis, 36 Mo. 546;

Jeffersonville etc. v. Estrele, 76 Ky. 667;

Southern Ry. Co. v. Perfection Laundry, 38 F.(2d) 74.

"Thus in actions for the recovery of the value of the land taken and in actions for damages where the *past, present and future damages are recoverable,* a satisfied judgment operates to transfer such title or proprietary right as defendant is entitled to."

20 *C. J.* 1216, Sec. 574;

U. S. v. Lynah, 188 U. S. 445, 470, 47 L. ed. 539;

U. S. v. Portneuf-Marsh Valley Irr. Co., 205
   Fed. 416;

Organ v. Memphis etc., 51 Ark. 235, 11 S. W.
   96;

Hickerson v. Mexico, 58 Mo. 61;

Omaha v. Redick, 61 Neb. 163, 85 N. W. 46;

East Dallas v. Barksdale, 83 Tex. 117, 18 S. W.
   329;

Huntington v. Kenower, 12 Ind. A. 456, 40 N.E.
   550;

Porter v. Railroad, 148 N.C. 563, 62 S.E. 741.

MEASURE OF DAMAGES THE SAME WHETHER THIS ACTION
   IS FOR "JUST COMPENSATION" FOR A WRONGFUL
   "TAKING", OR AN ACTION IN TRESPASS FOR DAMAGES
   FOR A PERMANENT INJURY.

Under the facts what is the measure of damages?
For permanent injuries to land there is no invariable
rule. The true rule is *compensation* which carries with
it the idea of an equivalent. This equivalent is the
measure of damages irrespective of the question as to
whether this is an action for "just compensation"
under the constitutional amendment for a wrongful
"taking" or whether it is an action in trespass for
damages for a permanent injury to real estate. In
either case the measure of damages is the same. (Opening Brief 25.)

The Supreme Court of the United States in *Hetzel
v. Baltimore & O. R. Co.,* 169 U.S. 26, 37 (42 L. ed.
648), quoted Rappallo, J., in *Baker v. Drake,* 53 N.Y.
211, 220, as follows:

"The inquiry is what is an adequate indemnity to the party injured, and the answer cannot be affected by the form of the action by which he seeks his remedy."

What is compensation? The Supreme Court of the United States in *Monongahela Nav. Co. v. U. S.,* 148 U.S. 312, 326, 37 L. ed. 463, said:

"The noun 'compensation' standing by itself, carries the idea of an equivalent * * * the natural import of the language would be that the compensation should be the equivalent of the property * * * what does full equivalent therefor demand? The value of property, generally speaking, is determined by its productiveness—the profits which its use brings to the owner."

The Supreme Court applied this rule in determining the value of a mining lease arbitrarily terminated in *Anvil Mining Co. v. Humble,* 153 U.S. 540, 550, 38 L. ed. 814, which case we cited and quoted from in our answering brief, page 37, on defendant's appeal.

> *Empire Gravel Co. v. Bonanza etc.,* 67 Cal. 406, 7 Pac. 810;
>
> *Northern Light Min. Co. v. Blue Goose Min. Co.,* 25 Cal. App. 282, 143 Pac. 540, 545;
>
> *Shoemaker v. Acker,* 116 Cal. 244, 48 Pac. 64.

PLAINTIFF'S CONTENTION AS TO THE MEASURE OF DAMAGES.

The plaintiff contends that the full equivalent for the property taken is to be measured by its value at the time of trial, in November, 1934, at which time the value of gold was $35.00 per fine troy ounce; and, that the gold content of the Gartner property should be determined on that basis. That plaintiff is entitled to recover all damages past, present and future, proximately caused by defendant's malicious tort, and that that means its value at the time of trial. (Opening Brief 27.)

The lower Court agreed :

"That this is such a case as entitles the plaintiff to all damages to him, past, present and future which the proof shows were or *will* be proximately caused by the silt, deposition and the filling of Goldstream channel, yet I cannot agree that the increased price of gold is more than fortuitously connected with plaintiff's damage. This case might well have been tried before the price of gold raised, which was over a year after the commencement of the action." (R. 53.)

The latter part of this statement is well answered by the Supreme Court of Illinois, as follows:

"As we have seen, the right of action accrued upon the happening of the injury, occasioned by the completion of the road, to recover not only present but prospective damages. It by no means follows that, because the right of action then accrued, it must then be enforced. The action may, under the statute, be brought at any time within five years from the accrual of the right, and the *railroad company or appellants, can not complain that the plaintiff has waited until his damages have become susceptible of absolute proof before bringing his action, instead of resorting to proof of prospective damages.* The authorities are that whenever the suit is brought, if within the statute of limitations, proof may be made of the permanent damages to the property, and the recovery being once for all, may include all damages flowing from the location and ordinarily skilful operation of the road." (Italics ours.)

    Suehr v. Sanitary Dist. etc. (Ill.), 90 N.E. 196, 198, quoting from *Weiss* case, 141 Ill. 35, 31 N.E. 138.

We do not agree with the statement of the trial Court that the increase in the price of gold was no more than fortuitously connected with plaintiff's damage. The increased price of gold was not **the result** of chance or accident or without a known cause. It was the result of an Act of Congress passed in May, 1933, empowering the President by executive order to reduce the gold content of the dollar by not more than 50%. The President on January 31, 1934, devalued the dollar 40.94% to its present level of 59.06% of its old parity. In other words when the Government fixed the price of gold at $35.00 per fine troy ounce, it also fixed the gold contents of the dollar at 59.06% of what it was when gold was $20.67 per ounce, which was the price of gold at the time this action was brought. After January 31, 1934, the gold miner received for his new mined gold $35.00 per ounce, but the net value of the gold in the dollars received is the same.

AS A RESULT OF THIS CHANGE IN THE PRICE OF GOLD THE
GOLD CONTENT OF THE GARTNER PROPERTY WAS IN-
CREASED IN TERMS OF DOLLARS FROM $320,994.64 TO
$543,470.55 AND THE PROFIT TO BE DERIVED BY MINING
WAS INCREASED FROM $38,686.65 TO $261,162.10, A DIF-
FERENCE OF $222,475.95. (Opening Brief 6.)

DAMAGES SUFFERED BY REASON OF INCREASED VALUE OF GOLD WAS A LOSS DIRECTLY RESULTING FROM THE INJURY COMMITTED BY DEFENDANT.

The lower Court in holding that the increase in
the price of gold was no more than fortuitously connected with plaintiff's damage, and not recoverable,
failed to distinguish between (a) the terms "injury"
and "damage"; (b) that in cases of wanton and malicious trespass the defendant is responsible for all
damages proximately caused by the injury, no matter
how unlikely, unusual or unforeseen they may have
been; and (c) that all damages accruing since the action was commenced which are the consequence of the
acts done before and constitute a part of the cause of
action are recoverable.

(a) Distinction between "injury" and "damage".

There is a material distinction between "injury"
and "damage". An injury in a legal sense is a wrong,
a violation of a right; while damage is the harm,
detriment, or loss sustained by reason of the injury.
(1 C. J. 964, Sec. 56.) In cases of trespass the cause
of action is the wrongful act of the defendant and
the injury is merely the measure of damages. (5 Am.
& Eng. Enc. Law, 16, cases cited in note 2.) Injury

means something done against the rights of the party, producing damage and has no reference to the fact or amount of damage.

>*City of North Vernon v. Voegler,* 2 N.E. 821, 824, 103 Ind. 314;
>
>*West Virginia Tran. Co. v. Standard Oil Co.,* 50 W.Va. 611, 615, 40 S.E. 591.

(b) Consequential loss in property by malicious tort recoverable.

In estimating the injury caused by reckless, willful or malicious acts, the Courts go beyond the proximate consequences. In such cases the law regards as the consequences of the tort the losses of the defendant in property. (5 Am. & Eng. Enc. Law, 11.) In cases of willful or wanton tort a wrongdoer is responsible for the direct consequences of his act, however unlikely, unusual or unforeseen they may have been.

>17 *C. J.* 750, 751, Sec. 81;
>
>*Eisle v. Oddie,* 128 Fed. 941, 949;
>
>*Garrison v. Sun Printing Co.,* 207 N.Y. 1, 100 N.E. 430.

(c) Damages accruing after action commenced.

Damages which have accrued after the action is begun may be allowed when they are the consequences of the act done before the beginning of the action and constitute a part of the cause of action declared on.

>"Where, however, the same wrongful act produces both *immediate and future injury,* independent of any subsequent wrongful acts, all damages resulting before or after commencement of the action may be recovered therein."
>
>17 *C. J.* 1085, Sec. 395.

Said the Supreme Court of the United States in
Lawler v. Loewe, 235 U.S. 522, 536, 59 L. ed. 341:

"Damages accruing since the action began were
allowed, but only such as were the consequence
of acts done before and constituting part of the
cause of action declared on. This was correct.
New York, L. E. & W. R. Co. v. Estill, 147 U.S.
591, 615, 616, 37 L. ed. 292, 304."

The United States Circuit Court of Appeals, Eighth
Circuit, said:

"It frequently happens that the consequences
of an act are not at once apparent, and that a
litigant on the day of trial is able to show that
certain damages have been sustained as the proximate result of a wrongful or a negligent act, which
could not have been proven if the trial had occurred at an earlier day; *but in such cases no
court has ever as yet intimated that the right of
recovery was limited to such damages as became
manifest immediately after the wrongful act was
committed.* On the contrary the rule is that a
plaintiff is entitled to recover compensation for
*such damages as he can establish on the day of
trial, provided they were the proximate results
of the alleged wrongful or tortious acts.* Hayden
v. Albee, 20 Minn. 159 (Gil. 143); Fort v. Railway Co., 2 Dill. 259, 268, Fed. Cas. No. 4,952;
Hagan v. Riley, 13 Gray 515, 516; Filer v. Railroad Co., 49 N.Y. 42; Suth. Dam. (2d Ed.) Sec.
113, and cases there cited."

Jones v. Allen, 85 Fed. 523, 527, certiorari denied, 171 U.S. 687.

16

Sedgwick on Damages (9th ed.), Sec. 86a, page 146, says:

"It is now universally recognized that a loss that happened after the action brought, as a direct consequence of a wrong for which the action was brought, may be compensated, though it had not happened or could not be foreseen when the action was brought."

See also:

Sutherland on Damages (4 ed.) Sec. 113.

In 24 Cal. Jur. 695, Sec. 35, it is said:

"All damages of which the trespass was the proximate and efficient cause may be recovered, and it is immaterial that part of the loss was suffered subsequent to the trespass, or that it could or could not have been anticipated."

Hawthorne v. Siegel, 88 Cal. 159, 25 Pac. 1114.

The rule in California is that damages may be awarded in a judicial proceeding for a detriment resulting after the commencement of the action. This rule is now founded on a statutory provision and under it proof of damage may extend up to the time of verdict and prospective damages may also be allowed.

8 *Cal. Jur.* 754, Sec. 22;

Bryson v. McCone, 121 Cal. 153, 53 Pac. 637;

Hicks v. Drew, 117 Cal. 305, 49 Pac. 189;

Shoemaker v. Acker, 116 Cal. 244, 48 Pac. 64.

The rule in California prevailed, however, prior to the adoption of the Code.

DeCosta v. Mass. Min. Co., 17 Cal. 613, 618;

Hicks v. Herring, 17 Cal. 566.

In *Hicks v. Herring*, supra, Field, C. J., cited with approval Greenleaf Ev., Sec. 268a (14th ed.), as follows:

> "The natural results of a wrongful act are understood to include all the damages to the plaintiff of which such act was the efficient cause, though in point of time the damage did not occur until some time after the act done. * * * And it is further to be observed, that the proof of actual damages may extend to all facts which occur and grow out of the injury, even up to the day of the verdict."

TEST OF RIGHT TO RECOVER PROSPECTIVE DAMAGES.

In our opening brief, page 27, we cited cases in support of the well established rule that for a permanent injury to real property caused by trespass all damages past, present, and prospective are and must be recovered in one action. This is also the rule in California.

> 24 *Cal. Jur.* 677, Sec. 19;
>
> *Williams v. So. Pac. R. Co.,* 150 Cal. 624, 89 Pac. 599;
>
> *Kafka v. Bozio,* 191 Cal. 746, 218 Pac. 753, 755;
>
> *Parks v. Hubbard* (N.Y.), 91 N.E. 261.

In *Ridley v. Seaboard etc.,* 118 N.C. 996, 24 S.E. 730, 32 L.R.A. 708, the Court said:

> "The right to recover prospective as well as existing damages in an action depends usually upon the answer to the test question whether the *whole injury results from the original tortious act, or 'from the wrongful continuance of the*

state of facts produced by those acts'. Troy v.
Cheshire R. Co., 23 N.H. 83, 101, 55 Am. Dec.
177.''

The direct and immediate results of defendant's
acts were the ''destruction'' of plaintiff's gold, a
''taking'' of his property. In *Jones v. Allen,* supra,
the Circuit Court of Appeals held that the plaintiff
is entitled to recover compensation for such damages
as he can establish on the day of trial, *provided they
are the proximate result of the alleged wrongful or
tortious act. The loss of the gold content of the Gartner property is the direct result of defendant's tortious
acts. If it were not for its acts—and its acts alone—
the Gartner property at the time of trial and today
would have a value of $261,162.60.* When the action
was commenced plaintiff was entitled to recover—the
injury to the property being permanent—all damages
past, present and *prospective.* The damages that were
prospective when the action was commenced were
actual present damages at the time of trial and it is
immaterial that this damage was the result of an increase in the price of gold. Gartner as a result of defendant's malicious conduct has suffered a loss of
$261,162.60. This is a consequence that has resulted
from defendant's malicious tortious acts. Defendant
should be compelled to pay therefor. Certainly Gartner should not suffer the loss.

Defendant (Brief 33) says: ''The plaintiff still
owns the land, with whatever increase in its value the
rise in the price of gold has brought about''. This is
not true. The land without the gold is valueless and

plaintiff can never *own the gold*. It has been taken away from him by the defendant, as effectually as though it had mined it and carried it away. It has deprived plaintiff of the means of mining it—has destroyed its removability—by taking away all water and drainage.

DAMAGES FOR RESTORATION NOT BEFORE THIS COURT.

The defendant has taken up considerable space in its brief in discussing the measure of damages when property can be "restored". This question is not before this Court nor was it raised or presented in the lower Court. No evidence was introduced as to the cost of removing the silt and debris deposited by defendant on the Gartner ground and in the channel of Goldstream Creek. The restoration of the Gartner claims to their former condition is out of question. The removal of the silt deposited will not restore the property. That is only a small part of the injury done by the defendant. The principal wrongs are the taking away of the water that formerly flowed through Goldstream Creek, and the destruction of the channel so essential for drainage. Defendant has diverted and appropriated the water and has destroyed the channel for a distance of three miles, in each direction, above and below the Gartner claims. How can the plaintiff restore the property?

> "Under some circumstances, the proper measure of damages may be the cost of restoring the property to its original condition, as where the injury is susceptible of remedy at a moderate

expense and the cost of restoration may be shown with reasonable certainty. * * * Such a measure cannot be adopted * * * where restoration is impracticable.''

17 *C. J.* 882;

Troy v. Cheshire R. Co., 23 N. H. 83.

Space does not permit an analysis of defendant's cases or is it necessary. The cases cited to support defendant's contention that damages are to be measured by the value of the land at the time of the injury was completed, irrespective of any damage that may have subsequently developed, do not support that doctrine. The cases cited relate to property that had a ''market value'' and was such that under the facts the rule adopted afforded adequate relief—cases where the property destroyed or damaged could be readily supplied in the open market or an equivalent given. None of them relate to ''prospective profits'' which in this case are fixed and certain and the amount thereof, upon the record, only a matter of calculation. In cases for future profit, the Supreme Court of California, in *Shoemaker v. Acker,* supra, said:

''Parties are entitled to the benefit of any facts transpiring subsequently to the bringing of the action which shows more clearly the gains prevented by the breach complained of, or the damages sustained from such a cause of action, or any other, *the injurious effects of which extended into the future.* Suth. Dam. Sec. 107. See also Tahoe Ice Co. v. Union Ice Co., 109 Cal. 248, 249, 41 Pac. 1020.'' (Italics ours.)

Defendant, in its attempt to make the cases cited by it applicable, has presumed that mining claims could be purchased in the open market or as *cordwood is sold upon the streets of Fairbanks.* (Defendant's Brief 30.) It says Gartner could have purchased for that sum ($30,000.00) land of equal value for mining purposes. Defendant knows better. It has had a vast mining experience. Besides where was Gartner to get the $30,000.00, the value of the property, according to defendant, that it had destroyed. The rule for measuring damages must not only be just but also practicable. Plaintiff is entitled, as we have stated, to recover for the "loss of profits" he has suffered. The amount thereof is certain. The findings show the quantity of gold; the cost of mining; and, the price of gold per ounce was and is fixed by law. The loss suffered, therefore, is a mere matter of calculation.

We respectfully submit that the judgment of the lower Court should be reversed, without retrial, and that this Honorable Court direct that the judgment to be entered in the lower Court be calculated at a gold value of $35.00 per fine troy ounce instead of $20.67.

Dated, San Francisco, California,
    January 17, 1938.

                    Respectfully submitted,
                        ROBERT L. MCWILLIAMS,
                        JULIAN A. HURLEY,
                        JOHN L. MCGINN,
                            Attorneys for Appellant.

No. 8168

United States
Circuit Court of Appeals
For the Ninth Circuit.

WILSON MOTOR COMPANY, a Washington
Corporation, Claimant of one Ford Coach
Automobile, Engine No. 181128696, tools and
accessories,

Appellant,

vs.

UNITED STATES OF AMERICA,

Appellee.

Transcript of Record

**Upon Appeal from the District Court of the United States
for the Western District of Washington,
Northern Division.**

FILED

MAY 4 - 1936

PARKER PRINTING COMPANY, 545 SANSOME STREET, SAN FRANCISCO

No. 8168

United States
Circuit Court of Appeals
For the Ninth Circuit.

WILSON MOTOR COMPANY, a Washington Corporation, Claimant of one Ford Coach Automobile, Engine No. 181128696, tools and accessories,

Appellant,

vs.

UNITED STATES OF AMERICA,

Appellee.

Transcript of Record

Upon Appeal from the District Court of the United States for the Western District of Washington, Northern Division.

PARKER PRINTING COMPANY, 545 SANSOME STREET, SAN FRANCISCO

INDEX

NAMES AND ADDRESSES OF PROCTORS.

Mr. FRED C. BROWN,
Proctor for Appellant,
418 Joseph Vance Bldg.,
Seattle, Washington.

Messrs. J. CHARLES DENNIS and
OWEN P. HUGHES,
Proctors for Appellee,
222 Post Office Building,
Seattle, Washington. [1*]

In the District Court of the United States for the
Western District of Washington, Northern
Division.

No. 13613.

UNITED STATES OF AMERICA,
Libelant,

vs.

ONE FORD COACH AUTOMOBILE, Engine
number 18-1128696, its tools and accessories.

LIBEL.

Under Secs. 306 and 1181, Title 26, and
Section 88, Title 27, U. S. C.

To the Honorable Judges of the District Court of
the United States for the Western District of
Washington, Northern Division:

Now comes J. Charles Dennis, United States
Attorney for the Western District of Washington,

---

*Page numbering appearing at the foot of page of original certified
Transcript of Record.

and Gerald Shucklin, Assistant United States Attorney for said District, who prosecutes for and on
behalf of the United States of America, and exhibits
this information of libel against ONE FORD
COACH AUTOMOBILE, Engine number 18-112-
8696, and its tools and accessories, and against all
persons lawfully intervening for their interest therein, which automobile was duly seized near Enumclaw, State of Washington, and within the Western
District of Washington, on the 6th day of April,
1935, by T. L. Chidester, a duly authorized, appointed and acting investigator of the Commissioner
of Internal Revenue of the United States, and is
still held in custody within said District in pursuance of said seizure, and thereupon the United
States Attorney doth allege and give the Court
to understand and be informed as follows:

I.

That on or about the 6th day of April, 1935,
near Enumclaw, Washington, and within the Northern Division of the Western District of Washington, the said ONE FORD COACH AUTOMO-
BILE, Engine number 18-1128696, with its tools
and accessories, was personal property found in an
inclosure and yard connected with and used with
the premises of a distillery for which no bond as
distiller had been given as required by law, and [2]
without having complied with the Regulations pro-
mulgated by the Commissioner of Internal Revenue
under the authority of Sec. 13, Title III of the

National Prohibition Act, which said distillery was used and intended to be used with intent to defraud the United States of the tax on the spirits distilled, imposed by Sec. 600 of the Revenue Act of 1918 as amended by Sec. 2 of the Liquor Taxing Act of 1934, contrary to the form of the statute in such case made and provided and against the peace and dignity of the United States of America.

II.

That on or about the 6th day of April, 1934, near Enumclaw, Washington, and within said District and Division, there were removed, deposited and concealed in said ONE FORD COACH AUTOMOBILE, engine number 18-1128696, certain articles and raw materials, to wit, solder and eight hundred pounds of sugar, used and intended to be used in the manufacture of certain goods and commodities, to wit, alcohol, distilled spirits, upon which there were then imposed certain taxes by Section 600 of the Revenue Act of 1918 as amended by Section 2 of the Liquor Taxing Act of 1934, with the intent to defraud the United States of the said taxes on the distilled spirits so manufactured, contrary to the form of the statute in such case made and provided, and against the peace and dignity of the United States of America.

WHEREFORE, the United States Attorney for the Western District of Washington, who prosecutes as aforesaid for the United States, prays that due process of law may be awarded in this behalf to enforce the forfeiture of said automobile and its tools and accessories, so seized, as aforesaid, and to give notice to all persons concerned to appear on the [3] return date of said process to show cause, if any they have, why said forfeiture should not be decreed; and that, for the causes aforesaid and others appearing, the said motor vehicle be condemned by sentence and decree of this Honorable Court as forfeited to the United States of America, to be disposed of in accordance with law.

J. CHARLES DENNIS
United States Attorney
GERALD SHUCKLIN
Assistant United States Attorney

United States of America
Western District of Washington—ss.

I, Gerald Shucklin, being first duly sworn, on oath depose and say:

That I am a duly appointed, qualified, and acting Assistant United States Attorney for the Western District of Washington, and as such make this verification to the foregoing libel; that I know the contents thereof, and that the matters and things therein stated are true to the best of my knowledge, information and belief.

GERALD SHUCKLIN

Subscribed and sworn to before me this 30th day of July, 1935.

[Seal]   TRUMAN EGGER
Deputy Clerk, U. S. District Court,
Western District of Washington.

[Endorsed]: Filed Jul. 30, 1935. [4]

[Title of Court and Cause.]

WARRANT OF ARREST AND MONITION.

The President of the United States of America to the Marshal of the United States, Western District of Washington, Greeting:

WHEREAS, an information has been filed in the Western District of Washington, on the 30th day of July, 1935, by the United States of America against one FORD COACH AUTOMOBILE, Engine number 18-1128696, its tools and accessories, and against all persons lawfully intervening for their interest thereon, which said FORD COACH AUTOMOBILE, hereinabove described, is in the possession of W. E. Burke, Acting Supervisor of the Alcohol Tax Unit, Seattle, Washington, and an officer of the United States, claimed as forfeited to the United States in that the said FORD COACH AUTOMOBILE, hereinabove described, was found in an enclosure and yard connected with and used with the premises of a distillery for which no bond as distiller had been given as required by law, and in that there were removed, deposited, and concealed in said automobile certain articles and raw materials used and intended to be used in the manufacture of alcohol and distilled spirits, with intent to defraud the United States of the tax imposed under the Internal Revenue Laws of the United States upon said distilled spirits, praying the issuance of [5] process in due form of law to enforce the forfeiture of the said FORD COACH, above described, its tools and accessories, so seized as set forth in said information, and to give notice to all persons concerned to appear on the return date of said process, to show cause, if any they have, why said forfeiture should not be adjudged.

YOU ARE, THEREFORE, HEREBY COMMANDED, to arrest and attach the aforesaid FORD COACH, above described, its tools and accessories, and to hold the same in your custody until the further order of this Court, respecting the same, and to give due notice of such seizure to all persons claiming the same or knowing or having anything to say why the aforesaid FORD COACH, above described, its tools and accessories, should not be condemned and forfeited, pursuant to the prayer of the said libelant, and that they be cited to file answer to said information, setting forth their interest in or claim to said property libeled, if any they have, with the Clerk of this Court, in the City of Seattle in the Northern Division of the Western District of Washington, on or before the 19th day of August, 1935, by causing the substance of said information and order to be published in the Daily Journal of Commerce, a newspaper of general circulation, published at Seattle, in the Northern Division of the Western District of Washington, and near the place of the seizure, and by posting up the same in three of the most public places in the City of Seattle, [6] in the Northern Division of the Western District of Washington, for a period of fourteen (14) days, and what you shall have done in the premises do you then make return thereon, together with this Writ.

WITNESS THE HONORABLE John C. Bowen, Judge of said Court at the City of Seattle, in the Western District of Washington, this 30th day of July, 1935.

> EDGAR M. LAKIN
> Clerk, United States District Court,
> Western District of Washington.

[Seal]   By S. COOK,
> Deputy Clerk, U. S. District Court,
> Western District of Washington. [7]

MARSHAL'S RETURN.

I hereby certify that I received the within Monition on the 30th day of July, 1935 at Seattle, Washington and thereafter on the 2nd day of August, 1935 I did cause to be published in the Daily Journal of Commerce, a newspaper of general circulation, a copy of this notice. Copy of Affidavit of Publication is attached herewith and made a part hereof. Notices were posted on the first and third floors of the Post Office Building, Seattle, Wash.

> A. J. CHITTY,
> > U. S. Marshal

By B. E. JOHNSON,
> > Deputy.

Returned this 7th day of August, 1935.

[Endorsed]: Filed Aug. 7, 1935. [8]

AFFIDAVIT OF PUBLICATION.

State of Washington,
County of King.—ss.

M. F. Brown, being first duly sworn, on oath deposes and says that he is the business manager and one of the publishers of The Daily Journal of Commerce, a daily newspaper. That said newspaper in a legal newspaper and it is now and has been for more than six months prior to the date of the publication hereinafter referred to, published in the English language continuously as a daily newspaper in Seattle, King County, Washington, and it is now and during all of said time was printed in an office maintained at the aforesaid place of publication of said newspaper.

That the annexed is a true copy of Notice of Seizure #13613, as it was published in the regular issue (and not in supplement form) of said newspaper on the 2nd day of August, 1935, and that said newspaper was regularly distributed to its subscribers during all of said period.

M. F. BROWN

NOTICE. Whereas, on the 30th day of July, 1935, the United States of America by J. Charles Dennis of Seattle, Washington, Proctor of Libelant, filed a libel of information in the District Court of the United States for the Western District of Washington, against One Ford Coach Automobile, engine number 18-1128696, its tools and accessories, in a cause of condemnation, numbered 13613. And whereas, by virtue of process in due form of law,

to me directed. I :
automobile an:
tice is hereby
same, or known:
the same should :
prayer of the s:
before the said
Western Dis:
of August. A. D.
noon of the same
Jurisdiction, o:
diction thereaf:
claim for the s:
in that behalf A.
shal. (5707)
Subscribed an:
of August, 1935.
[Seal]
Notary Public in
residing at S
tioned by V:
tion.) [9]

Form No. 252

RETURN
United States of
Western Distri:

I hereby cer:
annexed Warran:

to me directed, I have attached the above described automobile and retain the same in my custody. Notice is hereby given to all persons claiming the same, or knowing or having anything to say why the same should not be condemned pursuant to the prayer of the said Libel, that they be and appear before the said Court to be held at Seattle, in the Western District of Washington, on the 19th day of August, A. D. 1935, at ten o'clock in the forenoon of the same day, if that day shall be a day of Jurisdiction, otherwise on the next day of Jurisdiction thereafter, then and there to interpose a claim for the same, and to make their allegations in that behalf. A. J. CHITTY, United States Marshal. (5707)

Subscribed and sworn to before me this 2nd day of August, 1935.

[Seal]   L. J. BROWN

Notary Public in and for the State of Washington, residing at Seattle. (This form officially sanctioned by Washington State Press Association.) [9]

Form No. 282

RETURN ON SERVICE OF WRIT.

United States of America,
Western District of Washington.—ss.

No. 13613.

I hereby certify and return that I served the annexed Warrant of Arrest and Monition on the

therein-named ONE FORD COACH AUTOMO-
BILE Engine No. 18-1128696, its tools & etc., by
handing to and leaving a true and correct copy
thereof with A. E. Ekins, Mgr. Grand Central Ga-
rage, personally, (and posting copies of notice at
three (3) public places in the City of Seattle) at
Seattle, in said District on the 30th day of July,
A. D. 1935.

Marshal's fees: $2.87.

                    A. J. CHITTY,

                              U. S. Marshal,
          By JAMES BRIDGES

                              Deputy.

[Endorsed]: Filed Aug. 7, 1935. [10]

[Title of Court and Cause.]

AMENDED ANSWER AND CLAIM OF WILSON MOTOR COMPANY, a corporation.

To the Honorable Judges of the District Court of
    the United States for the Western District of
    Washington, Northern Division:

I.

Answering the allegations of the Libelant herein,
the Wilson Motor Company, a corporation, admits
that said automobile was seized and that the same
is being held by the officers and agents of Libelant,
and denies each and every other allegation therein
contained.

Further answering said Libel and by way of an affirmative defense, Wilson Motor Company, a corporation, alleges:

I.

That during the times hereinafter mentioned, the Wilson Motor Company, was and now is a corporation organized and existing under and by virtue of the laws of the State of Washington with its last annual license fee paid, and doing business at Seattle, King County, Washington.

II.

That on the 4th day of October, 1934, the Wilson Motor Company sold to Amico Chiola 1 Ford Model 40 Tudor Sedan Motor No. 1128696 and all accessories thereon, and at said time, as part of the consideration of the purchase of said automobile, the said Amico Chiola made, executed and delivered to petitioner a chattel mortgage, a copy of said chattel mortgage being hereto attached and marked Exhibit "A" and hereby made a part of this Answer, the original of which was duly filed with the County Auditor of King County, State of Washington on the 5th day of October, 1934 under recording No. 1155160; that as part of said transaction the [11] said Amico Chiola made, executed and delivered to the Wilson Motor Company his promissory note, a true copy of said note being hereto attached and marked Exhibit "B", and hereby made a part of this Answer.

III.

That there is now due and unpaid on said automobile the sum of Five Hundred Fifty-Five Dollars ($555.00).

IV.

That said mortgage provides that if there is default by the mortgagor in payment of any moneys due to said mortgagee; or encumber his interest in said property or in case said mortgagee shall, at any time, deem said debt insecure, all sums then remaining unpaid shall, at the election of the mortgagee, become immediately due and payable.

Said mortgage further provides that in the event of a default of the mortgagor in complying with the terms of payment hereof, the mortgagee may take immediate possession of said property and for this purpose the mortgagee may enter upon the premises where said property may be, and may foreclose said mortgage either by suit therefor or by the statutory procedure of notice and sale without action.

V.

That the said Amico Chiola is in default in the monthly payments on said note and mortgage; that the Wilson Motor Company has declared said debt insecure and has elected that all sums remaining unpaid have become immediately due and payable.

VI.

That on the 6th day of April, 1935 the said automobile was seized by the United States Internal Revenue Department under an alleged charge against the said Amico Chiola of using said automobile in the operation of a still and a violation of the United States Internal Revenue Laws, and is still in the possession of said Officers under a proceeding for forfeiture. [12]

VII.

That prior to making the sale to said Amico Chiola and receiving a note and mortgage for the balance due of the purchase price, the Wilson Motor Company caused an investigation to be made and that he was referred to the Wilson Motor Company by one of its regular customers to whom it had sold several cars, and that the Wilson Motor Company found that the said Amico Chiola was a resident of Seattle for twenty-four years and in the sewer and concrete contracting business; its information from the credit bureau indicated that Amico Chiola was a limited credit user, and with no derogatory information and no criminal record or criminal reputation. He had lived at the same address for several years. That the said Wilson Motor Company had no information that the said Amico Chiola was engaged or intended to engage in a violation of any law, and the officers of the Wilson Motor Company were led to believe him to be a reliable person. That at no time had the Wilson Motor Company any knowledge or reason to believe that said automobile would be used in violation of the laws of the United States or of any state relating to liquor.

VIII.

That petitioner states that it has a valid, paramount and superior lien against said Tudor Sedan.

That the amount of claimant's claim is in excess of the amount of the appraised value of said automobile.

IX.

That the Wilson Motor Company, a corporation herewith offers to pay the Government their costs and disbursements by reason of said seizure.

WHEREFORE, the Wilson Motor Company, a corporation, respectfully requests that said Ford Model 140 Tudor Sedan be returned to it, free from any seizure proceedings.

FRED C. BROWN
Attorney for the Wilson Motor Company, a corporation.

[13]

State of Washington,
County of King.—ss.

HARRY WILSON, being first duly sworn, on oath deposes and says: That he is the President of the Wilson Motor Company, a corporation; that he has read the foregoing Answer and Claim, knows the contents thereof, and believes the same to be true.

[Seal, Wilson Motor Co.]   HARRY WILSON.

Subscribed and sworn to before me this ...... day of September, 1935

[Seal]   E. M. DENTON
Notary Public in and for the State of Washington, residing at Seattle.

Received a copy of the within Amended Answer this 26 day of Sept. 1935.

J. CHARLES DENNIS
Attorney for Libelant.

[Endorsed]: Filed Sep. 26, 1935. [14]

INSTALLMENT NOTE COVERING CHATTEL MORTGAGE.

Seattle, Washington, October 4, 1934.

FOR VALUE RECEIVED I promise to pay to the order of WILSON MOTOR CO., payments as follows: $39.33 on the 5th day of November, 1934, and $39.05 on the 5th day of each month thereafter for seventeen months until said payments are fully paid. Any delinquent payments will draw interest at the rate of 1% per month from date of maturity. In case of default, in the payment of any installment, principal or interest, or if any proceeding in bankruptcy, receivership or insolvency, be instituted against the purchaser or his property, all payments then remaining unpaid, shall, at the election of the holder hereof, become immediately due and payable without notice.

The venue of any suit in Court or summary action on this note shall be in King County, Washington, and the parties to this note so expressly consent. In any action on this note an attorney's fee in a reasonable sum may be included in any judgment or proceeding hereon.

(Signed) AMICO CHIOLA

Motor Number 1128696 [15]

COPY

CHATTEL MORTGAGE

THIS MORTGAGE made the 4th day of October, 1934, by Amico Chiola, 1420 30th Ave So., Seattle, of the County of King, State of Washington, Mortgagor, to WILSON MOTOR CO., of Seattle, Washington, Mortgagee:

WITNESSETH: That said Mortgagor mortgages to said Mortgagee all that certain personal property situated and described as follows: 1 Ford Model 40 Tudor (standard) Sedan, Motor No. 1128696, and all accessories thereon, as security to said Mortgagee for the following payments: $39.33 Dollars on the 5th day of November, 1934, and $39.05 Dollars on the 5th day of each month thereafter, for seventeen months. Any DELINQUENT payments will draw interest at the rate of 1 per cent per month from date of maturity, according to the terms and conditions of one certain promissory note of even date herewith, payable to the order of said Mortgagee.

1. This mortgage is also given to secure any and all moneys hereafter due for work done or for material, supplies or accessories furnished or sold to mortgagor by said mortgagee and if there is default by the mortgagor in payment of said moneys due to said mortgagee, then the mortgagee may add such demand to the amount then due under this mortgage and use any means herein provided for the collection of security thereof.

2. In case said mortgagor shall default in the payment of any installment of principal or interest

under said note and mortgage as and when the same shall become due; or in case said mortgagor shall remove said property or cause or permit same to be removed from the county of King without the written consent of the Mortgagee, or encumber his interest in said property or suffer the same to become encumbered by any laborer's, mechanic's or artisan's lien; or a proceeding in bankruptcy, receivership, or insolvency be instituted against the said purchaser or his property; or in case said Mortgagee shall, at any time, deem said debt insecure, all sums then remaining unpaid shall, at the election of the mortgagee, become immediately due and payable.

3. In the event of a default of the Mortgagor in complying with the terms of payment hereof, the Mortgagee may take immediate possession of said property, and for this purpose the Mortgagee may enter upon the premises where said property may be and remove the same, and may foreclose said mortgage either by a suit therefor or by the statutory procedure of notice and sale without action.

4. The Mortgagor shall give immediate written notice to Mortgagee of any and all loss of or damage to said property. The proceeds of any insurance paid by reason of such loss or injury shall become subject to the terms hereof to be applied toward the repair and replacement of the property or protanto payment of the above obligations as the Mortgagee may elect.

5. The Mortgagee has the privilege of insuring said personal property against fire, theft, embezzle-

ment and confiscation, and the Mortgagor hereby agrees to pay the premium for such insurance. The Mortgagor waives and assigns to Mortgagee, all right in and claim to any dividends which may be earned on insurance premium.

6. The venue of any suit in court or summary action on this mortgage shall be in King County, Washington, and the parties to this mortgage so expressly consent. In any action on this mortgage an attorney's fee of 15% of the amount due may be included in any judgment or proceeding hereon.

In TESTIMONY WHEREOF, the Mortgagor herein named hereunto set his hand and seal the day and year first above written.

(Signed) AMICO CHIOLA

Witness:

State of Washington,
County of King.—ss.

Amico Chiola, the Mortgagor in the foregoing mortgage named, being first duly sworn, on oath deposes and says that aforesaid mortgage is made in good faith, and without design to hinder, delay or defraud creditors.

(Signed) AMICO CHIOLA

Subscribed and sworn to before me this 4th day of October, 1934.

(Signed) E. M. DENTON

Notary Public in and for the State of Washington, residing at Seattle.

State of Washington,
County of King.—ss

On this 4th day of October, 1934, before me, a
Notary Public in and for the State of Washington,
personally came Amico Chiola to me known to be
the individual described in and who executed the
within instrument, and acknowledged that he signed
and sealed the same as his free and voluntary act
and deed for the uses and purposes therein mentioned.

WITNESS my hand and official seal, the day and
year in this certificate first above written.

(Signed) E. M. DENTON

Notary Public in and for the State of Washington,
residing at Seattle. [16]

Amount of Contract $703.18
Dated Oct. 4, 1934.

[Title of Court and Cause.]

DECREE OF CONDEMNATION
AND FORFEITURE.

This matter having come on duly and regularly
to be heard on this 28th day of February, 1936, the
libelant appearing by its attorneys, J. Charles
Dennis, United States Attorney for the Western
District of Washington, and Owen P. Hughes, Assistant United States Attorney for said district, and
it appearing to the Court that due notice of the
seizure of said Ford coach automobile, engine number 18-1128696, its tools and accessories, and the

time and place of trial and hearing upon the information filed herein, has been given both by publication and posting of the same, in accordance with the statutes and laws in such cases made and provided, all of which is shown by the files and records herein, and it further appearing that the Wilson Motor Company, a corporation, filed or caused to be filed a petition for remission or mitigation under and by virtue of Section 204, Title II of the Act of Congress of August 27, 1935, and it further appearing that all other persons having or claiming an interest in said Ford coach automobile, engine number 18-1128696, its tools and accessories, have failed and neglected to enter their appearance herein and are in default for such failure to appear and defend herein, and that after due proclamation made, an order of default has been heretofore [17] duly and regularly entered herein against all said other persons, and the Court being duly advised in the premises, and it appearing to the Court that the above mentioned Ford coach automobile, engine number 18-1128696, its tools and accessories, were personal property found in an enclosure and yard connected with and used with the premises of a distillery for which no bond as distiller had been given as required by law and which said automobile was used for the removal, deposit and concealment of certain articles and raw materials used and intended to be used in the manufacture of certain goods and commodities, to wit, alcohol and distilled spirits, with the intent to defraud the United States of the tax on the distilled spirits so manu-

factured, all in v...
Title 26, U.S.C.A
U.S.C.A., as th...
therefore

It is hereby OR...
CREED that sa...
number 18-112...
before in the cus...
be and the same is...
unto the United S...
Done in open c...

J(

Presented by:
OWEN P. HU(
Asst. U...

Approved:
FRED C. BR(
Atty. for ...

[Endorsed]: FI...

[Title of Court an...

ST...

IT IS HEREBY
tween J. Charles I...
for the Western D...
P. Hughes, assist...
said district, and F...
Wilson Motor C...

factured, all in violation of Sections 306 and 1181, Title 26, U.S.C.A., and Section 88, Title 27, U.S.C.A., as the libel of information alleges; now, therefore

It is hereby ORDERED, ADJUDGED and DE-CREED that said Ford coach automobile, engine number 18-1128696, its tools and accessories, heretofore in the custody of the United States Marshal, be and the same is hereby condemned and forfeited unto the United States of America.

Done in open court this 28 day of February, 1936.

JOHN C. BOWEN

United States District Judge.

Presented by:

OWEN P. HUGHES

Asst. United States Attorney

Approved:

FRED C. BROWN

Atty. for Wilson Motor Co.

[Endorsed]: Filed Feb. 28, 1936. [18]

[Title of Court and Cause.]

STIPULATION.

IT IS HEREBY STIPULATED by and between J. Charles Dennis, United States Attorney for the Western District of Washington, and Owen P. Hughes, assistant United States Attorney for said district, and Fred C. Brown, attorney for the Wilson Motor Company, that the above mentioned

automobile was seized by the United States officers·
of the Internal Revenue Department in King County, Washington, and within the jurisdiction of the
District Court for the Western District of Washington, Northern Division, on the 6th day of April,
1935, while being used in connection with the operation of a still in violation of the Internal Revenue laws of the United States; that on or about·
the 30th day of July, 1935, a libel of information
was filed against said automobile by the United
States Attorney for the Western District of Washington; that on or about the 26th day of September, 1935 the Wilson Motor Company filed herein
its petition for remission or mitigation of forfeiture
of said automobile under and by virtue of Section
204, subsection a of Title 2 of the Act of Congress
of August 27th, 1935. This cause having come on
regularly for trial, the witnesses for the Government and the Wilson Motor Company being in at-
tendance, it being agreed in open court and by this
stipulation that [19] said libel was in proper form
and that said automobile was used in violation of
the Internal Revenue laws of the United States in
the operation of an unlawful still in King County,
Washington; that the Wilson Motor Company on
the 4th day of October, 1934, sold said automobile
to Amico Chiola, one of the defendants, who was arrested for the operation of said still; that the Wilson Motor Company sold said automobile in good
faith; that its officers had at no time any knowledge
or reason to believe that said automobile would be

used in the violation of the Laws of the United
States or any state relating to liquor. That its officers made inquiry at the headquarters of the Sheriff of King County, and the headquarters of the
Chief of Police of the City of Seattle, and the principal Federal Internal Revenue Officer engaged in
the enforcement of the liquor laws of King County,
as to the character of the said Amico Chiola, and
was informed by said officers that he had no record
or reputation of violating the liquor laws of the
United States or of the State of Washington. That
at the time of the sale of said automobile, the said
Amico Chiola made, executed and delivered to the
Wilson Motor Company a chattel mortgage on said
automobile and made, executed and delivered to
the Wilson Motor Company his promissory note;
that the amount due and owing to the Wilson Motor
Company from the said Amico Chiola, at the time
of seizure, was in excess of the appraised value of
the automobile made by the officers of the Internal
Revenue Department; that the said mortgage was,
on the 5th day of October, 1934, recorded in the
Auditor's Office of King County, Washington and
has not been satisfied; that the Wilson Motor Company has offered to pay the United States Government all expenses incidental to the seizure and forfeiture incurred by the United [20] States Government. That said Petition of the Wilson Motor Company is in proper form and the facts therein stated
are confessed. That upon these statements of respective attorneys, the cause was, by the Court, taken

under advisement and thereafter forfeiture was decreed.

DATED this 3d day of March, 1936.

J. CHARLES DENNIS

United States Attorney

OWEN P. HUGHES

Assistant United States Attorney

FRED C. BROWN

Atty for Claimant.

[Endorsed]: Filed Mar. 3, 1936. [21]

[Title of Court and Cause.]

ORDER.

This matter having come on duly and regularly to be heard on this 13th day of March, 1936, upon the petition of claimant Wilson Motor Company, a corporation, for remission or mitigation of the forfeiture in the above entitled cause under Section 204, subsection (a), Title II of the Act of Congress of August 27, 1935, and upon the stipulation filed herein March 3, 1936 the libelant appearing by its attorneys, J. Charles Dennis, United States Attorney for the Western District of Washington, and Owen P. Hughes, Assistant United States Attorney for said District, the Wilson Motor Company, a corporation, appearing by its attorney, Fred C. Brown, and it appearing to the Court that the seizure of the said Ford coach automobile, engine number 18-1128696, its tools and accessories, was made on or about April 6, 1935, in the Northern Division of the Western District of Washington, and it further appearing to the Court that a libel of information was filed by the United States of America on July 30, 1935 against the above described motor vehicle under Sections 306 and 1181, Title 26, U.S.C.A., and Section 88, Title 27, U.S.C.A., and that said petition of Claimant Wilson Motor Company was filed herein Sept. 26, 1935; it further appearing to the Court that the said Ford coach automobile, engine number 18-1128696, its tools and accessories, was condemned and forfeited unto the United States [22] of America on February 28, 1936, and the Court being fully advised in the premises,

It is hereby ORDERED, ADJUDGED and DECREED that the petition of the claimant Wilson Motor Company, a corporation, be and is hereby denied upon the ground and for the reason that Section 204, subsection (a), Title II of the Act of Congress of August 27, 1935 is not retroactive and does not apply to pending cases.

Claimant, Wilson Motor Company, a corporation, excepts to the denial of its petition for remission or mitigation and said exception is hereby allowed.

Done in open court this 13th day of March, 1936.

JOHN C. BOWEN

United States District Judge

Presented by:

OWEN P. HUGHES

Asst. U. S. Attorney

[Endorsed]: Lodged Mar. 12, 1936.
[Endorsed]: Filed Mar. 13, 1936.   [23]

[Title of Court and Cause.]

NOTICE OF APPEAL AND PETITION FOR ALLOWANCE THEREOF.

The above named claimant, Wilson Motor Company, conceiving itself aggrieved by the order made and entered in the above entitled cause on the 13th day of March, 1936, denying its claim for a remission or mitigation of the forfeiture of that certain automobile, one Ford Coach automobile, Engine No. 18-1128696, its tools and accessories, does hereby appeal from said order to the United States Circuit Court of Appeals, in and for the Ninth Circuit, for the reason specified in the Assignment of Errors, which is filed herewith, and it prays that the Appeal may be allowed and that the transcript of the record, proceedings, and papers, upon which the said order was made, duly authenticated, may be sent to the Circuit Court of Appeals for the Ninth Circuit.

FRED G. BROWN
Attorney for Claimant and
Appellant

Office and Post Office Address:
418 Joseph Vance Building
Seattle, Washington

Received a copy of the within Notice and Petition this 30 day of Mar. 1936.

J. CHARLES DENNIS
Attorney for Libellant.

[Endorsed]: Filed Mar. 30, 1936. [24]

[Title of Court and Cause.]

ASSIGNMENT OF ERRORS.

Comes now the Wilson Motor Company, the Claimant in the above cause, by and through the undersigned, its attorney, and specifies the following Assignment of Errors to be relied upon on its Appeal herein.

I.

That the Court erred in holding that Section 204, sub-section (a), of Title Two of the Act of Congress of August 27, 1935, is not retroactive and does not apply to pending cases.

II.

That the Court erred in entering its order denying the claim of the Wilson Motor Company's right to have the forfeiture of said automobile remitted or mitigated.

III.

That the Court erred in refusing to hear and determine and exercise its discretion as to what terms it would impose on the Wilson Motor Company to remit or mitigate said forfeiture.

IV.

For errors appearing upon the record.

For the above and foregoing errors apparent on the face of the record, petitioner prays that the Order herein rendered be reversed and that said Order be set aside and that the Hon. District Court proceed to hear and determine the petition of the Wilson Motor Company, claimant, for a remission or mitigation of the forfeiture.

FRED C. BROWN
Attorney for Claimant,
Wilson Motor Company.

Received a copy of the within Assignment of Errors this 30th day of March, 1936.

OWEN P. HUGHES
Asst. United States Dist. Atty.

Received a copy of the within Assignment of Errors this 30 day of Mar. 1936.

J. CHARLES DENNIS,
Attorney for Libellant.

[Endorsed]: Filed Mar. 30, 1936. [25]

———

[Title of Court and Cause.]

ORDER ALLOWING APPEAL.

This cause having come on regularly to be heard and it appearing to the Court that the above named claimant has presented herein his Notice and Petition for Appeal in the above cause, together with Assignment of Errors, and the Court being fully advised,

IT IS HEREBY ORDERED that said Appeal be allowed as prayed.

DONE IN OPEN
March, 1936.
Jul

Presented by:
FRED C. BROWN.
Attorney for C.
Wilson Motor (

Received a copy
day of Mar. 1936.
J. (

[Endorsed]: Filed

Title of Court and
ST.

It is hereby stip.
Dennis, United St.
District of Wash
Owen P. Hughes, A.
for said District, c.
necessary part of t
the question involv
1. The Libel.
2. The Stipula
United State
Brown.

DONE IN OPEN COURT this 30th day of March, 1936.

JOHN C. BOWEN,
Judge.

Presented by:

FRED C. BROWN,
Attorney for Claimant,
Wilson Motor Company.

Received a copy of the within Order this 30th day of Mar. 1936.

J. CHARLES DENNIS,
Attorney for Libellant.

[Endorsed]: Filed Mar. 30, 1936. [26]

Title of Court and Cause.]

STIPULATION.

It is hereby stipulated, by and between J. Charles Dennis, United States Attorney for the Western District of Washington, Northern Division, and Owen P. Hughes, Assistant United States Attorney for said District, and Fred C. Brown, that the only necessary part of the record for a determination of the question involved herein shall be

1. The Libel.
2. The Stipulation of March 3, 1936, between United States District Attorney and Fred C. Brown.

3. Order Denying Claim of the Wilson Motor Company for remission or mitigation of forfeiture.
4. Notice of Appeal and Petition for Allowance Thereof.
5. This Stipulation.
6. Assignment of Errors.
7. Order Allowing Appeal.
8. Citation.
9. Certificate of Clerk of U. S. District Court to transcript of record.
10. Bond.
11. The Praecipe.
12. Decree of Forfeiture, Feb. 28, 1936.
13. Warrant of Arrest and Monition and Marshal's Return.
14. Amended Answer, Sept. 26, '35.

Dated this 30th day of March, 1936.

J. CHARLES DENNIS,
United States Attorney.
OWEN P. HUGHES,
Asst. United States Attorney.
FRED C. BROWN,
Attorney for Wilson Motor Co.

Received a copy of the within Stipulation this 30 day of Mar. 1936.

J. CHARLES DENNIS,
Atty. for Libellant.

[Endorsed]: Filed Mar. 30, 1936. [27]

(Title of Cour: .

BOND ON .
I.

WHEREAS, a
in this Court ':
against one Fc: !
its tools and a:...

appeal to the C. '
an Order made .
such appeal an!
Co. give bond c: :
DRED FIFTY &
NOW, THERE!
agreed for the '..
the undersign.!
of Two Hundre! F.
that the Wilso.. \
appeal and all s:.:
until final deter:.
entitled cause an!
mobile truck in ..
costs as may be a:.
to the appellate ...
Dated this 25:b .

V.
B.
N.

[Seal]

(Title of Court and Cause.)

BOND ON APPEAL FROM ORDER OF DISTRICT COURT.

WHEREAS, a libel and complaint was duly filed in this Court by the United States of America against one Ford Coach Automobile, E#18-1128696 its tools and accessories; and whereas on .... an order was entered allowing an appeal to the U. S. Circuit Court of Appeals from an Order made herein and in said order allowing such appeal and requiring that the Wilson Motor Co. give bond on appeal in the sum of TWO HUN-DRED FIFTY & NO/100 ($250.00) DOLLARS

NOW, THEREFORE, it is hereby stipulated and agreed for the benefit of whom it may concern that the undersigned shall be and are bound in the sum of Two Hundred Fifty & No/100 Dollars conditioned that the Wilson Motor Co. will pay all costs on appeal and all storage costs from March 13, 1936, until final determination of the appeal in the above entitled cause and also storage charges on auto- mobile truck in cause #13612, together with all costs as may be awarded against it on such appeal to the appellate court.

Dated this 25th day of March, 1936.

WILSON MOTOR COMPANY,

By Fred C. Brown, its attorney.

NEW AMSTERDAM

CASUALTY COMPANY,

[Seal]   By M. A. Reese, Attorney in Fact.

O. K. as to form.

    ·OWEN P. HUGHES,

       Asst. U. S. Dist. Atty.

Approved this 30th day of March, 1936.

      JOHN C. BOWEN,

       U. S. District Judge.

[Endorsed]: Filed Mar. 30, 1936. [28]

————

[Title of Court and Cause.]

PRAECIPE FOR TRANSCRIPT OF RECORD.

To the Clerk of the above entitled Court:

You will please prepare a typewritten transcript
of record in the above entitled cause and file the
same in the United States Circuit Court of Appeals,
for the Ninth Circuit, the said record to comprise
the following papers:

1. The Libel.
2. The Stipulation of March 3, 1936, between
   United States District Attorney and Fred C.
   Brown.
3. The Order Denying the Claim of the Wilson
   Motor Company for remission or mitigation
   of the forfeiture, March 13, '36·
4. Notice of Appeal and Petition for Allowance thereof.
5. Stipulation between United States Attorney
   and Attorney for Claimant as to papers on
   Appeal.
6. Assignment of Errors.

7. Order Allowing Appeal.
8. Citation.
9. Certificate of Clerk of United States District Court to transcript of record.
10. This Praecipe.
11. Bond.
12. Decree of Forfeiture, Feb. 28, 1936.
13. Warrant of Arrest and monition and Marshal's Return.
14. Amended Answer, Sept. 26, 1935.

FRED C. BROWN,
Attorney for Claimant,
Wilson Motor Company.

I hereby acknowledge service of the above and foregoing praecipe and waive the right to request the insertion of any other matters than those in that prayed in said Praecipe and Stipulation; that the proceedings, papers, orders, and documents, included in said Praecipe constitute a full and sufficient record on appeal.

Dated this 30th day of March, 1936.

OWEN P. HUGHES,
Asst. United States Dist. Atty.

Received a copy of the within Praecipe this 30th day of Mar. 1936.

J. CHARLES DENNIS,
Attorney for Libellant. [29]

(Title of Court and Cause.)

CERTIFICATE OF CLERK U. S. DISTRICT COURT TO APOSTLES ON APPEAL.

United States of America,
Western District of Washington.—ss:

I, Edgar M. Lakin, Clerk of the United States District Court for the Western District of Washington, do hereby certify that the foregoing typewritten transcript of record, consisting of pages numbered from 1 to 29, inclusive, is a full, true, correct and complete copy of so much of the record, papers and other proceedings in the above entitled cause as is required by praecipe of counsel filed and shown herein, as the same remain of record and on file in the office of the Clerk of said District Court, and that the same constitute the apostles on appeal herein from that certain order filed March 13, 1936, denying claim for remission or mitigation of the forfeiture of the hereinabove described one Ford Coach automobile, to the United States Circuit Court of Appeals for the Ninth Circuit.

I further certify the following is a true and correct statement of all expenses, costs, fees and charges incurred in my office by or on behalf of the appellant for making record, certificate or return to the United States Circuit Court of Appeals for the Ninth Circuit, in the above entitled cause, to-wit: [30]

Clerk's fees (Act of Feb. 11, 1925) for making record, certificate or return, 63 folios at 15c .... $ 9.45

Appeal fee (Sec. 5 of Act) 5.00

Certificate of Clerk to Apostles on Appeal .50

Total $14.95

I certify that the above cost for preparing and certifying record, amounting to $14.95 has been paid to me by the proctor for Claimant.

I further certify that no citation on appeal was issued in this cause.

IN WITNESS WHEREOF, I have hereunto set my hand and affixed the official seal of said District Court at Seattle, in said District, this 6th day of April, 1936.

EDGAR M. LAKIN,

[Seal]   Clerk United States District Court, Western District of Washington.

By Truman Egger,

Deputy. [31]

[Endorsed]: No. 8168. United States Circuit
Court of Appeals for the Ninth Circuit. Wilson
Motor Company, a Washington Corporation, Claim-
ant of one Ford Coach Automobile, Engine No.
18-128696, tools and accessories, Appellant, vs.
United States of America, Appellee. Transcript of
Record Upon Appeal from the District Court of
the United States for the Western District of
Washington, Northern Division.

Filed April 8, 1936.

PAUL P. O'BRIEN

Clerk of the United States Circuit Court of Appeals
for the Ninth Circuit.

In the United States Circuit Court of Appeals For the Ninth Circuit

No. 8168

WILSON MOTOR CO., a Washington Corporation, Claimant of One Ford Coach Automobile, Engine No. 18-1128696; Its Tools and Accessories,

Appellant.

vs.

UNITED STATES OF AMERICA,

Appellee.

APPEAL FROM AN ORDER *of the* DISTRICT COURT OF THE UNITED STATES FOR THE WESTERN DISTRICT OF WASHINGTON, NORTHERN DIVISION

Honorable John C. Bowen, *Judge*

BRIEF OF APPELLANTS

FILED

FRED C. BROWN,
Attorney for Appellant.

In the United States
Circuit Court of Appeals
For the Ninth Circuit

No. 8168

WILSON MOTOR CO., a Washington Corporation, Claimant of One Ford Coach Automobile, Engine No. 18-1128696; Its Tools and Accessories,

Appellant.

vs.

UNITED STATES OF AMERICA,

Appellee.

APPEAL FROM AN ORDER *of the* DISTRICT COURT OF THE UNITED STATES FOR THE WESTERN DISTRICT OF WASHINGTON, NORTHERN DIVISION

Honorable John C. Bowen, *Judge*

BRIEF OF APPELLANTS

FRED C. BROWN,
Attorney for Appellant.

STATEMENT OF THE CASE

In King County, Washington, on April 6th, 1935, United States Federal Internal officers seized the above described automobile in an enclosure wherein a still was being operated, in violation of the United States Internal Revenue laws. The automobile had solder and sugar therein intended to be for use in operation of the still.

On the 4th day of October, 1934, the claimants, Wilson Motor Company, sold said automobile to Amico Cheola.

That at the time of sale, said Amico Cheola executed and delivered to claimant, his note and mortgage on the automobile as security for the unpaid balance.

That the mortgage was recorded, in accordance with law, with the Auditor of King County, Washington, on the 5th day of October, 1934.

That at the time of said seizure, the amount due claimant was in excess of the appraised value of the car.

That on the 30th day of July, 1935, the United

States District Attorney filed in Court, the libel against the automobile praying for forfeiture by Court Decree.

On the 26th day of September, 1935, claimant filed therein his petition as mortgagee praying a remission or mitigation of the offence, said petition being in proper form and in pursuance of the provisions of Section 204, sub-section (a) of the Act of Congress of August 27, 1935, hereinafter described.

That at the time of the sale of said automobile to Amico Cheola, he had no record of conviction or reputation as a violator of any law of the United States or of the State of Washington, relative to liquor.

That the case came on regularly for trial, the facts were stipulated in Open Court and reduced to writing and filed March 3, 1936. Transcript of Record, page 21. That on 28, Feb., 1936, Court declared the forfeiture. Transcript of Record, page 19.

That the Court denied the claimant's right to a remission or mitigation for the reason that the above described act was not retroactive and did not apply to pending cases. Transcript of Record, page 24.

ERRORS

1. The Court erred in holding that Section 204, sub-section (a), of Title Two of the Act of Congress of August 27, 1935, is not retroactive and does not apply to pending cases.

2. The Court erred in entering its Order Denying the Claim of the Wilson Motor Company's right to have the forfeiture of said automobile remitted or mitigated.

3. The Court erred in refusing to hear and determine and exercise its discretion as to what terms it would impose on the Wilson Motor Company, to remit or mitigate said forfeiture.

AUTHORITIES AND ARGUMENT

Congress, on August 27, 1935, enacted "The Liquor Law Repeal and Enforcement Act."

The act repeals Title I and II of the National Prohibition Act, reenacted certain provisions of Title II, amended and repealed various liquor laws, and for other purposes.

Section 204 (a) of Title II is as follows:

> Whenever, in any proceeding in court for the forfeiture, under the internal-revenue laws, of any vehicle or aircraft seized for a violation of the internal-revenue laws relating to liquors, such forfeiture is decreed, the Court shall have exclusive jurisdiction to remit or mitigate the forfeiture.

Then follow Sections (b), (c), and (d), that provide the conditions under which the Court shall remit or mitigate the forfeiture (under the Stipulation the Appellant has met all of those conditions). Transcript of Record, page 21.

Congress, by the enactment of this law, changed the policy of the law by giving the Court additional jurisdiction so as to protect the rights of lien holders in automobiles and airplanes, and as pertinently said in *"United States vs. One Paige Automobile,"* 277, Fed. page 524: And this protection has been conferred by Congress in such unmistakable language that an attempt to apply against a bona fide lien-holder the old judicial theory of destruction of his lien under the custom laws would be inexcusable judicial legislation to set aside the intention of Congress not to apply those laws in the enforcement of the Prohibition Laws.

In *Oakland Motor Car Company vs. United States,* 295 Fed., page 626, the Court said "the purpose of the provision of the National Prohibition Act, respecting forfeiture of vehicles * * * is to penalize only the law breaker and protect the interest of innocent persons in the vehicle and the vendor under a conditional sale contract who proves he has a lien * * * is entitled to payment of his lien." Citing *Vickmore vs. United States,* 295 Fed. 620.

United States vs. Sylvester, 273 Fed., 253: In interpreting and construing forfeiture under the National Prohibition Act the Court says: "But it is apparent that the Congress intended to penalize only the wrong-doer. This is accomplished in two ways—first, by imposing a penalty for the offence of transporting intoxicating liquor; and, second, by confiscating his interest in the vehicle. The Government, therefore, is to determine how to bring into practical operation the provisions of Section 26 to the Act, that the wrong-doer may be properly punished while the innocent parties may be protected against loss, as far as possible."

This Act does not change the procedure or modify any right of forfeiture in the Government up to the time that the Court decrees the forfeiture; after which time the Court has jurisdiction to remit or mitigate the forfeiture upon the petition of a lien holder in the automobile or aircraft, when such interest has complied with the conditions required therein.

III.

We contend that it is remedial legislation, to correct a harsh, unjust law.

II.

The Act applies to methods of procedure of cases pending before the Court.

// I.

It enlarges the jurisdiction of the Court in the hearing of judicial condemnation.

IV.

It does not by implication repeal any law.

Before its passage no power existed in any department of the Government to recognize the rights of conditional sale vendees, or mortgagees of automobiles or aircraft.

The Government will contend, we believe, that the

only remedy of the claimant for relief would have been under Section 158, Title 26, U. S. C. This section provides as follows:

> "158. *Compromises.* The Commissioner of Internal-Revenue, with the advice and consent of the Secretary of the Treasury, may compromise any civil or criminal case arising under the Internal-Revenue Laws instead of commencing suit thereon; and, with the advice and consent of the said Secretary and the recommendation of the Attorney General, he may compromise any such case after suit thereon has been commenced. Whenever a compromise is made in any case there shall be placed on file in the office of the Commissioner the opinion of the solicitor of Internal Revenue, or of the officer acting as such, with his reason therefor, with a statement of:
>
> 1. The amount of the tax assessed.
> 2. The amount of additional tax or penalty imposed by law in consequence of the neglect or delinquency of the person against whom the tax is assessed, and
> 3. The amount actually paid in accordance with the amount of the compromise."

Of course the libel in this case was not to collect a tax but was an action *in rem* to condemn the automobile as an instrument used in violation of law.

We believe that the opinion of the Attorney General as cited in Vol. 23 of the Opinions of the Attorney

General, page 507, does not give the Internal-Revenue Collector the power to compromise a case of this kind.

This opinion says: "This section provides for the settlement of cases arising under the Internal-Rvenue Laws, but it is obvious that the section refers to suits commenced by the Government to recover taxes."

If, however, the Commissioner of Internal Revenue, with the consent of the Attorney General, had the power to compromise a case of this kind, the Government was not injured by transferring that power to the Court.

If a seizure conferred a vested right in the Government then the same vested right is in the Government for a seizure taking place after August 27, 1935.

If there was the right to compromise before, that right has been transferred from the Internal Revenue Department to the Judicial Department.

It cannot be said that the government would be deprived of a compromise offer, for the Act gives the Court the power to impose terms (mitigate) against the lienholder, the same as under Sec. 158.

So how can it be contended that the Government is

injured by transfer of authority from one department of the Government to another?

The Court was of the opinion that the Act of Congress of August 27, 1935, did not apply to pending cases for the reason that it was an implied repeal of said Section 158.

Section 158 was not repealed as Section 8 of Title 1 provides:

> "It shall be unlawful to have or possess any liquor or property intended for use in violating the provisions of this title, or of Title III of the National Prohibition Act, or the internal-revenue laws, or regulations prescribed under such title or laws, or which has been so used, and no property rights shall exist in any such liquor or property. A search warrant may issue as provided in Title XI of the Act approved June 15, 1917 (40 Stat. 228; 18 U.S.C., Secs. 611-633), for the seizure of such liquor or property. Nothing in this section shall in any manner limit or affect any criminal or forfeiture provision of the internal-revenue laws, or of *any other law.* * * *"

LAW APPLIES TO PENDING CASES

The case of *Hollowell vs. Commissioners*, 239 U. S. 506, was a suit affecting the title to an allotment made under the Act of August 7, 1882. The Court said:

"It is unnecessary to consider whether there was jurisdiction when the suit was begun. By the Act of June 25, 1910, c. 431, 36 Stat. 855, it was provided that in a case like this of the death of the allottee intestate during the trust period, the Secretary of the Interior should ascertain the legal heirs of the descendant and his decision should be final and conclusive; with considerable discretion as to details. This Act restored to the Secretary the power that had been taken from him by Acts of 1894 and February 6, 1901, c. 217, 31 Stat. 760. *McKay vs. Kalyton,* 204 U. S. 458, 468. It made his jurisdiction exclusive in terms, it made no exception for pending litigation, but purported to be universal and so to take away the jurisdiction that for a time had been conferred upon the Courts of the United States. The appellee contends for a different construction on the strength of Rev. Stat., 13, that the repeal of any statute shall not extinguish any liability incurred under it, *Hertz vs. Woodman,* 218 U. S. 205, 216, and refers to the decisions upon the statutes concerning suits upon certain bonds given to the United States, 209 U. S. 306. But apart from a question that we have passed, whether the plaintiff even attempted to rely upon the statutes giving jurisdiction to the courts in allotment cases, the reference of the matter to the Secretary, unlike the changes with regard to suits upon bonds, takes away no substantive right but simply changes the tribunal that is to hear the case. In doing so it evinces a change of policy, and an opinion that the rights of the Indians can be better preserved by the quasi-paternal supervision of the general head of Indian affairs. The consideration applies with the same force to all cases and was embodied in a statute that no doubt was intended to apply to all, so far as construction is concerned."

The Government may contend the right of forfeiture became absolute at the time of seizure.

If this is true it would be so for a seizure after the Act of Congress of August 27, 1935. Still title is not perfected until judicial determination.

The acquittal of a defendant in a criminal charge is a bar to the forfeiture.

> *United States vs. One Ford Sedan,* 5 Cir. 297
> Fed. 830.

We believe the principle in *Larkin vs. Saffarans,* Vol. 15, Fed. Reporter, page 147, is decisive of this case.

The syllabus read as follows:

"Statutes which are remedial will be given a retrospective effect, unless they direct to the contrary. Where, therefore, an act of Congress enlarges the jurisdiction of the Circuit Court, it will be construed to apply to the cases pending and undetermined at the passage of the act unless excluded by its terms or necessary implication from the language of the act."

Page 149:

But, under the act of 1875, there can be no doubt of our jurisdiction, if the first section applies to cases pending in the courts at the time of its passage. And why does it not apply? Counsel say it is

because it is giving that act a retrospective operation, without any words directing that it shall so
operate, and because it interferes with vested
rights. The first obvious suggestion here is, can
the statute, in conferring jurisdiction over suits
then pending, be said to act retrospectively in any
proper sense? It acts immediately on a thing then
in existence, and from that moment gives the court
a power to act on that thing which it did not before have; but the idea that it acts retrospectively
is founded on the assumption that the question of
jurisdiction is to be determined as of the date when
the suit was brought, and not as of the date when
the decision is made, it being argued that the proceeding was void in the beginning, and cannot be
made valid by subsequent legislation. That Congress has the power to bestow jurisdiction over a
pending suit there can be no doubt whatever, if the
act says so in terms; and, in this connection, it
must be remembered that there are no constitutional restrictions upon Congress in the matter of
retrospective legislation as there are in some of the
states. *Satterlee vs. Mathewson,* 2 Pet. 380; *Sinking Fund Cases,* 99 U. S. 700.''

* * *

The case of *Sampeyreac vs. U. S.,* 7 Pet. 222; S.
C. Hempst, 118, is a direct authority for the power
of Congress to do what the plaintiff claims has been
done here; and it will be found that it has been
sometimes ruled in the state courts that such legislation interferes with no vested right, since one
can have no vested right to any particular remedy,
or to sue or be sued in any particular court or to a
defense growing out of mere remedial legislation.
For example, a party cannot complain if the legis-

lature enlarges the statute of limitations, if this be done before the bar actually attaches under the old statute. And it will be found that, both in the civil and common law, the repugnance to retrospective legislation was not understood to extend to remedial legislation of that character. In Tennessee we have a constitutional provision 'that no retrospective law, or law impairing the obligation of contracts, shall be made,' and yet at a very early day it was construed to apply only to the impairment of contracts, and not prohibitory of the large class of legislation affecting remedies, remitting penalties, etc. 'In short,' says the Supreme Court, 'so many are the past transactions upon which the public good requires posterior legislation, that no government can preserve order, suppress wrong, and promote the public welfare without the power to make retrospective laws.' *Townsend vs. Townsend,* Peck, 1, 17; 1 Tenn. Code, (T. & S.) 79, and notes; 2 Meigs, Dig. (2d Ed.) Paragraph 727, p. 886.

* *

Page 150:

And the rule is that 'where the enactment deals with procedure only unless the contrary be expressed, the enactment applies to all actions, whether commenced before or after the passing of the act.' *Broom, Legal Max.,* 35; *Wright vs. Hale,* 6 Hurl. & N., 227; *Kimbray vs. Draper,* L.R. 3 Q.B. 160.

This is only in accordance with the general rule that all remedial legislation shall be liberally construed, and particularly should this be so where new remedies are given, and with reference to the bestowal of jurisdiction on the courts. Strictly speaking, it may be that this statute is not an act relating only to procedure in the purview of the last above cited cases; but it takes away from these defendants no right of action, or defense to this action on its merits, if indeed an objection to the jurisdiction can be called a defense at all. The plea protests against the power of the court to act in the premises; it says this suit should not be entertained here because this court has not been empowered to try it. But the very non-existence of the power to hear it may be the strongest reason why the legislature should determine to confer it, and render this defense, if it may be called so, nugatory. Certainly nothing could be more appropriate than for Congress to confer on its own courts power to hear controversies arising out of its own laws as the Constitution has expressly authorized it to do; and I cannot see how any citizen can acquire a vested right in any omission of Congress to do this, nor why the rule of construction should not be, by analogy to that above mentioned, to apply the act to pending cases, unless there be an express direction to the contrary, as in *Good vs. Martin,* 95 U. S. 90, 98, where the question was whether a change in the law of evidence applied to pending cases.

Page 152:

* * * The interesting case of *Fisher's Negroes vs. Dabbs,* 6 Yerger, 118, illustrates the prevalence of the beneficient principle upon which I base this judgment, and which I find pervading the authorities everywhere, namely, that statutes are, in the absence of directions to the contrary, retrospective in their operation wherever they are remedial, as

where they create new remedies for existing rights, remove penalties or forfeitures, extenuate or mitigate offenses, supply evidence, make that evidence which was not so before, abolish imprisonment for debt, enlarge exemption laws, enlarge the rights of persons under disability and the like, unless in doing this we violate some contract obligations or divest some vested right. And I cannot see why this principle should not apply to statutes *enlarging the jurisdiction of the courts so as to embrace suits then pending and not ended.* In the case last cited one statute was retrospectively construed in favor of a legislative remedy for establishing the freedom of a slave, and the legislature was not permitted by a subsequent act to forbid that retroactive operation, and thereby jurisdiction over a pending suit was saved.

It is proper that I should refer to a class of cases in the Supreme Court of the United States which hold that laws repealing those acts of Congress which confer jurisdiction on our courts, operate on suits then pending to take away the jurisdiction, and I am unable to see why the same principle should not apply here. These cases show that Congress has the power to legislate retroactively in such instances;''

This case was cited with approval in *United Wall Paper Factories, Incorporated, vs. Hodges, et al,* 70 Federal, 2nd Series, Page 243.

To illustrate how far the courts have gone to relieve parties in condemnation cases, we call Your Honor's attention to *The Schooner Rachel vs. The United States,* U. S. Supreme Court Reports, 3 Law Edition, Page 239. This was a sentence of condemnation of a vessel for trading contrary to a temporary act of Congress; the vessel had been sold and the proceeds paid over to the government while the law was in force. Pending an appeal the act was repealed. Held forfeiture could not be affirmed after law was repealed. The money was given back.

The case of *United Wallpaper Factories vs. Hodges, supra,* on page 244:

> "(5-7) There can be no doubt that the amendment applied to pending cases; it was a mere change in procedure, of far less consequence then the amendments held to apply presently in *Lockhart vs. Edel,* 23 Fed. (2d) 912 (c.c.a. 4); *Royal Indemnity Company vs. Cooper,* 26 F. (2d) 585 (C.C.A. 4); *in re Carter,* 32 F. (2d) 186 (C.C.A. 2). It is the general doctrine that amendments touching only procedure apply to pending actions. *Pease vs. Wilson,* 186 N. Y. 403, 79 N. E. 329; *Sackheim vs. Piqueron,* 215 N. Y. 62, 109 N. E. 109; *Larkin vs. Suffarans* (C.C.) 15 F. 157."

We contend that this Act is remedial legislation. *Maxwell on "Interpretation of Statutes,"* page 401, says, "Procedure legislation is retroactive unless there is some good reason against it."

Remedial laws liberally construed.

Fisher vs. Harvey, 6 Col., page 16;

Haskell vs. City of Berlingtun, 30 Iowa, 232.

We believe that appellant's constituted right of the protection of its lien was entitled to greater protection than that of claimant in *United States vs.* 1—1935 *Ford Standard Coach Automobile,* Vol. 13—No. 2 Fed. Supplemental Advance Sheets, page 104, wherein there was filed a petition of a finance corporation to reclaim the automobile after forfeiture, under the provisions of this Act of Congress of August 27, 1935. The automobile had been used in connection with the operation of a still. The car was seized in June of 1935. A libel was instituted and the question at the trial after Aug. 27, 1935, was the interpretation of Section B of said Act. The Court said:

> "The Government contends that the requirement of Sub-division 3, with respect to inquiry of the designated county, city, and said officers as to the character and financial standing of Thornton in the instant case, is mandatory.
>
> In the opinion of this Court, such interpretation is too rigid; and, if made to apply to the exclusion of a consideration of Sub-divisions 1 and 2 in connection therewith, Sub-division 3 would be an unconstitutional provision, violative of the Fourth

Amendment to the Constitution, securing the citizen against unreasonable seizure of his effects.

The business of financing the purchase by individuals of automobiles, through the medium of discounting purchase-money notes secured by conditional sales agreements, has become a wide-spread activity in this country. Through such finance, individuals of moderate means have been able to purchase, use, and enjoy automobiles which would have been beyond their means to acquire except for such financing. The salutary aid given to enforcement of the Internal Revenue Laws by the enactments of Congress authorizing the seizure of automobiles engaged in unlawful liquor traffic must not be construed to have the effect of requiring unreasonable and generally unnecessary inquiry to be made of designated public officials, as to whether such individual in every instant case purchasing an automobile with the aid of an automobile purchase-money note finance company is a bootlegger or internal revenue law violator. To the average purchaser of an automobile, it would be, to say the least, insulting should a finance company inquire of the chief of police, the sheriff, and the local chief of the alcohol tax unit whether such person is stigmatized as a bootlegger or liquor law violator.

When the validity of an Act of Congress is drawn in question, and even if a serious doubt of constitutionality is raised it is a cardinal principle that the Court will first ascertain whether a construction of the statute is fairly possible by which the question may be avoided.''

We respectfully contend that the language of the

Act indicates that Congress intended the Act to apply to pending cases.

It says: "Whenever, in any proceeding in court for the forfeiture," etc.

If Congress did not intend the Act to apply to pending cases it would ~~would~~ have read:

"Whenever, in any proceeding in court, *hereinafter commenced* for the forfeiture, etc."

Respectfully submitted,
FRED C. BROWN,
Attorney for Appellant.

IN THE

UNITED STATES CIRCUIT COURT OF APPEALS

FOR THE NINTH CIRCUIT

No. 8168

WILSON MOTOR COMPANY, a Washington Corporation,
Claimant of ONE FORD COACH AUTOMOBILE, Engine
No. 18-1128696, its tools and accessories,

Appellant,

vs.

UNITED STATES OF AMERICA,

Appellee.

HONORABLE JOHN C. BOWEN, *Judge*

BRIEF OF APPELLEE

J. CHARLES DENNIS,
United States Attorney.

OWEN P. HUGHES,
Assistant United States Attorney

Office and Post Office Address:
222 Post Office Building,
Seattle, Washington.

BRIEF PRINTING CO.

SUBJECT INDEX

TABLE OF CASES CITED

IN THE
UNITED STATES CIRCUIT COURT OF APPEALS
FOR THE NINTH CIRCUIT

No. 8168

WILSON MOTOR COMPANY, a Washington Corporation, Claimant of ONE FORD COACH AUTOMOBILE, Engine No. 18-1128696, its tools and accessories,

Appellant,

vs.

UNITED STATES OF AMERICA,

Appellee.

HONORABLE JOHN C. BOWEN, *Judge*

BRIEF OF APPELLEE

STATEMENT OF THE CASE

On April 6, 1935, at Enumclaw, in the Northern Division of the Western District of Washington, agents of the Alcohol Tax Unit seized a Ford coach automobile, engine number 18-1128696, within an enclosure containing a distillery being operated without the bond as required by law and in violation of the laws of the United States. (Tr. 22). The said Ford coach automobile contained a quantity of solder and eight hundred pounds of sugar, used and intended to be used in the manufacture of tax-unpaid illicit liquors at said distillery. (Tr. 22) One Amico Chiola was one of the defendants arrested for the operation of said illicit distillery. (Tr. 22).

On July 30, 1935, the United States of America, by J. Charles Dennis, United States Attorney for the Western District of Washington, filed a libel of information in the United States District Court for the Western District of Washington, seeking the condemnation and forfeiture of the said motor vehicle under Sections 1184 and 1441, Title 26 U. S. C. A. (Tr. 1-4)

On September 26, 1935, the claimant Wilson Motor Company, a corporation, filed its petition seeking the remission or mitigation of the forfeiture under Section 204, subsection (a), of the Act of August 27, 1935 (Section 40a, Title 27 U. S. C. A.). The petition alleged that on October 4, 1934, the claimant, Wilson Motor Company, sold the said Ford Coach automobile to Amico Chiola who executed and delivered to claimant his note and mortgage on the automobile as security for the unpaid balance of the purchase price. (Tr. 11) It was further alleged that claimant was the present owner and holder of said note and mortgage upon which there was an unpaid balance in the sum of $500.00; that the said Amico Chiola was in default in the monthly payments on said note and mortgage. (Tr. 12) It was further stated in the petition that the claimant sold the automobile in good faith and had no knowledge or reason to believe that the said Ford coach automobile would be used in violation of the laws of the United States or that the said Amico Chiola had a record or reputation for violating the liquor laws of the United States or of the State of Washington. (Tr. 13)

On February 28, 1936, the District Court entered a decree of condemnation and forfeiture. (Tr. 19-20) Thereafter and on March 13, 1936, the petition of claimant Wilson Motor Company, seeking a remission or mitigation of the forfeiture came on for hearing upon a stipulation of the facts. (Tr. 21) The Honorable John C. Bowen, United States District Judge for the Western District of Washington, denied the petition of the claimant Wilson Motor Company upon the ground that Section 204, subsection (a), of the Act of August 27, 1935 was not retroactive and did not apply to pending cases. (Tr. 24-25)

This is an appeal from that order made and entered on March 13, 1936.

ASSIGNMENTS OF ERROR

The assignments or specifications of error set forth in appellant's brief present but one question, to wit, whether Section 204, subsection (a), of the Act of August 27, 1935 (Section 40a, Title 27 U. S. C. A.) is retroactive and applies to pending cases.

ARGUMENT

SECTION 204, SUBSECTION (a), OF THE ACT OF AUGUST 27, 1935, IS NOT RETROACTIVE AND DOES NOT APPLY TO PENDING CASES.

The seizure in this case was made by members of the Alcohol Tax Unit on April 6, 1935, (Tr. 22) and on July 30, 1935 a libel of information was filed seeking condemnation and forfeiture of the motor vehicle in question under Section 3450 Revised Statutes (Section 1441, Title 26 U. S. C. A.) and Section 3281 Revised Statutes (Section 1184 Title 26 U. S. C. A.). (Tr. 1-4)

Prior to August 27, 1935, the date upon which Section 204, subsection (a) of the Act of Congress of August 27, 1935 became effective, and at the time the seizure was made and the libel of information filed in this case, the claimant, Wilson Motor Company, as mortgagee, could have obtained no relief from the forfeiture under its petition.

As was said by Mr. Justice Brandeis in the leading case of *United States v. One Ford Coupe,* 272 U. S. 321, 47 S. Ct. 154, 155, 71 L. Ed. 279, 47 A. L. R. 1025:

> "If a forfeiture may be had under section 3450 for such use of a vehicle to evade a tax on illicitly distilled liquor, the interests of innocent persons in the vehicle are not saved."

In numerous other decisions it has been held that automobiles or other vehicles used in removing liquor on which the tax had not been paid, with intent to defraud the United States Government of such tax, would be subject to forfeiture under the Internal Revenue laws as against the mortgage taken by the seller of the vehicle, who voluntarily gave possession to the purchaser but who had no knowledge of its unlawful use.

> *Goldsmith-Grant Co. v. U. S.,* 1921; 41 Sup. Ct. Rep. 189; 254 U. S. 505; 65 L. Ed. 376.
>
> *U. S. v. One Saxon Automobile,* 1919; CCA (4th); 257 Fed. 251
>
> *U. S. v. One Chevrolet Four Door Sedan Automobile,* Dist. Court Okla. 1930; 41 F. (2d) 782.
>
> *U. S. v. One Chevrolet Truck,* Dist. Court, Wash. 1925; 4 F. (2d) 612.
>
> *U. S. v. One Ford,* Dist. Court Nebr. 1927; 21 F. (2d) 628.
>
> *U. S. v. One Ford V-8 Sedan,* Dist. Court Ky. 1935; 11 Fed. Supp. 515.

U. S. v. One Plymouth Coupe, Dist. Court Me. 1935; 10 Fed. Supp. 164.

The Act of August 27, 1935 entitled "An Act to repeal Titles I and II of the National Prohibition Act, to reenact certain provisions of Title II thereof, to amend or repeal various liquor laws, and for other purposes" was approved and became effective August 27, 1935.

Section 204, Subsection (a) of Title II of the above mentioned Act (Section 40a, Title 27 U. S. C. A.) provides as follows:

"Section 204. (a) Whenever, in any proceeding in court for the forfeiture, under the internal-revenue laws, of any vehicle or aircraft seized for a violation of the internal-revenue laws relating to liquor, such forfeiture is decreed, the court shall have exclusive jurisdiction to remit or mitigate the forfeiture.

"(b) In any such proceeding the court shall not allow the claim of any claimant for remission or mitigation unless and until he proves (1) that he has an interest in such vehicle or aircraft, as owner or otherwise, which he acquired in good faith, (2) that he had at no time any knowledge or reason to believe that it was being or would be used in the violation of laws of the United States or of any State relating to liquor, and (3) if it appears that the interest asserted by the claimant arises out of or is in any way subject to any contract or agreement under which any person having a record or reputation for violating laws of the United States or of any State relating to liquor has a right with respect to such vehicle or aircraft, that, before such claimant acquired his interest, or such other person acquired his right under such contract or agreement, whichever occurred later, the claimant, his officer or agent, was informed in answer to his inquiry, at the headquarters of the sheriff, chief of police, principal Federal internal-revenue officer engaged in the enforcement of the liquor laws, or other principal local or Federal law-enforcement officer of the locality in which such other person acquired his right under such contract or agreement, of the locality in which such other person then resided, and of each locality in which the claimant has made any other inquiry as to the character or financial standing of such other person, that such other person had no such record or reputation."

It will be noted that there is no provision contained in this Act which would make it applicable to pending cases or pending litigation.

The court in denying appellant's petition for remission or mitigation, held that Section 204, subsection (a), of the Act of August 27, 1935, did not apply to cases pending at the date it became effective and in an oral opinion referred to Section 13 Revised Statutes (Section 29, Title I U. S. C. A.) which provides as follows:

"Repeal of statutes as affecting existing liabilities. The repeal of any statute shall not have the effect to release or extinguish any penalty, forfeiture, or liability incurred under such statute, unless the repealing Act shall so expressly provide, and such statute shall be treated as still remaining in force for the purpose of sustaining any proper action or prosecution for the enforcement of such penalty, forfeiture, or liability."

It was further indicated by the court that Section 204, subsection (a), of the Act of August 27, 1935, by conferring exclusive jurisdiction on the district court to remit or mitigate forfeitures operated as an implied repeal of those strict forfeiture provisions of Sections 1184 and 1441, Title 26 U. S. C. A., and for that reason became subject to the provisions of Section 29, Title I U. S. C. A. There is a scarcity of authorities on this particular question but the following cases tend to sustain the court's contention:

U. S. v. Four Cases of Lastings, Dist. Court, S. D., N. Y. 1879; 25 Fed. Cas. 1176.

Maceo v. U. S., CCA5, 1931; 46 F. (2d) 788.

In the case of United States v. Four Cases of Lastings, 25 Fed. Cas. 1176, an information was filed to enforce a forfeiture of certain goods alleged to have been entered by means of a false invoice in violation of Section 2864 Revised Statutes which was in force at the time the alleged unlawful entry occurred. After the entry, but before the seizure or commencement of the suit, the Act of February 18, 1875 was enacted which amended Section 2864 Revised Statutes by providing as an alternative to the forfeiture of the goods, the forfeiture of "or the value thereof." The court held the Act of February 18, 1875 was not retroactive and did not apply to the case at bar which had been pending at the time of its enactment. The court used the following language:

> "It is true that the act of 1875 contained no saving clause, and it may well be held to have operated as a repeal of section 2864 within the meaning of this rule; but it was subject to the provisions of section 13 of the Revised Statutes, which is as follows: 'The repeal of any statute shall not have the effect to release or extinguish any penalty, forfeiture or liability incurred under such statute, unless the repealing act shall so expressly provide, and such statute shall be treated as still remaining in force for the purpose of sustaining any proper action or prosecution for the enforcement of such penalty, forfeiture or liability.' Motion denied."

Appellant has contended in its brief that the Act of Congress of August 27, 1935 did not effect an implied repeal of Section 158, Title 26 U. S. C. A. by reason of Section 8 of Title I of the Act of August 27, 1935, which provides in part as follows:

> "Nothing in this section shall in any manner limit or affect any criminal or forfeiture provision of the Internal Revenue laws or of any other law."

Although the appellant's contention is based upon a misconception, i. e., that the lower court held there was an implied repeal of Section 158, Title 26 U. S. C. A., it might be urged that the above mentioned portion of Section 8 of Title I of the Act of August 27, 1935 negatives an implied repeal of Sections 1184 and

1441, Title 26 U. S. C. A. A perusal of the foregoing section, however, will disclose that the quoted portion of that section is limited to that section, namely, Section 8 of Title I, and does not include within its provisions Section 204, subsection (a), of Title II.

The appellee adopts the foregoing reasons advanced by the court in holding that the Act of August 27, 1935 can be given no application to the instant case.

In addition, the general rule is urged that all statutes are to be considered prospective and will not be given a retroactive effect unless there is express language to the contrary or there is a necessary implication to that effect.

> *Caha v. U. S.*, 152 U. S. 211.
>
> *City Railway Co. v. Citizens' Street Railroad Co.*, 166 U. S. 557.
>
> *U. S. v. St. Paul, Minncapolis & Manitoba Ry Co.*, 247 U. S. 310; 38 Sup. Ct. Rep. 525.
>
> *Cox v. Hart*, 260 U. S. 427; 43 Sup. Ct. Rep. 154.

In its brief appellant contends that Section 204, subsection (a), of the Act of August 27, 1935 (Section 40a, Title 27 U. S. C. A.) is remedial and goes only to the method of procedure and jurisdiction and for that reason the Act should be given a retroactive effect, and should be given application to the instant case.

It will not be denied that statutes which affect only the remedy or procedure and do not impair a vested right will be given a retroactive effect, but where a statute, though it purports to affect only the remedy, in reality impairs the value of a right or an obligation incurred, it will not be given a retroactive application as to pending litigation.

> *Cameron v. U. S.*, 231 U. S. 710.
>
> *Plummer v. Northern Pacific Railway Co.*, 152 Fed. 206.
>
> *In re Chavez et al.*, 149 Fed. 73.
>
> *S. W. Coal and Improvement Co. v. McBride*, 185 U. S. 499; CCA (8th)
>
> *Hoyt Metal Co. v. Atwood*, 289 Fed. 453.

Prior to August 27, 1935 and at the time of seizure and at the date the libel of information was filed in the instant case, the only remedy available to those holding chattel mortgages or conditional sales contracts on automobiles or vessels seized for a violation of the Internal Revenue laws, was under and by virtue of Section 158, Title 26 U. S. C. A., which provides as follows:

> "Compromises. The Commissioner of Internal Revenue, with the advice and consent of the Secretary of the Treasury, may compromise any civil or criminal case arising under the internal revenue laws instead of commencing suit thereon; and, with the advice and consent of the said Secretary and the recommendation of the Attorney

General, he may compromise any such case after a suit thereon has been commenced. Whenever a compromise is made in any case there shall be placed on file in the office of the commissioner the opinion of the Solicitor of Internal Revenue, or of the officer acting as such, with his reasons therefor, with a statement of the amount of tax assessed, the amount of additional tax or penalty imposed by law in consequence of the neglect or delinquency of the person against whom the tax is assessed, and the amount actually paid in accordance with the terms of the compromise. (R. S. Sec. 3229)."

Under this statute, offers in compromise could be made to those officials named in the statute, for the return of the offending motor vehicles or vessels. This right of compromise on the part of those holding chattel mortgages or conditional sales contracts on motor vehicles seized for violation of the Internal Revenue laws has been recognized in the following cases:

U. S. v. One Reo Speed Wagon, Dist. Court, W. D. Wash., S. D., 1925; 4 F. (2d) 284.

U. S. v. One Chevrolet Roadster Automobile, Dist. Court. W. D. Wash., S. D., 1926; 13 F. (2d) 948.

At the time of the seizure and at the date the libel of information was filed, and up to the 27th day of August, 1935, the only relief available to appellant, as an innocent mortgagee, was by way of making an offer in compromise to the Commissioner of Internal Revenue in settlement of the liability to forfeiture of the Ford Coach automobile. The United States was in possession of said automobile and vested with the right to acquire absolute title under and by virtue of the forfeiture provision of Sections 1184 and 1441, Title 26 U. S. C. A. upon the entry of a decree of condemnation and forfeiture. Upon judicial condemnation the title acquired by the United States became indefeasible and could not be defeated by the claims of those holding chattel mortgages or conditional sales contracts on the condemned chattel.

With the Act of August 27, 1935 the court was given exclusive jurisdiction to remit or mitigate the forfeiture. Such a remission or mitigation of the forfeiture in the case at bar would divest the United States of its absolute right to acquire title to the said Ford coach automobile and would further deprive the United States of any compromise offer made under Section 158, Title 26 U. S. C. A. If a retroactive effect be given Section 204, subsection (a), of Title II of the Act of August 27, 1935, there would be an impairment of a right vested in the United States prior to the effective date of said Act of August 27, 1935.

In the case of *Hoyt Metal Co. v. Atwood*, 289 Fed. 453, the Circuit Court of Appeals for the Seventh Circuit stated the rule applicable to the instant case in the following language:

"By a long line of decisions, courts have uniformly held that a statute is presumed to operate prospectively only, unless an intent to the contrary clearly appears, and this is especially applicable where the statute, if given a retroactive operation, would be invalid as impairing the obligation of contracts, or interfering with vested rights.

* * * * *

"The statute in question, though purporting to affect only the remedy, in reality impairs the value of a right. Though it may be true that parties to actions have no vested right in rules of procedure, and the latter may be changed at will by the Legislature, without affecting the right, yet if statutes, professing only to affect procedure or remedy do in fact thereby affect or impair the right, they are void or inapplicable as against the enforcement of the right thus impaired. Whatever belongs merely to the remedy may be altered according to the will of the state, provided the alteration does not impair the obligation of the contract. But if that result is produced, it is immaterial whether it is done by acting on the remedy, or directly on the contract itself. In either case, it is prohibited by the Constitution. *Tennessee v. Sneed*, 96 U. S. 69, 24 L. Ed. 610; *Bronson v. Kinzie et al.*, 1 How. 311, 11 L. Ed. 143; *McGahey v. Virginia*, 135 U. S. 695, 10 Sup. Ct. 972, 34 L. Ed. 304."

CONCLUSION

In conclusion it is urged that Section 204, subsection (a) of Title II of the Act of August 27, 1935, should be given no application to the case at bar and it is respectfully submitted that the court committed no error in denying the appellant's petition for remission or mitigation of the forfeiture.

Respectfully submitted,

J. CHARLES DENNIS,
United States Attorney

OWEN P. HUGHES,
Assistant United States Attorney.

In the United States 7
Circuit Court of Appeals
For the Ninth Circuit.

PATRICK H. BRAY and CATHERINE PATRICIA
MARQUETTE,

Appellants,

vs.

HOFCO PUMP, LTD., and D. W. HOFERER,

Appellees.

Transcript of Record.

Upon Appeal from the District Court of the United States for the
Southern District of California, Central Division.

FILED

APR 10 1936

PAUL P. O'BRIEN,
CLERK

Parker, Stone & Baird Co., Law Printers, Los Angeles.

No.

In the United States
Circuit Court of Appeals
For the Ninth Circuit.

PATRICK H. BRAY and CATHERINE PATRICIA
MARQUETTE,

Appellants,

vs.

HOFCO PUMP, LTD., and D. W. HOFERER,

Appellees.

Transcript of Record.

Upon Appeal from the District Court of the United States for the
Southern District of California, Central Division.

Parker, Stone & Baird Co., Law Printers, Los Angeles.

INDEX.

[Clerk's Note: When deemed likely to be of an important nature, errors or doubtful matters appearing in the original record are printed literally in italics; and, likewise, cancelled matter appearing in the original record is printed and cancelled herein accordingly. When possible, an omission from the text is indicated by printing in italics the two words between which the omission seems to occur.]

PAGE

ii.

iii.

INDEX TO TESTIMONY.

vi.

INDEX TO EXHIBITS.

PAGE

Names and Addresses of Solicitors.

For Appellants:

ALAN FRANKLIN, Esq.,

114 West Third Street,

Los Angeles, California.

For Appellees:

BURKE & HERRON, Esqs.,

MARK L. HERRON, Esq.,

Chapman Building,

Los Angeles, California.

UNITED STATES OF AMERICA, ss:

To HOFCO PUMP LTD., and D. W. HOFERER—
    GREETING:

You are hereby cited and admonished to be and appear at a United States Circuit Court of Appeals for the Ninth Circuit, to be held at the City of San Francisco, in the State of California, on the 22nd day of AUGUST, A. D. 1935, pursuant to an order allowing appeal filed and entered in the Clerk's Office of the District Court of the United States, in and for the Southern District of California, in that certain cause numbered X-50-J Equity, filed and tried in the Central Division of said District, wherein Patrick H. Bray and Catherine Patricia Marquette are plaintiffs and you are defendants to show cause, if any there be, why the errors set forth in the Assignments of Error, accompanying said order allowing appeal in the said cause mentioned, should not be corrected, and speedy justice should not be done to the parties in that behalf.

WITNESS, the Honorable WILLIAM P. JAMES United States District Judge for the Southern District of California, this 23rd day of July, A. D. 1935, and of the Independence of the United States, the one hundred and *ninth*.

Wm. P. James
U. S. District Judge for the Southern
District of California.

Service of a copy of the above Citation acknowledged this 23 day of July, 1935.

Burke & Herron & Russell Graham
By Mark L. Herron
Attorneys for the above-named Defendants.

[Endorsed]: Filed Jul. 23, 1935.

IN THE UNITED STATES DISTRICT COURT
SOUTHERN DISTRICT OF CALIFORNIA
CENTRAL DIVISION

PATRICK H. BRAY,

Plaintiff,)

: Equity No. X-50-J.

-vs-)

HOFCO PUMP, LTD., D. W.)
HOFERER, JOHN DOE and :
RICHARD ROE,)

Defendants. :

BILL OF COMPLAINT FOR INFRINGEMENT OF
LETTERS PATENT NO. 1840,432

Now comes the plaintiff and for cause of action against the defendants and each of them, alleges

I.

That defendant Hofco Pump, Ltd, is a corporation duly organized and existing under and by virtue of the laws of the State of California, and is a citizen of that State, having its office and principal place of business in the City of Long Beach, County of Los Angeles, State of California.

II.

That defendants D. W. Hoferer, John Doe and Richard Roe are, and each of them is, a citizen of the State of

California and resident of the City of Long Beach, County of Los Angeles, State of California and plaintiff is ignorant of the true names of the defendants John Doe and Richard Roe but upon ascertaining the same plaintiff will ask leave of Court to amend his complaint herein by inserting the true names of said defendants in lieu of said fictitious names.

III.

That this is a suit arising under the patent laws of the United States and the Court has jurisdiction thereunder.

IV.

That prior to the 4th day of February, 1929, Patrick H. Bray, plaintiff herein, a citizen of the United States, residing in the City of Los Angeles, County of Los Angeles, State of California was the original, first, and sole inventor of a certain new and useful invention, to-wit, A certain oil pump and parts thereof and said invention is fully described in Letters Patent No. 1840,432 hereinafter referred to, which said invention had not been known or used by others in this country before his invention or discovery thereof, and had not been patented or described in any printed publication in this or any foreign country before his invention or discovery thereof, or more than two years prior to his application for United States Letters Patent therefor, and had not been in public use or on sale in this country for more than two years prior to his application for United States Letters Patent, and had not been abandoned and had not been first patented or caused to be patented by said inventor or his legal representative or assign in any foreign country upon an application filed more than twelve (12) months prior to the filing of said application for patent in the United States of America.

V.

That on the 4th day of February, 1929, plaintiff herein, Patrick H.. Bray, as the inventor of said invention, applied for Letters Patent of the United States thereon, and the plaintiff herein has complied with all of the laws of the United States and the rules and regulations of the United States Patent Office concerning such application and such proceedings are had that on the 12th day of January, 1932, Letters Patent of and in the name of the United States of America for said invention, under the seal of the Patent Office, signed by the Commissioner of Patents in due form of law and duly executed and recorded, and numbered 1840,432, were duly issued, granted and delivered for said invention to the plaintiff herein, whereby there was granted and secured to the plaintiff, his successors or assigns the exclusive right to make, use and vend the invention for seventeen (17) years from February 4, 1929, in the United States and territories thereof.

VI.

That plaintiff has been, from and since the date of issuance of said Letters Patent No. 1840,432, and he now is the owner of the entire right, title, and interest and to said Letters Patent.

VII.

That plaintiff is informed and believes and upon such information and belief alleges that the defendants have jointly infringed said Letters Patent No. 1840,432 within the Central Division, Southern District of California and possibly elsewhere, since the issuance of said Letters Patent and prior to the filing of this Bill of Complaint, by manufacturing, selling, or using, or causing to be manufactured, sold, or used, without license or consent of the plaintiff, devices embodying the invention disclosed in said Letters Patent and set forth in the claims thereof.

VIII.

That the invention patented in and by said Letters Patent No. 1840,432 was and is of great value and commercial use and that the public in the United States of America has generally recognized and acquiesced in the novelty, utility, value and patentability of said invention, and has recognized and acquiesced in the validity of said Letters Patent and the exclusive rights of the plaintiff thereunder.

IX.

All persons making or vending devices embodying the invention in said Letters Patent, without the authority of the plaintiff herein, have given sufficient notice to the public that the same was patent by fixing thereon the word "Patented", together with the day and year that the patent was granted; that the defendants and each of them have been duly notified in writing of said Letters Patent, and the infringement thereof by the same defendants, have wilfully continued and still continue, and, as plaintiff is informed and believes, and therefor alleges, threaten further to continue to infringe the same.

X.

That by reason of said infringing acts the defendants have, and each of them has, profited, and plaintiff has been irreparably damaged.

WHEREFORE, plaintiff prays:

1. That the defendants, and each of them, their officers, agents, servants, employees, and attorneys, respectively, or those in active concert or participating with them, or any of them, be enjoined, both permanently and pending this suit, from further infringing upon plaintiff's said Letters Patent, and upon the rights of the plaintiff thereunder.

2. That the defendants, and each of them, be required to account and pay to plaintiff defendants' profits and plaintiff's damages, and a sum in excess thereof not to exceed three times the actual damages and profits.

3. That the defendants be required to pay the costs and disbursements of plaintiff in this suit.

4. That plaintiff have such other and further relief in the premises as to the Court may appear proper and agreeable to equity.

Delmar W. Doddridge
Attorney for plaintiff

STATE OF CALIFORNIA

)

COUNTY OF LOS ANGELES) ss.

PATRICK H. BRAY, being first duly sworn, on oath says:

That he is the plaintiff in the foregoing and above entitled action; that he has read the within Bill of Complaint and knows the contents thereof; and that the same is true of his own knowledge except as to the matters and things therein stated on his information or belief, and that as to those matters and things he believes it to be true.

Patrick H. Bray

Subscribed and sworn to before me this 18th day of April, 1932.

[Seal]   Mary O. Terpenning
Notary Public in and for the County of Los Angeles, State of California.

[Endorsed]: Filed Jul. 13, 1932.

[TITLE OF COURT AND CAUSE.]

AMENDED ANSWER OF DEFENDANTS, HOFCO PUMP LTD. AND D. W. HOFERER TO BILL OF COMPLAINT

Come now the defendants, Hofco Pump Ltd. and D. W. Hoferer and answering the plaintiff's Bill of Complaint for themselves alone, admit, deny and allege as follows, to-wit:

I.

Said defendants admit the allegations contained in paragraph I of plaintiff's Bill of Complaint.

II.

Said defendants admit that the defendant, D. W. Hoferer is a citizen of the State of California, and a resident of the City of Long Beach, County of Los Angeles, State of California, and allege that they have no knowledge, information or belief as to the residence or citizenship of the defendants, John Doe and Richard Roe, and basing their denial upon that ground deny that the said defendants, John Doe and Richard Roe are, or either of them is, a citizen of the State of California, or a resident of the City of Long Beach, County of Los Angeles, State of California.

III.

Said defendants admit the allegations contained in Paragraph III of said Bill of Complaint.

IV.

Deny that prior to the 4th day of February, 1929, or at any other time, Patrick H. Bray, plaintiff herein, was

the original, first, sole or any inventor of the alleged
invention described and claimed in the alleged Letters
Patent referred to in paragraph IV of the Bill of Complaint; deny that said invention was either new or useful;
deny that said invention had not been known and/or used
by others in this country before plaintiff's alleged invention or discovery thereof; deny that the same had not
been patented and/or described in any printed publication
in this or any foreign country before plaintiff's alleged
invention or discovery thereof, or more than two years
prior to his application for United States Letters Patent
therefor; deny that the same had not been in public use
or on sale in this country for more than two years prior
to his application for United States Letters Patent; deny
that the same had not been abandoned and likewise deny
that said alleged invention had not been first patented or
caused to be patented by said inventor or his legal representative or assign in any foreign country upon an application filed more than twelve months prior to the filing
of said alleged application for patent in the United States
of America.

V.

Allege that these defendants are without knowledge and
upon such ground deny that, on the 4th day of February,
1929, said Patrick H. Bray, as the inventor of such
invention or otherwise, applied for Letters Patent of the
United States thereon; and for want of knowledge also
deny that plaintiff herein has complied with all or any
of the laws of the United States or the rules or regulations of the United States patent office concerning such
application; and likewise deny that such or any proceedings were had that on the 12th day of January, 1932, or
at any other time Letters Patent No. 1,840,432, of the

United States were duly or at all issued or granted or delivered for said or any invention to the plaintiff herein. Deny that there was thereby or at all granted or secured to the plaintiff, his successors or assigns the exclusive or any right to make or use or vend the said alleged invention for seventeen years from February 4, 1929, or for any other time or at all.

VI.

Defendants allege that they are without knowledge and therefore deny that plaintiff has been from or since the date of issuance of the said Letters Patent No. 1,840,432, or that he is now the owner of the entire or any right, title or interest in or to the said Letters Patent.

VII.

Defendants deny that they have jointly or otherwise infringed said Letters Patent No. 1,840,432 within the Central Division, Southern District of California, or elsewhere, since the issuance of the said alleged Letters Patent or prior to the filing of this Bill of Complaint, or at any other time, by manufacturing or selling or using, or causing to be manufactured, sold or used devices embodying the invention disclosed in said Letters Patent, or set forth in the claims thereof, or in any manner whatsoever.

VIII.

Defendants allege that they are without knowledge and therefore deny that the invention alleged to have been patented in or by said Letters Patent No. 1,840,432, was or is of great or any value, or of any great or any commercial use, and deny that the public in the United States of America has generally or at all recognized or acquiesced in the novelty or utility or value, or patentability of said alleged invention, or has recognized or acquiesced in the validity of said Letters Patent or the exclusive or any rights of the plaintiff thereunder.

IX.

Defendants allege that they are without knowledge and therefore deny that all or any persons making or vending devices embodying the alleged invention in said Letters Patent, either with or without the authority of the plaintiff herein, have given sufficient or any notice to the public that the same was patented by fixing thereon the word "patent," together with the day and year the said Letters Patent were granted. Deny that said defendants have wilfully, or at all, infringed the said Letters Patent at any time or at all, and deny that defendants threaten to continue to infringe the same; and allege that plaintiff has no reason whatever to fear that defendants will commence or continue any such alleged infringement as described or referred to in paragraph IX of the Bill of Complaint herein.

X.

Deny that by reason of said alleged infringing acts the defendants have or either of them has profited, and also deny that plaintiff has been or is being irreparably damaged or damaged in any degree, or at all, by reason of the defendants acts in the premises.

XI.

Further answering, and as a separate defense on their behalf, defendants allege that the said Patrick H. Bray was not the original, true and first inventor or discoverer of the alleged invention purporting to be covered by said Letters Patent No. 1,840,432, or any material or substantial part thereof, but that the same and all material or substantial parts thereof have been invented by, and patented to others prior to the date of the alleged invention thereof by the said Patrick H. Bray, in divers prior Letters Patent of the United States as follows:

United States Patents:

No.	Date		Patentee
211,230	January	7, 1879	William H. Downing
575,498	January	19, 1897	Wm. J. Wright
586,524	July	13, 1897	James S. Thompson
840,919	January	8, 1907	Philip H. Deis
879,166	February	18, 1908	Henry D. Haven
1,139,396	May	11, 1915	Oliver E. Barthel
1,309,738	July	15, 1919	Milton N. Latta
1,323,352	December	2, 1919	William Floyd Cummings
1,338,907	May	4, 1920	William J. Coulson
1,372,031	Mar.	22, 1921	William H. McKissick
1,378,268	May	17, 1921	Thomas A. Northrup
1,391,873	September	27, 1921	James Andrew Whitling
1,407,493	February	21, 1922	Roscoe W. Stephens
1,452,004	April	17, 1923	Clinton L. Savidge
1,454,400	May	8, 1923	W. A. O'Bannon
1,456,727	May	29, 1923	Aldo M. Franchi
1,474,718	November	20, 1923	Granville A. Humason
1,482,141	January	29, 1924	Emerson M. Parks
1,486,180	March	11, 1924	James C. Dickens
1,503,652	August	5, 1924	Charles W. Howe and Jesse J. Robertson
1,513,699	October	28, 1924	Warren E. Ellis
1,543,918	June	30, 1925	Robert W. Gunn and Willsie A. S. Thompson

1,559,766 November 3, 1925 Walter A. O'Bannon
1,625,230 April   19, 1927 Charles B. Thurston
1,163,435 December 7, 1915 Carl Mercer
1,710,581 April   23, 1929 George F. Le Bus
1,717,619 June   18, 1929 Karl P. Neilsen and
                             Daniel E. Byers
1,755,990 April   · 22, 1930 Roy George Hawley
1,720,979 July   16, 1929 Willard R. Tierce

and to others whose names and numbers and the dates of whose patents defendants pray leave hereafter to append hereto by amendment or otherwise when they have fully ascertained the same.

XII.

Further answering, and as a separate defense on their behalf, defendants allege that the alleged invention purported to be covered by said Letters Patent No. 1,840,432, and every material and substantial part thereof, prior to the alleged invention or discovery thereof by the said Patrick H. Bray and/or more than two years prior to his application for said Letters Patent, were described and shown in various printed publications and patents, to-wit: the several patents specifically enumerated in Paragraph XI, hereof, and in the following publications:

Catalogue of Frick and Lindsay Company, describing and illustrating the "O.F.S. Working Barrel" at pages 20 and 21 thereof, published by said Company in 1915, the address of which Company was at the corner of Sandusky and Robinson Streets, N. S. Pittsburgh, Pa., and the successor to the said Company is the Frick Reid

Supply Corporation, the address of which Corporation is at 111 Sandusky Street, Northside, Pittsburgh, Pa.

Catalogue of Frick and Lindsay Company, describing and illustrating the "Bramo Working Barrel," at page 43 thereof, published by said Company in May, 1919, the address of which Company was at the corner of Sandusky and Robinson Street, N. S. Pittsburgh, Pa., and the successor to the said Company is the Frick Reid Supply Corporation, the address of which Corporation is at 111 Sandusky Street, Northside, Pittsburgh, Pa.

Catalogue of Bradford Motor Works, describing and illustrating "Bradford Plunger liner" at page 7, and "O.F.S. Working Barrel" at page 8, and "Bramo Working barrel" pages 14 and 15 thereof, published by said Company in 1921, the address of which Company is Bradford Motor Works, Bradford, Pa.

Single sheet catalogue of "B.M.W. Products" manufactured by Bradford Motor Works, describing and illustrating "Admore liner Barrel" published by Bradford Motor Works in March, 1926, the address of which Company is Bradford Motor Works, Bradford, Pa.

Single sheet catalogue of Charles N. Hough Mfg. Co. of Franklin, Penn., describing and illustrating the "Huf Duo Pump," the address of which Company is Franklin, Pa.

Catalogue of the National Supply Companies, No. 30, describing and illustrating "Bramo Working Barrel" at page 393 thereof, published by said Company in 1921, the address of which Company is Toledo, Ohio.

Catalogue of the National Supply Companies, No. 35, describing and illustrating "Bramo Working Barrel" at page 401 thereof, published by said Company in 1925, the address of which Company is Toledo, Ohio.

and in other*s* patents and publications not now known to defendants, and which when ascertained defendants pray leave to insert by amendment or otherwise.

XIII.

Further answering, and as a separate defense on their behalf, defendants allege that the alleged invention or improvements purported to be covered by said Letters Patent No. 1,840,432, and every material and substantial part thereof, prior to the alleged invention or discovery thereof by the said Patrick H. Bray and/or more than two years prior to his application for said Letters Patent, had been known and/or used by the following named persons at the places indicated, to-wit:

Bradford Motor Works,
Bradford, Pa.,
used by said Company at or near
Bradford, Pa.

Fay L. Ingleright,
31 Pleasant St.,
Bradford, Pa.,
used by said person at or near
Bradford., Pa.

George F. Iverson,
177 Williams St.,
Bradford, Pa.,
used by said person at or near
Bradford., Pa.

George B. Morris,
123 Kennedy St.,
Bradford, Pa.,
used by said person at or near
Bradford., Pa.

William P. Kolbe,
Gilmore, Pa.,
used by said person at or near
Gilmore, Pa.

Carl Perkins,
R. F. D. #2
Bradford, Pa.,
used by said person at or near
Bradford., Pa.

Marcellus M. Seeley, 589 E. Main St., Bradford, Pa., used by said person at or near Bradford., Pa.

Raymond C. Flickinger, 410 E. Main St., Bradford, Pa., used by said person at or near Bradford., Pa.

Jas. G. McCutcheon, Jr. 22 Hobson Place, Bradford, Pa. used by said person at or near Bradford, Pa.

Sirvertus H. Johnson, 40 Melvin Ave., Bradford, Pa. used by said person at or near Bradford., Pa.

Gilbert T. Eidson, 280 South Kendall Ave., Bradford, Pa., used by said person at or near Bradford., Pa.

Leo H. Storms, 39 Sanford St., Bradford, Pa. used by said person at or near . Bradford, Pa.

Frick Reid Supply Co.,
Tulsa, Oklahoma.
used by said Company at or near
Tulsa, Okla.

Harland Oil Company
Wolco, Oklahoma.
used by said Company at or near
Wolco, Okla.

Forest Oil Corporation
Bradford, Pa.,
used by said Company at or near
Bradford, Pa.

Emery & Mason.
Bradford, Pa.,
used by said person at or near
Bradford, Pa.

South Penn Oil Co.
Bradford, Pa.,
used by said Company at or near
Bradford, Pa.

Oil Well Supply Co.,
Bradford, Pa.,
used by said Company at or near
Bradford, Pa.

Sloan & Zook Co.,

Bradford, Pa.,
used by said Company at or near
Eldred, Pa.

Carter Oil Co.,

Casper, Wyo.,
used by said Company at or near
Casper, Wyo

A. L. & L. M. Lilley,

Rixford, Pa.,
used by said persons at or near
Rixford, Pa.

E. L. Adams

Bradford, Pa.,
used by said person at or near
Bradford, Pa.

T. P. Thompson, Jr.,

Bradford, Pa.,
used on T. P. Thompson lease
near Bradford, and elsewhere
near Bradford, Pa.

Charles N. Hough Mfg. Co.,

Franklin, Pa.,
used by said Company at or near
Franklin, Pa.

Continental Supply Co.,
St. Louis, Mo.,
used by said Company at or near
St. Louis, Mo.

Prairie Oil & Gas Co.,
Drumright, Oklahoma,
used by said Company at or near
Drumright, Okla.

Lyle Travis
Bradford, Pa.,
used by said person at or near
Bradford, Pa.

Walter O'Bannon,   400 South Rockford.
Tulsa, Oklahoma.
used by said person at or near
Tulsa, Okla.

Walter O'Bannon Co.,
Tulsa, Oklahoma,
used by said Company at or near
Tulsa.

Dewey Nichols,
Tulsa, Oklahoma,
used by said person at or near
Tulsa, Okla.

and by others whose names and addresses together with the places of knowledge and use defendants pray leave to insert by amendment or otherwise when they shall have been ascertained.

XIV.

Further answering, and as a separate defense on their behalf, defendants allege that the alleged invention or improvements purported to be covered by said Letters Patent No. 1,840,432, and every material and substantial part thereof, had been in public use or on sale in this country by the persons or concerns referred to in Paragraph XIII, hereof at the places there indicated and elsewhere in the United States, and by others, for more than two years prior to the application of said Patrick H. Bray for said Letters Patent, and that the same had been abandoned to the public.

XV.

Defendants are informed and believe and therefore allege, that said Letters Patent No. 1,840,432 was and is invalid and void because the alleged invention purported to be patented thereby did not constitute patentable knowledge or invention within the meaning of the patent laws in view of the prior state of the art, and in view of what was common knowledge on the part of those skilled in the art, all prior to the time of the alleged invention of the said Letters Patent No. 1,840,432, by the applicant therefor, which prior art and prior knowledge defendants are ready to aver, maintain and prove.

Defendants are informed and believe and therefore allege, that while the application for said Letters Patent No. 1,840,432 was pending in the United States Patent Office, the applicant therefor so limited and confined the claims of said application under the requirements of the Commissioner of Patents, that the plaintiff herein can not now seek for or obtain constructions for such claims sufficiently broad to cover any apparatus made, used or sold by these defendants.

WHEREFORE, defendants deny that the plaintiff is entitled to the relief prayed for in the said Bill of Complaint, or to any relief, and pray to be hence dismissed with their costs in this cause sustained, and for such other and further relief as to the Court may seem just.

Joe C. Burke

Russell Graham

Attorneys for defendants Hofco Pump, Ltd., and D. W. Hoferer

UNITED STATES OF AMERICA
Southern District of California

} SS.

Southern Division
COUNTY OF LOS ANGELES

D. W. HOFERER, being by me first duly sworn, deposes and says: That he is one of the defendants in the above entitled matter, that he has read the foregoing Amended Answer and knows the contents thereof; and that the same is true of his own knowledge, except as to the matters which are therein stated upon information or belief, and as to those matters that he believes it to be true.

D. W. Hoferer

Subscribed and sworn to before me this
18 day of April 1934

[Seal]   Raymond Hoyt
Notary Public in and for the County of
    Los Angeles, State of California.

[Endorsed]: Filed May 1, 1934

(Testimony of Patrick H. Bray)

[TITLE OF COURT AND CAUSE.]

NARRATIVE STATEMENT OF THE EVIDENCE UNDER EQUITY RULE 75.

PATRICK H. BRAY,

one of the plaintiffs, called as a witness on his own behalf, testified:

My name is Patrick H. Bray. I am the plaintiff in this action, and the Patrick H. Bray to whom letters patent, No. 1840432, were issued, involved in this suit. This (indicating) is the original letters issued to me by the Patent Office.

This patent is offered and received in evidence as Plaintiff's Exhibit No. 1., and appears at Volume II, page 1 hereof.

At the time of the commencement of this action I was the owner of the letters patent in this case, and still continue to be the owner, with the exception of an assignment made to my daughter Catherine Patricia Marquette, who has been joined as plaintiff in this case. No other assignments have been made.

On or about the 30th day of January, 1932, I showed the original of this patent to the defendant, Hofco Pump Company. It was shown to Mr. Hoferer.

At this point it was stipulated by counsel for the respective parties that in the early part of February, 1932, the defendants in this case were notified by letter that the letters patent in the suit had been issued to Mr. Bray, that they were infringing and requested to cease.

Shortly after the notification I had a conference with Mr. Hoferer, the President of the Hofco Pump Com-

(Testimony of Patrick H. Bray)
pany at his office in Long Beach. Mr. Maxwell and the
lawyer that we had at that time by the name of Dodridge,
were present. He, Mr. Dodridge, had the case and he
was called away and had to give the case up, and I
turned it over to McAdoo and Neblett. I showed them
the original copy when I notified him that he was infringing; that is, my daughter notified him that she
had the original copy of this.

Here's a model of a pump; it is a standing column. On
top here is a plunger and this pump works down. We
will say this is in the pump and it don't go down to the
bottom unless you want it down there to unscrew or
screw up rods. Rods have to be screwed in a well very
often especially in these deep wells. If the plunger don't
drop over very free we unscrew it and the consequence
is it will save from 7 to 10 hours. In this way you can
do it without pulling everything out and make two round
trips in 15 minutes. This lower casing has a puller nut
and this nut goes up against the bottom of the plunger
so it pulls this lock. This lock is fastened into a shoe.
That is the Bray pump, a pump of the class having a
standing guide column. Right here is the standing
guide column. All of this up here at the top. That is
a standing guide column with a working valve. The working valve is in there. As this pump goes up this valve
is closed and this valve opens with the suction of the
pump and lets the fluid in between these two and confines it there. This goes open again and this closes and
this one opens and the oil is displaced. In other words,
the plunger is not working through a volume of oil all
the time. That has a valve, a traveling barrel. The

(Testimony of Patrick H. Bray)

traveling barrel and the valve is up at the top. It is running on the standing column, and the standing column is this inner structure. Everything in here.

In the operation of this pump the standing column does not move. The barrel is attached to the rods at this point, hook the first rods on and keep going down until they find the shoe. This is the plunger. This here is what we call the pulling tube. It is fastened on the plunger at the bottom of the bushing and lock nut and also fastened on this adapter on here. This is the pulling tube and standing column. It has a bottom guide. This is that guide.

I will explain this to you if I can. This guide here, this nut here, if that wasn't on there you would have quite a play in here with the barrel keeping going up against the tubing all the time and wearing the tubing out. You have got to have a guide to guide the column here evenly, because if you don't have it there you would have too much play, working in a crooked hole especially. Now, a loose guide of course will do the same work. I have an old one there that you might see.

That is the bottom guide. It is on the barrel, the bottom of the barrel. It screws on the bottom of the traveling barrel, and that has a bottom face forming a coupler. The bottom face is on the extreme bottom end of the guide, and there is an anchor for the column and a top face form a complementary coupler. That is the anchor or mandrel that goes in here, locks itself in a shoe and when you pull this out you have an open hole. You take your mandrel and pump all out on the rods and return them in the same way. In other words

(Testimony of Patrick H. Bray)

in this kind of a pump it eliminates the pulling of the tubing which means a whole lot, saves you sometimes two or three days to pull tubing, and this bottom anchor has a face forming a complementary coupling, and when you do that that is perfectly locked. Now you can screw your rods on or unscrew them without any trouble. You can get all the tension you want on it. You can snap it on or anything else from the top and it saves days, as I say, in making two round trips. The combination of these factors is the combination covered in claim 1.

Now, then, coming to claim 2, this bottom guide has a series of venting ducts which open directly towards the face. This model which I have here don't have venting ducts; it is just a model, but I have some others that do. I make pumps like this model is made here at the bottom. The oil from the barrel gets out of the barrel down here on the up stroke of the pump because it is large enough here to let the oil down around the standing column. Here is the idea, when this barrel is down this way this is only that long in the pump and as the pump moves up the oil cannot go past your plunger proper here, you see, because that is what it is for. That is a suction pump and it cannot go by any further and the consequence is it is driven up against that and a stream of oil is coming out of here and keeping the sand away.

Now coming to claim 3, this model contains a guide being apertured for the ejection of liquid from the barrel. That guide operates as I described it before, the guide comes down here and keeps the sand washed away from here. As it goes up it works the stream of oil. It does

(Testimony of Patrick H. Bray)

that on account of the clearance between the inner part of the bottom guide and the outer surface of the standing column, and that guide is running on the column.

I am engaged in the oil pump business and have been engaged in that business 11 years, I guess. I am actively in the field seeing oil pumps every day. I am very familiar with the type of pump which I have seen manufactured by the defendant Hofco Pump Company. This piece of metal is the adapter.

Model of Bray pump offered and received in evidence as plaintiffs' Exhibit No. 2.

It is a piece of metal covered with blue paint. I know of my own knowledge that one similar to that is used in the Hofco Pump Company pump. I guess it is the same thing. I don't know where I got it.

MR. SOBIESKI: I am very familiar with these, will you examine them and see if they are in accordance with the model you make?

MR. HERRON: Surely we will and if they are we will admit it. We will stipulate that they have the general appearance of the parts that we make. We are unable to tell, however, without measuring the inside diameter and that, of course, is important.

MR. SOBIESKI: With that exception.

MR. HERRON: They look the same, but we are not positive about the diameter without measuring.

BY MR. SOBIESKI: Now, what is this piece here, which is not painted blue?

A. That is what I would call a guide. That is what we would call a guide or puller nut, some call it a puller nut and some call it a guide. This model which I have

(Testimony of Patrick H. Bray)
been describing is about an inch and a quarter. Mr.
Hoferer works on a larger tube than I do. Now going
back to this model, this is a model of a pump which I
make. The pump of which this is a model is 12 or 15
feet long. The plunger in my pump proper is 56 inches
from the bottom of the plunger, where it starts, to the
top, 56 inches long. The pull tube is 7 feet long. This
inside here is just about half the size of a guide, 11
inches long, and this here about five and a half. Now
this piece of metal which I have described as a guide, or
is called a puller nut, is used in the Hofco pumps which
I have seen for the last 5 years. They are one of my
competitors. I am familiar with their products, and I
have examined their pumps down at their factory. The
factory is at 3001 Cherry Avenue, Long Beach.

MR. SOBIESKI: We will not offer this device as
being accurate as to the interior diameter of the pump.

MR. HERRON: Offer them as merely illustrative
of the general practice.

MR. SOBIESKI: Yes.

Device received in evidence as plaintiffs' Exhibit No. 3.

MR. SOBIESKI: Will you stipulate that this catalog
which we presented to the Court is the catalog which
is being issued at the present time by the Hofco Pump
Company?

MR. HERRON: I will stipulate that it has been
issued by us and we are issuing one of a similar type
to that.

Catalog received in evidence as plaintiffs' Exhibit No.
4., and appears at Volume II, page 5 hereof.

MR. SOBIESKI: Calling your attention to the
catalog which is in evidence as Exhibit 4, being a catalog

(Testimony of Patrick H. Bray)

similar to those being issued by the defendant, the Hofco Pump Company, does that catalog show a pump?

A. Yes. And that pump as shown there is of the class having a standing guide column. The standing guide column runs from here. It would be about 7 feet, I suppose. I suppose they have got that from 7 feet up to 10 feet, depending on the length of the barrel. It screwed into here, into this adapter. This part here is screwed into the adapter. This is screwed into the barrel. The other part is screwed in the bottom of the adapter. This coupling here shows at this point. I will mark on that the part which is the standing guide column. I will write "1" at the side there. Now, the standing guide column is marked 1, and this has a standing guide column with a working valve.

I will point out the working valve and mark it 2. That has a standing guide column with a working valve and a valve traveling barrel. I will mark the valve traveling barrel 3. The working valves and the valve traveling barrel are running on the column. The column is marked 1. That has a bottom guide on the barrel.

I will mark the bottom guide on the barrel No. 4. That bottom guide is on the barrel running on the column, and I have marked the column as No. 1. This bottom guide has a bottom face forming a coupler.

I will mark the bottom face forming a coupler No. 5.

Q. And an anchor for the column presenting primarily a coupler face and a top face forming a complementary coupler to be non-rotatively interlocked with the lower guide coupler, is there such a piece in there?

A. That is an adapter. From here down is that lock, do you want that marked?

(Testimony of Patrick H. Bray)

Q. We want the anchor. We want to mark that part of the pump as shown in the catalog which shows an anchor for the column presenting directly to said coupler face a top face forming a complementary coupling to be non-rotatively interlocked with the lower guide coupler.

A. Below the adapter is an anchor that locks into a shoe and makes it stationary. This has a cone that fits in here and is called a seat that is made out of babbitt.

Q. The anchor is marked—that is, is the anchor more clearly shown elsewhere in the catalog?

MR. HERRON: We don't raise any question about that, we have an anchor.

Q. BY MR. SOBIESKI: Is this the anchor here?

A. No, that is an adapter.

Q. Taking up claim No. 2, does the catalog show a pump of the class having a standing guide column?

A. Yes. I have already marked the standing guide column as 1. With a working valve—I have marked that 2. Your plunger has got a working valve, too. Your barrel valve is marked 2. Your barrel is marked No. 2 and your top valve in the pump is marked No. 5. You are getting to the plunger now and you haven't asked about the plunger before. For the sake of clarity, the standing guide column, that is marked No. 1, and the working valve is right up here. The working valve is No. 7 and the valve traveling barrel is No. 5, and the bottom guide on said barrel is No. 4, and having a bottom face forming a coupler. And an anchor for the column, that anchor would be No. 6. That is right, this is 3. It looks like a 5 and I couldn't tell very well.

Q. Now taking up claim 2, does that have a pump of the class having a standing guide column?

(Testimony of Patrick H. Bray)

A.   Yes, sir.   I have previously marked the standing guide column No. 1 with a working valve. The working valve is No. 7.   The valve with the traveling barrel running on said column is No. 3.   The bottom guide on said barrel is No. 4.   That bottom guide is running on said column and it has a bottom face forming a coupler. That is 4, which would be that whole thing.

Q.   And an anchor for the column presenting directly to said coupler face a top face forming a complementary coupling to be non-rotatively interlocked with the lower guide coupler?

A.   No. 5 would be the adapter and No. 6 would be—

Q.   Using the language of the claim, and an anchor for the column, where is that anchor, what is the number of the anchor?

A.   The anchor is a shoe.   Now, this is cased in with the tube.   It takes those two parts to make an anchor. When this one goes down here it locks in an offset in this shoe.   The anchor is a large piece.   The anchor has no top face forming a complementary coupling.   I find such an anchor in there.   The top face of the anchor forming a complementary coupler to be non-rotatively interlocked with the lower guide coupler is this adapter up here, No. 5, and said guide has a series of venting ducts which open directly towards the anchor coupler face.   It would be No. 4.   I did not see anywhere on this (catalog, Pltfs. Ex. 4) any ducts in this lower guide.

Taking up claim No. .... the barrel has no holes in the sides to allow liquid to come out during the pumping process, no holes in the barrel at all.

Taking up claim No. 3, I see in this catalog a combination of a pump of a class having a standing guide column

(Testimony of Patrick H. Bray)
with a working valve, and valve traveling barrel running on said column. Those items I have marked, No. 3. The standing guide column is No. 1 and the working valve in the standing column is No. 7. The traveling barrel is No. 3, and those all run on the standing guide column.

The catalog shows a bottom guide on said barrel. I have numbered the bottom guide on said barrel No. 4. The bottom guide of the barrel is operated on said column. It has a coupler part on its bottom end, and that is also No. 4. An anchor for said column presenting a top face forming a complementary coupling is on the catalog. I have marked that No. 5.

There is an anchor for said column having a top coupler part complementary to the coupler part of the said guide, whereby the engaged part may be meshed in non-rotative interlock. That guide is apertured for the ejection of liquid from the barrel directly towards the effective face of the lower standing coupler. I don't think the aperture shows here.

THE COURT: He is talking about the drawing now.

Q. BY MR. SOBIESKI: Will you mark the place on the drawing where the aperture would be?

A. Up there. That would be 8. These two pieces of metal represent, the top piece represents the bottom of the traveling barrel of the Hofco pump, and the lower piece represents the top of the anchor of the Hofco pump. I have seen many of these Hofco pumps operate, and I am familiar with them. I can tell the Court how the liquid gets out of the barrel in the operation of that structure. As this is down it is full of oil for the length of this pulling tube. This is on the bottom of the barrel, and as it goes up, according to the stroke, whatever

(Testimony of Patrick H. Bray)

strokes in pumping a well, according to the inches of the stroke, it will go up and compress itself against the bottom of the plunger proper and force the oil out here, coming down onto the adapter.

Q. That is the anchor?

A. Yes.

Q. Well, the oil, how does the oil get out of the barrel on the stroke of the Hofco pump?

A. The upstroke. The oil comes out on the upstroke, right up here on the lower side. There is nowhere else to come out. It comes out alongside of the tube, the pulling tube. It is in there under pressure, the pressure shoots the oil down and keeps the sand away. It goes right down straight.

Q. When you say it shoots it straight down, in the Hofco pump, is the effective face of the lower standing coupler directly underneath this moving part of the traveling barrel at the time of the operation?

A. The object of that is to get the two as close together as you can and eliminate your gas. In my opinion the operation of this guide on the bottom of the traveling barrel results in sending the liquid towards the top face of the anchor coupling in the Hofco pump. It is the only way you can get it down to the pump, going down straight. Some of these wells are very crooked. You have sometimes to move your tubing, it will bind at the two ends; sometimes they go below that place, and sometimes they raise up to get as straight a proposition as they can, but if it is a little crooked it will wear on the side.

Q. I show you here a metal ring—is this stipulated to be similar?

(Testimony of Patrick H. Bray)

MR. HERRON: It is stipulated to be one of the rings.

MR. SOBIESKI: We accept the stipulation.

Q. Can you explain to the Court how this ring is made in the Hofco pump?

A. Yes, it is sliding on the pulling stem, it is put on that way. Between this and that, see, two parts like that. It would set on like that and the standing column is right in here, do you understand—this is a smooth face and the other one is smooth faced, too. With respect to this bottom and top face coupler valve, this ring would be put on in the Hofco pump before the adapter was put on, which would be this way. It would be put on first. I will mark in the catalog that part of the pump which I have been referring to when I speak of the adapter. The adapter is marked No. 5. At the top of the lower coupler face this metal ring which I have been describing would be placed right on top of the adapter, between the adapter and the bottom guide. The metal ring is of harder metal than the adapter or the bottom guide. That stuff is generally made out of cold rolled steel.

MR. HERRON: We will stipulate that the ring is of harder metal than the other.

MR. SOBIESKI: We will accept the stipulation. Mr. Herron, will you stipulate that the interior diameter of the ring is of the same as the bottom guide on your traveling barrel, is that practically the same?

MR. HERRON: With respect to size it is.

MR. SOBIESKI: It has been stipulated that with respect to size the interior diameter of the metal ring is of the same size as the lower guide on the traveling barrel which you have described as the guide.

(Testimony of Patrick H. Bray)

A. The one goes in between.

Q. And it is of the same diameter. In other words, Mr. Bray, from your experience with oil pumps can you tell us if the presence of this metal ring where you have described it would make any difference with the way the oil comes out of the barrel on the stroke of the pump? Would it come out the same way if it was a sand well?

A. In my opinion I figure that this would get a lot of sand. It wouldn't be as free as the other. The oil would come out just the same, and would be directed straight down. I think the use of this ring would enable a person to twist up the rods. I am not sure. It has a tendency, if this is soft metal and this is hard, it probably hasn't strength enough to screw on and bring the guide up. We will say at 5,000 feet you would have 15,000 pounds of rods on there and that would make a pretty good weight on there, but there is no reason why it would keep the barrel from rotating.

THE COURT: Providing the teeth took hold?

A. Yes, that is soft metal and this is soft metal.

MR. HERRON: Could I ask Mr. Bray what you mean by rotation on one direction, you don't mean it would stop it in both directions?

A. No, not both ways. It is not made that way, it couldn't be both ways, because the teeth are all the same way. It is more for screwing on than off.

MR. HERRON: Thank you.

A. Screws both ways, you can screw on or screw off.

Q. By MR. SOBIESKI: Mr. Bray, in your pump as illustrated by the design, does the bottom guide actually touch—how close is the bottom guide to the standing column?

(Testimony of Patrick H. Bray)

A.   The standing column is on the bottom guide.

Q.   The bottom guide on the traveling barrel, how far is it?

A.   About 6 inches.   I will illustrate it on the model, about 6 inches, that gets the balls on top closer together. We place them as close as we can.   Some might be 4 and some might be 10.

Q   Going back to this model Exhibit 2, will you tell us how much clearance there is between the standing guide column and the guide at the bottom of the traveling barrel?

A   Pardon me, do you mean in that model or in the structure?

THE COURT: In your working pump.

A   Well, I believe I will have to call for my blueprint.

Q.   BY MR. SOBIESKI: There is a clearance of 3/32nds of an inch?

A   Yes.

THE COURT: You don't plan to have oil go through in that clearance?

A   Yes, a little bit, not much, though.   You have to run it pretty loose, anyway.   You cannot afford to have it too tight, if you do it is not going to work very good.

THE COURT: It is more, therefore, a guide?

A   Yes, the clearance is all right for making the hole.

Q   BY MR. SOBIESKI: Before you applied for your patent you made a pump?

A   I made it like that.

THE COURT: Out of your device, your patent, you conceived the idea of perforating and making additional jets?

(Testimony of Patrick H. Bray)

A   Yes. I found it more practical than the other. I now make them both ways. I make them with this and make them the same as this is; make them both ways and I find them very substantial. They give the man the worth of his money when he buys them. You see the difference here, this wears out a lot easier than this one.

MR. HERRON: He mentioned the diameter there, is that all free or merely on the side?

A   It is inside, I think.

MR HERRON: In other words, the total difference would be twice that; is that correct?

A   Come and look at it, the inside diameter is one and nine-sixteenths and the outside diameter is one and three-eighths. It is not necessary for this bottom nut to fit closely on the tube in order to guide the traveling barrel. It can guide it by leaving quite a space there, as far as that is concerned.

Q   Mr. Bray, in your experience in the oil business can you tell us if this device of a lock on the bottom of the traveling barrel is widely used in the oil business in this district?

A   As near as I can figure, probably 80 per cent, we will say, of the pumps used to pump oil *with* use this type of this kind of a pump, so it is used with most all of the companies.

MR. HERRON: We will stipulate it is generally used on all types.

Q   BY MR. SOBIESKI: Your letters patent which you described in this action have not been rejected or modified by the Patent Office?

A   No. They are now in effect.

(Testimony of Patrick H. Bray)

Q   And have you at any time licensed the defendants or either of them to use it?

MR. HERRON: We will stipulate that we are not licensed.

MR. SOBIESKI: We will accept the stipulation.

Q   Has the word "Patented" together with the number and date and year of the patent been stamped on the bottom part of your traveling barrel on all pumps sold by you, or others to be sold by you?

MR. HERRON: I think we can shorten it by saying we are raising no question and that the parts have been properly marked.   We will stipulate that they have.

MR. SOBIESKI: We will accept the stipulation, that the parts covered by this patent have been appropriately marked by the plaintiff.

MR. HERRON: That is, the parts you contend *is* covered.

MR. SOBIESKI: Correct.

CROSS EXAMINATION

I testified in relation to this model that the guide here has no series of vents and venting ducts. I testified likewise, that I found in that model a guide apertured for the ejection of liquid from the barrel directly towards the effective face of the coupler. I will point out what I call the aperture. It is an opening big enough here to let the oil through. Of course this is not that big. It is bigger than that. I mean the clearance between the standing guide column and the guide is greater in the pump than it is shown to be in the model. The aperture is all around this column. In other words, it is the annular space between the guide and the standing column.

(Testimony of Patrick H. Bray)

Looked at another way, it is the hole bored through the guide by which it is made from a block of metal into a guide. Prior to the issuance of this patent I had built certain of my pumps; actually built them. I don't know how many I built prior to the filing date of my patent. I never kept any track of them. I sold them when I could and sold what I could. They worked very good and I improved on them as I went along and made it probably 500 per cent better pump than it was when I first started, but they all worked from the first one I ever put in work. I probably had made 100 pumps prior to the filing of my application for patent.

The guide that was on my pump was put on first. When I first started making pumps in 1926 I had a guide with a smooth face just the same as this is, with holes bored in here, half holes, half of the holes. That was made first. I have a model of the locking device on the first guide I ever put on a pump, and I got into trouble, and then I conceived this idea of locking the barrel, and I got into trouble with the well at Huntington Beach.

Q I am asking you to describe the type of guides that you had made prior to the date of the application for patent.

MR. SOBIESKI: I object to that on the ground that it is immaterial.

THE COURT: He may testify as to the type of guide.

A Have you got one of them there, that little one— I will show you the one, fetch both of those up here. I had not made the type with ducts there like that at that time.

(Testimony of Patrick H. Bray)

Q You had made the type with borings under those parts and half holes cut out around them?

A Yes.

MR. SOBIESKI: I object to that on the ground that it is incompetent, irrelevant and immaterial; no abandonment or anything of that sort, your Honor.

THE COURT: Well, I don't know whether it was part of a prior art.

MR. HERRON: You understand, if the Court please, that he testified as to certain types of pumping guides in answer to a question by the Court on direct.

THE COURT: He may answer the question. Overruled.

MR. SOBIESKI: May we have an exception.

Q BY MR. HERRON: Now, then, the next type of guide you made was one with holes of that sort?

A Yes. I had not made any other type before I filed my application. I made one when I tried it without this to see how it worked. It worked very good. It worked just the same as Mr. Hoeferer's does. And that was prior to the time I made application for the patent.

Q Now, does it make any difference, whether you have ducts in the sense of holes bored through the guide or whether you have channels such as are indicated in this model you have here produced, (Defendants' Exhibit A) bored around the surface of the guide, or whether you have simply a hole larger in diameter than the outside diameter of the standing column in so far as the production of what you term is your patent, jet action, is concerned? Do you understand the question, Mr. Bray?

A Well, I figure—whether it would be the same, working the same in all of these—they probably would let out

(Testimony of Patrick H. Bray)

the oil, some quantity of oil, which is necessary to come out, anyway. They would all work the same. The reason I changed from this—of course I knew I had to have that and the reason I changed from this was I found out the tube running through this guide wore these out. There is no solid foundation enough for it so I changed to a solid foundation. When the guides wore out those ribs. It gave it a little more space in there. That probably jetted a little too much. In other words, I think the greater the clearance the greater the jet.

Q Do I understand that you say that you believe from your knowledge of the action of your pump that the greater the clearance between the standing guide column and the guide the greater your jet will be?

A No, it will be a smaller jet, but you can put more oil out, you wouldn't have the pressure to throw it down. If you have got a jet like that, or a smaller place, you have got to figure on how much oil goes in. In other words, to get a jet you have to have a smaller opening through which the dropped oil is passed; I think that is it. I said that prior to my application for patent I had built a pump with clearance like Mr. Hoeferer's. That clearance in the pump I built is on the blueprint there, you have got it. That guide is the same as the one I built that time. The dimensions are on there. That is not the blueprint I made before I made my application, but it is the same figures. In other words, the dimensions there are the same as the device I made. I never changed the dimensions any.

I changed from a guide with ducts to the guide without ducts because it was easier to make, and in putting holes through here I sometimes find that they would run out.

(Testimony of Patrick H. Bray)

There is a place right there. You can see where the drill run out, that hole right there. It wouldn't make much difference, you know, as long as the threads hold, but sometimes it gets crooked and gets pretty close to the threads.

I don't know how great the aperture can be between the standing column of the pump and the guide and yet produce the jet action which I mention in my patent. The only thing I can figure on is my practical experience and my pump, the way it drops, the way the plunger drops free, and I found out that I have no trouble, sand trouble; that my standing valve pulls free without any interference from sand, and I figure I have got about what I wanted. I think any greater clearance than that indicated on my blueprint would, in the light of my experience, teach me that I wouldn't get a jet action with any greater clearance.

I intend to imply by my answer, that I wouldn't get any jet action if the clearance was larger. The larger it is the less pressure you would get. According to my practical experience for ten years—I think I have had this kind of pump since 1926, and I was one of the first fellows that worked this kind of a pump in the field, and I found out I had to do a lot of preliminary work, and I have got it as I think I want it.

My pumps do not sand up occasionally unless it is a heave. In other words, I think my jet action is so effective that my pumps are kept free from sand except in the case of a heave. I haven't had a pump sand up where they have had to pull tubing to clear it, I think, in four years. There was one down at Lomita, there was a flowing mud down there and I pulled—they pumped all the oil out and then pumped the mud up—

(Testimony of Patrick H. Bray)

Q Well, in reference to your troubles in the Lomita field, with that exception your pumps don't sand, and the reason they don't sand is because of the diameter of the pumps as you make them and the construction as you have your guides, and the jet action which washes off the face of the clutch and hence enables you to go down and engage the portions of the clutch and engage your rods?

A It has a tendency to do that.

Q Now, what happens with the oil, just how does your jet work?

A Well, the jet you have got—my pulling tube is 7 feet from the top of this to the bottom of the plunger proper where it is fastened to the bottom of the plunger. My oil cannot go past that and as the barrel comes up this is filled with oil when it is down and as it comes up it compresses itself against the pump and it has got to go out here—it has got to go somewhere.

In other words, it is the spray which it gets because of the pressure through that restricted point. You rely upon the spray to exert the desired force to the liquid in the well and wash off the clutch face. Just as soon as that starts here that starts the jet and it has a greater tendency to keep the sand moving. As the barrel comes down again everything moves up.

Q You testified, did you not, that in the designing of the pump for which you applied for patent you were designing a pump which would furnish the maximum amount of jet action consistent with a reasonably free operation of the pump; is that correct?

A Well, I answered that question before. I will answer it again. I figured through my practical experience—that is how I manufactured them, through the

(Testimony of Patrick H. Bray)
practical experience with the pumps I worked with. I
have probably worked 500.

Q BY MR. HERRON: It was that type of pump
which you sought to construct, was it not, one which
would procure the maximum jet action consistent with an
operative pump?

A Yes.

Q Mr. Bray, you gave us a little while ago some
figures as to the interior and exterior diameters of certain
portions of your pump. Referring to that same pump,
what is the interior diameter of the barrel of the pump?

A One and fifteen-sixteenths.

Q Well, What is the length of the stroke of that
particular pump?

A 56 inch stroke. That is the longest stroke you can
get on the pump. That is the normal operating stroke,
that is the average. They will run from 26 to 56. I have
had them run as high as 26 strokes a minute and I have
had them run as high as 30 strokes a minute, but they
don't last very much. The average pump should be
pumped, in my opinion, well, about 20. If it is real
heavy oil, about 16. It depends on the gravity of the
oil how fast you can run a plunger and make it drop.
You have got to have your plunger drop free, if it don't
it will crystallize and break your rod. If it drops free it
don't make any difference because you have got weight
enough on the rod to force the oil up. I find that the
ordinary speed is about 20. Probably more people run
at 20 than any other stroke, especially in the deep wells.

Device produced by Mr. Bray offered and received as
Defendants' Exhibit A.

(Testimony of Patrick H. Bray)

REDIRECT EXAMINATION

Q As I understand it, Mr. Bray, in the operation of this pump it is necessary for the oil which has accumulated inside of the barrel to go somewhere; is that right?

A Yes, sir, it has got to go somewhere. And in my pump and in Mr. Hoeferer's pump the oil is let out exactly the same, they both work the same. It is done through having room enough here. This is O. D. and this is I. D., what I would call the guide to let the oil out through, or there are ducts. There are two ways to get it out, but I figure that Mr. Hoeferer's pump, take it all the way through, is as near like mine as is possible. They both have the same results, anyway. Neither of them use the device of letting the oil out through the barrel. In both of them the barrel may be turned end for end, to utilize both ends of the barrel.

In working the barrel you only work half way on a barrel. The plunger works just according to what the stroke is. Sometimes it is 26, sometimes 32, sometimes 40, sometimes 48 and sometimes 56, and the other end of the barrel does not wear, and this guide keeps it from wearing. There is nothing in here. This don't touch the barrel at all, so the barrel ought to be as perfect as it was, outside of a little further rust or something in there when you turn this, as it was when it was new. That is the idea of not putting holes in the barrel. The Hofco pump ejects its oil straight down along the barrel.

RECROSS EXAMINATION

Q Mr. Bray, in any type of pump in which you have a traveling barrel and a stationary column, the oil which is trapped below the lower packing and the bottom of the

(Testimony of Patrick H. Bray)

barrel will be forced down around the bottom of the barrel, will it not, unless channels are provided elsewhere for its removal?

A   When your tubing is always full of oil, or gas and oil, supposed to be full of oil, and it is filled down here to the shoe, that fills it until it cannot go any further. After the oil is by-passed—now, this down here, you understand, that 7 feet is full of oil below the plunger. Now, your displacement of this oil shows that this pump pumps with the weight of the rod. It pushes the oil up in place of pulling it up like a traveling plunger will. With a pump of this kind at 5,000 feet there are probably 15,000 pounds of rods, you understand, but as I say, the displacement is on the outside of this wall. We will say this is full; it is two and a half inches in diameter and there is an eighth clearance between the tubing and the pump and of course that is full of oil. As this goes up, that oil comes down here and fills this place. There is your displacement. In other words, your barrel, as that goes up, don't lift none of that oil. it runs down here and fills this here as it comes down. This barrel has got to have room and this pushes it up—in any type of pump in which you have a standing tube and traveling barrel, as the barrel moves up the liquid which is in the barrel, between the packing and the bottom of the barrel, will flow out of the bottom. It has got to.

Q   You did see many pumps in the prior art before you built yours?

A   No; oh, no, there wasn't out here, only two. There were only two out here. In the two that were here the liquid, as the barrel moved up, came out through holes here, holes bored in the barrel.

(Testimony of Patrick H. Bray)

Q And if the holes were stopped it would come out through the bottom?

A I don't know. If they had it big enough it would, I suppose. Anything would come out if it was big enough.

Q BY MR. SOBIESKI: In the pumps you saw prior to the pump you built for this patent, allowing oil to come out through holes in the sides of the barrel, bored in the barrel—

A Two of them, and some had four holes bored in the barrel, on the bottom end of it.

THE COURT: If you didn't have the holes and you didn't have the clearance something would break or it would stop?

A It couldn't work.

MR. SOBIESKI: I would like to call Mr. Hoeferer.

MR. HERRON: I think we can save time possibly if we know what you want to show. If you just mention the things we can stipulate and save a lot of time.

MR. SOBIESKI: The first thing I would like to have is Mr. Hoeferer, manager and officer of the defendant Hofco Pump Company, and the pumps which are manufactured which we claim infringe our patent, are manufactured under his supervision and direction; is that a fact.

MR. HERRON: It is so stipulated.

MR. SOBIESKI: And is it also true that the defendant Hofco Pump Company has been manufacturing pumps since the notification similar to those described in the catalog which has been introduced in evidence and is still manufacturing them, and unless restrained will continue to manufacture them or similar pumps?

(Testimony of Patrick H. Bray)

MR. HERRON: The answer to that is this: Prior to the receipt of notice, we stipulate that we were manufacturing pumps of which this catalog is roughly illustrative. We don't stipulate that it is intended as a working drawing. It is just a rough illustration of what we are making. We also assert it to be a fact that from the date of the notification to to the first day of January of this year we made no pumps of the type or any of the types so illustrated which had any form of clutch other than the ring clutch, on the first of January of this year, when we again began to manufacture the line of pumps including the jaw clutch.

MR. SOBIESKI: May we consider that statement as a stipulation?

MR. HERRON: Oh, yes, surely.

MR. SOBIESKI: And unless you are now manufacturing pumps of which the catalog is an accurate general description,—

MR. HERRON: Well, an inaccurate general description. We will be glad to show the structure.

MR. SOBIESKI: You are doing that at this time?

MR. HERRON: Yes, and will continue to do it unless this Court finds that it is an infringement on your rights in so doing.

MR. SOBIESKI: And the Hofco Pump Company manufactures these and sells them?

MR. HERRON: The Hofco Pump Company manufactures and sells pumps of the sort I have described in the stipulation we have just heretofore entered into.

MR. SOBIESKI: We accept that stipulation. After receiving notification, whether they sold any of the pumps theretofore manufactured, and how long?

(Testimony of Patrick H. Bray)

MR. HERRON: Not to exceed, he says, in October. We sold none except with your permission. You allowed us a few days, I understand, in which to decide whether we would pay royalty or cease making them, and starting in with a period of a few days after notification you gave us two or three days, and between that time—that roughly speaking, subject, of course, to correction that our books may show.

MR. SOBIESKI: The only other thing we wish to show in our case is the dimensions.

MR. HERRON: We can save time if you will indicate what parts you want at the next session of court.

MR. SOBIESKI: The points I wished to ask Mr. Hoferer were with respect to the dimensions of the barrel traveling barrel at the middle and at the bottom.

MR. HERRON: If you will be so good as to indicate on this diagram which they are, as much as you can.

MR. SOBIESKI: We would like the interior dimensions of the traveling barrel—I think that is marked C, the inside diameter of C, and the inside diameter of D, and I take it that C is the traveling barrel and D is your bottom nut on the barrel, and then we would like the outside diameter of A, which is the standing column, and then also we would like the dimensions of the maximum stroke of the traveling barrel.

MR. HOFERER: What size pump are you referring to, two and a half pump, that is the standard.

MR. SOBIESKI: Surely. Running two and a half tubing, that is the common pump.

MR. HERRON: The inside of C is two inches. That means the interior dimension, inside diameter.

MR. SOBIESKI: And C is the traveling barrel.

(Testimony of Patrick H. Bray)

MR. HERRON: C is the part indicated on your sketch entitled Hofco Pump, figure 1.

MR. SOBIESKI: That is two and a half inches?

MR. HERRON: Two inches.

MR. SOBIESKI: The inside diameter of D.

MR. HERRON: We have employed two diameters of puller nuts. The size of the puller nut which we used was one and twelve-sixteenths or one and three-quarter interior diameter. Later, following a conference of various pump manufacturers with the code authorities, it was suggested about three months ago that we agree upon a tentative diameter of one and eleven-sixteenths. A few were made of that diameter but we are making them and intend to make them of the one and twelve-sixteenths diameter, that is, one and three-quarters. The outside diameter of A, or the puller tube, is one and one-half inches.

MR. SOBIESKI: We accept that stipulation. With those measurements, then, in evidence—this is on the two inch pump; is that correct?

MR. HOFERER: What we term two and a half inch pump, two and a half inch tubing.

MR. SOBIESKI: And is this pump typical of the general types of the Hofco pump?

MR. HOFERER: Yes, sir.

MR. SOBIESKI: That is the large size, they just change proportionately?

MR. HOEFERER: Yes.

MR. SOBIESKI: We will accept those statements as stipulations and with that we rest.

(Testimony of Patrick H. Bray)

MR. HERRON: If the Court please, we will now offer in evidence the file wrapper and contents of the application of Mr. Bray for this patent.

MR. SOBIESKI: If the Court please, we object to that, first, on the ground that it is incompetent, irrelevant and immaterial. We are standing right squarely on the patent and I don't see how there is any materiality in the thing here. As I understand it, it is only in the case that they plead for some claim of abandonment, and there is no evidence before the Court of abandonment.

THE COURT: Sometimes it aids in interpreting the patent.

MR. HERRON: We maintain that the patentee as shown in the file wrapper clearly indicated the change.

MR. SOBIESKI: It is our contention, if the Court please, that the file wrapper is useful in interpretation only in the event the Court should feel there is an ambiguity in the meaning of the patent. It is our contention there is no showing of ambiguity here. In the showing of an abandonment claim that would be material, but statements or argument and that sort of thing has no bearing on the case. The patent speaks for itself unless there is doubt as to what the patent means, and if there is then the file wraper and contents may aid in construing the doubt.

MR. HERRON: We offer it.

MR. SOBIESKI: Our contention is that the patent is plain on its face.

THE COURT: Will you claim that providing apertures is duplicating the function of the perforating holes?

(Testimony of Sirvertus H. Johnson)

MR. SOBIESKI: Our claim No. 3, your Honor, makes no mention of perforated holes at all.

THE COURT: I think I will file it, subject to your exception. If it doesn't appear to be relevant we will disregard it.

File wrapper and contents received in evidence as Defendants' Exhibit B., and appears at Volume II, Page 9 hereof.

SIRVERTUS H. JOHNSON,

witness on behalf of defendants gave his deposition as follows:

My name is Sirvertus H. Johnson.

MR. HERRON: I take it, your Honor, it is not necessary to read the preliminary matters.

THE COURT: No.

MR. SOBIESKI: We at this time wish to object to it as incompetent, irrelevant and immaterial, has no bearing on this case, does not affect any of the elements involved or any of the contentions involved in this patent.

THE COURT: It is rather an anticipation of a prior art—

MR. SOBIESKI: In the aid of time, may it be stiuplated that we have an objection to all of these things and an exception to the ruling?

THE COURT: Yes.

MR. SOBIESKI: All right. Proceed with the reading of the deposition and that will serve.

I reside at 40 Melvin Avenue, Bradford, Penna. My business is Secretary of the Bradford Motor Works. I have been secretary since 1928. Prior to the time that I

(Testimony of Sirvertus H. Johnson)

became secretary I was employed by the Bradford Motor Works. I entered their employ May 1st, 1920.

Q At that time (and for the purpose of shortening the record I will refer to the Bradford Motor Works by the term, 'Company') did the Company make a pump known as "Bramo"?

A It did. This cut is a true and correct representation of that pump.

(Cut received in evidence, and marked Exhibit 1. It appears at Volume II, Page 29 hereof.)

The Bramo is an oil well pump commonly known as a liner barrel. The pump has a reciprocating plunger tube, designated by "A" on Exhibit No. 1, the outside tube, which is designated by "B" on Exhibit No. 1 remaining stationary. Part C shown on Exhibit No. 1 is a stuffing box gland in which a groove is milled to engage a milled section in the upper connection, which is designated as "D" on Exhibit No. 1. Part C is at the upper end of an assembly at the upper end of Part B.

At the time I entered the employ of the Company, the Bramo at that time was being constructed as illustrated in Exhibit No. 1. This Frick & Lindsey Company catalog, dated May, 1919 was furnished us by the Frick & Lindsey Company. I first saw it, to the best of by knowledge, in this office when I came to work here. Page 43 thereof illustrates the device concerning which I have been testifying.

(Catalogue marked Exhibit No. 2 received in evidence. Page 43 thereof appears at Volume II, Page 30 hereof.)

(Testimony of Sirvertus H. Johnson)

MR. HERRON: I might say, if the Court please, that this is the other type of pump, that is the straight pump. That is the one with the fixed barrel and the moving plunger, and the object or purpose in refering to these structures is to prove the development and the history—

MR. SOBIESKI: It is not the type of pump which is claimed in the patent.

MR. HERRON: No, the predecessor. Mr. Herron: (Reading from the deposition): That catalog remained in the files of the Company the time I entered its employ until the present time. We still manufacture the Bramo device. This was being used and was upon sale at the time I entered the employ of the Company. It has been in use and upon sale since that time. I would say not less than ten thousand Bramo pumps have been manufactured and sold by the Company since I entered its employ. During the time I have been with the Company, the Company made pump parts for the Walter O'Bannon Company. We have our cost records pertaining to those parts. Those records were made under my direction and supervision. They are true and correct. There are nine photostatic copies of Bradford Motor Works specification sheets numbered consecutively from 1 to 9, inclusive. I caused these photostatic copies to be made of the original records of the Bradford Motor Works specification sheets.

(Nine sheets received in evidence and marked Exhibit No. 3. They appear at Volume II, Page 31 hereof.)

These specification sheets were made up, as of the date which each bears, in our planning department, being based on blueprints or samples furnished by our customer,

(Testimony of Sirvertus H. Johnson)

the specification sheets to be used as guides in manufacturing the parts to which they apply, and from which job orders are made up and given to the operators who actually made the parts.

These photostatic copies of job orders numbering 1 to 8, inclusive, are photostatic copies of the original record made and kept by the Bradford Motor Works. The records from which these photostats as well as the preceding ones were made, were made in the regular course of business at the time of the transactions which they depict, and they are true and correct.

(Photostatic copies received in evidence and marked Exhibit No. 4. They appear at Volume II, Page 40 hereof.)

These job orders, Exhibit No. 4, sheets 1 to 8, were made up as orders were received from the Walter O'Bannon Company, the information as to machine operations being taken from such specification sheets as Exhibit No. 3.

This document entitled, 'Work in Process Detail', consisting of sheets 1 and 2, are part of the original records of the Company, kept in the due course of business and under my general direction and control, and made at the time of the transactions which they depict, and are true and correct.

(These sheets received in evidence and marked Exhibit No. 5. They appear at Volume II, Page 48 hereof.)

Exhibit No. 5, sheets 1 and 2, are a cost record of the operations performed in the manufacture of parts for the Walter O'Bannon Company, and show the job order

(Testimony of Sirvertus H. Johnson)
numbers and the operations which were performed in
making these parts.

This cut, represents a device made by the Company.

(This cut received in evidence and marked Exhibit No.
6, and appears at Volume II, Page 52 hereof.)

The device there illustrated was first made early in the
year 1926. The Admore is an oil well pump commonly
known as a liner barrel, which is designed to seat in a
special seat or common working barrel. The standing
valve, comprised of parts 14 to 19, inclusive, to which is
attached the plunger tube, part 2, and plunger fitting,
which is made up of parts 13-B-8-13-11-12-7-9 and 10,
remains stationary during operation, and the outside tube,
designated as part 1, together with parts 3-4-5 and 6,
reciprocate with the pumping action. Part 15, which is
known as a closed crown, is designed with milled tongues
which engage in part 6, which is known as the tube guide,
part 6 being milled with slots which engage with the
tongues on part 15, part 6 being assembled at the lower
end of the outside tube, shown as part 1 in the illustration.

This one sheet catalog contains an illustration of the
device concerning which I have been testifying. That
catalog was published in March of 1926. It was distributed generally to the trade by mail and by salesmen
of the Bradford Motor Works.

(Catalog received in evidence and marked No. 7, and
appears at Volume II, Page 53 hereof.)

These sheets No. 1 to 14, inclusive, are photostatic
copies of the original records of job orders and were
made under my general supervision and direction in the

(Testimony of Sirvertus H. Johnson)
regular course of business of the Company at the time of the transactions which they record. The original records were true and correct.

(Photostatic copies received in evidence and marked Exhibit No. 8, and appear at Volume II, Page 54 hereof.)

Exhibit No. 8 disclose the same information in general with relation to the Admore, that Exhibit 4 disclosed in general with reference to the parts made for the Walter O'Bannon Company. We do not have work specification sheets for the Admore pump similar to those relating to the Walter O'Bannon Company, marked Exhibit No. 3. We never had such sheets, because most of the parts used in the assembly of the Admore liner barrel had been previously made and used in other types of pumps, and new specification sheets were, therefore, considered unnecessary.

These photostats, sheets No. 1 to 10 inclusive, are copies of sales sheets upon which we billed our customers for purchases of Admore liner barrels. The original sheets, of which those are photostatic copies, were kept under my general direction and control in the usual course of business of the Bradford Motor Works, and they are true and correct. They were made at the time of the transactions which they record.

(These photostatic copies received in evidence and marked Exhibit No. 9, and appear at Volume II, Page 68 hereof.)

Sheet No. 1 of Ex. No. 9, is a record of a charge, dated May 4th, 1926, for six 2-inch by 6-foot Admore liner barrels, ordered by the Frick-Reid Supply Company, Tulsa, Oklahoma, to be shipped to the same company at

(Testimony of Sirvertus H. Johnson)

Bartlesville, Oklahoma, and is a record of the fact that that device was so shipped. The same would be true of each of the succeeding sheets, that is, that the set-up of the sheet and the information disclosed thereby is the same in character. The original sheets, of which these are photostatic copies, were the original records of the transaction.

To the best of my knowledge this catalog being catalog of Frick & Lindsey Company, bearing notation "Copyright 1915" was in the office of the Company when I entered its employ. The writing on the fly leaf pasted therein is mine. That fly leaf was pasted in the catalog by me on or about October 8, 1923.

(Catalog received in evidence and marked Exhibit No. 10, and a photostatic copy of the pertinent matter appearing therein appears at Volumn II, page 78 hereof.)

I helped make up this catalog entitled "Pumping Supplies for Oil Wells—Bradford Motor Works" and caused it to be printed about August, 1921

(Catalog received in evidence and marked Exhibit No. 11, and a photostatic copy of the pertinent matter appearing therein appears at Volumn II, page 79 hereof.)

Catalog, Exhibit No. 11, was distributed to all our customers either by mail or through our salesmen who called upon them. The mailing of the catalogs was done under my general direction. The number of copies that were distributed by mail or otherwise was not less than 5,000.

I first saw this part of a pump on the lease of T. P. Thompson, at Red Rock, Penna. It is a plunger assembly and tube guide of an Admore liner barrel

(Testimony of Sirvertus H. Johnson)

(This plunger assembly and tube guide received in evidence and marked Exhibit No. 12)

The circumstances under which I first saw Exhibit No. 12 were these. We were looking for samples of Admore barrels which had been manufactured among the first which we put out, and we discovered this Exhibit No. 12 among the junk which had accumulated in a barn on this lease. Mr. William Kolbe and Mr. W. H. Maxwell were present when I discovered it. Exhibit No. 12 is in the same condition now as it was when I discovered it in the barn.

I have also seen this part of a pump before. I first saw this one also on the lease of T. P. Thompson at Red Rock, Penna., at the same time that Mr. Kolbe, Mr. Maxwell and I discovered Exhibit No. 12. It is a plunger assembly and tube guide of an Admore liner barrel.

(This plunger assembly and tube guide received in evidence and marked Exhibit No. 13.)

This part of a pump was also discovered at the same time as Exhibits No. 12 and No. 13. At that time Exhibit No. 13 and this exhibit were assembled as one unit.

(The part referred to received in evidence as Exhibit No. 14)

Exhibit No. 14 is the outside tube, upper connection, crown, ball and seat. About 308 Admore pumps were made and sold in the year 1926. About 14,000 have been made and sold since that time.

(Testimony of Fay L. Ingleright)

Depósition of Sirvertus H. Johnson and Exhibits 1 to 14 inclusive referred to therein offered and received in evidence as Defendants' Exhibit C.

FAY L. INGLERIGHT,

witness on behalf of defendants, gave his deposition as follows:

My name is Fay L. Ingleright. My residence 31 Pleasant Street, Bradford, Penna. My occupation is machinist and superintendent of the Bradford Motor Works. I first entered the employ of the Bradford Motor Works between 4:00 and 5:00 o'clock on Thursday, February 25, 1926, and I have been continuously employed by the Company since that time.

I saw an exact copy of Exhibit No. 11 in February, 1926. It was sent to me at Grand Rapids, Michigan, by Mr. Morris. That was immediately prior to the time I entered their employ. The catalog which was sent to me contained those illustrations of the Bradford plunger liner, shown on page 7, and the O. F. S. working barrel, shown on page 8 of Ex. No. 11.

MR. HERRON: That, your Honor, is the other type.

MR. HERRON: (Continuing the reading of the deposition):

When I entered into the employ of the Bradford Motor Works, they were making and selling the structures illustrated on the two pages I have just named, and my attention being called to the catalog at the point where the letter A is placed in ink on the cut on page 7, and the letter B on the cut on page 8, the structures which the

(Testimony of Fay L. Ingleright)

Bradford Motor Works was then manufacturing contained the clutch which is there illustrated.

At the time I entered the employ of the Bradford Motor Works the Company was constructing another additional pump known as the Admore liner barrel. A structure similar to Exhibit No. 6 had been designed by the Company when I entered their employ. The first such pump was actually built under my direction. This drawing accurately represents the construction of that pump built by me with certain exceptions.

(Drawing received in evidence and marked Exhibit No. 13-A, and appears at Volume II, Page 81 hereof.)

I have compared this drawing with Exhibit No. 13. It is a drawing of that device.

Now, the difference between Exhibit No. 13-A and the first Admore liner barrel pump which I built after I entered the employ of the Company is:

The original pump built had one lock nut at the point marked A, and there was a lock washer at the point D between the nut and part E. The illustration has missing one piston ring at B and at C, which rings were present at those points on the pump originally built. With those exceptions, Exhibit No. 13-A represents the structure of the first Admore built by me.

Referring to Exhibit No. 6, I will briefly explain in what respect the pump which the Company was building in 1926 was similar to the cut of Exhibit No. 6, and in what respect it differed from that cut, and likewise explain any similarities or differences in operation. The pump which was being built at the time that I entered

(Testimony of Fay L. Ingleright)

the employ of the Bradford Motor Works had a standing valve, similar to the one shown on Exhibit No. 6, which is represented by part numbers 14-15-16-17-18 and 19, a plunger tube, similar, represented by part No. 2, an upper valve assembly which was similar, represented by parts No. 3-4 and 5, an outside tube, exactly similar, represented by part No. 1. There was a difference between the pump then built, and the one illustrated in Exhibit No. 6, the difference being in the plunger fitting. The early plunger fitting, as shown in Exhibit No. 13-A, consisted of a body, 'F', four cups, 'G', three cup rings, 'H', one lock nut on top at A, and grooves cut in the body to receive four piston rings, 'I'. The plunger fitting shown on Exhibit No. 6 consists of a body, No. 10, upon which are mounted four cups and three cup rings. Directly above that, another member is attached, carrying a female coupling which in turn has screwed into its upper end another tube upon which is mounted a spring, part No. 12, a packing follower, part No. 11, a coil of packing, part No. 13, a top nut, part No. 8, and a lock nut, part No. 13-B. The early pump had a shake proof lock washer mounted on the upper end of the plunger fitting. In running a liner barrel of the type shown in Exhibit No. 6 and in Exhibit No. 13-A the lower valve, parts No. 14 to 19, inclusive, form a standing valve which is run down in the top of an old working barrel or special seat, seating at the lower edge of part No. 15. The upper end of part No. 15 is milled to form a tongue at its upper end, which engages a female slot in part No. 6, same being screwed into the lower end of part No. 1. When the well is being pumped a string of sucker rods attaches to the upper end

(Testimony of Fay L. Ingleright)

of part No. 3 to give the reciprocating pumping action. In the event of rods becoming unscrewed, the upper, movable member, consisting of parts No. 1-3-4-5 and 6 drop down by their own weight. The operator at the surface of the ground may then lower the rods until they engage, and a slight twist of the rods from above causes the notch or female groove in part No. 6 to become engaged with the tongue on part No. 15. This locks the lower end of the string of sucker rods, and because of the lower valve, namely, parts No. 14 to 19, inclusive, being held in a fixed position in the working barrel or special seat, allows the operator to tighten up the loose joints of sucker rods. After hooking onto the power the movable members, parts No. 1-3-4-5 and 6 are spaced in their proper pumping relation. This disengages the lock between parts No. 6 and No. 15 of Exhibit No. 6. The design of the lock between No. 6 and No. 15 of Exhibit No. 6, and of part No. 6 and No. 15 of Exhibit No. 13-A, has been unchanged since the early manufacture, except for some minor details of dimensions. The position of parts No. 6 and No. 15, in respect to the other parts of the pump has been unchanged.

Exhibit No. 12 is a plunger fitting assembly of an Admore liner barrel, practically the same as the one illustrated in Exhibit No. 13-A, and comprising a standing valve, No. 15 on Exhibit No. 13-A, a tube guide, No. 6, a plunger fitting, parts A to I, inclusive, and a plunger tube No. 2.

I am able to tell from the design of that exhibit approximately when it was manufactured. This assembly was assembled during the summer of 1926. It is not of the first lot of manufacture—it was of the second. I say that

(Testimony of Fay L. Ingleright)

because of the fact that this Exhibit No. 12 has two lock nuts on top. The earliest plunger fitting made had one lock nut. The plunger fittings which had two lock nuts had no threads in the member directly below the two lock nuts, that is, the member which carries the two piston rings. In the first lock the upper member, carrying the piston rings, had a tapped hole. The upper member with piston rings, corresponding to E on Exhibit No. 13-A, was made to be slideably mounted on a plunger fitting body. The shake proof lock washer between the two lock nuts on Exhibit No. 12 was found by experience to be faulty, as it had a tendency to break up, scoring the working barrel tube and the piston rings. The last shake proof lock washers were assembled in the latter part of 1926, and none were used thereafter. The piston rings were also discontinued during the latter part of 1926. The plunger tube, corresponding to part No. 2, Exhibit No. 13-A, is made from black iron pipe, the length of same being 3 feet 6 inches. The material was afterwards changed to seamless steel tubing. The closed crown, corresponding to part No. 15-A, Exhibit No. 13-A, measures one and seven-eighths inch O. D., which is of a size that was discontinued after the second run of Admore closed crowns. The tube guide, corresponding to part No. 6 of Exhibit No. 13-A, is of a design used in making up the first two lots of Admore pumps, and is identified by the shortness of the length of the large outside diameter, this being afterward lengthened to provide a better wrench hold.

Q Calling your attention to Exhibit No. 13, can you from its design fix the approximate date of its manufacture?

(Testimony of Fay L. Ingleright)

A   The approximate date of its manufacture would be the same as of Exhibit No. 12. The fact that Exhibit No. 13 does not have a shake proof lock washer simply means to me that it was on there and was lost off. The length of the tube, Exhibit No. 13, is for a 6-foot Admore, while that for Exhibit No. 12 is for a 5-foot Admore. The first lot of 5-foot Admore plunger tubes, similar to that on Exhibit No. 12, were cut from the length of that shown on Exhibit No. 13. Because of the fact that no great sales were anticipated in the 5-foot Admore, none of the proper lengths of plunger tubes were purchased.

Q   Referring to Exhibit No. 14, can you from its structure fix the approximate date of its manufacture?

A   Exhibit No. 14 was made during either early 1926, or along in the summer of that same year, because the part corresponding to part No. 3 of Exhibit No. 6 is what is known as a long-pattern valve cage. The two earliest runs of Admore liners were equipped with this length valve cage.

Q   Directing your attention to the characteristics by which you have fixed the date of the structure concerning which you have testified, and now calling your attention to Exhibit No. 8, can you, with the aid of those records, fix the date of the manufacture of the devices more closely?

A   Yes, I can. Sheet No. 3, dated April 30th, 1926, being completed June 25th, 1926, specified 250 of the Admore closed crowns, corresponding to No. 15-A of Exhibit No. 13-A, to be made of one and seven-eighths inch O. D. round cold rolled steel. Sheet No. 6 specifies 400 of the same parts to be made from the same size steel, and was completed August 27th, 1926. That is the last

(Testimony of Fay L. Ingleright)

date of record of any Admore closed crown being made having the same outside diameter as that member on Ex. No. 13, and Ex. No. 12. The next lot of Admore closed crowns specified an outside diameter of one and twenty-seven-thirty-seconds inch round cold rolled steel which were started January 7th, 1927, and were completed January 28th, 1927, as shown by sheet No. 14. Sheet No. 7, Exhibit No. 8, is a record of the last job order of Admore plunger fitting bodies of the piston ring type, corresponding to part F on Exhibit No. 13-A and found on Exhibit No. 13 and Exhibit No. 12. The design referred to was changed at the completion of this run of parts, and a new run of a totally different design was started and completed October 11th, 1926, as shown by the record in our job order files. The document from which I was reading is the original record.

(This original document is received in evidence, marked Exhibit 19, and a photostatic copy thereof made and attached to this deposition as Exhibit 19, and appears at Volume II, Page 82 hereof.)

Sheet No. 10, Exhibit No. 8, is a record of the mating part for the Admore plunger fitting body, corresponding to "E" on Exhibit No. 13-A, and is a record of the last run of those parts. The Admore cup nuts were started August 9th, 1926, as shown by sheet No. 10, and were completed August 13th, 1926. On succeeding Admores this part was replaced by parts of standard design and carried regularly in stock, and a search of our records shows that no more of the Admore cup nuts referred to were made. The Admore plunger fitting bodies were completed August 13th, 1926, as shown by sheet No. 7.

(Testimony of Fay L. Ingleright)

Deposition of Fay L. Ingleright and Ex. 13-A and 19, referred to therein, offered and received in evidence as Defendants' Exhibit D.

FAY L. INGLERIGHT

being called as a witness on behalf of defendants gave his deposition as follows:

DIRECT EXAMINATION

I have referred to the plunger tube as a feature enabling me to fix the date of Exhibits No. 12 and No. 13. What I had in mind was that the plunger tube of Exhibit No. 12 and of Exhibit No. 13 corresponds to Part No. 2, and is made from 3/4" extra heavy pipe of which sheet No. 2, Exhibit No. 8, is a record of the second run of same. The last run of this material was completed August 27th, 1926, as shown by job order No. 4679 in our job records.

(The document referred to is received in evidence marked Exhibit No. 20, and a photostatic copy thereof made and attached to this Deposition, and a copy appears at Volume II, Page 83 hereof.)

When our supply of these 3/4" pipe plungers was exhausted our design was changed to specify a seamless steel tube 1-1/16" O. D. x 13/16" I. D. x 4'6" long. The first run of these tubes was completed November 19th, 1926, as shown by our job order record, No. 5022.

(This record referred to is marked Exhibit No. 21 and a photostatic copy thereof made and attached to this Deposition as Exhibit No. 21, and a copy appears at Volume II, Page 84 hereof.)

Deposition of FAY L. INGLERIGHT and Exhibits 20 and 21 referred to therein was offered and received in evidence as Defendants' Exhibit D1.

(Testimony of George B. Morris)

GEORGE B. MORRIS,

witness on behalf of defendants, gave his deposition as follows:

My name is George B. Morris. My residence 123 Kennedy Street, Bradford, Penna. My business is manufacturing. I am the President of the Bradford Motor Works. I first became connected with that company at a time when it was a partnership—at the organization of the partnership. That was approximately 1911. This sketch represents a pump which the company has manufactured. That pump was known as the O. F. S. working barrel. It was being manufactured by the predecessor of the Bradford Motor Works, and when the Bradford Motor Works started in business in 1911 the manufacture of this pump was contined.

(Sketch received in evidence and marked Exhibit No. 15, and appears at Volume II, Page 85 hereof.)

The O. F. S. working barrel is of the inserted type, the lower end of which is seated in the top of a conventional or standard working barrel. The tube, part 1, is stationary. Reciprocating within the tube is a plunger fitting, attached to a valve stem, part 2, which in turn is connected to the sucker rods by a substitute, part 3. Fluid is discharged from the working barrel into the tubing through holes in the crown, part 4. The crown is fitted at its upper end with a notch, or groove, part 5. A corresponding tongue, No. 6, is formed on the substitute, No. 3. By the engagement of tongue No. 6 in slot No. 5 rotation of part 3 is prevented, thus making it possible

(Testimony of George B. Morris)

to screw up loose or disconnected sucker rods. This was a well recognized and valued feature of the unit.

I first saw catalog Exhibit No. 10 about 1915 or 1916. The device, concerning which I have just testified, is listed therein on page 20. I have seen other copies of that catalog, in various supply stores throughout the country.

I was instrumental in the preparation of this catalog Exhibit No. 11 and saw the first shipment from the printer in 1921. The device concerning which I have been testifying is illustrated in that catalog, on page 8. We are not manufacturing O. F. S. pumps in quantities and advertising it, but continue to furnish repair parts, and at intervals have been supplying complete units. It is so long since we worked aggressively on this pump that I don't recall the volume with exactness, but I am certain that more than 1,500 were made and sold prior to 1927.

We made a pump of which this sketch is a rough illustration. We began the manufacture of it about 1912. We termed that pump the Bradford plunger liner, or sometimes the I. X. L. plunger liner.

(Sketch received in evidence and marked Exhibit No. 16, and it appears at Volume II, Page 86 hereof.)

Referring to Exhibit 16, this is a pump of the inserted type, and is operated without the customary cups. The suction is provided by reciprocation of a plunger tube, No. 1, through a stuffing box, No. 2, fitted interiorally with packing. The groove No. 3, and tongue No. 4, operate in exactly the same manner as the groove, No. 5, and tongue No. 6 on Exhibit No. 15, and for the same purpose. That structure is illustrated in catalog Exhibit

(Testimony of George B. Morris)

No. 11, on page 7. I am familiar with the extent to which that catalog was circulated. Between 5,000 and 10,000 copies of this catalog were printed, and an effort was made to put at least one copy in every oil well supply store in the United States, and also one copy in the hands of important production men in all of the oil companies in the United States. That was done both by mail and personal distribution through a salesman. That mailing was done under my personal direction or supervision. With relation to the circularization by means other than mail, salesmen were furnished with a supply of catalogs and asked to deliver them wherever no copies were to be found.

Referring to my records, I can give you the date those catalogs were purchased—several shipments in the year 1921. They were paid for by us on September 19th, 1921, as shown by cash disbursements, page 30. My recollection, after the passage of many years, may not be extremely accurate, but I am quite sure that at least 500 of these Bradford plunger liners, or I. X. L. pumps, were made and sold.

The Bradford Motor Works made devices of which Exhibit No. 1 is an illustration. The first one was completed and installed in a well in March, 1917. The structure of this pump embodies a combination of the essential features of the pumps on Exhibit No. 16 and Exhibit No. 15. It has the cup structure and lower valve like Exhibit No. 15, and the plunger tube and stuffing box, like Exhibit No. 16. It also has the lock feature common to both Exhibits No. 15 and 16, and indicated by letters C and D on Exhibit No. 1, the tongue appearing on part D and the groove on part C.

(Testimony of George B. Morris)

The Bramo device is illustrated in Exhibit No. 2, on page 43. I first saw that catalog sometime during 1919. I have seen copies of that catalog elsewhere than in our plant. I have seen them in various supply stores. We are still making the Bramo pump. Not less than 15,000 were made and sold by us. I have during the time I have been in charge of the Company, had occasion to furnish parts to the Walter O'Bannon Company for a pump distributed by that Company. Exhibit No. 3 and No. 4 disclose the manufacture to which I have testified.

I saw the pump after it had been assembled by O'Bannon. At periodical intervals when I visited his office and shop at Tulsa, Oklahoma.

My company manufactured the device pictured in these photographs, three sheets numbered 1, 2 and 3 inclusive. We called it the inverted Bramo working barrel.

(Photographs referred to are received in evidence as Exhibit No. 17, and appear at Volume II, Pages 87 hereof.)

About a year ago we shipped an inverted Bramo pump to Mr. William H. Maxwell in Los Angeles, California. In the original Bramo working barrel the stuffing box and main tube were stationary while the pump was in operation. The plunger fitting was reciprocated within the tube and connected by a plunger tube to the sucker rods. In the inverted Bramo barrel the main tube, No. 1, and stuffing box, No. 2, reciprocate over a stationary plunger fitting within the tube. This plunger fitting is connected to the standing valve, No. 3, by means of a plunger tube, also stationary. The tongue and groove assembly, indicated at No. 4, is the same as that used on the original

(Testimony of George B. Morris)

Bramo working barrel, but is located at the lower end of the pump, due to the inverted construction of the pump. As a matter of fact, we have used this construction on every type of inverted pump we have ever manufactured, and the only deviation from a general standard is due to the necessity for thread changes in order to adapt the features to different types of pump. We made approximately not more than 25 of those devices. They did not function to our satisfaction.

Q In the design of this pump no satisfactory provision was made for the escape of oil trapped between the main tube and the plunger tube, and, consequently the stuffing box principle was abandoned, and a pump built of the type later given the name of Admore. Exhibit No. 6 is the type of device to which I just referred. Referring to that illustration, the Admore pump is of the inverted type, involving a stationary plunger tube, part No. 2, a stationary plunger fitting located at the upper end of said plunger tube, and a standing valve screwed to the lower end of said plunger tube. It also involves the reciprocating parts, No. 6-1-3-4 and 5. The reciprocation of tube, No. 1, creates a suction above the plunger fitting. The locking device is positioned at the same place as in Exhibit No. 17, and in construction and purpose identical with Exhibit No. 17.

The device concerning which I have just testified is illustrated in Exhibit No. 7. The company ordered and received 5,700 of these circulars. They were distributed in the usual manner to the supply companies and oil companies. I know of my own knowledge that that distribution was made, and it was made under my own direction.

(Testimony of George B. Morris)

I recall some of the earliest ones made of the Admore pump. One of the first pumps installed was in the month of March, 1926, on a property near our factory, and owned by Emery & Mason. I personally saw this working barrel put in the well, and watched it operate at periodic intervals for a month or two. We have a record of this installation. The record of this installation is shown on sheet No. 4 of Exhibit No. 9. The words "Sent out on trial March 30th, 1926," mean that the pump was taken by us from our shop to the property on that date. The invoice is dated August 6th, 1926. The pump was sold on a trial basis, and invoice was not mailed until the customer expressed approval of the pump and willingness to pay for same. That expression of approval occurred about the 1st of August, 1926. Following the early uses, which I have testified began some time in March, 1926, the Company began widespread sales efforts in relation to the Admore pump. All during the year 1926 we made consistent sales efforts both in the Eastern fields and in the Mid-Continent fields, and by the end of the year had a rather wide distribution and a consistent demand. Since that date we have continued to make and sell the Admore pump. We are at the present time manufacturing and selling that pump. We have made no changes in the general construction of the locking device from the original Admore. Its general construction is identical with that used on the first pump made.

(Deposition of George B. Morris, and Exhibits 15, 16 and 17, referred to therein is offered and received in evidence as Defendants' Exhibit E.

(Testimony of William P. Kolbe)

WILLIAM P. KOLBE,

witness on behalf of defendants, gave his deposition as follows:

My name is William P. Kolbe, I reside at Gilmore, Penna. My business is pumper—oil field worker. I am now employed on the Thompson Lease, Red Rock, Pennsylvania. On the 11th day of last month I have been employed there five years. I know Mr. W. H. Maxwell and Mr. S. H. Johnson. I had occasion to examine this physical Exhibit No. 12 in the presence of Mr. Maxwell and Mr. Johnson, about the middle of August, 1933. Exhibit No. 12 was at that time in the barn located at the Thompson Lease at Red Rock, Penn. It was in substantially the same condition as it is now, a little dirtier, if anything.

I examined Exhibit No. 13 and Exhibit No. 14, the two parts of the other pump, at the same time and place and in the presence of Mr. Johnson and Mr. Maxwell. They are in substantially the same condition now as they were then. A little bit cleaner than they were then—otherwise the same.

Deposition of William P. Kolbe offered and received in evidence as Defendants' Exhibit F.

(Testimony of Gilbert T. Eidson)

GILBERT T. EIDSON,

witness on behalf of defendants, gave his deposition as follows:

My name is Gilbert T. Eidson. I reside at 292 South Kendall Avenue, Bradford, Pennsylvania. I know what an Admore pump is. Exhibit No. 13 and Exhibit No. 14 when assembled, make up the pump known to me as an Admore.

I installed, or caused to be installed, an Admore during the year 1926 in wells No. 140 and No. 145 on the Melvin Lease. I have a record book which shows such installation. This is it. That record appears on page 62.

(Record book received in evidence and marked Exhibit No. 18, and a photostatic copy of the pertinent matter appearing therein appears at Volumn II, page 90 hereof.)

MR. HERRON: That is the little book attached to Exhibit 18.

MR. HERRON: (Reading from deposition): The pumps which I installed worked. Page 62 of that exhibit shows the installation in wells No. 140 and 145.

Deposition of Gilbert T. Eideson and Exhibit 18, referred to therein, offered and received in evidence as Plaintiffs' Exhibit G)

(Testimony of Sirvertus H. Johnson)

SILVERTUS H. JOHNSON,

a witness on behalf of the defendants, being recalled, gave his deposition as follows:

These four yellow sheets (attached together) on stationary of the Bradford Motor Works are sales sheets taken from our original record of sales.

(Four sales sheets received in evidence and marked Exhibit No. 22, and appear herein at Volumn II, page 91 hereof.

I have heretofore testified that the records of the Bradford Motor Works were made under my direction and control in the usual course which they depict, showing the sale of the Admore Liners sold to the companies mentioned upon the sheets. Referring to Exhibit No. 22, these sales sheets are numbered in consecutive order when placed in our records. Sheet No. 123 covers the sale of 2 Admore working barrels to the Carter Oil Company, Caspar, Wyoming, on April 12th, 1926. Sheet No. 128 shows the sale of 1 Admore liner barrel to A. L. & L. M. Lilly of Rixford, Pennsylvania, on April 15th, 1926. Sheet No. 308 shows a sale of 1 Admore liner to the Sloan & Zook Company, Bradford, Pennsylvania, on April 26th, 1926. Sheet No. 327 covers the sale of 1 Admore liner barrel, together with other material, to the South Penn Oil Company of Bradford, Pennsylvania, on April 27th, 1926.

The deposition by Silvertus H. Johnson, so recalled, together with Ex. 22 referred to therein was offered and received in evidence as Defendants' Exhibit H.

(Testimony of Raymond C. Flickinger—Carl M. Perkins)

RAYMOND C. FLICKINGER

a witness on behalf of defendants gave his deposition as follows:

My name is Raymond C. Flickinger. I reside at 410 East Main Street, Bradford, Pennsylvania. I came to work here for the Bradford Motor Works in February, 1913. I left here December 12th, 1925, and returned in the latter part of January, 1927. I am familiar with the type of pumps manufactured in the plant at the time of my return. Exhibit No. 6 represents the type of device then being constructed. Referring specifically to the locking device on parts No. 6 and No. 15 on Exhibit No. 6, I recall specifically that the pump which was being manufactured at the time of my return included that locking device. All of the Admore pumps manufactured since that time have carried that locking device.

Deposition of Raymond C. Flinckinger, offered and received in evidence as Defendants' Exhibit I.

CARL M. PERKINS,

a witness on behalf of defendants gave his deposition as follows:

My name is Carl M. Perkins. I live at 26 High Street, Bradford, Pennsylvania. I was employed in the year 1926 at the South Penn Oil Company at Bradford, Pennsylvania. I had occasion during the year 1926 to install an Admore pump in a well No. 149 of the South Penn Oil Company. In the year 1926 I installed the pump to which I refer around the forepart of the summer. I was assisted in that installation by Louis Rounds. He is not living now. He died on the 18th of January, 1927. The

(Testimony of James G. McCutcheon, Jr.)

Admore pump which I installed in well No. 149 had four cups on the bottom, and a plunger which goes up into the barrel. The barrel works on the outside of the plunger, and it has a plunger which has four cups and piston rings. The standing valve has four cups and a Ball & Seat. There is a guide bushing on the bottom of the outside tube that works over the standing plunger tube, and when the outside tube is set down the guide bushing locks with the standing valve crown so the rods can be tightened. The plunger in Admore pumps now has a spring and no piston rings.

Deposition of Carl M. Perkins offered and received in evidence as Defendants' Exhibit J.

JAMES G. McCUTCHEON, JR.,

a witness on behalf of defendants, gave his depsition as follows:

My name is James G. McCutcheon, Jr. I live at 22 Hobson Place, Bradford, Pennsylvania. My business is printer. My business in 1926 was printer. I printed Exhibit No. 7 the day the job was given to me on March 16th, 1926. I have a record of doing that work on page B-35 of Ledger No. 1. Our job number was No. 2567, the order was placed on March 16th, 1926, and the order was for 5,700 folders—Admore liner barrel. The price paid was $55.00. I kept this sheet, Exhibit No. 7, as a retained or filed copy of the document I printed. The time elapsed between the time the order was placed with me and the time I printed it was approximately two weeks.

Deposition of James G. McCutcheon, Jr. offered and received in evidence as Defendants' Exhibit K.

(Testimony of Lyle D. Travis)

LYLE D. TRAVIS,

a witness on behalf of defendants, gave his deposition as follows:

My name is Lyle D. Travis. I reside at No. 1 North Bennett Street, Bradford, Pennsylvania. I worked for the Bradford Motor Works from March, 1923, till July or August of 1926.

Referring to Exhibit No. 3, I recognize these documents as being photostatic copies of records of the Bradford Motor Works. I kept the original records, of which those are photostatic copies, in the regular course of business of the Bradford Motor Works, under the general direction or supervision of Mr. S. H. Johnson, during the time I was employed there.

Referring to Exhibit No. 3, sheets No. 1 to 9, inclusive, I kept the originals, of which those sheets are copies, under the general direction or supervision of Mr. S. H. Johnson, during the regular course of business of the Bradford Motor Works. These records, which I kept, truly and accurately reflect the transactions which they purport to record.

Referring to sheets No. 4-5-6-7 and 8 of Exhibit No. 4, I recognize those sheets as being photostatic copies of original records of the Bradford Motor Works. I kept the originals, of which those are copies, in the due course of business of the Bradford Motor Works, under the general direction and control of Mr. S. H. Johnson. The original records, of which those are photostatic copies, truly and correctly depict the transactions which I recorded upon those sheets. All these records, concerning which

(Testimony of George F. Iverson)

I have testified, were kept by me in the due course of business of the Bradford Motor Works, and as part of my regular duty, and were made at the time of the transactions which they record.

Deposition of Lyle D. Travis, offered and received in evidence as Defendants' Exhibit L.

GEORGE F. IVERSON,

a witness on behalf of defendants, gave his deposition as follows:

My name is George F. Iverson. I reside at 177 William Street, Bradford, Pennsylvania. I was first employed by the Bradford Motor Works on March 2nd, 1925. I have worked continuously for them ever since. My work has been the first year billing clerk, and from then on planning clerk. Among the duties of planning clerk, it has been my duty to keep the original records of which sheets No. 2 to 14 inclusive, of Exhibit No. 8, are photostatic copies. I kept those sheets in the regular course of business of the Bradford Motor Works. The entries which are made thereon truly and correctly record the transactions which they depict. I kept those records under the general direction and control of Mr. S. H. Johnson.

My attention being directed to the point A on Exhibit No. 13-A and to the lock washer on Exhibit No. 12, the last date that the lock washer was used in Admore pumps, was

A   Approximately November 11th, 1926.

(Testimony of George F. Iverson)

Q When was the type of plunger assembly found on Exhibit No. 12 and No. 13 discontinued?

A The last run of piston ring type was finished on August 13th, 1926, as shown by sheet No. 7, Exhibit No. 8. The first run of cup-and-packing type was finished on October 11th, 1928, as shown by this book record marked Exhibit No. 19, another photostat.

Q When you refer to "piston assembly" what do you refer to, identifying the parts on Exhibit No. 13-A?

A Parts A-D-E-B-G-H-C-F and I.

Q. Referring to Exhibits No. 12 and No. 13, and to the plunger tubes illustrated by Fig. 2 in Exhibit No. 13-A, when was the use of plunger tubes, such as those, discontinued?

A The tubes in these exhibits—by which he is referring to this part (indicating)—are iron type, and the use of iron pipe was discontinued August 27th, 1926, as shown by the book record marked Exhibit No. 20. In place thereof we use seamless steel tubing. The first run of plunger tubes from seamless tubing was finished November 19th, 1926, as shown by job order No. 5022, marked Exhibit No. 21. I remember this change on account of the tubing we received being soft and hard to thread, and we put the ends of it into the furnace to harden it some so that we could thread it easily.

Q Referring to the closed crown of Exhibits No. 12 and No. 13, shown as No. 15-A on Exhibit No. 13-A, when was the use of the closed crowns of the dimensions shown in Exhibits No. 12 and No. 13 discontinued?

A The last run of closed crowns made from 1-7/8" stock was finished August 27th, 1926, as shown on page 6, Exhibit No. 8, and the first run made from 1-27/32"

(Testimony of William H. Maxwell)
stock was finished January 28th, 1927, as shown on page 14, Exhibit No. 8.

Q From your knowledge of the structures built by the Bradford Motor Works, and from your examination of the records, can you state when Exhibits No. 12, No. 13 and No. 14 were made?

A I can't state the exact date, but I know they were made during the year 1926.

Deposition of George F. Iverson was offered and received in evidence as Defendant's Ex. M.

WILLIAM H. MAXWELL

called as a witness on behalf of defendants, being first duly sworn, was examined and testified as follows:

DIRECT EXAMINATION

My name is William H. Maxwell. My business, patent attorney. I was employed by Mr. Hoeferer to investigate the state of the prior art with relation to this pending litigation. In the course of that investigation I received by express from the Bradford Motor Works of Bradford, Pennsylvania, the device which you have just carried up here with you. Following its receipt, I caused certain photographs to be taken of that device.

Referring to Exhibit No. 17, consisting of three photographs, they are the photographs which I caused to be made of that device. I proceeded to Bradford, Pennsylvania, in company with Mr. Herron and attended the taking of the depositions which have just been listed. In the course of those depositions those photographs were referred to.

(Testimony of William H. Maxwell)

MR. HERRON: Yes. You see, we did not want to carry that heavy thing back, so we photographed it and took the photographs.

THE WITNESS: The device is in exactly the same condition now as it was when I received it from the Bradford Motor Works. That device there has been described in the depositions. I was there when the depositions were taken.

MR. HERRON: I would suggest it be given the number 17-A, and then it will correspond with the photographs.

MR. SOBIESKI: I understand we have a general objection to all these depositions; is that correct? If not, I wish to object to the introduction of this on the ground it shows to be a different structure from that which is covered by our patent in this action, therefore immaterial.

THE COURT: That will be open for argument. The objection is merely overruled and exception noted.

The device is received in evidence, as Defendant's Ex. 17-A and 17-B.

MR. HERRON: At this time, if the Court please, I wish to introduce certain patents which we have pleaded as constituting anticipations, and likewise as illustrating the prior state of the art. Under stipulation, I am introducing office copies.

MR. SOBIESKI: We have no objection, of course, to the copies. We do object to each one of these patents as being immaterial, not bearing on the action or anticipating in any way any of the features covered by our patents. In the interests of time we make a general objection to the whole.

(Testimony of William H. Maxwell)

MR. HERRON: It may be understood that that objection which counsel has just made goes to each of the patents which I will now introduce.

THE COURT: The objection will be overruled and exception noted, then.

Thereupon, the following patents were offered and received in evidence:

A patent to Deis, No. 840,919, as Defendant's Ex. N.

A patent to Cummins, No. 1,323,352, as Defendant's Ex. O.

A patent to Whitling, No. 1,391,873, as Defendant's Ex. P.

A patent to Downing, No. 211,230, as Defendant's Ex. Q.

A patent to O'Bannon, No. 1,454,400, as Defendant's Ex. R.

A patent to Thurston, No. 1,625,230, as Defendant's Ex. S.

A patent to Thompson, No. 586,524, as Defendant's Ex. T.

A patent to Wright, No. 575,498, as Defendant's Ex. U.

A patent to Northrup, No. 1,378,268, as defendant's Ex. V.

A patent to Ellis, No. 1,513,699, as Defendant's Ex. W.

A patent to Dickens, No. 1,486,180, as Defendant's Ex. X.

A patent to Neilson, No. 1,717,619, as Defendant's Ex. Y.

(Testimony of Edward T. Adams)

A patent to Savidge, No. 1,452,004, as Defendant's Ex. Z.

A patent to Tierce, No. 1,720,979, as Defendant's Ex. AA.

A patent to Franchi, No. 1,456,727, as Defendant's Ex. BB.

A patent to Stephens, No. 1,407,493, as Defendant's Ex. CC.

A patent to Barthel, No. 1,139,396, as Defendant's Ex. DD.

The above patents appear in the order in which they were introduced, at Volume II, pages 95 to 172 inclusive hereof.

EDWARD T. ADAMS,

called as a witness on behalf of defendants, testified:

My business is that of a mechanical engineer. I have been a mechanical engineer for close to half a century. My initial training was that of a mechanic. I afterwards had a college training at Stanford and Cornell; was instructor at Stanford for two years in machine design; instructed at Cornell at Ithaca in machine design for two years. I was afterwards engineer for Hi-Speed Engine Company, and from that went to the Edward B. Allis Company of Milwaukee as engineer. After a year or two I was made engineer of the pumping engine department and with them was successively the sales engineer for that department, manager of pumping engine department, manager of the pumping engine and steam engine department, manager and chief engineer of all their heavy machinery

(Testimony of Edward T. Adams)

departments and pumping engines, heavy machinery of all sorts. And that takes up the major part of my active business life. After that I was engaged in consulting engineering with various firms and with my own independent offices. I have been greatly interested in oil well pumps, and one of the prominent oil well pumps now made on the market is built from my patents and designs.

I am a member of the American Society of Mechanical Engineers. I have made a careful study of each of the patents which have just been introduced in evidence. I took practically all of these prior art patents and colored them so as to represent the parts in the various structures which move and those which stand still. In the system I used, I colored the moving parts red. The standing parts blue, the valves green, and in cases illustrating a reversal of a structure, shown any additional fittings which would be necessary to tie together the structure as reversed, in pencil.

MR. HERRON: At this time, if the Court please, I hand to the Court and likewise to counsel a copy of the Deis patent, so colored.

This colored copy is offered and received as Defendants' Ex. N^1.

(Note: At the trial, office copies of patents marked "Defendants' Exhibit N to DD" inclusive were received in evidence as heretofore noted. Another office copy of certain of these patents, colored to indicate those parts which moved, those which stood still, and the valves of the structure, were offered in evidence. Each such patent bore the same letter as the corresponding non-colored copy

(Testimony of Edward T. Adams)
followed by the prime sign, that is, the uncolored copy of
the patent to Deis is marked N and the colored copy N'.
In order to avoid encumbering the record, in each such
case only the colored copy of the patent so introduced is
set out in Volume II hereof and is there marked with both
the numeral and the numeral followed by the prime sign,
as for example, the patent to Deis is marked "Exhibit N
and N'."

THE WITNESS: I caused to be made under my direction a large drawing of the Deis design. That is the
drawing to which I refer.

This drawing offered and received in evidence as Defendants' Exhibit EE, and appears at Volume II, Page
173 hereof.

Referring to the Deis patent and to this diagram, this
patent is for a traveling barrel pump of the same type
in that respect as the patent in suit. It differs in form
somewhat, due to the fact that this is a balance type. In
the ordinary pump the load of the fluid is carried alternately by the pump and by the tubing surrounding
the pump. Deis wanted to balance his pump, therefore
it was necessary that the load should all of the time be
carried by the pump, otherwise a fixed weight, such as
he shows depending from the handle of his pump, would
not balance a weight which was carried alternately by the
pump and by the tubing. The diagram EE has been
simplified by depicting those parts which go to make it
a single-acting pump and eliminating the double-acting
feature. The action of the pump as a single-acting
pump, eliminating the double-acting feature, is the same
in so far as the single action goes, with that elimination

(Testimony of Edward T. Adams)

as without it. The lower or inlet valve of the pump, instead of being stationary as it would be in an ordinary pump, therefore, has to travel with the barrel. That does not in any way affect the action of the valve as an inlet valve. So we come with the change made by eliminating the balancing, to a pump that in every way acts as the ordinary oil well pump would act.

Figure 4 of Ex. EE shows the parts of the pump that are stationary, standing parts, as they are called. We have a standing column 1 and 6, being the same lettering as that of the patent, and sliding over this, running on it, is the running part shown in red in Figure 5, and in section in Figure 3. So that when the pump is assembled we have the traveling barrel J of Figure 3 lowered or sliding down over the standing part 6. In operation when this barrel is lifted up, the fluid within the pump barrel itself in the space where the valve u is located and the hollow space inside of the tubing above that, which we will label A-A is discharged out through the valve n shown in green. As the barrel is moving up while the plunger 6 is stationary, the volume of the pumping chamber is diminished. Therefore, this fluid is discharged out of these discharge valves, which consist of an annular ring surrounding the barrel J and seated upon the openings there shown. This fluid goes up through the tubing in the usual manner, just as it would in a pump of any type.

There is a space above that plunger where one might think air or fluid would be compressed but compression is prevented by the sucker rods being hollow so that air can flow back and forth in that space to prevent a vacuum being formed or fluid being trapped in it.

(Testimony of Edward T. Adams)

On the downward stroke of the plunger the space A-A increases in volume because 6 is stationary and the barrel travels down over it. There will, therefore, be a drop in pressure, and fluid from the well will flow in from the bottom through holes, indicated by circles 5, at the bottom of the stationary tubing, up through the annular inlet valve u which is shown in green and which opens at that time and allows the fluid from the well to flow back into the pumping chamber A-A so that we have completed the stroke. The chamber is again filled with fluid which, on the following upward stroke of the plunger, will be discharged from the discharge valve n in the usual manner for pumps of this type.

This structure has a clutch of the jaw type. Figure 7 is an enlargement of it. It is formed of the male projection z, of the member not numbered but part of y, and the female member part 4 of the tube v. When it is desired to tighten the rods in the pump, the traveling barrel is lowered, this female member 4 engages the tongue z and is thus positively interlocked against movement in either direction. The lower end of the rods, therefore, being fastened, the upper part may be tightened as desired. The object of locking that clutch so as to prevent relative rotation of the parts is specifically mentioned in the patent, page 2, lines 16 to 25.

In that structure I find a standing guide column, No. 1, with working valves u and 10; a valved traveling barrel, marked i, j. r. The valve is marked n. I find that valved traveling barrel running on the column 6, and 1.

(Testimony of Edward T. Adams)

I find a bottom guide *s* and *v* on the barrel. This bottom guide very clearly runs on the barrel, that is parts *s* and *v* run on 1. That bottom guide has a bottom face forming a coupler. I find in that structure an anchor for the column marked *w* and *y*, *y* being specified as being locked stationary. That anchor presents directly to the coupler face a top face *z*, forming a complementary coupling or jaw clutch between *z* and 4. The structure is so designed that the anchor for the column is designed to be non-rotatively interlocked with the lower guide coupler in either direction.

MR. HERRON: We have no objection either to counsel deferring all his cross examination until we have explained all the structures, or asking any questions he may wish with relation to the structure as to which the witness is testifying.

THE COURT: I think you had better cross examine as you go along on each exhibit.

CROSS EXAMINATION

Q BY MR. SOBIESKI: Mr. Adams, with respect to this structure here that you have testified about, this lock to which you testified on the bottom of the barrel here, how is that connected with the bottom of the traveling barrel? Is that part of the traveling barrel?

A Speaking now of the upper lock, that is formed on the part *v* and *s* which is attached by screw threads to the part of your traveling barrel marked *r*. That is attached to the lower end of the traveling barrel. The four holes that are marked BB on Ex. EE and marked *p* on the drawing, are the openings through the traveling

(Testimony of Edward T. Adams)

barrel to allow the fluid to flow up through the discharge valve *n*.

Q   Referring to the patent, I notice that you see eight holes on the patent.   Do you know how many holes there are intended by the inventor?   Four here on this drawing.

A   Oh, this is only a single row shown on the drawing, but that is simply illustrative of the drawing.   The holes in the drawing are shown as much larger size than those shown in the patent.   There is not any attempt in any way to reduce area there, to change it in any form. It is simply illustrative of the drawing.   The drawing does not specify how large they are, but it is evident they are ample for the purpose both in the illustration we have made and in the patent drawing.   During the upper stroke of the plunger the fluid flows out through the valve.   The traveling tube at that point is practically double and these holes simply attempted to connect the inner part of the tube to the outer part.

MR. HERRON: That is from space *AA* to this little space marked *XX*.

MR. SOBIESKI: Now, the part in green marked *n*, that is a discharge valve.   At this space marked *u*, that is the inlet valve, at the bottom of the traveling barrel. Now what sort of a valve is that?   Do you know how it works?

A   Yes.   It is a plate valve guided by a central stem, with a flat seat, a flat valve and flat seat.   There is the valve 10 at the top of this barrel.   These are the valves *n* and *u*, annular flat valves going around the entire surface.   There are thus three valves on the traveling barrel.   There are no other valves in the pump.

(Testimony of Edward T. Adams)

Q   Is there anything in this patent which refers to a lock to enable a person to tighten up the rods?

A   Page 2 line 16 of the specification reads, "stem v has a cross-cut 4"—that is the jaw clutch we are speaking about—"to receive the key z of stud y. The stud whereby the spindle is anchored to the plug w is adapted to be screwed into the latter by introducing down into the well-pipe the assembled plunger, sleeve, spider, and stem with the key and cross-cut engaged and then turning the stud (thus interlocked with these parts against relative rotation) in the plug w. Above the plug w the well-pipe has orifices," etc.   Now, in the operation of these pumps this entire barrel, that I colored red, moves up and down.

Q   Now, where is the spindle in this barrel, as described in the patent? You refer to a spindle here; which piece is the spindle?

A   It refers to the spindle 1, the spindle being in that case part of the standing tube.

Q   Starting on page 2 line 11 of the specification, "This stud receives the lower end of a spindle 1, held permanently therein by a pin 2 and penetrating and snugly fitting the stem v of spider s, which stem may have a gland 3 to prevent liquid escape between the spindle and stem." Does not the patent, as thus described, prevent the escape of liquid along this spindle?

A   Yes.   By "spindle" in that case we mean what, in the rest of the testimony, would be called a standing column, and for that purpose it is right.

Q   And it prevents the egress of liquid along that standing column?

A   Yes.

(Testimony of Edward T. Adams)

Q Yes. And this bottom guide which locks has nothing in it to allow the flow of oil down, does it?

A It is packed against the flow of oil. It is a close fit.

DIRECT EXAMINATION

resumed

MR. HERRON: I now call your attention to the patent to Cummins, No. 1,323,352; you have made a careful study of that device have you, Mr. Adams?

A I have.

MR. HERRON: I hand to the Court a photostat copy of the drawing of Cummins, photostated for the reason that the Patent Office was out of printed copies, and hand one to counsel. This, I suggest, be marked O' and be admitted, together with the office copy of the same patent. I have handed counsel a colored copy of that. I take it, Mr. Adams and Mr. Sobieski, that it may be understood that the witness, if asked whether he had caused all of these colored copies to be made, will give the answer he gave in relation to Deis?

MR. SOBIESKI: That he made all these drawings?

MR. HERRON: Caused them to be made—made them all together in that office so he is familiar with them.

MR. SOBIESKI: All right; I will accept that.

MR. HERRON: And that the same colors would indicate the same relative action of the parts.

MR. SOBIESKI: I am supposed to accept that statement.

MR. HERRON: I also caused to be made an enlarged drawing of the Cummins device. This is the drawing.

(Testimony of Edward T. Adams)

This drawing offered and received in evidence as Defendants' Ex. FF and a colored copy of the patent drawing offered and received in evidence as O'. Exhibit FF appears at Volumn II, page 174 hereof.

This pump as shown in the patent is of the stationary barrel and traveling plunger type. The barrel is divided in two parts as a means of introducing a lubricant in which the patentee was interested. Another feature is that at the top of the standing barrel, as is shown very clearly in the enlarged drawing at 14, there is a packing which is introduced between a nut and the barrel, and is compressed out against the tube as shown by a dotted line in the enlarged drawing and in the patent, to prevent the sand which settles out of the liquid in the column from dropping below the top of the standing barrel. There is a jaw clutch, or, as it is called in the Bray patent, a coupler, shown very clearly in all the drawings at the top of the standing barrel and at the bottom of the traveling plunger. The action of the pump is the same as any other pump. When the pump moves upward fluid is drawn from the well into the pump chamber; on the downward stroke the plunger settles down through that and the hollow plunger is then filled with oil, and on the upper stroke again the inlet valve of the plunger, traveling valve it is called, holds the fluid in place and the fluid is lifted.

MR. HERRON: Now, Mr. Adams, have you considered whether or not a pump of that type could possibly be reversed in such fashion as that the barrel could move and the plunger remain stationary?

(Testimony of Edward T. Adams)

MR. SOBIESKI: I object to that question, your Honor, on the ground there is nothing to show that this type of device ever has been reversed; and I think, regardless of what might be done or what could be done, if they are showing anticipation or priority, they should show what has been done.

THE COURT: He may answer and exception may be noted.

I have caused a diagram to be made of the pump as so reversed. I have it here.

This diagram is offered and received in evidence as Defendants' Ex. GG and a colored copy of the patent drawing offered and introduced in evidence as O^2. (Exhibit GG appears at Volumn II, page 175 hereof.)

This diagram GG marked "Cummins reversed" is the diagram of such change as I have testified could be made. Such a change would involve nothing but mechanical skill, as distinguished from invention. Absolutely nothing.

MR. HERRON: Referring now to the drawing GG, will you explain to the Court just what changes are there indicated and how the pump would work in the reverse form as there shown?

A   In the pump shown as reversed the barrel 2 which, in its form as shown in the patent, is the stationary part, now becomes the traveling part. I think it would be clear if we consider that the pump chamber is the vital part of the pump. As it increases in area or in volume fluid is drawn into it and as it decreases in volume fluid is expelled from it. Therefore, it does not make any difference which part moves, whether the plunger moves and the barrel is stationary, or whether the barrel moves and

(Testimony of Edward T. Adams)

the plunger is stationary, or if both move. Any time that the pump chamber between the discharge and the inlet valve is increased fluid is drawn in, and when it is decreased fluid is pushed out, regardless of how it is done. So that, reversing it did not make any difference in the basic operation of the pump at all, or it would not have made any difference if both parts moved instead of one. The parts, in their reversed position, are the same. All that was necessary to do was to leave that top cage and its valve in the position which it originally occupied and to leave the mandrel and shoe, which hold the standing parts in place, in the position which they occupied, just the same in both cases. Those parts are 23, referring to the top cage, and parts 8, 9, 10 and 11, referring to the mandrel and shoe,—the shoe not being shown. The mandrel fits into a shoe, which would be a part of the completed structure in the well. Then the barrel, to become a traveling barrel, is itself then simply reversed. First, we take from the top the nut 19 with the female member
                              female
21, which is the ~~male~~ member of the lock and which in the reverse position is attached to parts 8, 9, 10 and 11, or the standing mandrel. Now, because we are using the parts just as they are shown in the patent, just as if they were an actual pump, made up and present here in court, when we make a reversal we find that we have two male threads to put together. It is necessary therefore to introduce a small adaptor, as it is called in the trade, simply a short piece of metal threaded on the inside at both ends to receive the male members of the threads that exist in the pump as it stands. So this little short piece shown in black on the outside, this adaptor, is in-

(Testimony of Edward T. Adams)

troduced at that end, and a short adaptor is introduced at the top. The only reason for that is this: In the pump as shown in the patent before it is reversed, we have an inside and an outside thread joined together between the top cage 23 and the nut 19, and we have the same condition at the top. When we reverse that instead of having a male and female fit here, we bring together two female fits which makes the necessity of introducing this extra piece. Except for the way the parts happen to have been fitted together, no change whatever would be necessary in the pump. The reversal brings the jaw clutch to the bottom instead of at the top. As originally made, the jaw 21 in the part 19 matched up and interlocked with the male part of the jaw 22 in the nut 14, and when those were dropped down into position, the standing tube being firmly locked in the well, the traveling plunger then became locked in the tube and it was then possible to tighten up the sucker rods if they had become loosened. When you reverse the pump you have just exactly the same condition, the only difference being that, when the barrel is stationary the jaw clutch is at the top, and when the barrel is the moving part and the plunger is stationary the jaw clutch comes to the bottom, but the action is just the same as before. Now what I have done in this suggested reversal is to literally take the part which was the plunger of the pump, turn it over and created it into the stationary part, which fits into the anchor. That plunger, then, instead of moving is held immovable with the anchor, which puts the valve 17 and the former bottom part which produces the pumping action at the top of the plunger and in the relation shown on this sketch at points 16, 17 and 18. I have literally

(Testimony of Edward T. Adams)

turned the barrel over so that the bottom of the barrel has become the top, and I have simply moved the jaw clutch down from its former position and placed part 14, (which we will label), at the bottom of the barrel and the cooperating part, part 19 on the top of the anchor, that is that part which cooperates with the anchor, the spindle or mandrel, part 5, Ex. GG.

Further

CROSS EXAMINATION

This green in there marked 6, is a valve. As shown by this patent, there is a close fit between the plunger and the barrel, up here, for instance, and I have preserved that close fit in turning the same parts around.

Q   Now, if, then, this were, as you say, turned around and put into a well, and if that operation was attempted how would you have the escape of oil; how would you take care of the necessary escape of oil in this region when that barrel goes up in the pumping process?

A   By "this region" I understand you to mean the escape of oil between the traveling barrel 2 and the standing tube 15, is that right?

Q   That is what I have in mind, yes.

A   That is no way changed. We leave it just as the patentee of the pump designed it in the first place. In other words, there would be no way in which the oil here would escape in the upward motion except by ordinary leakage of any pump.

Q   Except for ordinary leakage there would be no escape of the oil in the lower part of the traveling barrel on the pumping motion, is that correct?

(Testimony of Edward T. Adams)

A Well, I wonder if I understand you right. You understand that the part 15, which is the standing tube, is a hollow tube, do you not, and the fluid goes up through it? Were you mixed up about that? The standing column 15, as the patentee shows it and as it is shown in sectional figure in the original drawing is hollow. That is what you were mixed up about. The normal flow of the pump goes up through this hollow tube 15. It is marked in blue to show it is a stationary piece but it is still hollow, exactly the same form as shown in the patent and as shown in section in our drawing which is a copy of the patent. The Bray pump and the Hofco pump involved in this suit have the hollow tube I have been describing. Exactly the same thing,

Q They have the hollow tube. With this close fit, would there not be considerable friction in the operation of this pump? What I mean to say is, without allowing some method of escape for this fluid in the pumping motion other than this natural leakage through this close fit, would that not be a very inefficient type of pump?

A No. The method of clearance to make a fit between a standing part and moving part is extremely old in the art and there is a tremendous amount of data regarding it. That clearance depends upon the type of pump you have. Sometimes they are packed, as shown in this drawing, and you have stationary cups making a mandrel fit into a hole in the shoe, in which the leather cups form the packing. This man here has shown a metal to metal fit; that is, the standing tube is metal and his traveling tube is metal. The amount of clearance he would need for the amount of leakage he would get, which would depend upon the clearance between that trav-

(Testimony of Edward T. Adams)

eling barrel and the standing tube, and would depend upon the metal. If you had two soft steel parts you would need a large clearance, a good deal of clearance; and if you have the type that Mr. Bray specifies very clearly in his patent in which he uses a bronze, which is notoriously good metal to work with steel, then you would make a lesser clearance, a lesser clearance and less leakage. But that is purely a mechanical matter. One type of pump, properly designed, would be as efficient as another as to the amount of leakage you have.

Q Is there anything in this pump, as shown in the patent, which allows the escape of oil over the anchor face?

A The anchor face is buried in oil. The pump sets down at the bottom—

Q What I mean to say is, from the barrel over the anchor face in the pumping motion?

A There is a very distinct washing of the anchor face by oil coming from the tubing. In this case, as originally designed, the barrel was stationary. At the threaded top of the barrel there was the leather which packs the barrel in the tubing and prevents sand from dropping below the nut 14 which is the face of the jaw clutch, and the moving part above that which is the plunger 15, and immediately above the moving member the jaw clutch 19. Now, as these members move up the fluid in the tubing above that would flow very, very rapidly because of the packing member at 14, and wash the face of the standing member of the jaw, but there is nothing which causes the fluid from within the barrel to go over the lock face. It does not come from the inside of the barrel in that case. It comes from the out-

(Testimony of Edward T. Adams)

side of the traveling plunger, and by this leather device here just below 14, they prevent the tubing from being sanded up so it could not be withdrawn from the well. This is a lift type pump. It is made to be withdrawn from the well without withdrawing a tubing.

Further

DIRECT EXAMINATION

In the reversal I have made, the valve marked 6 and the valve up in part 23, which we will mark AA, on Ex. FF remain in the same position when you have reversed your structure, and the valve marked 17 is turned over, so that it seats from the other side. It does not involve invention to turn over a valve so it seats with the other end. That is old in the art. In my opinion, there would be nothing at all involving invention in making that part of what becomes the standing column in the reversed pump, designated on the chart by part 15 of greater or less dimension in order to give such clearance as would, in point of fact, be there desirable to permit trapped oil to escape, or for any other reason.

Further

CROSS EXAMINATION

BY MR. SOBIESKI: The change of dimension to permit trapped oil to escape is an efficient change to make, isn't it? It adds to the efficiency of the pump, does it not?

A If I understand you correctly, you are saying that if there is too close a fit and the oil is trapped in this annular space and has a hard time getting out, it will limit the motion of the pump. If that is what you are getting at, that is true.

(Testimony of Edward T. Adams)

Q The change of dimension of the standing part to give a greater clearance so that the trapped oil may escape down here past tube 2 would be a change which would cause the pump to operate more efficiently, would it not?

A You are pointing to the traveling barrel itself and the change in the seal would be made in accordance with the type of metals that you had. That would be the primary consideration. Mr. Bray makes that very plain in his patent, in which he speaks of the use of bronze that is especially adapted to operate with steel. So the amount of clearance at this point between 2 and 15 is dependent on the type of metals which you use to form the seal.

Q Sufficient clearance at that point to allow for the escape of the trapped oil, or, in other words, the allowance of the escape of trapped oil adds to the efficiency of the pump, does it not?

A I doubt I understand what you want. I am not quite sure that I do. The seal wants to be designated by consideration of a seal. Are you speaking of the trapped oil in the reduced area or in the enlarged area of the traveling barrel? I do not know what trapped oil you mean.

Q Of course, you are a pump man and, as I understand it, when the traveling barrel goes up it is necessary for some of the oil in the barrel to escape in order for efficient operation; is that correct?

A Well, yes. In this reversal which was made I was not concerned with the trapped oil at all. I was simply concerned in showing to you, the Court, and everybody, that it was perfectly simple to turn this pump end for

(Testimony of Edward T. Adams)

end; and I wanted to turn it end for end using the pieces just as they were, not as I would design the new pump necessarily, but simply show that if you take the pieces, and turned it end for end, the jaw clutch shown at the top, might just exactly as well have been at the bottom. That brings the trap area here at the top below 16, and the clearance and escape from the trapped area, would be designed necessarily for whatever clearance you wanted. If I had been after that part of it I would have changed the pump more, so that trapped area would have been at the bottom of the pump barrel instead of at the top of the plunger. I did not try to do that. All I wanted to do was just to show it was perfectly easy, and that has been done.

Q The point I wish to make is, that changing the pump end for end that you have performed, does not make a proper provision for the trapped oil.

A No; I have not changed the provision that was made for the trapped oil in the first place. I have not touched it. Whatever provision Mr. Cummins made we have just left. Mr. Cummins evidently makes provision for the trapped oil. That is, by the clearance between the part 16 and the standing barrel. There it is (indicating); here is your clearance. He does not specify what that is. That would be well known. Anybody who operated pumps at all, would know what clearance he would have to make. This provision for the escape of the trapped oil is at a part of his pump quite removed from the locking device, in the initial pump before it is turned over. The provision for the escape of the trapped oil was utterly disconnected with the locking means. After

(Testimony of Edward T. Adams)

it was reversed I made no change of that. I did not touch that at all.

Further

DIRECT EXAMINATION

I not only did not turn over valve 6 but I left it in the same relation in both the inverted pump and in the other, absolutely untouched in any way. The only valve that is turned over is the valve 17, which is necessarily turned over due to its changed position. In my opinion, with my knowledge of the art, it would not have represented any act of invention, as distinguished from the exercise of mechanical skill, to have turned this pump over in the year 1926.

Q Would it have involved any exercise of inventive genius, as distinguished from mere mechanical skill, to have made the dimensions of that portion of the standing column represented by part 15 to an appropriate dimension to allow for proper clearance between the barrel, in the light of the condition encountered in the actual pump, in 1926?

MR. SOBIESKI: We object to that question, your Honor, as not proper expert testimony in view of the evidence that Exhibit 17, which has been introduced with relation to these exhibits, is somewhat of a turn-over of this pump. I understand that is their contention, and it failed to make such an allowance for that clearance, and that was done by the oil motion itself; and my argument is that the evidence already introduced by these defendants shows that ordinary practice and ordinary skill would not anticipate the turning over of this pump and the proper allowance for the escape of the trapped oil.

(Testimony of Edward T. Adams)

MR. HERRON: I have asked Mr. Adams for his opinion and, of course, Mr. Sobieski is entitled to his, but we differ. I submit my question.

THE COURT: I understand you to say, examining this patent, you do allow it to escape by reason of the clearance, allow the escape of the oil?

A You are now referring, your Honor, to the original patent?

THE COURT: To the original patent, yes.

A In the drawing, as shown in the patent there is an annular space made necessary by the device for pulling the pump, which lies above the part 16 and below the upper section of the stationary barrel 1. In this annular space fluid is trapped; and in the design as shown in the patent, anybody skilled in the art of making pumps at that date would have made proper allowance there. I do not believe that it was probably necessary to specify what it was. Pumps have been made for many years and when you have a trapped volume I would know what clearance I would have to make, that is, a practical man would know that. The upper part is a close fit, so I feel in the initial drawing there was a clearance space for that oil to escape.

Q Is there any different arrangement in the reversed structure?

A I have reversed it without making any difference whatever. If, as an engineer, I had been asked by Mr. Cummins to reverse it, I would have then changed it to the extent that I would have made more of a fit at the top and less at the bottom, and would have made the discharge flow out at the bottom and wash that space there. But in showing to your Honor that it was perfectly simple

(Testimony of Edward T. Adams)

to reverse the pump, I reversed it, making no change whatever, so it would be simply a straight-forward change. I was not attempting to design a pump. I was simply reversing it and showing how easy it was done, and that there was no invention or special skill required in order to reverse it.

MR. HERRON: And the testimony you have given, you would give, had you taken that action in the year 1926?

A   In the year 1926.

Further

CROSS EXAMINATION

This provision for the escape of trapped oil is utterly disconnected with the locking device in the pump as shown in the patent before being reversed, and I have reversed it just the way it was, without making any change in the locking device, simply to show it could be reversed. By the same token, when the pump is reversed, any provision made for the escape of the trapped oil would have inevitably resulted in a flow down across the clutch?

Q   May not provision for the trapped oil also be made by holes in the barrel?

A   That would have the same result as far as the washing is concerned.

Q   But I am asking you if that provision for the trapped oil would not necessarily mean a flow through the lock?

A   Yes; either provision would.

Q   If they had a hole through the barrel that would not be a provision by way of a flow through the lock, would it?

(Testimony of Edward T. Adams)

A   A flow around the lock, washing the top of the nut of the locking device. That is, you can't consider the pump without considering the tubing in which it is confined.

Q   When you get rid of your trapped oil would it not have occurred to you, Mr. Adams, to put some holes in the side?

A   No; it would not. I would not have done it. That is in the seal and I would not think of it. Whatever clearance there is down there I would leave just as it is. To bore holes in the side would simply be, in effect, to cut off a portion, to shorten the length of the tube.

MR. SOBIESKI: What I mean to say is, that one making allowance for the escape of the trapped oil would not inevitably have had to do it by means of a hole in the lock down here; but could have done it by means of a hole in the barrel, couldn't they?

A   They could if they wanted to. There is no reason for doing it.

MR. HERRON: Would you have considered it involved inventive skill to have used either of those two contrasted means?

A   No; I would not.

Further

DIRECT EXAMINATION

A   I find in Cummins as so reversed a standing guide column 15, with a working valve 16, 17 and 18. I find a valved traveling barrel 1, 2, and 23. I find that valved traveling barrel running on a column that is 1 and 2 running on 15. I find a bottom guide 14 on that barrel run-

(Testimony of Edward T. Adams)

ning on the column 15. The bottom guide has a bottom face forming a coupler 22. I find an anchor for that column, parts 8, 9, 10, 11 and 19. The anchor for the column presents directly to said coupler face a top face on 19. That forms a complementary coupling 21, to be non-rotatively interlocked with the lower guide coupler 22.

The number in the Cummins patent of this interlocking part at the end of the barrel is 14, and that of the part on the top of the anchor is 19. The notches in those parts are 21 for the anchor and 22 for the barrel.

In the Deis patent there is an apertured guide t.

Further

CROSS EXAMINATION

In the Deis patent, what I call the apertured guide is the seat for the plate valve. The valve covers the apertures, which apertures are marked t in the large sectional drawing. The apertures are below the plate valve.

On the down stroke of the traveling barrel oil flows into the inside of the barrel, and on the up stroke, through the aperture and plate valve u. And on the up stroke of the pump, by reason of the plate valve, no oil escapes from the bottom end of the traveling barrel. The flow is in the opposite direction at that time, washing through the holes 5 in the side of the standing barrel through which the fluid flows in across the lock nut 4, and up through the passages t. No oil flows from the barrel down onto the coupler face, through the apertured guide I have described. It flows upward, past the coupler face through the aperture t.

Q Does that allow for the escape of trapped oil through the aperture you have described, or is that rather an intake valve for the barrel; that is, an intake valve for

(Testimony of Edward T. Adams)

the barrel and not a means for the escape of the trapped oil?

A   It is a means of inlet for all of the oil which comes into the trapped area everything that comes in flows by there.

Q   And that is the intake valve in the barrel?

A   That is the intake valve of the entire flow of the pump.

MR. HERRON: Referring now to the large chart of the Deis patent, Exhibit EE, the apertures to which you refer are marked *t* in Figure 6?

A   They are marked *t* in Figure 6, in the drawing and marked in the patent with the same letter.

Further

DIRECT EXAMINATION

MR. HERRON: I now hand to the Court a colored Patent Office copy of the patent of Whitling, No. 1,391,873.

This colored drawing offered and received in evidence as Defendants' Exhibit P^1.

THE WITNESS: I have made an examination of that Whitling patent and have also made a drawing of that device.

This drawing is offered and received in evidence as Defendants' Exhibit HH, and appears herein at Volumn II, page 176 hereof.

I have also considered the question of reversing that device. This sketch represents in Figure 1, Whitling as shown in the patent, somewhat simplified for the purpose of easy illustration, and in Figure 2 the same device re-

(Testimony of Edward T. Adams)

versed. As shown in the patent it has a traveling plunger
with a stationary barrel. It has at the top a lock 25a
in the form of a jaw clutch, which locks against rotation
in either direction. That pump as shown has an annular
space between the stationary barrel 5 and the traveling
plunger 22, in which fluid is trapped, provision is not
made for the escape of a measurable amount of leakage
because at the upper end of that space a .packing unit
number 31 as shown. That shows a packing in the stationary barrel sealing the barrel between the member 29
and the traveling plunger. As the plunger is lifted fluid
is drawn in through a standing valve 12 and up through
the lower end of the standing barrel 5—between the
standing valve 12 and the lower end of the traveling
plunger, due to the increase of the volume between these
two valves. On the return stroke of the plunger the
volume between the standing valve 12 and the traveling
valve 23 is diminished and fluid is therefore expelled up
through the hollow tubing 22, out through the top valve
27 and so up the tubing to the top of the well.

Q Now, have you an opinion as to whether in the
year 1926 that pump could have been reversed in such
fashion as to make the moving parts stand still and the
parts which stand still in this pump move, that is, turning
it into the type of pump with the standing plunger and
the movable barrel, without the exercise of any inventive
faculty, as distinguished from the exercise of mere mechanical skill?

MR. SOBIESKI: We object to that question on the
ground it is immaterial whether it could be done or not.

THE COURT: I will allow the answer, and exception
shown.

A Yes.

(Testimony of Edward T. Adams)

MR. HERRON: Have you an opinion on that?

A It could have been done and without any inventive skill whatever, and is shown so reversed in Figure 2 of Ex. HH. I will explain briefly how that reversal has been accomplished. In making that reversal you would first make a break at the point between the traveling cage 26 and the part immediately below it, 25. At the point where that threaded connection appears, which we will call AA I would unscrew that connection. The pump is now disconnected up to that point from the operating mechanism, which is left just as it was before, carrying the valve 27 with it. At the bottom we would make another disconnection. Here is the stationary part 3, which is a seal or a block or a portion of the standing tube marked 3, and I would simply unscrew the part 5 from the point where it is screwed into 3 which I am marking now, BB. That being done, we would reverse these two parts just as they stand. We have now the barrel of the pump completely disconnected at the top from the mechanism that operates it and at the bottom from the mechanism which holds it in place. You take these two parts just as they are and turn them end for end. So that part 17 shown at the bottom in the patent, becomes part 17 at the top as shown in Ex. HH Fig. 2, and parts 25 and 25a here shown at the top becomes part 25 at 25a at the bottom. You would couple the parts together in that position.

Q Now, then, what would be the operation of the pump so coupled?

A The part 5 which was previously the stationary barrel, now becomes the traveling barrel and part 22 which was previously the moving plunger, now becomes part 22, the stationary tubing. Therefore, when the pump is operated it would still operate by the sucker rods

(Testimony of Edward T. Adams)

31 and the barrel would travel on what was formerly the plunger, which now stands still.

Q Now, with relation to the valve, what change, if any, would be made?

A The valve 12 at the bottom of the standing column and the valve 27 at the top in the traveling cage 26 also remains the same. When you make the reversal it is necessary for valve 23 to seat on the other side. When it is reversed it needs to seat at the opposite side of the same pocket. In other words, you would simply have it seat at the other side of the cage.

Q And is it a usual operation in any mechanism to change the seat of valve?

A It don't amount to anything. It can be done very easily. It does not involve any invention or particular knowledge. Now, then, as so turned over, you would find a clutch at the bottom in which the parts 25 and 25a can be made to engage by lowering the plunger until they come into mesh.

Q It would be true in this case, would it not, that there would be no downward fluid passing over the face of the clutch member between these parts and the standing column?

A There is a packing there, and in the event that packing prevented any flow, there would be none. You would have then a closed air chamber.

In Whitling as reversed, to-wit, Figure 2 of Exhibit HH, I find a standing column with a working valves 12 and 23. I find a valved traveling barrel 5 with its valve 27. The valved traveling barrel runs on a column 22. I find a bottom guide 25a on that barrel, which runs on the column 22. The guide has a bottom face forming a coupler which is marked 25a. I also find an anchor for

(Testimony of Edward T. Adams)

the column. The anchor for the column is 3, there being a joint between 3 and 25. That anchor presents directly to said coupler face a top face, the top face of 25, forming a complementary coupling between 25 and 25a, to be non-rotatively interlocked with the lower guide coupler. Technically speaking, packing indicated at 16-17 on the column 22 runs in the barrel 5. The part 5, really runs over part 22 without touching it. The packing 17 is on part 22, and part 5 comes in contact with the packing.

Further

CROSS EXAMINATION

BY MR. SOBIESKI: I understand that this reversal does not show in the patent but is your conception of what it would be like if it were, is that correct?

A I have taken the parts as shown in the patent drawing and reversed them without any change.

Q I believe you stated the purpose of the packing was to prevent the oil which is trapped in the barrel from escaping downward toward the coupler face; is that correct?

A I did not intend to say "the purpose," but such a packing at that point would prevent a flow of liquid from the trapped space past it—and also prevent a flow of liquid into it, looking at it technically either way. The liquid check is outside. It cannot go through this portion designated as 25a, if the packing is tight.

Q Now, I notice the lock up here in this situation is right beneath this ball here. Is the principle of the operation the same when you place it down below and have two balls above it instead of only one ball above it?

A So far as the lock is concerned, it would make no difference.

—115—

(Testimony of Edward T. Adams)

Q   Well, doesn't that make a different type of pump out of the reverse structure as you have it, with these balls up here close together instead of being down here with this distance apart?

A   I do not know the distance they are apart. They are coming out with a design of pumps. It is considered good practice to have the standing valve and the upper traveling valve close together.

Q   What I was wondering is, would not this make a different type of pump than it is, when you have in one case the balls close together and in the other case whatever the distance be, this ball being at the bottom of the standing barrel?

A   No.

Q   Ball 23 being at the bottom?

A   No, the essential thing in a pump is to have the two working valves that are on the opposite sides of the pump chamber near together. Now, the pump chamber is the space between the standing valve and the nearest traveling valve. This is not the pump chamber up through the valve. The pump chamber in Whitling as shown in the patent is between the valve 12 and the valve 23. When that plunger moves up its volume is increased, and increasing the pump chamber draws fluid in, just as decreasing throws fluid out. And so when the pump is reversed the pump chamber is also reversed, and thus instead of being at the bottom it comes out of the top of the combined structure, and the pump chamber then lays between the valve 23 and the valve 27. So the two valves are close together in both cases.

Q   And this valve down here marked No. 12, that is a valve merely for the purpose of preventing the oil from flowing down the hollow tube out into the oil sands?

(Testimony of Edward T. Adams)

A   This is a standing valve, as called in the art, or an inlet valve, if you wish to call it that.   It is the inlet to the pumping chamber but because it is commonly stationary in oil well practice, is spoken of as a standing or stationary valve.

Q   Yes.   And this packing here prevents the flow of oil from the barrel onto the coupler face; that is correct, is it?

A   Well, there is no flow of oil from the barrel.   The packing to which I refer is the packing 31 on standing column 22, as reversed.   That would prevent the flow downward of the fluid trapped between the outside of the standing tube and the inside of the barrel.

Further

DIRECT EXAMINATION

Either of those two structures, in my opinion, would be an operating pump.

Assuming that in the reversed structure the barrel was let down so that parts 25a and 25 come into union, and that to the string of sucker rods attached above this portion 31, a rotary motion is imparted, the effect of that rotary motion upon any joints which there were in the string of sucker rods, would be to tighten the threads in the sucker rod, if the rotary motion was in the right direction to tighten them and to tighten any joint which was loose.   The twisting being done at the top comes down through the sucker rods to the lock between 25a and 25, which fastens the barrel to the standing portion. That motion would be resisted by the fastening of 25 into 3 which is the stationary part, held in the tubing at that point.   In other words, when you attempt through the

(Testimony of Edward T. Adams)

sucker rods to impart rotary motion to the barrel, that rotary motion would be stopped by the engagement of the clutch face on the bottom of the traveling barrel with the clutch face on the top of the anchor. That stops the bottom end of the sucker rod against rotation, and as the sucker rods at the top are rotated it would of necessity follow that any loose joint in the rods would be tightened, if the turn was in the direction in which the threads would tighten.

Further

CROSS EXAMINATION

That packing 31 prevents the trapped oil from escaping through the coupler face?

A   That is technically correct.

Further

DIRECT EXAMINATION

BY MR. HERRON: At this time, if the Court please, I pass up a colored copy of Downing, No. 211,230, and ask that it be marked Q'.

This colored copy is received in evidence as defendants' Ex. Q'.

The Downing patent is a traveling barrel and a stationary plunger type. It is a removable type. The stationary part B and C being made to engage in a taper at the bottom of the tubing A so that by pull on the sucker rods the entire structure can be removed from the well. The operation is the same as practically all of the other pumps considered. The letter G appears in what is the pumping chamber, which increases in volume as the sucker rods are lifted, thereby lifting the traveling barrel and

(Testimony of Edward T. Adams)

filling that space with fluid. On the down stroke, the space in G, which is between the valve O and the valve B, dimishes in volume and the fluid is ejected. I find in that structure a standing guide colume provided with a working valve B. The structure has a valved traveling barrel G with valve O which runs on the column, that is G runs on the packing at the upper end of C. There is a bottom guide H on the barrel which runs on the column C. I find an anchor for the column mounted on a taper at A in the tubing.

Further

CROSS EXAMINATION

BY MR. SOBIESKI: Do you find anything that enables the bottom of this traveling barrel to lock with the—

MR. HERRON (interrupting): We make no contention there is any locking means or clutch there.

Further

DIRECT EXAMINATION

MR. HERRON: We now pass to the Court a colored copy of the O'Bannon patent No. 1,454,400 and ask it be marked R'.

This colored copy is received in evidence as Defendants' Ex. R'.

I have likewise examined that patent.

This again is a traveling barrel removable type of pump, removable meaning that the whole structure may be withdrawn by means of the separate lining. The operation is the same as the other pumps discussed, in which the pump chamber lies between the traveling valve 14 and the valve which is inside and concealed in the part marked 13. When the barrel is lifted fluid is drawn into

(Testimony of Edward T. Adams)

this pumping chamber and when the barrel descends it flows out through the valve at 14 into the standing tube.

I find in that structure a standing guide column P. provided at the top with a working valve at 13. It is hidden but it is specified that it is there on page 1 of the patent, line 89: "As is well known the upstanding valve 13 if provided", etc.

I find a valved traveling barrel B, running on 15 and 16 which is the packing part of the standing column P. I find a bottom guide 12 on the barrel which runs on the column P. There is an anchor for the column at 10. I find that that guide has a series of venting ducts, marked 16'. I have considered the effect of the liquid issuing through those venting ducts.

Q. What, in effect, would be the result?

MR. SOBIESKI: We object to that as incompetent, irrelevant and immaterial. The patent shows on its face that these ducts do not direct it towards the bottom nut, but that there are holes in the side of the barrel which go outside those.

THE COURT: That is true, but we will see what he says about it.

MR. HERRON: I am not claiming identity. I am merely asking what the result would be of the structure there shown.

A Essentially, the action would be in any practical sense the same as if the holes were vertically downward through the nut 12, because the clearance between the inside of the traveling barrel B and the standing tube is

(Testimony of Edward T. Adams)

very small, and as the volume of this annular space, which is vented by the ducts 16′, is diminished, the flow out of that, in so far as it is radial through this section of the metal of the traveling barrel itself, is immediately deflected down and not up to any extent at all.

Q THE COURT: Why is it deflected down?

A Because the barrel is moved upward. Presume it is moved one inch, then it would have left what would be a vacuum at one inch below it, and the fluid can't move upward because the column from there to the surface of the earth is completely filled with fluid. There is only one place to which this fluid could flow, and that is to the space immediately below the nut 12 which is vacated by its upward motion. No fluid flows out of this annular space except at a time when the traveling barrel is moving. If it has moved up and nothing has flowed into the space where it was just an instant before there would temporarily be a vacuum immediately below 12 and the fluid which flows out of 16′ has no other place to go. That is the only thing it could do, just flow down past the outside, which would be with quite considerable velocity, into the vacuum that was made by the upward motion of the plunger. There is no such space, there is no vacuum in the motion of the plunger upward. There is nothing up there. There is no space vacated into which that fluid could flow.

MR. HERRON: By the way, calling your attention to that structure, that valve is located at the upper end of the column, is it not?

A The valve is located at the upper end of the column at 13.

(Testimony of Edward T. Adams)

Further

CROSS EXAMINATION

These holes 16′ are in the side of the traveling barrel. I think I said that the flow out of these holes 16′ would have the same effect in washing the top of the structure 10 as if they were through the guide 12. I did not mean to say that there were any holes in the guide.

Q And, as a matter of fact, the guide has a tight fit, packing to actually prevent the flow of oil downward through the guide, hasn't it, and the standing column, is that correct?

A No.

Q What is correct?

A The packing to which you refer, I think you refer to the spring 19 that compresses the packing above it.

Q. Yes.

A Well, that is simply the seal of the plunger. It has nothing to do with the annular space. The annular space is below that packing. But provision is made in this patent for escape of the trapped oil through a hole in the side of the traveling barrel. In this patent I do not find a lock between the bottom of the traveling barrel and the base. There is no clutch.

Q Does the patent contain any specification as to how close the traveling barrel is to the inside of the tube?

A It does not specify it exactly. It says that it passes through it. And this patent shows that the holes for the escape of the trapped oils allowed them to escape through the traveling barrel in a direction horizontal to it.

(Testimony of Edward T. Adams)

Further

DIRECT EXAMINATION

MR. HERRON: I now have passed to the Court a colored copy of Thurston, No. 1,625,230, and ask that it be marked S'.

This colored copy is received in evidence as Defendants' Ex. S'.

Thurston is again a removable type of pump with a traveling barrel and stationary plunger, the pumping chamber lying between the valve 7 in the stationary part and the valve 26 in the traveling barrel. When the barrel is lifted this volume will increase, drawing in fluid; when the barrel is reciprocated downward fluid will be discharged upward through the valve 26. I find there a standing guide column. No. 11, Figure 2b provided with a working valve No. 7; a valved traveling barrel 19 with valve 26, running on the column 19, on 9 and 11. There a bottom guide on that barrel 20, referred to in the patent as a combined puller nut and guide. It says that it is a slidably fitted guide, and that guide runs on the column 11, 20 on 11. I find an anchor 4 for the column. I find a venting for the fluid, as the guide is slidably mounted, which would give a vent between the guide and the tubing 11. (I find no vents but I feel that there would be some escape around the sleeve which I have mentioned). When the patentee says in line 95 page 1 "slidably mounted" it gives the impression of being loosely enough mounted so there would be escape of fluid.

(Testimony of Edward T. Adams)

Further

CROSS EXAMINATION

BY MR. SOBIESKI: To what in the patent do you refer when you say "he says that it is slidably mounted"?

A I am reading on page 1, line 95: "slidably embraces"—"a combined pull nut and guide 20 which slidably embraces the extension 11 of the plunger," etc. I construe that to mean that it engages the extension of the plunger so that it can slide up and down. That is the only place in the patent where that feature is described, I think. And according to the patent, then, this bottom nut is engaged to the standing column so that it will slide up and down on it, act as a puller nut, part 20, in striking that abutment 10 of the standing tube, and it is slidably engaged with the standing column.

Q Is it not a fact that all of these pumps with the traveling barrel which have a bottom nut, have a bottom nut which slides up and down on the traveling barrel? Have you shown us a pump this afternoon which did not have a bottom nut slidably engaged to the standing column?

A Well, I would want to look at that again and see. The pump before this, O Bannon, shows a nut that is a loose enough fit to slide up and down. In Whitling, while the nut moves up and down there is a packing to prevent fluid leakage out through past that point. It is put in for that purpose. Whether it does it or not, I do not know. But the bottom nut in the Whitling patent also slides up and down.

Q And in the Deis patent, if that were turned around, would the bottom nut s slide up and down on the standing column?

(Testimony of Edward T. Adams)

A   Yes. That is combined with v. And v is a guide sliding on the fixed column 1.

Q   Now, as an expert, can you tell us whether you could have a bottom nut on the bottom of the traveling barrel which did not slide up and down and the pump would work at all?

A   Oh, yes, if I understand. Your question really is: Well, it is not necessary for the nut to touch the column in order for the pump to slide up and down?

Q   No. My question is: Can you have a pump with the bottom nut on the traveling barrel which does not slide up and down; in other words, if the bottom nut would not go up and down could you have a pump at all?

A   Oh, yes. The nut, as a part, is attached to the barrel although it moves up and down. That does not necessarily say that the nut does or does not slide on anything. It is a part of the barrel and moves with the barrel. Whether it slides on anything or not depends on whether it touches it. My hand moves up and down relative to this lighting fixture (illustrating). I am not anywhere near the lighting fixture. The lighting fixtures now represents the standing column. My hand represents the traveling (barrel), my little finger represents the nut on this the traveling barrel (illustrating). But if it slides, it is slidably fixed on it, then I come to a point where my finger is actually touching the thing and sliding. I object to the word "sliding".

Q   I see. Isn't it true, however, when this bottom nut that you describe slides up and down when it is operating properly there is always a film of oil between the nut and the inner tube, otherwise it would burn itself out, wouldn't it? I mean, it does not actually touch?

(Testimony of Edward T. Adams)

A   A film of a few thousandths.   I am not trying to quibble about it.   What I mean is a very wide difference, one being a quarter of an inch away and one an eighth of an inch away, and a film of oil would be only a few thousandths of an inch.

Q   I am referring to the Thurston pump, which says that the bottom nut slidably embraces the standing column. If it did not slide up and down it would be impossible for the pump to operate, isn't that correct?

A   It slidably embraces it.   It slides on it and there ought to be, and probably is, room for a film of liquid.

MR. HERRON: In other words, the amount of oil which would be found between the standing column and the puller nut of that patent would depend upon the closeness of the embrace between the puller nut and the collar, is that correct?

A   And the amount of wear.   Whatever it was made when it was new would be one thing; after it had been in use for a time and had worn, it would be something else.   And in all events, the passage there would depend upon the fineness of the fit; and if it was so closely fitted as to run with nothing but this small film of oil I have mentioned there would, from a practical standpoint, be practically no seepage through it, and if it was more loosely fitted there would be more.

Further

DIRECT EXAMINATION

MR. HERRON: Now, let us pass to the Thompson patent 586,524.   I likewise pass up a colored copy and hand counsel one.   I ask it be marked T'.

(Testimony of Edward T. Adams)

This colored copy is received in evidence as Defendants' Ex. T'.

Witness. The Thompson pump is a reversible traveling barrel type, and interesting because it shows it both as a traveling barrel type and reversed as a traveling plunger type. It is shown as a traveling barrel pump in Fig. 1 and Fig. 2, and shown as a traveling plunger pump in Fig. 4. An interesting feature of this pump is that an object of it, which has led to a complication of design, if you may call it such, was that it would have a washing effect with the motion of the barrel. Coming now to Figure 1, when the barrel is lifted there will be an increase in volume between the valve i, which is the standing valve, and the valve O', which is the traveling valve, and fluid will be drawn in with the increase of that volume, and on the return stroke expelled with the decrease of the same volume, so you have an operative pump.

The lower standing valve is in the same location in Figure 4 that it is in Figure 1. There is a traveling valve and that would also be in its same location at the upper part of the traveling plunger, and when the plunger was lifted that would increase the volume between the traveling valve O and the stationary valve which is e^4. With that increase of volume fluid would be drawn in. When the plunger descends this volume would decrease and the fluid would flow out, causing a pumping action. I find in Figures 1 and 2 a standing guide column marked f, provided with working valves, i and e^4. I find also a valved traveling barrel k with the valve o. This barrel runs on the column k on f. The packing and the column are really one structure. That is true of all of these pumps. The packing part is the seal. I find a bottom

(Testimony of Edward T. Adams)

guide on the barrel marked k', shown in Figure 1. With respect to that bottom guide I find that it is a loose sleeve arrangement. K' is specified as a sleeve. We commonly call it a puller nut, but it is specified in line 53 on page 2 that it does not fit closely, so allows a tendency to discharge. The specification reads, "Fluid may also be admitted to this above-mentioned space through the sleeve k', which does not fit closely around the tube f." Thus allowing fluid to be admitted, lines 45 to 55 on page 2 are devoted to that. I find that that bottom guide runs over that column. Thompson specified a loose fit for the flow of liquid through there and that it runs over the column rather than on it. I find in that structure an anchor for the column indicated d.

Further

CROSS EXAMINATION

BY MR. SOBIESKI: What are these orifices that are described as k' in this page 2, line—the orifices described as k^2?

A They are shown more clearly in Figure 2 and shown very clearly in Figure 3, which is a cross-sectional drawing, where they are marked t'. Those orifices allow the escape of the trapped oil in a direction horizontal from the standing column. They are in there for entirely another purpose. The pump is designed to allow escape in there with very considerable force, carrying out a washing effect which the patentee had in mind. That was not the washing of the clutch. There is no clutch, as a matter of fact, in this patent, and there is no device for washing the clutch. These orifices to which I refer permit the escape of the trapped liquid in a horizontal direction from

(Testimony of Edward T. Adams)

the traveling column; They show the ejection of liquid for the purpose of washing in a horizontal direction in this case, and not downward along the standing column.

Further

DIRECT EXAMINATION

MR. HERRON: I now will hand the Court colored copy of Wright 575,498 asking that it be marked U'.

This colored copy received in evidence as Defendants' Ex. U'.

Wright is a traveling barrel removable type of pump and is also designed for the purpose of a washing action which the inventor had in mind. As the traveling barrel is lifted the pumping chamber above the valve e' and below the valve G of the traveling barrel increases in volume; fluid flows in and on the downward stroke of the plunger, valve $é$ closes and the fluid escapes upward through the traveling valve G, making an operative pump. I find in that structure a standing guide column b with b', provided with two working valves, E at the bottom and E' at the top. The structure has traveling barrel D with a valve G, which runs on the column b. I find a bottom guide on the barrel, which is the flanged bottom of the barrel. It is not given a specific number, but the opening through between the flanged bottom of the barrel and the lower part of the standing tube b' is marked e. That bottom guide or flange runs over the column, and is not in contact with it. I find an anchor for the column, b^2 shown more clearly Figure 1.

In the operation of that pump fluid passes downwardly through the space marked on the patent by $e,$ made for that purpose, directly over the anchor.

(Testimony of Edward T. Adams)

Q And what is the function which the patentee ascribes to that action?

A He says, in line 77 on page 1, "The lower end of which is apertured, as at *e*, for the passage of the pendent member *b'*, of the standing barrel B, which aperture *e* is, however, of a somewhat larger diameter than the said member *b'* to admit of a free entrance or exit of the fluid held within the space F,"—being what we have spoken of in all our discussion as the annular space. "the lower end *e'* of the working barrel operating as a piston for agitating and forcing the liquid in such space F through the passages *b³* out against the rock to wash same at all times during the pumping operation."

Q. Now, if you visualize that anchor as having upon its upper face such a clutch as has been referred to in the prior art, would that stream so referred to, or that ejection of liquid have a washing action against the clutch face, if there is any washing action in that type of structure?

A You would have that same washing action in every case. If there was a clutch there it would wash it, just exactly the same as it washes it through the holes.

Q Would you consider there would be any invention involved in providing a jaw clutch type clutch on the bottom of the working barrel and on the top of the anchor?

MR. SOBIESKI: I object to that question, your Honor, as the patent shows that that is impossible in this type of a patent, for the reason that the bottom part B here is curved out to permit the flow of—to cause the oil to flow against the sides of the wall, as I understand the patent, and no clutch could be inserted in this patent without destroying and changing the bottom construction

(Testimony of Edward T. Adams)

of the patented pump, so as to fail to accomplish the purpose for which this patent was issued.

BY THE COURT: Do you agree with that?

A No; I do not. I am sorry. It would be perfectly easy to put a clutch there without interfering in any way with the washing action that he has in mind. There are plenty of clutches that have been shown that could be used in this particular construction. The curvature would not have any effect at all.

MR. HERRON: Would you think there would be any invention involved?

A There would be no invention in doing it.

Q In view of the state of the art in 1926, in modifying that curved surface so much as might be necessary to there insert a clutch?

A There would not be any invention in doing it and it would not be necessary to modify the curved surface, for that matter.

Further

CROSS EXAMINATION

BY MR. SOBIESKI: There is no clutch in this patent?

A No clutch.

Q As I understand it, this base here *with* is marked b^3 is curved so that the fluid will be sent outwardly against the walls of the well, is that correct?

A Well, I think b^3, as I understand it, is the openings in the part b^2, and the washing action forces the fluid down through the holes in that upper face. If you imagine you had the lower face of a clutch or a coupler there, such as you had in the Bray patent, and holes in through that as

(Testimony of Edward T. Adams)

b^3, then the tendency would be to wash the fluid through that hole. The curve has nothing to do with it.

Further

DIRECT EXAMINATION

MR. HERRON: We will now pass to Northrup, No. 1,378,268, and I ask the Clerk to mark this colored drawing V'.

This colored copy is received in evidence and marked Defendants' Exhibit V'.

This is a traveling barrel removable type, and by "removable type" each time I have used that phrase in testifying I have meant a type with a puller nut striking against a projection on a standing column, enabling the barrel and column to be lifted as a unit. I find in this pump a stationary valve 21, a traveling barrel having a valve at 28. The pumping chamber proper is between the two valves, and increases and diminishes in volume with the motion of the barrel and, therefore, acts as a workable pump. The special feature of the pump is the side openings in the barrel for the purpose of agitating the sand and oil at the anchor face. That action is referred to in the specification on line 33, page 2. "The barrel 22 is provided with apertures 32 near its lower end for the purpose of permitting sand or other sediment to escape from the barrel. These openings also possess additional functions given below." Again at lines 59 to 67, "Upon ascent of the working barrel, this oil is forcibly ejected through the openings 32 so as to effectively wash all sand and the like therethrough, as well as agitating the oil in the tubing to prevent the accumulation of sand upon the extension 8 which supports the pipe 1"—etc.

(Testimony of Edward T. Adams)

It goes on to say that sand is held in suspension in the oil and discharges upward with the oil through the well tubing.

Q Now, in that pump likewise would you consider it would amount to the exercise of anything more than mere mechanical skill, not rising to the grade of invention, to insert at the top face of part 6 and on the lower face of part 24 clutch parts to operate to engage against rotation?

MR. SOBIESKI: I want to object to that, your Honor, on the ground the evidence here shows that that is not what they did.

THE COURT: I will allow him to answer and exception to appear.

A The introduction of a jaw clutch at that point would involve no invention. I understand that the specification of the patent would teach that one of the functions of agitation of the oil through the hole 32 would be the washing down over those faces and the keeping of it free from sediment. It so states. I find in that device a standing guide column No. 1, best shown in Figure 2, with a working valve 21. I find a valved traveling barrel 22, the valve 28 being at the top. This traveling barrel runs on the column 1 and its packing. I find a bottom guide 24 on the barrel which in Figure 2 runs on the column, and in Figure 1 is shown with a very material clearance. I would say, that as shown in that Figure, it runs over the column rather than in contact with it.

Q Now, in the event that the pump were operated with the clearance shown in Figure 1 would there be any ejection of the liquid downwardly through that aperture alongside of the standing column between the standing column

(Testimony of Edward T. Adams)

and the puller nut which would result in washing the flat top of 5?

MR. SOBIESKI: I object to that, your Honor, unless he points out wherein the patent they state that that is the result, or intended to be the result of anything here. There is nothing to show that that actual clearance, or if such be extra clearance, is not the extra amount that would merely result from a lettle freer fit rather than as he says.

THE COURT: What do you think about it, Mr. Adams? You are the expert. Tell us what you think about it?

A   The objects that Northrup had in mind, as he specifies, are the agitating of that sand and stirring that up to prevent it settling at the anchor face; and he has shown, as I would read the drawings, two methods of doing it; one in Figure 2, in which I would say that the puller nut 24 was to run in contact, and therefore, is running on the standing column and all of the fluid is ejected through the holes 32. In the drawing No. 1, the lower portion of the tubing reaching to the top of the well is shown, and the clearance between the puller nut 24 and the tubing is shown to be small; and just as I explained in a previous patent, when the plunger moves up the radial flow from 32 would flow downward towards the top of the stationary anchor and fill up the space vacated at the bottom by the upward motion of the plunger, which carries the puller nut with it. So there is a strong downward flow shown through holes 32. Then if we take the structure shown in Figure 1 with the holes at the side at 32 and also an opening, not given a number, between the puller nut 24 and the standing column 5, the flow would be divided.

(Testimony of Edward T. Adams)

Part of the fluid trapped in the annular space would flow out through the holes 32, and part would flow out through the clearance between the puller nut and the standing tube, and both flows would unite.

THE COURT: That would show that anchor was intended, then, as a working clearance?

A   No; I would not say that, your Honor, except I am just going from the drawing.   The drawing shows a considerably more than working clearance, and to what extent he is bound by that is beyond me.

Q   It shows, the fact he put the holes there, would show he intended the escape to be through the holes?

A   If I may disagree with your Honor—

Q   No; I am asking you.

A   I really think, if I had made the two drawings myself and was not explaining to you what I had done, I would say that, knowing that I had in mind initially the stirring up of the sand at the bottom and the preventing of sand settling therein, I would have shown both structures in my patent so that. I would be free to use either one and have the patent benefit of it.

MR. HERRON: At all events, Mr. Adams, would anyone skilled in the art of reading patent drawings who would look at those patent drawings, and as of the date of its issuance (1921) there find taught the making of a structure in which there would be the two cooperating flows you have mentioned?

A   I did not catch the word; find what?

Q   In the drawing a structure is shown in which there would be the two downward flows, namely, the one from 32 along the side of the tubing down across the flat face of 6, and the other the flow down around through the annular

(Testimony of Edward T. Adams)
space around the standing column between the standing column and the puller nut?

A   If you took these two drawings, cut them apart and put them in the shop to be built they would build Figure 2 with a close running fit and they would build Figure 1 with a very loose running annular space through which fluid would flow.   That would be the way a shop mechanic would interpret the drawings.

THE COURT: You would figure the play would interfere with the effect of the guide, I suppose, considerable play there?

A   If it is a guide; I do not think that this author specifies that as a guide.   He specifies it a bushing, or he does call it a bushing or guide.   If it is to be a guide I fully agree that it is necessary for it to be in contact. If it is to be a bushing, and he specifies both members, then it can be an opening through which the fluid would flow, as shown in Figure 1.

MR. HERRON: Examining that structure, then, I understand you would say that the question as to whether it showed the downward motion of fluid directly towards the flat surface 6 would depend on the question of whether the drawings and the specifications and the claims of the patent were intended to cover a guide running on the column, as contradistinguished from a puller nut running over it?

A   I would read that working of the patent as meaning both.   He says "a guide" and I think that the guide is Figure 2.   He says "a bushing" which is a plain puller nut which is shown in Figure 1; and in Figure 1 there would be a truly annular space in the bushing and direct downward flow on the face of the anchor.   The volume

(Testimony of Edward T. Adams)

through hole 32 would go to the same point so that the two join, and all the fluid goes to the same spot, no matter by which way it goes. I find in that structure an anchor for the columen No. 6, more plainly shown In Figure 2.

THE COURT: If you were providing for the washing effect, would you use the holes or would you provide for the play?

THE COURT: I mean as a practical thing, would you use the holes shown here and minimize the play between the column and the guide, or would you use—

A (interrupting) As far as washing is concerned, your Honor, I feel that there is very little, if any, practical difference. I can't answer that without coupling with my answer the fact that the pumps are made as large as they can be made, and be introduced into the tubing; that is, you have a tubing a certain size and the pump itself in order to get action the pump is made as large a size as it can be made. Therefore, the clearance between the nut 24 and part 1 is very small and there is a high velocity of flow through to the outside.

Q Does it make any difference to the well whether you would use a large barrel or not?

A I didn't get that.

Q If you had a 9,000 foot oil well would you use a large clearance or small?

A It is a question of whether the—the clearance between the outside diameter of the pump?

Q Is there any feature of the weight of the lift there?

A Of course, that is taken into account, but it is also taken into account in the size of the tubing. The size of the pump, however, is fixed more by this: If the well

(Testimony of Edward T. Adams)

is crooked you have to have a certain clearance. I think in most of the pumps we have been discussing we are talking very largely here of a standard type of what is called two and *on*-half inch pump, and the total clearance in there is something of the order of three-thirty-seconds of an inch on a side, that is, three-sixteenths total, is practically that, so the clearance is small and there is a fixed amount of fluid in that annular space—when you lift the plunger you squeeze that fluid out of there. When it is squeezed out of there, where can it go? It has got a column of fluid 5,000 feet above it, no holes in it anywhere, just a solid fluid. But immediately below it as soon as the tube has moved up there is a space made vacant by the upper motion of the bottom part of the plunger, into which that fluid flows. Therefore, there is a strong stirring action by the flow outside. There is also a flowing action vertically downward; and which would have the most stirring action on the sand I would not say. Personally, I do not think there is any material difference.

Further

CROSS EXAMINATION

BY MR. SOBIESKI: Do you find anywhere in the patent other than in the drawing anything indicating the inventor had in mind that the trapped oil should escape along the side of the standing column?

A Why, I think that is the reason why on lines 15 and 16, page 2, he specifies a bushing or a guide. Now, I interpret "guide" as something that is really a guide, such as Webster's Dictionary gives, or as we use it in mechanics, a guide of a locomotive, on which one part

(Testimony of Edward T. Adams)

is strictly guided on another, or a guide on a hydraulic piston in which the water is strictly guided along a path, that is carrying the idea of contact. Therefore, I think he means just what he says, that 24 as shown in 2 is a guide and 24 as shown in 1 is a bushing, which is something screwed in to make a stop whereby the pump could be lifted and have an opening around it.

MR. SOBIESKI: Now, you have seen these fishing poles that have guides for the line to draw along them?

A   Yes.

Q   Now, the inside diameter of these guiding means confining the activity of these lines is quite considerably larger than the outside diameter of the fishing line, isn't it?

A   Yes.

Q   Many times larger, isn't it?

A   Very much larger.

Q   And yet, you would not claim that those were not guides, would you?

A   No.

Q   The mere fact that there is a considerable amount of clearance does not prevent it from being a guide if it is close enough to accomplish the end in view, is that the idea?

A   No, that is not the reason. A fishing pole when it is in use is a bow, and the guiding is simply on one point. The line makes tangents between points in the curve, so it is guided simply by a point and not guided in the sense I mean. Take a locomotive guide, we all see the train going by and we have seen the crosshead working on the locomotive, and it is perhaps the best example you can fix. That keeps the motion of the piston in a

(Testimony of Edward T. Adams)

straight line, which is necessary in order to keep the piston from cutting the cylinder in which it works. Now, there is a reason for that and in here when the patentee says "a guide", and when he makes a guide, there is a reason for it act as a guide. But in the instance that you selected you have one point of a curve. You have a curved path. You are not trying to hold that fishing line in any one straight up and down motion. You are trying to keep it from getting too far away from the pole at any point between the butt and the tip. That is all that you are trying to do. It is not guiding in the same sense.

Q Well, you would not maintain that those agate rings whose inside diameter varies considerably in excess of the outside diameter were not guides, because there was considerable difference in the diameters?

A I would say they guide at a point in circumference of the agate, to keep the line close to the pole. They are not guides in the same sense. I think a more sane and honest interpretation is to call them "stops". I would not call them "a bushing". That would not do. But consider that the rod is bent in a semicircle so that we can all visualize it, just bent in the arc of a circle. If there is nothing to restrain it, the line from the reel to the tip would make a straight line. Now, when you put in these rings; what would the rings do? They limit the motion of the line. You allow it to move a certain distance away from the rod but not the two or three feet that it would move from the rod if it was not there. I think it is much nearer an honest interpretation to call that ring "a stop" than it is to call it "a guide". They are commonly known as a guide and is a guiding action, but not

(Testimony of Edward T. Adams)

the same type of a guiding action, by any means, involved in a pump or in a locomative, when the word "guide" is used for a like purpose.

Q   Then, as I understand it, your definition of a guiding action is where the ring or nut, or whatever be its word, limits the motion of some other article?

A   No, That I would call a stop.   You have a limiting of motion in a puller nut.   A puller nut is put at the bottom of the traveling barrel.   Now, that is a stop.   You can go down a certain distance and the puller nut would strike, and you go up a certain distance and it will strike again.   Take in your own pump, and probably those two parts are about 9 feet apart, so it could never move up above 9 feet or below 9 feet.   In that case it is a stop. Now, a guide implies the idea of contacting, the idea of control, rigid control of motion.   I am not talking about the fishing rods, of which you can speak loosely in all manner of ways.   I am talking of pumps; and when you use a guide as in a locomotive or in a pump, the idea of contact is there, a sliding contact in which the guide actually controls the motion from one end of the stroke to the other end of the stroke completely, and all the time such guide is the controlling factor.   The word "control" comes in as a synonym for "guide".   Now, when it is loose you have a part that is not a guide because the part is free to move within limits, just as your string is free to move within the limits of the diameter in that agate when you are fishing.   That is more of a stop than it is a guide. It is limiting the motion but it does not guide it rigidly. In these pumps when we think of a guide, we think of something that rigidly guides. But in fishing rods, for instance, they are known as guides, although it does not

(Testimony of Edward T. Adams)

guide it rigidly. It guides it loosely. In this patent the fact that the patentee says over here "having mounted thereon an externally threaded bushing or guide" leads me to believe that he had in mind the flow of trapped oil to some extent along the edge of the standing column. A bushing simply reduces the diameter. In the patent I think that the purpose for which the whole thing is done is fully described. The purpose was to agitate sand which is held in suspension at the bottom of the tube. On page 2, beginning at line 33, he says that the purpose of agitation is accomplished by the apertures 32 in the barrel Q. And is there anywhere in the patent any language which says that the purpose of agitation is accomplished any other way?

I do not know any place in which he mentions that flow other than to say "bushing". "Bushing" in mechanics is very clearly a reduction of area, as compared with "guide" which is very clearly a guiding action.

Q But he at no time indicates by any language that his purpose of agitation is accomplished by anything other than the holes in the side of the barrel?

A I don't know.

There is no lock in this pump. No clutch.

MR. SOBIESKI: Directing your attention to these pumps you have discussed, Mr. Adams, you stated awhile ago what the average clearance was ordinarily between the side of the barrel and the tubing in the ordinary operation.

A I made the statement that in a two and one-half inch pump about three-thirty-seconds on each side; that is two and seven-sixteenths approximate inside diameter of the tubing and two and one-quarter inch outside diame-

(Testimony of Edward T. Adams)

ter of the barrel. This Northrup patent which we were discussing, particularly that bottom guide or bushing known as No. 24, the cross hatched portion designated as C, to represent the tubing. I understand by hearsay that the largest pump that will go in a two and a half inch or two and seven-sixteenths tubing is one with two and a quarter outside diameter barrel. I have no way of knowing how much space is probably represented by the white space between No. 24 tubing marked C. There is no dimensions on that drawing. If that were of the same general dimensions as pumps in general that I have given you, that space would be three-thirty-seconds of an inch, both sides, for the largest pump that would go in that space.

Q Assuming that that was three-thirty-seconds of an inch, wouldn't you assume that this space between the nut 24 and the standing column which is left there in white would be approximately one-third, or one-thirty-second of an inch, judging from the diagram given in the patent?

A I made patent drawings for part of my living for a number of years, and in making those drawings things that were not material to the drawing itself were made simply in an illustrative manner. I never did—and by instructions of the attorney that I worked with would not if I was making a patent drawing pay particular attention to that outside dimension. It was illustrative that the pump was enclosed in the tubing in the well, and that is all it is there for. I would admit that the space outside of the plunger was more than the space inside the tubing. That is obvious on its face. That space is just merely illustrative of the fact that there is a space, I wouldn't

(Testimony of Edward T. Adams)

say it is illustrative of the fact that that space is considerably larger than this other little space between 24 and the standing column. I would not also call that merely illustrative. I really would not. I do not think that would be fair. The tubing on the outside had nothing to do with the patent on the pump. One making the drawing would make it any way he happened to draw it without much fuss about it, but the part 24, as I read the patent, is specified both as a guide and as a bushing. The guide is something that guides and a bushing is something that reduces the diameter in the tube and changes the size of the hole. In other words, in the absence of some particular description in the patent relating to either one of those spaces, I would regard them as merely illustrative. The patent gives no definition as to the amount of opening around the tubing through the nut 24, other than to say it is a bushing, made so for the purpose of directing the fluid downward into the bottom of that space. It does not say that the fluid is directed downward by means of the annular opening in the bushing 24.

Q And in the absence of something in the patent calling attention to the various items in the drawing, you would take it that the items which are not particularly called to our attention are drawn in an illustrative manner only?

A If you are speaking of the tubing, I feel that that is drawn in an illustrative manner. If you are speaking in your question of the clearance inside the bushing, you have words in the patent directing your attention to that.

Q And that depends upon your interpretation of the word "guide" as being something limited to a close fit and not merely something which limits the operation to

(Testimony of Edward T. Adams)
the extent that the designer wants it limited; is that a correct statement?

A   Well, in part, the other part being that the purpose of these holes and these openings wherever made is to agitate the sand that accumulates in the pocket above the anchor and below the barrel, and that such an opening through there with a direct flow through it would agitate it just as much as the flow upwardly and around the outside of the puller nut. The flow that goes through the holes from the inside flows around the outside of the puller nut. The flow that goes through the annular hole flows down through the puller nut, but both flow into the same space.

Q   But nowhere in the patent does the inventor mention the fact that there is a flow from down the sides of the standing column, but you gather that there is on account of the drawing and the fact that the mention of the bushing is used,

A   No. I gather that because it is a pump. I question whether he means that there is any flow upward in the column, but because it is a pump, it pumps or discharges fluid upward and this annular space downward is really in the nature of a pump itself, when it is diminished in size the fluid is squeezed out of it. It has to flow somewhere and it will flow through any openings that are made for it.

Q.   You think, then, that the distance between the tubing and the bottom nut there which is given as considerably larger than this annular space we have been describing, you think that that is put in there for illustrative purposes only?

A   There are no dimensions given.

(Testimony of Edward T. Adams)

Q   And there is no dimension given for the annular one, is there?

A   For any other space.

Q   And neither space is described in the specifications, either, is it?

A   No; it is not.

Further DIRECT EXAMINATION

As illustrative of the fact that non-essential portions of these drawings are intended to be merely illustrative, it is a fact that the presence of earth and rock, down around the anchor are illustrated by just a little cross-hatching.

Q   I ask that this diagram being the Whitling drawing heretofore marked HH, for identification be marked in evidence.   I am not quite clear from the stamp whether it is in evidence or merely for identification.

MR. SOBIESKI: I understand our objections to these patents also go to the digram as well.

MR. HERRON: I presume so, if your objection covers that.

MR. SOBIESKI: May it be understood that way?

MR. HERRON: Yes; as to this one, surely.   Now, if the Court please, we will pass to Ellis, No. 1,513,699, a colored copy of which I will ask be marked W'.

(Colored copy is received in evidence as Defendants' Exhibit W')

I have studied the structure shown in that patent.   Ellis shows a removable, traveling barrel type of pump.   A very interesting feature of the Ellis pump is that he is interested in and his patent largely relates to the velocity of flow through apertures in the puller nut.   He realizes, and

(Testimony of Edward T. Adams)

perhaps the other designers have done the same, but he is the first one who mentions the fact, that if there is a velocity downward through the puller nut there is on the opposite stroke equal velocity upward; and if you conceive that the pump is worn and that there is a material amount of leakage there is a greater velocity upward than there is downward. The pump has a valve at the top of the standing column and the pump chamber is between that and the traveling barrel, so that when the traveling barrel is moved up and down the volume of this pumping chamber increases and diminishes in volume and the pump operates.

In Ellis, I find a standing guide column, No. 12, provided with a working valve 18 at the top. That structure has a valved traveling barrel 19 the valve being 20. That barrel 19 runs on the packing which is a part of the column. There is a bottom guide 21 on that barrel referred to in the specification as bearing on 12. It is distinctly a guide, shown perhaps most clearly in the section Figure 3, and it is shaped as an apertured guide, that is the guiding surface is interrupted by grooves which are marked 22 in the drawing, making apertures in the guide through which liquid may pass. That apertured guide runs on the column because it is guided by the material left between the apertures as is illustrated by the part 23 in the sectional drawing Figure 3. I do not find a coupler in that pump. It has a series of venting ducts which are marked as 22 in Figure 3, and which are shown by the clearance in the side of 21 in Figure 2. Those venting ducts open directly toward the anchor face, the top of part 13.

It would not in my opinion involve the exercise of any inventive faculty, as distinguished from ordinary me-

(Testimony of Edward T. Adams)

chanical skill, to provide that top surface of part 13 with either the male member or female member of a jaw clutch, and to provide the bottom of a guide on the bottom of the traveling barrel with a cooperating clutch part. And it would not have done so in 1926, or '24, which is the date of the patent.

In that drawing I find that the guide is apertured, (the apertures being vertical grooves on the puller nut 21) for the ejection of liquid from the barrel directly toward the upper face of lower standing part, that is anchor part 13.

Q Now, I notice that there is a part illustrated by Figure 5 and Figure 6, which consists of a sort of a cage at the bottom of the apertured guide, in which a washer is inserted. What is claimed to be the purpose of that washer in the patent?

A Ellis was concerned about the upward velocity of flow through these apertures and the fact that the space above the anchor 13 would be filled with sand and silt in suspension, which would be carried by that upward velocity through the apertures to the packing cups 15 of the column 14, and he placed a ring a cup which acts as a valve. It is a fiber element. So that on the downward stroke of the plunger, when fluid would be flowing into the annular space above the puller nut, that flow would be restricted and sand would not be carried in at such high velocity. On the stroke in the opposite direction, that is the upward stroke of the plunger, the fiber valve 25, would drop down to the bottom of its retainer, which is shown as the end of 27, and allow free escape of the fluid.

In other words, Figure 5 and Figure 6 represent a cage and the cooperating ring which, together form a valve which would have the effect of preventing, according to

(Testimony of Edward T. Adams)

the teaching of this patent, the ingress or drawing in of sand upon the down stroke of the traveling barrel. Upon the upstroke, that is, when the traveling barrel is moving upward with relation to the standing column 13, there would be ejected through this series of ducts or apertures, illustrated in Figure 3, downward jets of oil. The downward direction of that ject would be interrupted by this ring in the cage and it would be deflected around the valve 25, some of it, perhaps, passing to the outside and some on the inside. That might very well be.

Q Is there anything in the specification of Ellis as to whether this cage and the cooperating washer, which together constitute the valve you have mentioned, are or are not an essential part of his pump?

A On page 2, beginning with line 37: "Under some conditions the valve ring may be omitted as the passages in the cap nut permit the free flow of liquid and prevents the formation of a dead chamber containing trapped liquid which will resist the free operation of the pump." If the structure illustrated in Figure 5 and Figure 6, which together constitutes the valve, is omitted, upon the upstroke of the pump, the guide shown in Ellis, being provided with a series of venting ducts or being apertured, would then be a guide so apertured or so provided with ducts, as to provide for the ejection of liquid from the barrel directly toward the upper face of part 13. In that event, if you assume that the upper face of part 13 has been provided with the lower part of a clutch, the guide would be apertured for the ejection of liquid from the barrel directly toward the upper face of the lower coupler part That would be a coupler part if it were present. And if a clutch were there, then it would be apertured for the ejec-

(Testimony of Edward T. Adams)

tion of liquid from the barrel toward the effective face of the lower standing coupler part. If in accordance with the teaching of the patent, I desired to do away with the floating ring valve illustrated in Figure 6, I would likewise do away with the cage part shown as the lower part of Figure 5; in other words, I would do away with the valve in its entirety if I was eliminating its function.

Any good mechanic would do that. It is exactly what you would do. It, of course could not by any stretch of the imagination rise to the grade of invention to so do. I find in Ellis a standing guide Column 12, provided with a working valve at its top; a valved traveling barrel 19, running on the packing of 12; a true bottom guide 21, running on the column. I do not find a bottom face forming a coupler, but do find anchor 13 for the column; and find a guide having a series of venting ducts 22, opening directly toward the upper surface of 13. That guide is apertured with the aperatures 22, so that fluid ejected from the barrel would be ejected directly toward the upper top face of part 13 directly and repeatedly in the event that the cage is removed in accordance with the teaching of the patent.

Further

CROSS EXAMINATION

BY MR. SOBIESKI: Do you refer to this bottom bushing as a guide, Mr. Adams?

A. Yes.

Q. Looking at the figure marked 21, is there not a space between that marked red there and the blue standing column just about as large as the space we were discussing in the Northrup patent?

(Testimony of Edward T. Adams)

A. I would not say whether it was or was not, but there is space there and that space is properly drawn. The space 22 is shown in Figure 3, which is a section and the section passes through the space. If you draw a horizontal line through Figure 3 then you will have the line on which the section in Figure 2 was taken and you will see that that passes through the aperture 22 and it is shown correctly.

Q Then, despite the existence of that space there, you think that it nevertheless is a guide?

A The space has nothing to do with it. It is a guide. Look at Figure 3 in your drawing, that section in the upper right hand corner. You look down and you see the number 23. 23 goes to a point that is in contact with the column, and those four points, of which one is numbered 23, are the guides. 22 are the apertures in the guides. There is no clutch in this patent.

Q Is it not true that when the trapped oil is ejected it goes out sideways through this opening in the bottom nut?

A That depends on the design of 25. If 25 (which is the fiber washer shown in Figure 6, and also shown at the bottom of the puller nut in Figure 1) fits very closely on the column 12, then there would be no flow through the inside. It does not teach whether it does or not, but in any event, no matter whether there is a downward flow past through the inside of the fiber washer 25 there will be a side flow somewhat radially, possibly wholly radially in extent, clear around that.

Q Don't you think when you say "possibly wholly radially" that if this washer is to accomplish its purpose it would have to be entirely and wholly radially, or else

(Testimony of Edward T. Adams)

the washer would not act as a valve, as is set out in the claims in the patent?

MR. HERRON: I think that is a fair assumption. We will stipulate it.

A I would not quibble about it. As a matter of fact, when the valve is some little distance away, as it is shown by the drawing, flow would be more or less at an angle and I should call it a radial flow.

MR. SOBIESKI: Well, then, you take it in this patent the liquid is not ejected directly towards the opposite face of the clutch, if that had been a clutch there, but instead is directed radially?

MR. HERRON: I think, Mr. Sobieski, you misunderstand the witness. My question and the witness' answer were directed to the situation where, in accordance with the teaching of the patent, the valve and the cage had been removed; not directed to the situation where it was working under conditions where the cage was desirable. In the patent he points out both circumstances and I specifically directed my question to the case where, in the teaching, the valve had been removed.

MR. SOBIESKI: I will re-ask the question then.

Q With the cage there, the nut sends the oil out in a radial direction, is that true?

A O. K.

Q It does not send it directly out toward the opposite face of the anchor, does it?

A When the nut is there it flows radially, or thereabouts. When the nut is not there the flow is not likewise radially. It is practically against the top face of the anchor 13.

(Testimony of Edward T. Adams)

Q Now, I notice the bottom of this part 26 of Figure 5 is specially drawn. That has no aperture in it. Don't you construe that to fit closely around the standing column and thereby require a radial flow whether the washer is there or not?

A Well, if you will look at drawing No. 4 you will see that 27 is drawn in section, 27 in No. 4 being the standing rib that holds the lower part in place, and next to the standing column 12 you will see dotted lines away from the column, indicating that there is free clearance between the portion of the cage 26 and the standing column 12.

MR. HERRON: May we ask that that dotted line be marked by let us say "T" on the Court's patent? Does that line where the arrow indicates, marked T point to the dotted line you have reference to in Figure 4?

A It does.

Q Wouldn't you say, considering the size of that clearance between the bottom of the cage 26 and the standing column, that most of the oil when the washer is out would have to go out in a horizontal direction through the nut?

A I feel that I would be constrained to answer in the words of the patent itself, appearing in line 36 on page 2,—"passages in the cap nut permit the free flow of liquid". So I feel that the design of the patentee intended free flow.

MR. SOBIESKI: Referring to Figure 5, Mr. Adams, to what does No. 27 refer?

A That refers to the vertical strip of metal which holds the ring 26 in position. Between the ring 26 and 21 there may or may not be inserted a fibrous washer 25.

(Testimony of Edward T. Adams)

Q Then, there is a lateral space between the part marked 26 and the part marked 21, is there not?

A There is a space between 26 and 21 to retain the washer 25 when it is present. That washer is not large enough to fill that space completely. So that when the washer is in there working, any oil that is ejected is ejected in a lateral direction.

Q Now, when the washer is not there, is it not true that on account of the fit of the part called 26 most of the oil that is ejected is likewise still ejected in a lateral direction?

A I have previously said that I thought not; that the author has already expressed his intention of having, to use his own words, "a free flow" of the liquid, and I would interpret that as meaning that the fit which we have marked T in Figure 2 was ample for the passage of the liquid.

Q Now, comparing the size of the space you have marked T with the size of the lateral opening we discussed between 26 and 21, isn't it true that the T is practically negligible in size when compared with that large lateral opening?

A That would depend upon the thickness of the valve. I could not answer. The inventor had in mind a free flow in each condition and I can't with my eye measure the height of the valve 25 at the bottom, and then with my eye measure the lateral opening, that is distance between the ring 26 and the body 21. The author specifies it would be enough for the fluid to flow out and he also states that there will be a free flow for that fluid when that valve 25 is not there. So we might say in each case there is room enough for free flow. When the valve 25 is

(Testimony of Edward T. Adams)

there I assume most of the oil would flow in a lateral direction. It just depends upon how close a fit the inside of the valve has with the tube, which I know nothing about.

Q Isn't it a reasonable construction of this patent that without the valve there the oil would still continue to flow out in a lateral direction, however, being a free flow this time because of the fact that the washer was not there?

A No. If the valve is out I see no reason why the flow should be lateral. I think that the intent of the patent is plain that there shall be a free vertical flow when the washer is not present. The word "vertical" is not used, but if the valve is not present then the flow is vertical and the flow through the nut is necessarily vertical. It goes through the nut. There is no other way to get out.

Q Would it not get out laterally, as it was able to do even with the valve there, only get out much better with the valve gone, laterally?

A No. Through 21, which is the nut, you have a series of orifices or vents or ducts, whatever you call them, 22, and through those there must be a vertical flow because of the position of the puller nut, therefore, the flow is through that much of it vertically. Now, then, the author specifies a free flow, shows the ring 26, as I read it, enough larger to allow that free flow, which is a vertical **flow.**

Q You believe, then, that the oil would be more likely to go through the narrow place you have marked T than it would be to go through the large opening which exists laterally between the part marked 26 and the part marked 21 on Figure 5?

(Testimony of Edward T. Adams)

A   We are getting into the line of speculation, but I think it would; and I will tell you why. If you go to the specification line 33 page 1, "in the upward movement of the barrel when the fluid is expelled and enters the dead chamber rapidly in the downward movement during which the dead chamber is created." We find in the reading of this patent that Ellis had in mind rather restricted openings and a high velocity of fluid. Now, then, if the fluid was flowing rapidly and is directed downward—I can only take it as he has it in his specification. I can't go outside of that and say what the sizes are—I read the specification as saying that there is a rapid flow, as he shows, through the small ducts, and that there is space for it to pass through. I am going to interpolate again just what I have said before, that in either case the fluid comes to the same place with, perhaps, equal violence of flow. I see no reason for believing that whether the washer is there or not, that all but a negligible portion of the fluid goes out in a lateral direction through the nut.

Q   Would you disregard the effect of that large opening, that large lateral opening that we can see clearly in Figure 5, and assume that the oil instead would go through this very small opening that you have marked T in Figure 4?

A   I do not know that it is small. The author says it allows free flow.

Q   And that opening does not, of course, prevent the bottom nut from being a true guide, does it, in either event?

A   Part 21 is the true guide. 26 is not a guide.

Q   And part 21 shows an opening between it and the traveling barrel about the same size as the opening you

(Testimony of Edward T. Adams)

described in the Northrup patent, which you said would prevent it from being a guide, is that correct?

A No. I don't know as I said that. '

Q Well, is the opening shown in both patents approximately the same size?

A I would have to look at Northrup, and I don't know as a looking at it fixes the size. I don't know.

Further

DIRECT EXAMINATION

Referring to Figure 3 of Ellis those portions of the guide which are marked 23 in the patent bear directly upon the standing column and therefore are a guide. The small opening which appears, referring now to Figure 1, at part 21 is due to the fact that that cross-sectional drawing is drawn as *through* you were looking at Figure 3 cut directly across the opening through the ducts, that is as though you draw a horizontal line through duct 22 to the duct shown on the opposite side of Figure 3, which is not lettered.

In the event that the washer and its seat, which is the ring supported by the vertical strips 27, making a cage— in the event that that washer and the cage which confines it, and which serves no purpose except to confine it, is removed from the patent, then in that event there is nothing whatsoever to impede straight downward flow of fluid through the ducts 22. My feeling is this: That the clearance in the ring was intended by Ellis to be sufficient to allow a free flow between the ring 26 and the standing column 12. Now, one reason for that is that he shows the dotted line which we have marked T, and the other reason for that is that he specifies a free flow. The

(Testimony of Edward T. Adams)

valve ring may be omitted. I understand by the "valve ring"—he may mean the "valve ring". I understood when I read that as meaning the "ring valve" with the words just reversed. In any event, if you remove the valve you would remove this operating seat, as a question of ordinary mechanics. You would not carry a valve seat if there was no valve, and good mechanics would dictate that it be gotten out of the way.

MR. HERRON: I now hand the Clerk a colored copy of the patent to Dickens, No. 1,486,180, and ask that it be marked X'.

(This colored copy received in evidence as Defendants Exhibit X')

I have studied that patent. This is a traveling plunger type with a stationary valve and the pump chamber lies between the inlet valve 18 and the discharge valve at the top of the plunger in the cage which is not lettered or numbered. With the operation of the plunger the volume of the pump chamber will increase and decrease; and that, with its valves, make a workable pump. The essential feature of the pump is a jaw clutch for the purpose of tightening the rods, referred to in lines 93 to 105 of the patent, on page 1. "In pumping the well sometimes a sucker rod 21 becomes unscrewed at one of its joints or from the valve 20. In such a case the traveling valve 20 will ordinarily descend by its own weight and rest on the head 11, the tongue 23 seating in the groove 12, and the traveling valve will thereby be held against rotation when it is attempted to reconnect the disconnected joint of the sucker rod 21, said traveling valve thus serving as an anchor to permit the tightening up of the joint or joints of the sucker rod." I find in the pump an anchor for the column marked 6. There is a coupler face on its top

(Testimony of Edward T. Adams)

part that is the female joint 12. That is the slot in the top of the coupler. The male coupler part 23 is on the bottom of the plunger. The anchor directly presents to said coupler face 23 a top face. When the parts are in engagement there is such a structure as to be non-rotatively inter-locked with the lower guide coupler. That would be 12 with 23, or if you take the entire coupler it would be 22 with 11.

Further

CROSS EXAMINATION

This is what is known as a traveling plunger type of pump. There is no trapped oil in that pump. That arises from the fact that it is not an essentially removable type. There is no trapped oil; there is nothing in there but the volume of fluid pumped.

THE COURT: It is not a barrel pump?

A It is not a barrel pump. It is a plunger type.

MR. HERRON: I now hand the Court a colored copy of Neilsen, No. 1,717,619, and ask that it be marked Y'.

This colored copy is received in evidence as Defendants' Exhibit Y'.

Further

DIRECT EXAMINATION

This is a traveling barrel type, the characteristic feature being a washing action. The pump chamber, as shown in Figure 1, lies between the inlet valve 12 and the discharge valve 20. In the operation of the pump this volume is increased and diminished, cooperating with the valves to make an operative pump. In this structure I find a standing guide column of Figure 2, which is provided with a working valve, 36 or 35—it is not clear which—at the top,

(Testimony of Edward T. Adams)

and a second valve at the bottom. I find a valved traveling barrel .37 with the valve 43. This barrel runs on the
column 32. I find in that structure a bottom bushing 39
on the barrel. It is not a guide, it does not touch the
column, but runs very loosely around or over and separated from it. I find an anchor for the column 30 with
3, perhaps. There is the same trapped liquid in this
pump that we have in so many that we have examined and
find in the Hofco and in the Bray types also, which lies
above the bushing 39 and below the packing of the standing column and inside of the traveling barrel; As that
volume is increased and diminished by the motion of the
barrel the fluid is washed from that space down around
the anchor 30 or 3 and through the holes 41 in that
anchor to wash the valve 31, the inventor's idea being
that he wanted to prevent the accumulation of sand at
that particular point. *Incidently,* it really washes the
face of the anchor, which is immediately below the valve
and has a seating of 3 in the part 2.

There is a washing action in the operation of that
pump. There is a direct downward flow vertically between
the bushing 39 and the standing column 32. This vertical
flow washes the face of the anchor 30; from there it
passes through the passage 40, to holes 41, washing the
valve.

If a clutch were interposed on the top of part 30 and
the cooperating part of the clutch on the bottom of part
39, there would be a direct flow of liquid towards the
effective face of the lower standing coupler part. The
patentee refers in several places in his specification to the
washing action which I have mentioned. On page 2,
beginning with line 2, the reference being apparently to
drawing Figure 1: "During the downstroke of the

(Testimony of Edward T. Adams)

plunger the liquid in the chambers 15, 17 is forced through the ports 18 into the cage 6, thus washing over the ball 7 and removing all sand and sediment therefrom. Thus it will be seen that this ball is always clean and all sand, and so forth, removed from the seat thus assuring that it will be properly cleansed at all times." On the same page in line 54, it says, apparently referring to Figure 2, that the liquid in the passage 40—"is forced through the ports 41, washing over the ball 31 to thoroughly cleanse the same in a manner previously described under Figure 1."

In other words, in the action of this pump the liquid is forced down past the bushing 39 out over the shoulder formed by the top of part 30, down the channel thus presented and between that part and the casing or tubing of the well and in through part 41 and over ball 31. It is forced through the bushing rather than around it. Thus there would be that washing over the shoulder formed at the top of 30. In my opinion, it would not have involved any invention in 1926 to have placed co-operating clutch parts on the top of part 30 and on the bottom of part 39, and it of course would not have involved invention in 1927, either.

Further

CROSS EXAMINATION

There are three tubes and I read the specification rather carefully and it seems to specifically avoid the terminology, at least, that would have made this into a fluid-packed pump. We will take Fig. 2 which shows a traveling barrel fitted with a traveling valve 43 and fitted at its lower end with a puller nut 39. This traveling barrel is

(Testimony of Edward T. Adams)

mounted on and runs over a standing tube 32, with its corresponding packing parts 33 and 35, and carries at its upper end a top standing valve 36, and its bottom, a standing valve 31. So far, the action of this plunger and traveling barrel are identical with any traveling barrel pump. Outside of the traveling barrel there is a combination of two parts. It consists of an outside connection numbered 1, Fig. 1, with a liner to make a wearing surface inside. The liner is numbered 42 near the top of Figure 2. This liner is outside of the traveling barrel, and is fitted thereon, and I now mark it "L". The traveling barrel, therefore, slides up and down between the standing column and the liner L. The object of doing this is apparently to secure a greater washing action than he would otherwise have, as he therefore forces the entire displacement of the lower part of the traveling barrel to flow down through the passage 40 and to wash the valve 41. The specification reads, page 1, line 25: "Fig. 1 is a fragmentary, longitudinal, sectional view of a 3-tube type of pump." The three tubes would be the liner, traveling barrel and stationary plunger. They are described by the inventor as three tubes. There is no clutch device in that patent.

Q No clutch. I understand, then, that the tubes are so arranged so that the liquid will flow between them, is that correct?

A Well, three tubes, your question is a little indefinite.

Q As I understand it, that patent is so arranged that the liquid flows between all of the three tubes which the inventor describes; is that a correct statement of the invention?

A Well, hardly. Supposing I state it as it looks to me. The fluid in the portion above the puller nut, which

(Testimony of Edward T. Adams)

we have always called the annular space in considering the prior art, flows from that space down through the bushing and after passing the bushing flows through the space between the standing tube and the liner and so through the opening 41 to the valve 7.

Q   And it then forces this oil down through the lower valve 31 into, you might say, the central tube, as described by the inventor, and through the pumping motion that goes out up into the top of the well, is that correct?   In other words, that oil is recovered which is forced down between parts 39 and 30; the oil that is forced down there goes through valve 31 and up through the tubing to the top of the well and is recovered, is that correct?

A   Whatever fluid is displaced by the thickness of the metal of the lower portion of the traveling barrel, as distinguished from the annular space, is forced down and in between the two valves, 31 and 36, and on the following stroke is carried out by the pump.   My distinction is this: That all of that fluid in that annular space is not pumped out at that time, is not pumped out at all.

Q   But part of it is pumped out and goes down through 31 and out through the top of the well, is that correct?

A   In part.

MR. HERRON: Did he say "through 31' or "over 31"?

MR. SOBIESKI: I said "through".

MR. HERRON: It does not go through that at all.

A   Over 31 it should be.   31 is the valve; through 41 and over 31.   The amount displaced is simply the volume occupied by the wall of the barrel, not that occupied by the puller nut or that contained in the annular space.   This pump has no clutch.

(Testimony of Edward T. Adams)

Further

DIRECT EXAMINATION

If the part which I have marked "L" in pencil on the drawing were removed the pump would still be an operative pump, in every sense.

MR. HERRON: The next patent, that of Savidge, No. 1,452,004, simply illustrates a different type of clutch.

Q I hand you this patent to Savidge, No. 1,452,004, Exhibit Z, I ask if it is not the fact that that patent shows a type of clutch a little different in design?

THE WITNESS: The clutch in Figure 1 is shown in the parts marked 11, 12, 22, 10, B, 10 and 20. They are of an interrupted thread type. When they are slipped into position they would lock the pump in non-rotative position.

Further

CROSS EXAMINATION

This pump is not a traveling barrel type of pump. It is a traveling plunger type. There is no provision here for the escape of trapped oil. There is no oil trapped by the operation of that type of pump.

Further

DIRECT EXAMINATION

BY MR. HERRON: I now call the witness' attention to the patent to Tierce, No. 1,720,979, Exhibit AA, which we offered because it was cited in the history and also because it shows a little different type of clutch.

THE WITNESS: This is a traveling plunger type, stationary valve, and the clutch consists of a stationary

(Testimony of Edward T. Adams)
part, shown very clearly in Figure 4 which is threaded
into the top of the standing valve cage and at the bottom
of the traveling plunger is a slot through which these two
pins 15 and 16 will enter, and then when the plunger is
turned this drops into a locked position, holding it in non-
rotative locking.

Further

CROSS EXAMINATION

It is not a traveling barrel type of pump, a traveling
plunger type; no trapped liquid.

Further

DIRECT EXAMINATION

BY MR. HERRON: The next patent is that of
Franchi, Exhibit BB. I think we can sufficiently dispose of it by saying that it illustrates a type of piston
packing. It has no significance other than it was part
of the history.

Further

CROSS EXAMINATION

MR. SOBIESKI: Have you looked at this Franchi
pump, Mr. Adams?

A I will. The patent evidently refers partly to a
grooving of an engine piston. It could be for a pump or
a compressor or an engine.

Q Is there anything there which indicates a traveling
barrel pump?

MR. HERRON: We stipulate there is not.

(Testimony of Edward T. Adams)

Further

DIRECT EXAMINATION

This patent to Stephens No. 1,407,493, Exhibit CC illustrates another type of clutch. This is a traveling plunger type pump, showing a clutch at this lower end, made up of the parts at the lower end of the plunger and the bottom end of the barrel in which the pins 11 on the stem 10, which forms a part of the standing valve cage, engages with the triangular shaped slots 12 at the bottom of the traveling plunger, forming a clutch.

Further

CROSS EXAMINATION

This patent is described as a "pump and method of packing the same". He is lubricating the same, I think. This is not a traveling barrel type of pump. There is no trapped fluid in this pump. Consequently, no provision is made for trapped fluid.

Further

DIRECT EXAMINATION

MR. HERRON: The next patent is that to Barthel, No. 1,139,396, Exhibit DD.

THE WITNESS: This, simply illustrates a different type of piston packing; and it is submitted simply because it was cited in the history. This clearly is a method of grooving a piston.

Further

CROSS EXAMINATION

This is a piston and not a pump patent. Of course, what we call a plunger in a pump is inferentially a piston

(Testimony of Edward T. Adams)

in an engine. The functions are the same. There is no provision for the escape of trapped fluid.

FURTHER
DIRECT EXAMINATION

I have made a careful study of the so-called jet action described in the Bray patent, and likewise of the action produced by such emission of fluid around the puller *not* as exists in the Hofco devices. I have caused this sketch to be made, illustrating broadly the different types of the pumps made by Mr. Hoeferer. I am familiar with his various structures.

This sketch is offered and received in evidence as defendant's Exhibit II, and appears at Volume II, page 177 hereof.

MR. SOBIESKI: Does this sketch embody all of the pumps which the Hofco Company makes of this type?

MR. HERRON: I will be glad to explain what the sketch means.

MR.SOBIESKI: Fine.

MR. HERRON: As I understand it,—and you may take this as a stipulation if you care to—Mr. Hoeferer makes one general type of pump, which is illustrated in Figure 1. It is made in various dimensions, some 70 to 80 per cent being of the 2-½ inch size. That general type of pump differs; first, in regard to the different types of packing which are placed in it, and which are illustrated by Figure 2, Figure 3, Figure 4 and Figure 5.

I take it that we are not here interested in the question of types of piston packings and, in consequence, I will say no more about them.

Another variation in the type of pump which Mr. Hoeferer has made, lies in the type of device used for the

(Testimony of Edward T. Adams)

purpose of screwing up rods. One such device is the jaw type of clutch, as illustrated in Figure 1 by the parts E and the cooperating part I, which are likewise shown in Figure 2 lettered the same.

We also employ in the different sizes of pump, with one or the other of the different types of packing, another type of clutch, which is illustrated in Figure 4 at Point Y. This consists of a flat surface on P, a flat surface on the top of H, and a notched ring. I offer in evidence Part P of Figure 4.

This part is received in evidence as Defendant's Exhibit JJ.

MR. HERRON: At the bottom below part P of Figure 4 we have a part H which I offer in evidence.

This part is received in evidence as Defendant's Exhibit KK.

MR. HERRON: Resting over the standing column A is a ring illustrated in Figure 4 by part Y, an exemplar of which I now ask be marked in evidence.

This part is received in evidence as Defendant's Exhibit LL.

MR. HERRON: Figure 3 is just intended to indicate the character of those surfaces at P and H before the ring Y is interposed. We make no pumps as finished pumps as set out in Figure 3.

MR. HERRON: We have heretofore given the interior diameter of the nut, part D in Figure 1 and 2. The outside diameter of Part D, the puller nut, in a 2-½ pump is 2-¼ inches.

(Testimony of Edward T. Adams)

The A. P. I. inside diameter of what is called 3-inch tubing is 2.441, which is approximately two and seven-sixteenths. When I say "A.P.I." I mean the American Petroleum Institute, which has standardized these sizes of various oil well supplies.

Q   I am talking about 3-inch casing, not 2-½.

A   I so understood you.

Q   Are you wrong about that?

MR. BRAY: 3-inch, did you say?

MR. HERRON: 3-inch tubing, what they call 3-inch tubing and I want the interior diameter of the 3-inch.

A   2-½ is what we have been discussing and 2.441 is the internal diameter of 2-½ inch.

Q   In other words, you consider the tubing you are considering is 2-½?

MR. SOBIESKI: We will consider any one, but previously we considered your dimensions to apply to 2-½ inch tubing, that is the only reason.

A   That is 2-½ inch.

MR. HERRON: As a matter of fact, we have not gone into the dimensions of the tubing.

A   All we are interested in is the internal diameter.

MR. HERRON: All right.   Now, if counsel cares to accept what I said, it may be deemed that witnesses familiar with the facts have been called to the stand and have testified as I have just stipulated the facts to be.

MR. SOBIESKI: We will accept that stipulation in saving time.

This sketch is a sketch of the device shown in the Bray patent made from the drawing of the Bray patent and represents it accurately.

(Testimony of Edward T. Adams)

This drawing is offered and received as defendant's Exhibit MM. and appears at Volume II, page 178 hereof.

The Bray pump is a traveling barrel type with a stationary plunger and operates in the manner common to traveling barrel pumps, in which the pumping chamber is at the top between standing valve S and the traveling valve above it. With the upward motion of the pump this pumping chamber increases in volume and fluid is forced in. On the return stroke, as this diminishes, fluid passes out through the tubing, making an operative pump. We are interested somewhat in the grooved packing AA at the bottom.

AA and BB, at the top, is, in the specification, specified to be made of a special metal particularly adapted to operating with a very close fit so that as the plunger runs over the standing column there will be very little leakage, and this is fitted with grooves to make a resistance to flow and scrape away any silt that may accumulate.

Between these parts there is a soft packing CC with a spring action to hold it tight making an effective seal, and for that reason a fairly short seal is shown. The part that we are particularly interested in is a jaw clutch at the bottom, its lower stationary *mail* member 2 being part of the anchor P, numbered and its female part being a part of the guide 3 on the bottom of the traveling barrel B. The jaw 4 of the guide 3 joins with the part 2 of the anchor P when they are brought into conjunction to make a non-rotative engagement whereby the traveling barrel is firmly locked with the stationary plunger P, as against rotative either to the left or to the right. The part 3 is clearly a guide. By "a guide" I mean it is a part which actually serves a guiding purpose and runs in

(Testimony of Edward T. Adams)

running contact between the guide 3 and standing column 11 at all times both in the normal operation of the pump and when it would be subjected to strain.

Q Now, the patent speaks of jet action. Will you explain the action of this pump in relation to the emission of fluid which Mr. Bray has term "jet action"?

MR. SOBIESKI: Just where in the patent does it speak of "jet action"?

The specification, page 1, lines 21 reads: "For this reason one object of the invention is to maintain the sand in such a constant state of agitation that it cannot pack down on the lock shoe box and lock shoe pin.

"To that end means are provided to cause a positive jet action of liquid over the lock means and keep the sand in a state of suspension so that it will be carried off in the discharge, and therefore prevent 'sanding up'."

I also considered the following language of Claim 2, "said guide having a series of venting ducts which open directly toward the anchor coupler face so that during the strokes of the barrel, ejected liquid is impelled toward the anchor top face to aid in prevention of sand accumulation thereof." And in the latter part of claim 3: "said guide being apertured for the ejection of liquid from the barrel directly toward the effective face of the lower, standing coupler part to aid in preventing sand settlement thereon." Also, lines 75 to 80 on page 1, which is perhaps the clearest of them all, reading: "Further, the bushing is provided with a series of ducts 5 from top to bottom so that as the barrel slides up from bottom position, liquid will be forced down in strong jets and wash sand from the top of the pin P so that this can be readily pulled from its seat in the lock box L."

(Testimony of Edward T. Adams)

Will you explain generally the movement of liquids in a well in their relationship to the washing action at or about the bottom of the traveling barrel.

This is a sketch which I have made, illustrating the lower part of the traveling barrel. It is purely a diagrammatic sketch but it is to correct a vertical scale and in horizontal scale is double size. It shows the lower end of the traveling barrel and the lower end of the tubing in diagrammatic form, dimensioned correctly, but dimensioned to double the scale of the Hofco Pump.

This sketch is received in evidence as Defendant's Ex. NN. and appears at Volume II, page 179 hereof.

It does not show the clutch. It simply shows the parts that we are interested in in showing the flow of liquid. This is a very brief general discussion of the fluid action at the bottom of the pump and as non-technical as it can possibly be made. In the first place when the pump is placed in the well and the barrel is in its lower position there is in good practice about 6 inches between the bottom of the barrel and the top of the face of the anchor. The top face of the clutch is offset above the bottom face by about 6 inches or thereabouts, so that between these faces there is a pocket into which sand or silt could settle. The object of the invention was to keep this sand stirred up and to wash the sand by the action of the traveling barrel away from the top face of the coupler clutch shown in the Bray patent thereof as Part 2.

Now, I think there is quite a lot of misapprehension as to what takes place. The sand in that space is stirred up and it is stirred up by several forces. The one to which Mr. Bray refers is often the minor one and a large part

(Testimony of Edward T. Adams)

of the time is entirely non-existent. But, first, you need a picture of it. It is very essential that we consider that this pump, as a whole, that is the traveling barrel, the plunger and all its parts are set at the bottom of a string of tubing. This tubing from top to the bottom is completely filled with the fluid being pumped and every part of the structure to which the fluid has access is filled with fluid. So from the top of the well the fluid comes down around the traveling barrel and between the traveling barrel and the tubing, completely filling that space, and comes down into the bottom and fills this pocket. I speak of that space below the traveling barrel and above the lock nut marked DD in Exhibit MM.

The fluid will also flow up through the orifices 5 in the guide and fill what we have continuously spoken of as the annular space, by which I mean the space above the puller nut below the packing on the standing tube and enclosed between the standing tube itself and the wall of the traveling barrel, which we will now mark EE, and the dimensions of which will vary with the movement of the pump. And this is also marked EE in the free hand sketch NN.

That colored pencil sketch which was introduced in evidence is made with an ordinary ruler and pencil, not with accurate drawing instruments, but it is very, very closely to double scale, that is, to twice the size of the parts. I mention that because I want you to see as you look at it that the clearances between the different parts are accurately shown. The interior diameter of the tubing is the API size 2.441 inches. The outside diameter of the traveling barrel marked 2 in this sketch is 2-$\frac{1}{4}$ inches. The inside diameter of the puller nut marked 4 is 1-$\frac{3}{4}$

(Testimony of Edward T. Adams)

inches. The outside diameter of the standing tube is 1 ½ inches. Now, it is clearly understood that in the drawing those dimensions are double those which I have named, simply for clearness, so that you can see them from here. At the beginning of the recess we were showing that the pump was at the bottom of a column of liquid in the tubing and completely immersed in fluid.

Q Does that column of fluid tend to exert any pressure; if so what and in what direction?

A In California conditions the net head,—the difference between the rock pressure and the total pressure. would run between 4,000 and upward feet of head, and this pump being at the bottom of that column would be under that pressure at all times, wherever that pressure had access to it. When the traveling barrel moves up, we think of liquid being ejected from the annular space above the puller nut through whatever openings there may be in the puller nut, as 5 in the Bray patent, and a jet being formed, flowing down for the purpose—of agitating sand. It is not like a jet. It is a different jet from anything that we are ordinarily accustomed to. A jet, as we think of it, would be a jet from a hose or a jet from a kitchen faucet, in which the orifice is stationary and the fluid flows from it under pressure, the fluid moving and the orifice being fixed. In the case here we have an entirely different situation. We have the entire orifice immersed in the fluid and such jet action as occurs, occurs by moving the jet through a fluid otherwise stationary. Therefore, while you can think of a downward flow from the jet 5 you must bear in mind that, accompanying that there is an upward movement which causes that flow through the guide 3. It is not a case of the guide standing still

(Testimony of Edward T. Adams)

and a jet gushing out, but a case of the ducts being carried through the fluid so that the motion of the guide through the fluid is the thing that makes a flow through the guide relative to the guide itself. The net motion of the fluid, therefore, altogether depends on the size of the opening and on the speed with which this guide is moved.

That is not the only force acting here. In order to get a picture of it I would now refer to Ex. NN. It seems to me that the clearest representation of what occurs is to think of the traveling barrel being immersed completely in the fluid and moving upward, and you have a case exactly the same as a ship moving through a canal. Say it has moved up a distance of just one inch. There is here a space now occupied by the puller nut FF, Ex. NN, considering it as a piece of steel immersed in fluid and when the puller nut is moved up one inch that space has been vacated. The puller nut no longer occupies it, and fluid must flow into that space to fill it up. In a tube that is completely full from top to bottom, fluid to move has got to have some space into which it can move. There has got to be an opening and when the plunger moves up and this puller nut FF has moved one inch there is such a space. Now, then, where does fluid come to fill that space? If we look at the Bray drawing MM, we see that the annular space is full of fluid and if we follow down through the grooves in the puller nut to the top of the anchor we see that that is also full of fluid and we could not take the fluid entirely out of the annular space because, if you took the fluid from there bodily through these groves we would leave space up there. So then, any displacement of the fluid that fills the vacated space must flow essentially from the unlimited supply,

(Testimony of Edward T. Adams)

with which the entire tubing is filled. The fluid which fills that space flows down through to occupy it as shown in the sketch. As the puller moves up an inch the fluid may not necessarily come from the annular space to fill that space vacated by the puller nut, because if I took it from the annular space I would have to create a vacuum in there in order to do it. Now, fluid or anything that is displaced by "displaced" meaning if it moves and leaves a space in the fluid which it formerly occupied—the fluid to fill that must come from around the tubing and flow down in there, and that is exactly what happens. It flows down as I show you between the tubing and the traveling barrel to fill the space formerly occupied by the puller nut; and that is why, in discussing some of the prior art, I made the statement that it did not make any serious difference whether the opening was through the puller nut or whether it was radial, because as soon as the fluid from the inside was displaced outward it came into this downward flowing current which flows in to fill the space vacated by the upward moving puller nut.

The only point I wanted to bring out at the present instant is that as the barrel moved up it left the space which it formerly occupied and that that space was filled by a flow outside of the tubing, between the tubing and the barrel, because that is the only place it could come from. Let us assume that the puller nut is removed and we have simply the bottom of the tubing without any inward projection such as is made by the puller nut; then as the tubing moves up and down—

MR. SOBIESKI: We will object to that, your Honor. There is no device here which does not have a bottom nut on it.

(Testimony of Edward T. Adams)

THE COURT: It may be for demonstration purposes.

A That is for demonstration, Mr. Sobieski, divided between the two parts it is simpler, that is all. If I removed the puller nut and if I advanced the tubing up and down, then move the tubing up and down I have—

MR. HERRON: Move the barrel of the pump up.

A Move the barrel of the pump up and down and I would have exactly the condition that you would have if you took a lead pencil and moved it up and down in a glass of water, for example. And then that makes it clearer than ever, because then it is absolutely evidence that when I do that the fluid stands perfectly stationary except for the amount that is displaced by the volume of the pencil itself. You see, as I move it up and down, as it moves down, sinking a lead pencil or the tubing, either one, the fluid moves sideways to let that down but is not displaced from its relative position except by the volume of the pencil that comes in there. And when the pencil or the tubing is moved up fluid flows in behind it, just as the reference I made, like the flow behind the sterm of a ship. Now, then, let us put the puller nut in place and we come then to a construction which is exactly such as we have in the Bray patent or in the other patents of the prior art. Then we have now added to the tubing an inward projection of the puller nut. This fluid above that projection, and there is fluid below it and there is fluid between that and the tubing, completely encased. Let us see, now that I move that up one inch from the position it now occupies what have I done? The amount of fluid above it is diminshed by the volume of that as moved upward in it and the volume below it is increased by that amount. In other words, the fluid has simply

(Testimony of Edward T. Adams)

flowed around the nut. Suppose I move it very slowly, the fluid in general would be stationary. The bottom of that column of fluid would be at the bottom, here at the face of the anchor and the top of it would be up here but the finger would not have moved. I simply move the puller nut through it gently. It flows from one side of it to the other side of it. That is all that it does. In general that fluid is stationary but if this opening is restricted then just relative to the puller nut it has a higher velocity as it is passing the nut, which immediately drops off into low velocity as it comes on the other side. It simply transferred liquid from one side to the other side.

MR. SOBIESKI: During that transference the liquid is not in stationary form and character, in my estimation.

A The liquid, as a whole, is stationary, but because of the restricted opening there is a velocity through there relative to the tube and that will persist, depending on the size and speed. I do not say there is not any jet. I admit that there is a jet and I will give you the volume of it in figures.

Q Then this liquid that is here on the upstroke, it does move down, is that correct?

A No; I could not say that.

Q It does not stay in there, does it?

A Yes; the tubing moves up. The liquid stays there and the tubing moves up into it. Take this sketch, as the puller nut is moved upward two inches higher, then part of the liquid is above. It passed through that liquid.

Q I confess, after it has quit moving it is stationary, but while this is in motion, isn't there a lot of liquid com-

(Testimony of Edward T. Adams)

ing out of here and going down around the end of this jet—around the end of this puller nut, the edge of it?

A   That happens but you said it just backwards.

Q   It does happen the way I say it?

A   No, this way: What really happens is we have moved the puller nut through the liquid. The thing I want you to get clear is that this jet which occurs is a jet formed by moving an orifice through the fluid, not an orifice that stands stationary and shoots the fluid out of it; and the net velocity it would have, would be the velocity by the nut minus the velocity with which the nut is moving away.

Q   Yes; but that would be more than the velocity of the nut, wouldn't it, in this case?

A   I wound't know the exact figures on that, not necessarily.

Q   As I understand the proposition—I think that all of us understand, the parties at least—before the nut moves up this liquid—there is considerable liquid, shall we say above the puller nut?

A   Yes; the annular space is full of liquid.

Q   And after the nut and the barrel have moved up the amount of liquid which was in that annular space above the puller nut is reduced, and the amount it is reduced depends upon how far up the nut has traveled, is that correct?

A   Again you and I state it just backwards.

THE COURT: You start with equal pressure of the liquid inside and outside; all space is filled and pressure is equal before you start to move the plunger, is that correct?

A   Yes, sir.

MR. SOBIESKI: Yes, sir.

(Testimony of Edward T. Adams)

A Yes, sir. The question counsel has asked me is in error, in this respect: That he has considered the tubing as moving and the puller nut as moved and the annular space, which is then diminished by that motion. He is thinking of two different volumes. I think the only clear way we can think of it is to forget the annular space just for the moment, and just as his Honor has stated it, here is liquid going from the anchor face at the bottom clear up through around the puller nut and clear up to the top of the annular space, which ends at the bottom end of the packing in the barrel. That is entirely full of fluid and is at equal pressure above and below. Then as the barrel is moved upward, carrying the puller nut with it, it is pulled up through this stationary column of fluid and the bottom and the top of the column of fluid remains exactly where they were before and they are not disturbed. Now, then, is there a velocity there? Yes. What is the velocity? The velocity is the velocity of the flow as you pull the puller nut through the tube of liquid, of the flow past the restricted area of the puller nut. In that sense there is a jet. It is a jet that is formed by an orifice that is moved through a fluid.

MR. SOBIESKI: Well, I think that is correct. I think we have the matter.

I said that this jet was only one of the forces. That flow through that orifice has a stirring motion. Another force that has a stirring motion, and that is equal, and many times much greater, is the downward flow between the traveling barrel and the stationary tubing due to the displacement of the lower end of the traveling barrel. It is entirely surrounded with fluid, as it moves up and down there is a displacement which is exactly the same as the

(Testimony of Edward T. Adams)

displacement when the pencil was moved up and down in the glass of water. That displacement must be filled up by the flow between the barrel and the tubing. This flow is vertically downward toward the anchor face and between the barrel and the tubing, and is of very great moment and of great importance, for the reason, that it is a constant on each stroke of the pump. It does not make any difference what the conditions are. If the barrel makes an upward stroke there is that flow and it is a fixed amount, equal to the volume of the tubing which is displaced. As that steel volume moves upward that fluid flows down around and fills that up again.

MR. HERRON: You mean the volume of the metal barrel?

A   I mean the well tubing, yes, the barrel—no; it is the volume between the standing tube and the barrel pocket. We have then two forces, one the jet action of Mr. Bray; two, the action which I have just pointed out which is a constant, that is the flow between the barrel and the tubing to take care of the displacement; and third, the purely physical action of the churning action of the barrel itself. It is danced up and down, and without any explanation it is perfectly obvious that that creates a churning action in there. All three forces stir up sand.

MR. SOBIESKI: One other factor I want to be considered along with these others which we are considering; that is, if this opening here in the puller nut did not exist and there were a perfectly tight fit between the puller nut and the traveling barrel would that not make it impossible for the pump to work, or, if it did work it would work quite laboriously,

(Testimony of Edward T. Adams)

A   In that you raise a purely theoretical question, and like any question of that sort, it rests entirely on the basis of what the physical facts are.   Now, there can be two sets of facts; one fact is, as you say, that this is a perfect fit so that there is no leakage past the puller nut between the puller nut and the tubing, and we will assume also that the packing on the tubing is a perfect fit and that there is no leak past it.   If the pump in that condition is introduced into the well the annular space will be filled with air. If there is no leakage that air will be compressed up and down and you will have a very, very lovely working pump from which we would have no sand or other troubles and it would work perfectly.   The air in the annular space would compress and expand with the motion of the pump and be very satisfactory.   Now, we know that we can't do that, because of the question of leakage.   We will say that there is a little leakage.   You can move that pump then at a rate, depending on the amount of leakage or venting.   If there is only a small amount of leakage you can move that pump slowly, and if there is a large amount of leakage—in other words, if the space is very great, you can move it rapidly, and there you are between those two conditions.   Legally, suppose that there is enough leakage so that the pump can be worked at some speed and some head, it would be, for legal purposes, a pump; but as a practical man, looking for production, you would not want to hamper yourself; you would prefer to have the opening large enough so that the plunger would make its full stroke.   Under those conditions a pump would work at a slow speed, but commercially I would not advise, and any engineer would not advise a pump which worked against itself in that fashion.

(Testimony of Edward T. Adams)

MR. SOBIESKI: So that it is necessary, for the proper operation of the pump, that in some manner the oil which is in this annular space that you have described get from above the bottom nut to below the bottom nut?

A   Yes.

MR. SOBIESKI: Assuming that some oil got in here or that the oil has got in there, it is necessary for any operation of the pump that the oil in this annular space, as the traveling barrel goes up, that that oil get out of the annular space somewhere?

A   That is true.   There are two theories under which a pump can be designed, and Mr. Bray in his design very skillfully worked on one theory and Mr. Hoferer in his design has worked on an entirely different theory; and we had in the art Ellis, who worked on a third theory.   But between Hofco and Bray there is the broad distinction.   I use this simply to illustrate all pumps.   It is necessary when the pump moves up for the oil in this annular space to go somewhere?

Now, let us consider not a pump just as it leaves the shop, but a pump which has been in use for some length of time.   Let us assume an average condition, that a pump has served perhaps a third of its useful life.   Then there will be a leakage past the packing because the pressure outside of the pump, between the tubing and the barrel and underneath the packing and up through into the annular space is the full pressure of the entire fluid column, while above the packing in the pumping chamber itself on the upward or pumping stroke, there is, naturally, the rock pressure.   Say the pump is set—a fair statement would be at 4500 feet and the fluid in the well extends 500 feet above the pump, thus leaving a net dif-

(Testimony of Edward T. Adams)

ference in pressure between the pumping chamber in the pump and the fluid column in the annular space of 4,000 feet of head of oil. That pressure is trying to make a leakage past the seal into the pumping chamber. That leakage, you will note, necessarily comes from the annular space. Therefore, if there is any leakage then the downward flow from the annular space would be reduced by that amount; that is, the annular space is full of liquid, and as the barrel moves up a part of it leaks upward into the pumping chamber and part of it flows downward through the puller nut. Now, that is a material amount at low speeds as the figures will show us. Let us assume one stroke per minute. We start this barrel on its upward stroke,—and we know that this is true—that for half of the time, no matter whether the pump travels fast or slow, for one-half of one minute there is leakage upwards past the seal and for the other half of the minute there is no leakage whatever; the pressures above and below and inside this annular space are equalized and are equal to the full pressure going to the surface of the well. Then if there is any material leakage, if the pump is worn, we will say, to a third of its life, in the one stroke it not only "can" but does happen that the leakage from the annular space, as the pump moves upward, is at a rate greater than the decrease in volume of the space, and so during that entire time the flow is upward through the orifices or openings in the puller nut into the annular space, and is not downward at any time, provided that rate of leakage past the plunger is greater than the reduction of area in the annular space. I want to modify that in a moment, your Honor, by showing with the model that this rate of motion is not uniform; that some-

(Testimony of Edward T. Adams)

times the barrel travels faster than it does at others, while the rate of leakage is constant; so I will speak now only of the average rate of leakage. I am going to assume 20 strokes per minute, 4,000 feet of head, 30 gravity, fluid of about 50 viscosity, and 85 per cent volumetric efficiency. By "volumetric efficiency" I mean that the amount discharged by the pump is 85 per cent of the volume in the pumping chamber. If the entire amount trapped in the pump chamber was discharged the volumetric efficiency would be 100 per cent, if a part of it leaks, you have a lesser efficiency. I have assumed 85 per cent as an average condition for a pump of this size which has used about a third of its useful life. I think that is a very fair assumption for a pump of that condition, and I back it up in my own mind by figures that I obtained from experimenting on leakage on pumps of this general type, made by some of the oil companies who are entirely disinterested and had no interest in this suit or any knowledge of it, the experiments being made a year or more ago.

With the average rate of leakage at 85 per cent volumetric efficiency, it will be found that up to somewhere about 6 strokes per minute, the average flow would always be upward through the orifices and in the Hofco design, upward through the opening in the puller nut, because the average rate of leakage from the annular space is as great as, or greater than the rate of displacement of that pump. So we can say broadly that up to a speed of about 5 or 6 strokes per minute, as pumps are operated on the present low production rate, you would have no jetting action from the orifices in the puller nut at all. You would have just the same stirring action from the flow

(Testimony of Edward T. Adams)

between the barrel and the plunger, and you would have just the same stirring action due to the churning action of the barrel itself. So a pump may be operated at a low speed and still keep this fluid and this sand stirred up.

As you increase the speed above 6 strokes per minute there begins to be a flow past the puller nut, and if you are moving very slowly the puller nut simply moves through the fluid, and it is displaced from one side to the other, and there is motion relative to the puller nut but not relative to the earth. I guess I had better keep out of those motions. We will strike those out. I don't want to be technical in it, but that is what really happens just the same. At any rate, up to 5 or 6 strokes per minute there is no jet action, not average jet action.

As we go beyond that, we have to consider another thing. This barrel does not move at a uniform rate. This is a model of pump and crank motion.

This model received in evidence as Defendants' Exhibit OO.

The foot of the model is marked A to represent the face of the anchor, marked C to represent the well tubing, marked B on the little inner strip to represent the standing tube, and marked D to represent the barrel or plunger. The connecting rod is marked E, and the crank is marked F.

Pumps of the type in suit are operated by a crank motion at the surface of the ground. This crank moves at a uniform rate, it has a uniform circular motion, and each degree of motion is made in an equal length of time, that is, through each unit of time it flows through a fixed distance, but due to the linkage of the crank and connect-

(Testimony of Edward T. Adams)

ing rod the pump, to which this is attached through the sucker rods, has a variable motion. As the crank is at or near its centers the motion of the pump is very slow. If you will move the crank through 20 degrees on one side of dead center or 20 degrees on the other side, you can see that the crank has moved through 40 degrees of the 360, but it has only moved a small distance at D. By dead center I mean at the end of motion. Now, if you move the crank to its central position, when the crank is moved up to this central position then the pump is moved more rapidly but has now moved a long ways away from the face A down here. I bring that up to show that the motion at the ends of the stroke is slow and that the motion at the center of the stroke is rapid.

We are dealing here with something on the order of a sucker rod going down in the well, and that is elastic. There has been a tremendous amount of work done on that, very interesting to the engineer but not necessary for this action at all. So I simply say that experimental figures have shown that the travel of the pump itself is very materially less than the stroke at the top. Take a little child's toys. Tie a rubber band around the finger with a little ball on the end and dance that up and down; you have exactly a representation of this. As I move my hand, due to the elastic, that ball falls in the same sort of coincident motion. I do not think anybody knows exactly what—and I am cutting that all out. I will assume that the stroke of the pump is 3 feet because if I assume that we have fixed things to do with. But at 20 strokes a minute on the data we have, the plunger or barrel will on say a 54 inch stroke of the polish rod move 36 inches.

(Testimony of Edward T. Adams)

The point I want to bring forth is this: As shown in this model, the pump stroke at the end of the stroke is slow and at the middle of it more rapid; and that due to the stretch of the rods, and the inertia and a lot of things that we know about but do not talk about, the stroke of the plunger in some degree follows the motion of the polish rod and one travels 54 inches plus or minus, and the other, I will assume, travels 36 inches, in order to get definite figures.

MR. HERRON: May I ask if that model has been made substantially to scale, that is, you have proper proportions between these various moving parts?

A I think that is perhaps scaled. The length of the crank and connecting rod, yes, essential parts are substantially correct. And the motion illustrated, therefore, would be substantially as the motion occurs, with such modifications due to such factors as I have just mentioned.

A Out of the model, we have developed that up to a modest number of strokes, say, 5 or 6 strokes per minute, the flow is upward through the puller nut instead of downward at all times, taking an average. At the end of the stroke the barrel moves very slowly. When the barrel starts upward the discharge valve is closed and the whole pressure or volume of the liquid then falls on the packing. Now, as long as that condition maintains and as the crank moves up. nearly to midstroke this pump is moved slowly and is under pressure, and there will be leakage out of the annular space past the seal and into the pump chamber because the leakage is a fixed thing, so much per unit of time. During the beginning of the stroke, therefore, and the end of the stroke it is true that the flow will be inward

(Testimony of Edward T. Adams)

into the annular space rather than outward out of it. Presently we come to a place where there begins to be a flow outward relative to the puller nut. That does not necessarily mean a velocity towards the face of the anchor, because what really is happening is that the puller nut is being moved away. If the puller nut is moving up at the rate of one inch a second and the flow through the puller nut is downward at one inch a second, the net result is that the fluid itself has not changed its distance from the face of the anchor. So we only get a downward flow from the jet, when we have moved up far enough in our stroke, and are moving upward fast enough, so that the flow past the puller nut or thru the restricted area of the orifices, is greater than the upward motion of the nut. The orifice is moving and you can only have a velocity relative to the anchor face, which is what we are considering, when the flow through the puller nut is at a greater rate than the upward motion of the nut itself.

We thus come to the two methods of design which I spoke of a moment ago. If you want to create a velocity through this puller nut relative to the puller nut you would make this orifice 5 as small as Mr. Bray has made it in his design, with the object of getting a jet action. When we come to see the figures I think we will find that this is not very likely to be called a strong jet, but whatever jet action you get is paid for by building up pressure in the annular space and retarding the drop of the barrel, so that the engineer of the Bray design very skillfully made a guide, which was tight on a standing tube. He had orifices, as shown by that puller nut shown in evidence with 8-3/16 inch holes. That would build up quite a material pressure in the annular space, and restrict the down-

(Testimony of Edward T. Adams)

ward stroke of the barrel. As you build up pressure there to gain a velocity through these orifices, you by that much restrict the capacity of your pump, because the plunger drops under gravity and the weight of the rods above it. Anything that retards that motion shortens the stroke and that in turn lessens the amount of fluid displaced and the quantity pumped.

We have therefore the Bray type of design in which this clearance is restricted; and another type of design, that of Hofco, which is shown to scale here, in which the opening through the puller nut is made the maximum amount consistent with having an upper surface left with which the parts may be pulled from the well. The difference in area is very great but the theory back of it is that if this is made a perfectly large opening, perfectly free, and if you disregard possible flow that you get, any possible jet action which you get from the interior of the puller nut, you would make that orifice as large as you could and with a low rate of flow through it.

I have made some figures on the basis of the assumptions that we have made and they are quite illuminating. I measured up the dimensions of the used Admore pump which was put in evidence the first day of the trial, I measured up the Hofco pump, the dimensions of which I have just given you, and I took the measurements of the Bray pump as given me by Mr. Bray, which I will now check to make sure I have them right. There are only three of them; the OD of the standing tube, 1-⅜ inches; the ID of the traveling barrel, 1.937; that is 1-15/16; the puller nut ID 1.516 or 1-9/16.

I think I have sufficiently explained the different forces acting to stir up that sediment at the bottom of the tub-

(Testimony of Edward T. Adams)

ing. We come to another thing of importance, which is the amount of liquid involved in forming this so-called jet action and the time during which it acts. In explaining that jet action I tried to keep entirely away from the technical side of the relative motion, which is apt to be confusing in oral testimony, but it comes in to this extent: By relative motion, I mean that if a man is walking towards the rear of a moving train, he would be walking backwards with respect to the train, but relative to the earth the train would actually be carrying him forward. So we have something of the same effect in this jet which is formed by an orifice moving backwards and a flow going past that. I do not want to carry the impression that there is absolutely no velocity there. There is a velocity relative to the puller nut when it is in motion, and if this velocity is of any great magnitude it will persist somewhat and be carried downward.

In order to get at the quantity of liquid involved we have to make some assumptions and I want to cut out entirely the mass of assumptions that would have to be made if we started with the motion of the crank, at the headworks or surface of the ground. I want to get rid of the effect of inertia on the mass of puller rods that transfers that pressure of the liquid from the pump on the upstroke back to the tubing on the backstroke, and the effect of the stretch of the rods. All of those things are of a theoretical and somewhat controversial nature, and I get rid of that by doing this: I am going to assume that this barrel, has a stroke of 36 inches and base my figures on that.

Now, I do not care what the motion is at the top of the ground, that is whether the polish rod moves through 36 inches or whether it moves through 64 inches. The polish

(Testimony of Edward T. Adams)

rod is the upper end of the line of sucker rods and has essentially the same vertical motion, as that imparted by the crank. It does not matter as far as this discussion is concerned what the stroke of the crank is. According to the best authorities, if we have 36 inches actual stroke of the plunger or barrel, that will correspond with a 54 inches plus or minus travel of the polish rod or stroke of the crank. I have computed the volume of the annular space above the puller nut. In the Admore, which was the old, worn pump introduced in evidence the first day, the gross displacement is 32.4 cubic inches; in Hofco the displacement of the annular space is 49.5 cubic inches; and in Bray, this displacement is 52.7 cubic inches. That is the gross displacement on the basis of no leakage whatever. I am taking average conditions all the way through and I have assumed that after a pump has served a third of its life, it would be in average condition. At that time we may reasonably expect a volumetric efficiency of 85 per cent, that is that the actual amount discharged is 85 per cent of the change of volume of the pumping chamber of the pump. Under those conditions we come back to those figures and we find that this leakage—

BY MR. SOBIESKI: Do you have any evidence that any of these Bray or Hofco pumps have a leakage of 85 per cent, or an efficiency of 85 per cent?

A I don't know of the Bray. There was a great deal of evidence of leakage that I have examined.

Q That is leakage in general, but it is not leakage with relation to this particular pump?

A Yes. I have evidence relating to this. I might say it will shorten the time if I make this statement right now, as you will see when I get through with my figures, that

(Testimony of Edward T. Adams)

it makes no real difference whether I ignore that leakage or not, because the velocities we are coming to are so small. If I ignored it, it would not make any difference. The figures all having been made the same way, I would prefer to read from my written record and I will make any correction to suit any condition that you want when I am through.

MR. SOBIESKI: Your Honor, the testimony assumes certain ideal conditions which are not shown to be related to the way this actual pump in question works. Therefore, I object to it as incompetent, irrelevant and immaterial, going too far afield into theoretical discussions of matters not directly related to the clutch.

THE COURT: Of course, the witness can proceed with the assumptions that he is making and give his deductions, and then modify them to suit the other conditions if they exist.

I don't know, your Honor, the efficiencies of the Bray type of pump, which has a soft packing at the center, which very likely would change the figures, as far as that is concerned. As you will see when I come to the end, if I allow the 15 per cent difference, which I can do readily, it will not make any material change in my deductions, but I have experimental figures made, as I said this morning, by one of the independent oil companies on the type of packing the same as that used by Hofco, that is, on a type of metal packing. But from the result of those figures, my assumption of 85 per cent volumetric efficiency for a pump that is less than half worn out is reasonable.

MR. SOBIESKI: Your Honor, I object to experiments made by somebody who is not here and the company

(Testimony of Edward T. Adams)

not even named who made these experiments. I think it is getting far afield from this simple clutch device on the bottom of the traveling barrel in allowing the oil to escape downward through the face of the bottom nut. It seems to me that those investigations and computations made by somebody unknown here are incompetent and immaterial in this case.

THE COURT: Perhaps you can tell us who made the tests and how you got the basic figures.

I took a carefully conducted test made for their own purpose by the General Petroleum Company some few years ago on leakage, carefully conducted. I looked over their apparatus and their methods of conducting the test and am satisfied of the accuracy of the work, which I have gone through with carefully. It makes very little difference, since the resultant flow is of the minor character, whether I make this deduction or do not make it. I will make the correction when I get to the end. I prefer to give you the figures as I have made them, because they are all in decimals and worked out carefully and it would take time to change them. I will correct them on any basis. I will even assume the pump does not leak a drop, which I do not believe is reasonable, and will correct them on that basis.

MR. SOBIESKI: We are objecting to the figures in their entirety on these computations he is going into because they are not shown to be directly related to the pump here in question, especially the Bray pump.

THE COURT: I think that the witness is proceeding to determine the general effect of general conditions. I think I will allow him to go ahead.

(Testimony of Edward T. Adams)

The gross displacement of the Bray pump, as figured from the dimensions given me by Mr. Bray's counsel, are 52.7 cubic inches. Now, at 85 per cent volumetric efficiency the cubic inches that leak past the seal or that leak out of the annular space upward, and therefore are not discharged through the orifice in the puller nut downward are appreciable amounts. In the Admore pump this amount is 9-½ cubic inches slip. In the Hofco pump, 17 cubic inches, and in the Bray pump 15.9 cubic inches. I have subtracted the slip, that is the amount that flows upward, from the gross displacement of the annular space we have left the amount that flows down past the puller nut producing the velocities under discussion. The net quantity that will flow down in the Admore pump is 22.9 cubic inches, in the Hofco pump, 32.5 cubic inches; and in the Bray pump, 36.8 cubic inches. Those are the net volumes. An ordinary table drinking glass would hold something on the order of 12 to 15 cubic inches; so that in the case of Admore pump the quantity flowing downward which is relied on to form a jet, if there is a jet, would be somewhat over two glasses of fluid. In the Hofco pump, the 49.5 cubic inches would be about three full glasses; and in the Bray pump, some little in excess of that.

Now, this may be the proper time to say that at the beginning of the upward stroke, as we have shown, the flow is upward through this puller nut, so that after the puller nut has moved some little distance there begins to be a downward flow, which continues through the point of rapid motion up toward the other end of the stroke when the motion slows down, at which time there is again no outward or downward flow but rather an inward flow to make up for the leakage.

(Testimony of Edward T. Adams)

So we come to this, that the actual discharge down through the puller nut, forming what is termed a jet, occurs somewhere in the middle of the stroke and not at the ends at all; and occurs after a considerable part of the liquid has leaked up in through the pump, and another and a great part has escaped at very, very low velocities at the time that the pump was just beginning to obtain its motion and after it has reached the top and is slowing down and coming to the end of its motion. If we say then that the quantity of fluid in, for instance, the Hofco pump, which is 49 cubic inches or something over three glasses of fluid —if we say that one or even one and a half glasses of fluid is the amount passing through the orifice at the time that it is moving to form any appreciable flow, I think that we have stated the case very fairly for all concerned, which is what I am trying to do. Further, that flow is as an intermittent flow.

We will assume here what I understood Mr. Bray agreed was an average conditions, namely, 20 strokes per minute and, therefore, this action occurs once in 3 seconds; and just to get a picture of what is occurring we mean that we are stirring up the sand down here with such velocity as occurs with the flow of a glass or a glass and a half of fluid once in 3 seconds. Just to get a mental picture of what is really happening, because I think we all do exactly what I do—I think of a jet as a flow from a hose, or from a faucet, that is as something that has a fixed and really an average velocity, whereas, this is a very, very different case. So that fixes the quantities and the times. I have computed from these figures average velocity, that is, the velocity of flow of this jet, if the flow was at a uniform rate from start to finish. That is not

(Testimony of Edward T. Adams)

true, your Honor. It varies. At the beginning of the stroke it is low; at the middle it is high; and at the top it is again low. But nobody can state accurately what the relation between those velocities are. We can figure the average with perfect accuracy because that is determined by quantity and time, and we know exactly what it is under the conditions we have assumed, and we know the limits. We know that the motion of this pump is a kind of a sine curve and we know from that that the average velocity would be divided by two, or approximately one and a half. I think 1.57 times the average would be the maximum velocity. So after the midstroke the maximum velocity would be one and one-half times that. Now, then, this pump only travels 36 inches; the plunger of the polish rod travels, say, 54 inches and just what relationship comes in we do not know, but I am safe in saying that the maximum could not be over two times the average. So I will place my figures somewhere between there on the average, and we can take double that or multiply it by one and a half or any other factor that we want to guess but it lays between those values.

Now, computing it in that way I have the following figures, which is the average velocity through the opening in the puller nut relative to the earth or relative to any fixed point. We will say relative to the tubing; that is, it is the actual velocity in the puller nut relative to the puller nut itself, less the movement of the puller nut in the opposite direction. The actual velocity in the orifice, disregarding the fact that the rod is moving in an opposite direction from the jet, is in the case of the Admore 3.24 feet per second; in the case of a Hofco 4.6 feet per sec-

(Testimony of Edward T. Adams)

ond; in the case of Bray as he actually builds it, 7.7 feet per second; and in the case of Bray of the patent, using as the opening in the guide eight 3/16 holes, which was the size of the nut shown here in court, 15 feet per second.

The actual velocity of the orifice in the direction opposite to that in which the jet is flowing is in all these cases 3.24 feet per second, which would need to be deducted from the above velocities of the jet, as they are figured relative to the orifice alone, (as tho it was not moving) and we then have the rate at which the flow from the orifice approaches the face of the clutch or the pocket at that point. That would mean that there is practically no change of velocity in the case of the Admore, that is, at the middle of the stroke the velocity of down flow would be very slight but the average would practically come to zero. It would mean that the average velocity of flow through the orifice in the Hofco pump is 1.36 feet per second, the velocity of Bray as built is 4.46 feet per second, and the velocity of Bray of the patent, that is, with the eight 3/16 orifices, is 11.76. Expressed in "miles per hour", in the Hofco pump, the average jet flows at the rate of a little less than one mile per hour, .93 or thereabouts miles per hour. I have reduced it to miles per hour for this reason: A very slow flowing river flows somewhere between two and three miles per hour, say, two miles per hour to be on the very conservative side; in other words, the average rate of flow past the puller nut in the Hofco pump, is that of a slow flowing stream. Very slow, sluggish flow, and tidal flows, in ordinary run much higher than that; and I reduced it to miles per hour, so we could visualize what occurs.

(Testimony of Edward T. Adams)

If I ignore the 15 per cent slip I have only raised my figures by 15 per cent, just a trifling amount. In other words, there is no jet. Let us compare it with an ordinary faucet in which you have, say 50 pounds pressure. There is 150 foot of head of water behind that faucet to make that flow, while to make the flow of the Hofco pump would take about 15/100 of one foot, that is about an inch and a half. The velocity of outflow is proportioned to the heads which make it, and you have therefore somewhere between an 800 and 1,000 to 1 difference. When we think of a jet that we have seen or think of the definition of "jet" as meaning a strong forced flow, you see that it simply is not present, that there is not anything comparable to it there.

In considering what really takes place in the bottom of that well, we must in thinking of the pump realize the magnitude of the flow around the outside. There are therefore three forces. Of these the least, in my estimation, is the velocity of flow through the orifice in the puller nut, for the reasons that the flow is intermittent, the quantity is small, and at low speeds is non-existent; it is non-existent at all speeds at both ends of the stroke. It only exists when the pump is at midstroke or thereabouts, at that time the orifice has been moved upwardly (it is originally set 6 inches above the top of the anchor) very materially farther away from the face of the coupler, which it is the stated purpose of the patent to keep free from sand. I do not dispute that the coupler may be kept free of sand. I simply say that of the three forces which keep it free from sand I feel that this one is perhaps the least important.

(Testimony of Edward T. Adams)

A second force which tends to keep this stirred up is the plain churning action of this metal of this tube. This 2-¼ inch piece of tubing, this 3 feet long barrel is danced up and down in that confined space making a physical displacement of the fluid, a churning action—that describes it.

The third force tending to disturb and stir up the sand and to keep the face of the coupler clear is the flow down the outside. The magnitude of that flow is equal to the displacement of the barrel, plus whatever leakage there is, that is, if any amount leaked up past the seal it must be added to the quantity flowing down the outside to make up for the displacement of the barrel.

These two forces, the churning action of the barrel itself and the strong flow down between the tubing and the pump barrel, are constant. They occur at each stroke; and if the stroke is 36 inches as I have assumed, that will be the constant to each stroke. If it is but one stroke per minute that displacement will be there, while the flow from the annular space at such speeds is totally non-existent.

I have repeatedly said that I felt that it did not make much, if any, difference, whether the flow was radial through the side of the barrel above the puller nut or whether it passed through the puller nut, for the reason that if a flow is to exist, and if it is to go to anywhere, it must exist by means of a difference in pressure. In other words, there must be a place for it to go and that place is made by the upward displacement as the barrel moves up, which leaves a space into which that fluid must flow because it is the only place that it can flow. It therefore flows outward and past the outside of the puller nut instead of past the inside; and I do not feel that it makes

(Testimony of Edward T. Adams)

any material difference as far as the stirring action is concerned which way it goes.

Q Now, Mr. Adams, you have referred to a flow of fluid through the puller nut and have given us the velocities which may be attained in the flow past that point, and have said that in considering the flow at the proper time of the upstroke to get the greatest velocity that the initial velocity will in some measure persist?

A Yes.

Q In your opinion, to what extent does that velocity persist, and what are the forces that tend either to cause it to persist or to cause it to become dissipated?

A The figures that I have given are based wholly on the theory that a flow persists, with the velocity given, in a vertical direction directly towards the face of the puller nut. Now, of course, that is not true. In the first place, there is formation of eddy currents. Any text book on hydraulics or almost any engineering book will give you the theoretical loss due to that and will give you, in tabular form, the experimental results showing that that theory is correct. That loss is proportional to the difference of the square roots of the area through the orifice and of the area at the bottom between the well tubing and the standing column, which is very, very great in this case. The factor found by experience used would average something to the order of 50 per cent. In other words, the loss by eddy currents, by which I mean the friction of this jet or flow against the fluid with which it comes in contact, is of such an order that, (at the velocities I have computed here and which actually exist at 20 strokes per minute) the entire energy of that jet would disperse within a very few inches of motion, even

(Testimony of Edward T. Adams)

if it was a clear fluid. The energy of that jet is proportionate to the weight of the fluid going through and the velocity with which it moves, and the weight, as we have shown, is only that of a glass or a glass and a half full of fluid, and the velocity is about a foot and a half or less per second. Now, we are not dealing with a clear fluid. We are dealing with a fluid of an assumed viscosity of about 50; that was something like Santa Fe Springs fluid, and that would be practically one and a half times the viscocity of water. Viscocity is the factor of stickiness or tendency of the particles of the fluid to adhere; and the eddy currents in such a case would be that much more severe. Added to that, we have by hypothesis, the whole bottom of the well filled up with a silt which we are trying to keep churned up. That would add to the weight of the fluid to be disturbed and use up the energy of the jet or flow that much quicker. In other words, there is not, in the Bray pump, in any of the other pumps that we have discussed, a jet action at any of the speeds which we have considered. Those are average speeds and the flow would not reach to the dignity of a jet. I might note that the jet from a faucet is about 80 to 90 feet per second as compared to 1.36 feet per second from the Hofco pump.

Further

CROSS EXAMINATION

My testimony with relation to the velocity of jet actions was not based on the assumption that this lower nut and traveling barrel were moving upward at a uniform speed. The shaft at the top of the well which carries the crank moves at a uniform speed, but gives this nut a variable

(Testimony of Edward T. Adams)

speed. The velocity with which this oil comes out of the bottom nut, varies at various times as the nut is going up.

Q Then you have given the velocity of the Hofco in a certain definite figure; what varying point did you pick out in which you gave the figure?

A I gave that as an average figure because that is the only way that it can be computed with accuracy. I also said that if you wanted to find the maximum you could multiply it by one and a half. It can't be less than one and one-half times that. Or you can multiply it by two, which is probably the maximum it could be. The reason that this can't be stated definitely is that the stroke of the pump is only 3 feet at the time that the stroke of the crank at the top is four feet and one-half, so that nobody knows exactly the ratio at which those two synchronize, but it lays between those limits, and that maximum is the maximum at the instant of maximum flow.

Q In other words, the maximum would be double the figures you have given us?

A That probably would be the maximum possible. It very likely would be less, but even at that I took 1.36, and twice that is 2.72, 1-¾ or up to two miles per hour. This is based on the assumption that the inside diameter of the bottom nut on the traveling barrel in the Hofco pump was 1-12/16 inches.

MR. SOBIESKI: Now, I understand that some of the pumps they make are 1-11/16 inside diameter. Is that correct, Mr. Herron?

MR. HERRON: We are not making them now. There was a suggestion made that that size should be adopted and some few were made, probably 15 pumps of that diameter made altogether. They have been discontinued and we

(Testimony of Edward T. Adams)

are not making them. It was an experiment under the code.

MR. SOBIESKI: Is that the type of pump which you have brought here in the court room?

MR. HOFERER: That is 1-11/16, I think.

MR. SOBIESKI: Now, if the inside diameter were 1-11/16 instead of 1-12/16 that would increase the velocity considerably, wouldn't it?

A I would not say "considerably". I can figure it for you in half a minute if you would like. Put down the figures I read you. I will read them from my slide rule. The tube is 1-½ inches in diameter and, therefore, the area of the tube is roughly 1.76. Now, 11/16 is 2.225. Understand, these are slide rule figures and not exact. Now, the difference between those two is .49, .49 multiplied by 6.38; that would change the velocities as follows: Using the slide rule figures made here in court only, if the puller nut was 1-11/16 in diameter instead of 1-12/16 internal diameter, the net average velocity of the jet at 20 strokes per minute would be increased from 1.36 feet per second to 1.77 feet per second.

Q Now, if we increase the operation from 20 strokes per minute to 40 strokes per minute that would more than double the velocity of the oil going through the puller nut, wouldn't it?

A No. You would increase the flow in direct ratio as you increased the speed, not more and not less. If you double the speed you double the velocity. It would be directly proportionate. 40 strokes per minute of a 54 inch pump would be considered an almost impossibly high speed, would it not?

(Testimony of Edward T. Adams)

Q   If there was a sand condition down there, I imagine for a temporary operation they very well might plan to increase the speed so as to get a greater churning motion, shall we say, or greater jet?

A   I took 20 strokes a minute partly on Mr. Bray's testimony and partly on my own belief that 20 strokes per minute was a fair average condition to consider.

Q   Now, you have previously said that your testimony disregarded the stretch in the rods. I understand that these pumps that we have been discussing are very frequently put in wells of four to five thousand feet depth, is that correct?

A   I have assumed here a 4500 feet depth and I possibly misled you. What I meant to say and what I really did was this: I took the stroke of the pump in the well at 36 inches because I did not want to go into all the theoretical discussion about the stretch of the rods. We do not care what the rods stretch because nobody cares what the stroke was up to the polish rod. I took 36 inches down here at the pump and I have neglected all these considerations. It is probably about a 54 inch stroke up at the polish rod so in selecting 36 I have taken an account of the stretch of the rods and I have not tried to figure the exact amount. I don't care whether it is 54 inches up there or 60 or what it is. I previously gave the illustration of a ball at the end of a rubber string.

Q   And that the ball at the end of that would not move in direct or in predictable ratio to the movement of pipe; is that a fair statement of what you had in mind?

A   I said that they move in the same sort of synchronization but I did not know what, and I don't think anybody knows. I just illustrated with that so that the

(Testimony of Edward T. Adams)

Court would understand the nature of the things that I was leaving out and why I took 36 inches as a definite stroke, in order to avoid all the theoretical discussion that we would otherwise get into due to the stretch of the rods, inertia, and all of that.

Q   When an oil operator operating at the top of the well sets his polish rod, at 56 inches, or whatever the case may be, does he have any way of knowing, that in fact he is getting the 36 inch stroke you have assumed?

A   Yes. There has been a great deal of work done on that. I don't know as he knows it exactly. They have had a lot of work done by measuring up the marks on the pump and there has been a great deal of theoretical figuring on it, all of which I wanted to keep out of, and I presume you yourselves would wish to, such as double shooting the stroke, what may be expected on the pump with different strokes of the polish rod, different weights of tubing, all things come into it, and all of which I tried to keep out of. I said that the type of packing which the pump uses, whether it be a soft packing or a metal packing or various other types known to the art, might affect the quantity and the velocity of the flow that goes through the nut on the bottom of the traveling barrel. I have to keep my figures that I make for the Court consistent. My figures were based on experimental results that I examined and that I knew were made of pumps with no soft packing whatever, so I made that reservation as to what might affect it because I wanted to keep my record clear.

Q   The type of the packing which is used in the pump up here does affect, in your opinion, the quantity and

(Testimony of Edward T. Adams)

the velocity of the flow of oil through the nut on the bottom end of the traveling barrel, is that correct?

A   I can't say that because I am not enough familiar with the packing that you use; in fact, I have had no experience, you might say, with cup packing. All my experience has been with metal packing. If you assume that this packing has the effect of preventing any leakage whatever, which I do not believe is possible, you would then raise the figures which I have made by 15 per cent.

Q   I understand that with these pumps down four or five thousand feet in the wells it is a practical impossibility to make actual measurements of how much flow you are getting anywhere along the pump, and particularly with respect to the flow through this bottom nut. That is to say, your figures here are based entirely upon calculations and not upon observations of how those pumps are operated, is that correct?

A   No.

Q   To what extent are they based on calculation and to what extent on observation?

They are based wholly on calculation, but on calculation of absolutely fixed conditions. We have specified the exact volume of the annular space. That is accurate. It is not a guess. It is exact to a fraction of an inch because we have assumed a stroke, which we may honestly do, as long as we do not at the same time assume a stroke at the top of the pump. I have assumed a certain stroke in the bottom of the well, in order to get rid of all that theoretical stuff above there. The inside diameter and all the dimensions are accurately taken, so the volume that I gave you and the average velocity that I gave you are actually not theoretical. They are accurate and actually

(Testimony of Edward T. Adams)

corerct. In the figures which I gave for the Bray pump
as built, I understand that instead of building a guide
Mr. Bray has built a bushing which has no guiding effect,
I took the dimensions of that bushing which you gave me,
and used the area between the bushing and the standing
tube which you gave me and computed the velocity of flow
in the pump as built. I also took eight 3/16 holes in a
guide which was properly a guide and ran on or with
rubbing contact or sliding contact with the standing tube,
and figured the velocity through the eight 3/16 holes.
Your annular opening in the present structure is practically double in area of the ducts shown by the patent,
and by the nut which you exhibited in the court room.

Q The patent contains no figures as to the actual
diameter of the standing column, the inside diameter of
the bottom nut, does it?

A I did not figure from that. I took the figures that
you gave me for the standing column, the figure which
you gave me for the inside diameter of the traveling
barrel, and took such a guide as you showed in court with
eight 3/16 inch holes. So I did not attempt to interpret
the patent at all, more than to say that the guide which
you exhibited in court with eight 3/16 holes, by your
testimony, was the nut that you built at the time that the
patent was applied for.

Q That was one of the nuts we were building at the
time.

A And you had an apertured nut also, that is, one with
grooves in the face of the guide, and I did not go into that
because I did not know what those areas were.

MR. HERRON: Mr. Sobieski, may I interrupt just
long enough to ask: Do you expect to introduce that

(Testimony of Edward T. Adams)

guide with the eight 3/16 holes? If not, I take it it may be stipulated that you did and it is a guide having those dimensions, and that that was the guide concerning which Mr. Bray gave testimony?

MR. SOBIESKI: Didn't we introduce that into evidence?

MR. HERRON: I do not think so. If you did not, the stipulation covers it.

The patent discloses no dimensions as to how this guide is to be constructed. The patent limits itself to the proposition of allowing the trapped oil to get out through the bottom nut by means of a jet action, but makes no claim to any particular velocity except that it says "strong jets".

I have said that I did not think that in any of these structures, including Bray, that the flow from that orifice rose to the dignity of a jet by any term that we can compare it with.

Q But the patent itself contains no specification as to measurements of any of these dimensions?

A Oh, if you want to be technical about it, the patent is drawn correctly in this drawing and the diameter of the orifice, less the thickness of the tubing, which is known to be one-eighth of an inch, so, figuring from that I would say the pump would never draw or get any production.

Q Didn't you the other day say that these patent drawings were descriptive?

A That is why I kept away from that, but when you kept beating me down, why, then I told you what I thought.

Q Am I to understand you to say in Mr. Bray's patent the drawing is an accurate representation and the

(Testimony of Edward T. Adams)

other patent we were discussing was descriptive merely, or are they descriptive in both?

A Descriptive merely. Absolutely, in all of them. And the patent contains no dimensions.

MR. HERRON: I will stipulate that. I think we can practically remove that from the case.

MR. SOBIESKI: Very well.

The figures I have given were based upon my actual measurements of the Admore structure.

I took an appropriate instrument and measured it. By the Admore pump, I mean the pump made up of the tube, Exhibit 14, and the plunger assembly, Exhibit 13, which parts I measured, and in which the standing pipe was commercial one inch iron pipe.

This Admore pump has four holes in the barrel near the lower end to allow a portion of trapped liquid to escape.

Q When you were stating that the oil had a downward motion you were referring to the oil which comes out of this barrel through the holes in a horizontal position, hits the side of the tubing and then goes down, is that correct?

A Well, it probably hits the side of the tubing. It flows radially from the holes in the side of the tubing into a strong downward flow, which is at that moment taking place, and is carried with it. At the instant it leaves the barrel it is going radially. It leaves the barrel at a point an inch and a quarter from the bottom of the barrel and it flows around to the space in the fluid vacated by the moving barrel about an inch and a quarter below it. It has to flow around there because that is the only place it can go.

(Testimony of Edward T. Adams)

Q   Did this barrel have a lock nut on the bottom of it?

A   Yes. The nut is shown here (indicating). There is possibly a little more than half of the opening in the holes, and the rest is through the lock nut.

Q   I understand then it would have to go downward on the outside of the barrel and the lower lock nut?

A   Part of the flow would be past the outer part of the barrel and lock nut into the space below the nut and part of the flow would be directly downward between the lifting nut and the tubing. At the time the oil is going out of the barrel through these holes it is not going in a downward direction, after it gets out it changes direction and goes downward. It can't change direction until it gets out. I refer to the oil coming through the holes, as distinguished from the oil in the annular space. The oil from the annular space is going directly downward.

MR. SOBIESKI: And I understand that, in your opinion, the traveling barrel reaches its greatest velocity about half way up and then slows down at both ends of the stroke, is that correct?

A   I don't know where it reaches its greatest velocity; somewhere in there and probably about half way up. I presume that is pretty nearly right. Nobody knows exactly. The figures I gave as to the amount of flow, are calculated on the average velocity and not on the basis of the greatest velocity with the maximum lying between 1.57 and 2 times the average.

Q   How much is the bottom nut ordinarily spaced from the bottom in the pump when it is at rest?

A   I accepted the figure which Mr. Bray gave of 6 inches. Some other operators gave me 8. I think it would vary with the nature of the well. If there was

(Testimony of Edward T. Adams)

gas present they would set it as low as they dared to and if it was a clean well they would be more careless about it.

Further

DIRECT EXAMINATION

BY MR. HERRON: Calling your attention to the four annular openings in Exhibit 14, what is the area of those four openings relative to the area between the puller nut and the standing column upon which the puller nut slides; that is the annular space between those two parts in Exhibit 13?

A   Well, roughly, in the nature of 3 to 4; that is, 3 representing the area through the puller nut and 4 representing the area through the 4 inch holes; that is, one would be 3/7 and the other would be 4/7.

Q   Mr. Adams, I now hand you Exhibit 1 attached to the depositions. What do you call this?

A   Exhibit 1 in connection with the deposition of Johnson, taken at Bradford, is an illustration entitled "Bramo", I have examined the drawing there and familiarized myself in a general way with the operation of that pump.

Q   Calling your attention now to Cummins, which has been introduced as one of the structures of the prior art, what is the difference between that pump, and the Bramo pump in its operation? Point out if there are any differences in the essential operation.

A   Both are traveling plunger types and are essentially the same, the difference being that in the Cummins pump, we have a metal plunger working in a metal traveling barrel, whereas in Bramo we have a cup leather pack-

(Testimony of Edward T. Adams)

ing traveling in a metal traveling barrel; that is, the essential difference lies in the nature of the packing used to form a seal. It makes no essential difference. The locking means in each case is a jaw clutch located at the top of the structure.

Further

CROSS EXAMINATION

This Bramo is not a traveling barrel pump. It is a traveling plunger pump. On account of that type of pump there is no trapped oil to escape. Therefore, there is no provision made for the escape of trapped oil.

Further

DIRECT EXAMINATION

A   This Inverted Bramo, Exhibit 17-A, is a traveling barrel pump which I have taken apart. Here is the barrel with the traveling valve at the top, and threaded above the top of that cage the connection to the sucker rods. It is of the traveling barrel type. This is a stationary plunger, which I have just removed from the barrel, and has a cup leather packing and a standing valve at the top of the plunger. At the bottom of the plunger there is a gland, carrying packing, set up with a spring. The nut which sets up the spring in the packing is formed with the female member of a jaw clutch adapted to engage with a male member at the top of the anchor. The type of packing is cup leather as shown. This structure concerning which I am now testifying is a Bramo pump reversed, that is, the moving parts are made to stand still and the standing parts made to move.

Q   With relation now to the structure of Cummins reversed, shown by Exhibit No. GG what, generally, are

(Testimony of Edward T. Adams)

the points of difference between Cummins reversed and that inverted Bramo structure?

A One has a cup packing as the seal and the other has the metal packing for the seal and that, in practical effect, is the essential difference. Each has a standing valve at the top of the standing column; each has a traveling barrel with a traveling valve at the top of the barrel. The Bramo pump has, however, a packing gland at the bottom immediately above the guide. That is a true guide as in the case of the Bramo it runs upon the column. The packing gland, spring and nut would tend to pack and prevent any downward motion of fluid escaping into the barrel on the upstroke or out of the barrel on the downstroke.

Q Now, from your knowledge of the structures in the pump art, would the inverted Bramo pump be an operative pump? I mean would it work and function under any conditions?

A We had that same matter up. It is simply a matter of packing. The packing closes off what we have called the annular space, and if it was inserted in the well and there was absolutely no leakage, then in the annular space you would have an air cushion on which the pump would alternatively compress and expand. That is not a practical condition because leakage will occur in any pump. This pump will operate successfully just depending on the rate of leakage. If the leakage is small the rate of motion must be slow; and if the leakage is greater the speed can be greater.

The essential difference between the Admore and the Inverted Bramo pump, is the outside annular space. In the Bramo which I have just examined, that space is closed in by packing.

(Testimony of Edward T. Adams)

Q That is to say, the packing gland at the bottom of the inverted Bramo has been removed and in lieu of the removed gland you have two things in the Admore, a vented barrel, and a loosely fitted puller nut so fashioned as to provide an annular space between the inside of the moving barrel and the outside of the standing column, is that correct?

MR. SOBIESKI: I object to that, your Honor, on the ground that the so-called loosely fitting puller nut which they here produce in evidence is obviously a worn puller nut and, as Mr. Adams has stated here, these things wear and, naturally, they may wear into a certain condition different. Now, there is no evidence that this Admore pump was constructed to permit the escape of oil downward. Therefore, the question is improper because it assumes a fact which we contend is not in evidence.

MR. HERRON: To which I make two answers. First, if I may ask one preliminary question:

Q In the event that a nut was loose by reason of wear, would the nut present an evenly rounded orifice at its top or would the wear make it egg-shaped or sloped? In other words, if wear was caused by sliding up and down a rod would it wear evenly all around the surface of the rod, or would it wear on a slope or egg-shape or otherwise?

A In other puller nuts which I have observed used on pumps of this type where wear had occurred you would find that one side of the puller nut was very materially finer than the other. This nut appears to be round, of uniform thickness, as you can see by this.

(Testimony of Edward T. Adams)

MR. SOBIESKI: Isn't it true that these barrels are constantly moving around a shaft as they go up and down also, which would cause them to wear more or less evenly?

A   Well, here it is in evidence, and I see absolutely no evidence of wear on the tubing, which would be necessary if there was to be wear on the nut. Also, I got my figures from this tubing, which is 1 inch rough iron pipe and I measured it with my micrometer at various places on it and I got essentially, oh, as close as I could make it, and it is a rough piece and there is no change in length from one end to the other.

MR. HERRON: And the deposition fixes, if the Court please, the size of this pipe as being the same diameter which the witness has just testified he found it to be by measurement, hence it follows that the wear could not have been by virtue of rubbing on the pipe, as counsel suggests. My impression is the ID of that puller nut is given in the deposition. I thought it did give it but I guess it did not. It gives the OD but not the ID because there was no change in it. Now, I submit my question if you can find it.

THE COURT: He may answer with the explanation given.

(Question read by the reporter)

A   It is.

MR. HERRON: Mr. Adams, in your judgment, was any exercise of inventive faculty shown in changing from the packed gland of the inverted Bramo to the vented barrel and the loosely fitting nut of the Admore?

(Testimony of Edward T. Adams)

A   There was not.   It is perfectly evident that as long as the rate of speed is controlled by the amount of leakage that that pump needed more space and needed to be vented.   In other words, any ordinary mechanic, skilled in the pumping art, if presented with an inverted Bramo in which the free drop of the pump was impeded by that packed gland at the bottom would discard the packed gland and use in place thereof either a loosely fitting puller nut or a vented barrel or a combination of both or any means to give a sufficient area for the fluid to escape.

Q   Would it involve any exercise of inventive faculty, and anything greater than the ordinary skill of a mechanic trained in the art, to change from that tight fitting guide on the inverted Bramo to the loose puller nut on the Admore?

A   It would not.

Further

CROSS EXAMINATION

BY MR. SOBIESKI: I understand, then this inverted Bramo, No. 17-A, has a packing means for the very purpose of preventing oil escaping through the bottom of the traveling barrel, is that correct?

A   It would have that effect.   The effect of the design was to keep oil from escaping through the bottom of the traveling barrel in any large amount.

MR. HERRON: Either that, Mr. Adams, or would it not be true, that conversely, to prevent oil from coming in, depending on how they looked at it?

A   Either way.   If he was trying to keep sand out, why, it would have that effect.

MR. SOBIESKI: And you mentioned that there might be an ideal condition existing if there was only

(Testimony of Edward T. Adams)

air inside the barrel. In the absence of some oil in there the pump would not be lubricated, would it?

A   It does not imply that there is oil inside of the annular space in order to have either the top or the bottom of it lubricated. In a case like this, when you have to answer regarding a theoretical condition, the fairest way to do is to take the two extreme limits and tell what they were, and say that you could operate between them. That is what I attempted to do. I do not consider that the pump would operate in a well with an air chamber in it; that is, that would not be the practical way it would be operated.

Q   An air chamber would not exist. Now, until this pump, Exhibit 17-A, wore some opening to permit the oil to get in and out it could be operated, if at all, only slowly and with difficulty, is that correct?

A   I don't know about the difficulty. It would have to operate slowly to give time for the fluid to leak in and to leak out again and that amount of leakage would govern the rate at which it could safely be operated.

MR. HERRON: I might say, if there is any question as to whether that Bramo is an operative pump, we will be able to set the doubt at rest. If there is any serious contention it is inoperative, in the light of the patent law, we will be glad just simply before the case is finally closed to go out and operate it, if you have any serious thought about it. I think you probably haven't.

MR. SOBIESKI: From your experience in patents on pumps, did you ever know of a pump of that design being operated in a deep well, such as we have in this district?

THE WITNESS: It has been done.

(Testimony of Edward T. Adams)

MR. SOBIESKI: Has such a pump been operative in deep wells such as we have in this district?

A In what is known as the Miller patent, there is an annular space not vented by any means whatever, and I know a good many of them, some 50 or more, were built and operated in the field.

The Miller pump is not what is called a fluid-packed pump, but it had an annular space of equal magnitude not vented. I have never heard of a Bramo pump, being operated in this district as a deep well pump. I simply said that the venting of the annular space would not prevent it being operated in some fashion. You would limit the drop of your pump.

Q It could only operate slowly?

A Well, I think these pumps were operated about 20 strokes per minute. The fluid-packed pumps I have been describing—

Q Calling your attention to that pump, until it became worn, in your opinion, could it be operated at 20 strokes per minute in deep wells such as we have in this district?

A I do not know. My best judgment would be that the stroke of the pump would be restricted so it would not get its full stroke. That does not mean to say that it would not operate.

MR. SOBIESKI: Well, could it be operated 20 strokes a minute in a deep well such as we have in this district before it became worn?

A What I had in mind, was the conditions that you have named, that the pump could operate. It could get 20 strokes but they would not be full length strokes, that might be quite sure. I do not attempt to say what, how long a stroke or how short a stroke.

(Testimony of Edward T. Adams)

MR. HERRON: Let me say, so we may not have a long controversy over this, we are not going to say that that is an advisable thing to do nor suggest it to be done at all. That is not the issue that we made. We are not intending to be understood as asserting it.

MR. SOBIESKI: Well, in your opinion, can you give us an opinion of how short these strokes would be?

A No. The size, at the rate of 20 strokes per minute, would depend entirely upon the friction, the resistance, the amount of the leakage that came in, how fast that leakage came in, how fast it could come out, and the length of time you operated it. It would be one thing when you started and be something else after you have run for a while.

MR. HERRON: I will stipulate that that obviously is not a commercial pump designed for use in deep wells; and if that will end the controversy, you are entitled to it and you have it.

MR. SOBIESKI: That is my honest opinion on it.

MR. HERRON: I think we both agree on that and have always.

MR. SOBIESKI: I understand the Cummins reversed makes no provision for the escape of trapped oil, does it?

A It does not—at the bottom.

Q On account of that, until it became worn, it would impede the free motion of the traveling barrel up and down after the pump had been reversed?

A No; not at all. The trapped oil in that case is trapped at the upper end of the pump and would be perfectly free to escape at that point. You see, as I explained before, in reversing this I attempted to show his Honor that taking the parts as they were and adding

(Testimony of Edward T. Adams)

two small bushings, one at the top and one down at the bottom, we could reverse it and make a successful operative pump. I did not attempt to re-design it. If I was re-designing that pump I would have made my packing at the top and my annular space at the bottom, but using the pump as I was, the annular space is at the top and there is supposedly free flow. He gives no dimensions. In Cummins reversed the way I reversed it the escape of trapped oil would be at the opposite end of the traveling barrel from the locking device. The escape of oil is around the locking device. There are no openings through it. Past the barrel there would be no downward flow except the leakage past the barrel.

Further

DIRECT EXAMINATION

It is always very desirable to provide for a free drop of the barrel, in a pump of the traveling barrel type. In proportion as you restrict the free drop of the barrel by trapping fluid between the downward side of the plunger and the upper side of the guide, you impede the free downward drop of the barrel. That space I have referred to in all of my testimony as the annular space. Whatever energy you use to force that fluid out of the annular space must come from the energy in the moving barrel and the movement restricted to that amount.

Q Now, if you restrict it to the extent shown in the Bray patent, thereby producing what he is pleased to call in the patent "jets", you, to the extent of that impeding impede the free drop of the barrel, is that correct?

MR. SOBIESKI: I object to that on the ground, your Honor, that the Bray patent contains no measurements at all and it is perfectly plain, of course, that if the

(Testimony of Edward T. Adams)

great weight of these rods—that there is a considerable force driving that thing down, so that you can have a jet action both going and coming without impeding the free operation of the barrel. It is obvious—it seems obvious to me that until your impediment, if there be one, exceeds that of the weight of these 4,000 feet of rods that the impediment, if any, is immaterial. So there first should be some showing that the Bray patent compels that impediment to a greater extent than the weight of the 4,000 feet of rods.

MR. HERRON: I submit the question.

THE COURT: You may answer, Mr. Adams, with any qualifications.

A The smaller the opening will be, the greater the pressure that you build up, the greater the resistance, and that resistance must be vitiated out of the energy in the barrel and will by that much reduce the drop of the barrel. And if that clearance is reduced to such a point as to be non-existent, you would have the case which I have described, where either there would be an air cushion in the ideal event that it was absolutely tightly sealed, or some leakage into the pump which would impede its operation.

Q Let me ask if, from your examination of the Hofco structure you believe that practically a maximum of barrel drop has been procured, considering now the possible limits of the structure, as distinguished from the production of jet action?

A I do.

Q Now, looking at the Bray patent, on the other hand, the structure taught by the Bray patent, would you say that that structure has procured, perhaps, the maxi-

(Testimony of Edward T. Adams)

mum of so-called jet or jet-like action, as distinguished from free dropping of the barrel?

MR. SOBIESKI: I object to that, your Honor, on the ground that there are no dimensions given in the Bray patent, and the Court must know that these 5,000, or, say, the 4,000 foot well—the 4,000 feet of rods, they falling down with this traveling barrel on the bottom of them, it does not take an expert to argue to your Honor that there is a considerable weight there and a considerable force; and, consequently, unless the making of the orifices in the bottom of the nut, considering that type of nut, unless those orifices create a pressure which is greater than the total of the weight of those 4,000 feet of rods, it is just plain common sense that orifice or no orifice, it is going to drop just the same. Now, we maintain that without taking those factors into consideration the question is improper and the Bray patent itself does not give measurements. It merely gives the number of the holes through the bottom nut on the traveling barrel, or the bottom nut on the traveling barrel be apertured so that the oil can go in and out in a jetting action, and beyond that general proposition it is improper to assume any particular dimensions which would or would not prevent a free flow—a free dropping of the barrel, especially when we are considering that it is a free dropping of a barrel on the end of these 4,000 feet of rods.

MR. HERRON: I submit the question.

THE COURT: Does the crank means that lifts and drops or allows to drop the pumping rods support the weight of the rods on the downstroke so that it has a uniform stroke? What I mean is: Does the weight of the pumping rods drop, adding their weight to the engine and causing the crank to move faster on the downstroke

(Testimony of Edward T. Adams)

or does it move uniformly; in other words, holding the weight of the rods on the downstroke to preserve uniform motion. In other words, if you do not do that you would have irregularity in the turning of your crank, I suppose?

A Pardon me. I thought it was addressed to Mr. Bray. In my answer to his question I had not taken into account the uniformity of motion of the crank at the top, although that would come in, or would be largely a question of counterbalance. I took the very much simpler and elementary scheme, which is this: That there is a certain amount of energy, a certain amount of resistance required to force fluid into strong jets through small openings.

MR. HERRON: I think, Mr. Adams, the Court's question was simply whether, in point of fact, there was a continuous operation, whether the crank supported the weight of the rods upon the downstroke or not, as a matter of fact, as I understand the Court's question.

A The crank supports the weight of the rods. We get into questions of inertia there, your Honor, and all the stretch of the rods that I wanted to keep out of. The crank does carry the weight of the rods.

THE COURT: That is the purpose of the counterbalance?

A Yes.

Q In other words, you do not get the force of the weight of the rods in the compression on the traveling barrel? What I mean is, you do not get the full weight; you may get some?

A When you get into the weight of the rods that is hung on the barrel, then you get into also the question of inertia and the speed with which you are traveling. If you are traveling down rapidly you not only have the

(Testimony of Edward T. Adams)

weight of the rod, but at the bottom of the stroke you have stored energy to overcome. By whatever amount that is, that is added to the weight. Now, then, a resistance such as trapping the oil and forcing it through orifices to get a strong jet would limit the stroke of the pump at the bottom; that would limit the stretch of the rods and to a certain extent the load on the crank, but it would be taken probably out of the stretch of the rods due to inertia. In other words, the rods are not a fixed thing, your Honor; they are like the rubber band and they have more or less stretch in them due to their own weight and due to the weight of the pump and due to inertia and due to the speed at which they move. But the eventual fact that I wanted to bring out was that any resistance to the free flow of the plunger at the top is without much tendency to limit your motion of the plunger, great or small, according to the circumstance itself.

MR. HERRON: You mean "free flow to the barrel", don't you?

A Free flow to the barrel, yes. Pardon me.

MR. SOBIESKI: May I ask a question?

THE WITNESS: Yes.

Q The thought that your Honor's question brought in my mind was: Does the operation of the engine at the top—is that calculated to impede the downward dropping of the rods and barrel so that they will go down in a uniform fashion?

THE COURT: The depth of counterweights on the wheel or on the crank means are there for that purpose, to equalize the downstroke?

A Yes, sir; to equalize the downstroke.

(Testimony of Edward T. Adams)

MR. SOBIESKI: Then, as I understand it, the weight of the rods and of the barrel going down creates a force that the engine has to be used to hold it back, is that correct?

A You are counterbalancing—you get into another problem. There are two factors; one factor is the physical weight of the rods and of the pump that is an amount of steel hung on this polish rod; in addition to that is the weight of the column of fluid of diameter of the outside of the traveling barrel and reaching from that to the surface of the well. Now, in pumps of this size the weight of the fluid is alternately picked up and lowered down again; that is, in the upstroke the weight of the fluid is carried by the plunger and in the downstroke the weight of the fluid is carried by the tubing. Now, then, when you get into counterbalancing you are more concerned with the dropping operation of he load than you are with anything else. Here is one time you have got practically double the load that you have at the other.

THE COURT: The engine is then lifting the counterweight on the downstroke?

A Yes, sir; giving a more uniform stroke.

THE COURT: When it gets on the other side the counterweight helps the uplift?

A Yes, sir.

Q Lifitng the weight up?

A Yes, sir. It was to avoid all of that, your Honor, that I fixed the stroke down here and left the other thing to be whatever it was.

MR. SOBIESKI: In view of this, then, that we have to have a counterweight to protect against the free down-

(Testimony of Edward T. Adams)

stroke, we maintain that it is immaterial to go into the question of whether the use of certain ducts down here at the bottom is an additional, you might say, brake on the downward movement, because until that brake arose to the dignity of being at least equal to the force of the downward drop of the barrel and all of the rods it is utterly immaterial, because it would be taken care of by an adjustment from your engine and your counter-balances.

MR. SOBIESKI: We urge our objection to the question which Mr. Herron had asked.

(Question read by the reporter as follows: "Now, looking at the Bray patent, on the other hand, the structure taught by the Bray patent, would you say that that structure has procured, perhaps, the maximum of so-called jet or jetlike action as distinguished from free dropping of the barrel?")

THE COURT: In any form that it might be constructed. You may answer.

MR. HERRON: That is, I take it the Court means within the limits of the patent?

THE COURT: Yes; within the teachings of the patent.

MR. HERRON: Do you get the question now, with the court's modification, which I adopt?

A Yes, I do. I feel that the structure, answering for the structure shown by the eight 3/16 inch holes through the tight-fitting guide, yes.

The Bray structure does not get the maximum of free drop of the barrel, it gets the maximum of jet action. The jet action which it does get is obtained at the expense of the free drop of the barrel.

(Testimony of Edward T. Adams)

I have examined Exhibit No. 7, being a part of the Bradford deposition, which is a sheet from a catalog and have examined the drawing there shown of the Admore liner barrel. A mechanic skilled in pump manufacture would be able to construct that Admore pump from the illustration there shown.

Further

CROSS EXAMINATION

BY MR. SOBIESKI: I understand that the operation of the pump, the barrel has to go down with a uniform motion—as uniform a motion as can be obtained; is that correct, Mr. Adams?

A No.

Q In the dropping of the barrel in ordinary pumping practice in this district, these deep well pumps, the engine and counterweights have to be used to retard the drop of the barrel and rods, is that correct?

A They have the effect of equalizing it.

Q Equalizing it. Now, to use that Bray pump, as made according to the patent, in deep wells, in your opinion—or, do you know whether the engine has to be used to slow up the drop of the traveling barrel?

A I did not testify regarding the engine in my direct testimony and I did not take that into account; and I do not see that there would be any difference between that in the Hofco pump or any other pump in that respect.

Q You mean to say, then, that the Bray pump would go down with practically the same speed as the Hofco pump—I mean the traveling barrel would go down?

A I did not say that. No; it would not.

(Testimony of Edward T. Adams)

Q   It would not.   Then, the engine would have to be used less on the Hofco pump than on the Bray, wouldn't it?

A   I haven't gone into the engine and haven't gone back to that end of it and it is confusing to try to answer a question bringing in stretch of the rods, the inertia of the rods and the counterweights and all the type design which I specifically left out in order to get a concrete case of facs that were really facts and that could be known by fixing a specific stroke at this end.

Q   The point I have in mind is to try and discover whether these facts that you have been discussing amount to anything, or whether, since you have to use the engine as a brake on the downstroke anyway, this slower downward motion of the Bray pump barrel which you describe, whether that amounts to anything.   That is the reason for introducing the question about the engine.

A   Very well, Let us consider the pump as standing in equilibrium so that we cut out the question of the inertia which comes in with the motion, and we will say that the plunger barrel has been lifted appreciably, say a foot or a foot and a half, so that the entire weight of the barrel, column of the fluid and the sucker rods clear up to the polish rod all hang on the walking beam and they hang in equilibrium at that time.   Now, then, the sucker rods are stretched; they are a spring, just as much as if they were a coiled spring; they stretch and that stretch is governed by the same laws that govern the stretch of a spring, one being tension and the other being torsion.   It hangs in equilibrium and is theoretically affected by any resistance that you put at the bottom, despite all that tremendous weight up there.   If you should increase that by an

(Testimony of Edward T. Adams)

amount that you could lift with your finger, theoretically, not, practically, you have by that much lessened the tension of that spring. Now, if you will just come to what we know practically, we know that any resistance down here shortens the stroke of the pump. I referred to fluid packed pumps which had an annular space which was not vented and said those pumps did not make their full stroke, but only made a portion of a stroke. Suppose this pump was sanded up, and suppose you got enough resistance in here so that the plunger is sanded up, it is well known that the sucker rods simply buckle in the tubing and that the plunger would remain practically stationary. Therefore, between the touching that would theoretically resist the free dropping in a certain amount, and the actual stopping of the pump by sand, you have all the degrees of resistance that lie between the two, and one of them is that degree which will hinder the free drop of the pump and will shorten the stroke. Now, does that answer?

Q   Then, as I understand it, your testimony to this point has been that, despite the weight of the rods and the counterbalancing of the engine, that any friction down here would have some effect?

A   I testified that any friction down here would shorten the stroke of the pump despite the 5,000 or 10,000 feet of rods going down. Any friction down there would shorten the stroke by the amount of that friction, that is, be proportional to it.

Q   If the strength of this friction, whatever the proper term for it is, was less than the force of the 5,000 feet of rods, which would win?

A   The 5,000 feet of rods are not on the pump. The 5,000 feet of rods are hung up at the top and you are just

(Testimony of Edward T. Adams)

touching the pump. If you have a weight down at the pump hung by a spring, the total weight carried by the spring is up here on the sucker rod a mile in the air; that weight is up there. That weight is up where you take hold of the walking beam, up 4500 feet or a mile away from where you have the pump. We are now talking about friction down at the bottom of the well. Any friction there would reduce the tension of the spring. The tension of the sucker rod is the spring which it has, and any friction there would reduce the tension of the spring. If you reduce the tension of a spring, as a rubber band, why, it slaps back and by that much you shorten the stroke.

Q By how much?

A I wouldn't give a figure for it, wouldn't begin to.

MR. SOBIESKI: Then, I move that this line of testimony be stricken out, your Honor, on the ground of it being entirely theoretical and being unable to give any definite figures what effect, if any, this friction has. That shows on its face to be negligible and not worthy of consideration.

THE COURT: I will leave that for the argument on the main case.

MR. SOBIESKI: If you say there is friction on account of the use of a nut with holes where you use such a nut, would it not be the fact that there would be more friction if the oil had to escape out sideways, say, through the barrel and then go down, rather than where the oil can escape directly into the position it will ultimately rest?

A You assume the same velocity of flow in each case?

(Testimony of Edward T. Adams)

Q The same conditions in both cases, the only change being that in one case you have that oil escaping directly downward through the bottom nut and in the other case you have that oil escaping outward through the side of the barrel, all other conditions remaining the same.

A There would be no difference. Oil escapes due to a pressure and the pressure doesn't care which way the oil goes. You said "the same conditions." By the same conditions I mean you have the same pressure inside of the annular space forcing the oil out, that pressure does not care what direction the oil goes.

Q No. But, as I understand it, you have here on this side the tubing which is perhaps a quarter of an inch away.

A Yes.

Q And this oil, if it goes out with any considerable velocity, will hit the tubing, won't it?

A You are talking now of what happens after it has left the hole. You asked me if I had uniform conditions, if I had fixed pressure in there whether it would make any difference in the frictional result as far as the pump was concerned, whether the oil went sideways or went the other way. If the pressure is the same, then the fluid escapes under that pressure and the pressure does not care which way it goes, and the resistance is the same because the pressure is the same. After that fluid is out and has a certain amount of energy in it what is done with that energy, that is another matter.

Q Let me ask a question. Perhaps it is the way I am asking the question. With the holes in the side would not the barrel be less free flowing than where the oil escapes out directly through the bottom?

(Testimony of Edward T. Adams)

A   If the pressure in the annular space causing the oil to flow is the same in each case the motion of the pump will be the same in each case.

Q   And can you answer the question "Yes" or "No", the one I asked?

A   No; I can't answer the question "Yes" or "No" because your question was not definite.   I tried to give it a definite answer but you do not fix the conditions under which I was to answer.   I supplied them for you.

Q   Wouldn't this oil going out against the side and hitting the side of the tubing and churning around on the side out here, wouldn't that tend to slow up the operations of the barrel more than if the oil had come out through the bottom nut on the traveling barrel?

A   No.

Q   I will call your attention now to this copy of the patent, Mr. Adams, the Ellis patent.   Now, I understand in this drawing, calling your attention to page 1 of the specification, reading at line 92:   "The interior surface of this nut is formed with a series of oil passages 22 intermediate of the bearing surfaces 23 so as to permit the free discharge of oil from the dead chamber 24 which exists between the nut and the valve at the upper end of the standard.   For the purpose of controlling the flow of oil through these passages an annular check valve 25 is provided"—now, with relation to that, I understand that they have a washer of some soft substance in there, fibre?

A   Preferably fibre.

Q   And that causes the oil to go out of the nut in a horizontal position, is that correct?

A   Presumably so.

Q   Now, how does it get it in the horizontal position?

(Testimony of Edward T. Adams)

THE COURT: 25 is not in thickness the full width of the aperture?

A Right. It does not say whether the fibre washer 25 is a close fit, is a running fit on the tube or standing tube or not, so I don't know whether there is any flow in that direction.

THE COURT: If it was tight there would be very little get out?

A That is it.

MR. SOBIESKI: Now, then, do I understand your testimony to be that if we take 25 out of there the oil would not go out in a lateral direction?

A That is correct. I said that if you took 25 out, and the teaching of the patent, beginning at page 1 lines 101, is that part 25 "may be mounted upon the nut in any desired manner, for instance by means of the valve cage comprising the ring 26 separated from the nut by spacers 27 and open at its sides so as to permit the free insertion or removal of the annular check valve 25". That is, part 26 is simply figurative, and I base my statement also upon the fact that there is a statement in the specification that there is a free flow. It is stated in line 37 of page 2. "Under some conditions the valve ring may be omitted"— that is, the ring 25—"as the passages in the cap nut permit the free flow of liquid and prevent the formation of a dead chamber" etc.

Q Then, am I to understand your testimony now to be that in order to get a free flow you do not have a passage out through the side, but when he says "free flow" he means that you have to have a passage downward, that is, the free flow is not a passage through the side?

(Testimony of Edward T. Adams)

A   No.   I did not say that.

Q   In other words, it is not your position that by the mere use of the words "free flow" he does not mean that the oil goes out through the side?

A   He simply says "free flow" and I took him at his word.   His drawing shows that that part 26 is larger in diameter than the tube that it encloses, and as a mechanic I think of how the thing is to be formed.   When it is shown it will be formed with a shaper on and while it is possible to spot the shaper in the slot that is cut at the side, it is more probable that in making it the man first turned out in 26 an area big enough so he could run the shaper straight through and not touch it.   That would leave enough opening so that a jet issuing with any velocity could flow down through that.

Q   Does Figure 5 show any of those things you have described on 26?

A   Figure 4 shows what I referred to.

Q   Does Figure 5 show any, though?

A   There are no dimensions on Figure 5 but there is an inferential measure on Figure 4 which we marked with the letter T which clearly shows that it is larger in diameter, and any mechanic who had to make the thing would infer it was enough larger to allow it to be finished with the shaper, which is what is evidently intended to be done; in fact, I don't know whether that it could be finished any other way, either that or broached.

Q   Well, isn't it true, Mr. Adams, that if this bottom 26 was very much smaller than the fibre it was supporting that the unsupported part of the fibre would be destroyed by the force of the oil coming down and, therefore, any

(Testimony of Edward T. Adams)

reasonable construction of the pump shows that 26 and the fibre thing are practically the same diameter?

A   No. I do not agree with that and the pressure that comes upon it is proportional to the velocity with which it flows; that is, the head is equal to the square of the velocity divided by 64.4, and we have previously shown that the head at the most is a very few feet and the pressure coming from that ring is very slight.

Q   You still maintain, then, that all of the oil would go out through the small opening, straight down, rather than through the large opening to the side, is that correct?

A   I have not maintained where the oil went. I have maintained that there was ample opening there for the oil to go and I saw no reason at all why it could not go there. It could go sideways, as far as I am concerned.

Q   I thought you told us the oil did go down, straight down?

A   Goes straight down. It has a vertical flow through the orifices in the nut and that gives a downward vertical flow towards the anchor at the bottom and there is by the teachings of the patent, space between the part 26 through which it could flow when that valve 25 is not present. The patent teaches it need not be there under certain conditions.

Q   When the ring is out do you maintain that the oil does not flow through there in a horizontal direction?

A   Ample room for it to flow in a vertical direction and the patent specifically says that that ring may be— he don't confine himself, as you will notice, to the top of the ring. He says of the support for the valve.

MR. SOBIESKI: Will you read the question, Mr. Reporter?

(Testimony of Edward T. Adams)

(Question read by the reporter)

A I don't see any necessity for it to do so.

Q But you believe it does not?

A I think it is a vertical flow.

Q That it goes through the smaller opening T rather than through the larger opening horizontally?

A It goes in the direction it is flowing, rather than making a turn.

REDIRECT EXAMINATION

BY MR. HERRON: Now, in the answers you have given to the questions with relation to the action of this pump when the ring is removed, you have assumed today, as you assumed yesterday, have you not, Mr. Adams, that when the ring was removed the cage which served as the cooperating part of the valve, together with the ring, may be likewise removed, within the teaching of the patent?

A Within the teaching of the patent, it might certainly be removed. I base my answer on the teaching of the patent which said there was a free flow and on the drawing it said it was inverted in direction.

Q Let us, Mr. Adams, contemplate two possibilities; one is that only the ring 25 is removed. In that event, as I understand it, you say that there would be some vertical flow downward and some outward deflection by reason of the cage part 26 which is left?

A I did not so testify. I confined myself to the wording of the patent which does not show how much is cut away. It simply shows it is removed, so I did not want to invent something of my own. I took the wording of the patent, which I have previously read, beginning at line 39, page 2, "the passages in the cap nut permit the free

(Testimony of Edward T. Adams)

flow of liquid and prevent the formation of a dead chamber containing trapped liquid". I took that to mean that that ring 26 was so formed that there was plenty of room for the fluid to flow vertically down past the ring. However, it is a perfectly natural thing if a man was going to do away with the valve 25 a man would throw away that ring 26 down there, and if he did that, of course there would be nothing to impede it.

MR. SOBIESKI: How does the patentee state, according to your opinion, that you divide up this solid piece—Fig. 5.

A It is fully explained in the patent.

MR. SOBIESKI: Isn't it true that the only thing that the patentee contemplates being removed is merely this fibre ring and then, with taking the fibre ring out, the oil goes out laterally more easily than when you had it in there?

A No.

Q Does the patent anywhere show how this solid piece, Figure 5, may be divided any other way?

A He describes that in page 1, beginning at line 90 and going through onto the other page; and states "At the lower end of the working barrel 19 a cap nut 21 is threaded therein" and then he goes on and describes that and what it does, and in line 98 states: "For the purpose of controlling the flow of oil through these passages an annular check valve 25 is provided and may be mounted upon the nut in any desired manner," The nut is the portion 21 in the Figure 5—"in any desired manner, for instance by means of the valve cage comprising the ring 26 separated from the nut by spacers 27 and open at its sides so as to permit the free insertion or removal of the

(Testimony of Edward T. Adams)

annular check valve 25 which may be formed of any desired material," etc. So he distinguishes between the nut 21 which has the vertical orifices, the apertured nut he might have called it, and part 26 and its supporting pieces 27, which is the retainer for the valve. I did not want to invent a scheme of my own and—as I read the patent there was room, as shown in Figure 4 and as specified in the patent, for the vertical flow of oil between the ring 26 and the plunger. If that is true, and if he means what he says by the "free flow of oil", then the oil would persist in the vertical direction it had, due to being guided by the openings 22 in nut 21, and it would continue in that direction rather than to turn sideways. However, I do not see any reason, if that valve 25 was taken out, why this flimsy thing which he describes really as an addition, may not and it probably would be removed at the same time. In line 37 on page 2 it says: "Under some conditions the valve ring may be omitted". I think the words valve ring refer to the fibrous part 25.

Q 25. That is the only thing he means as being removable, isn't it?

A That is all.

MR. SOBIESKI: He does not mention removing any other part of the cap nut, does he?

A I did not so read it.

Q Just removes the fibrous ring so as to get a free flow of liquid; that is all he mentions?

A I did not say he said "fibre ring".

Q No; he did not. "The valve ring", he says; and you consider the valve ring to mean this fibrous ring?

A I think so.

(Testimony of Edward T. Adams)

REDIRECT EXAMINATION

I cannot, in the light of my experience, think of any mechanical reason why a seat or cage for a valve should be retained after the valve has been removed. I testified yesterday, and am still of the same opinion that ordinary good mechanical practice would teach that that sort of a thing be done away with as an unnecessary appurtenance. He describes such removal as an act done to the main nut. And he furthermore states in the specification, that that valve may be mounted either in a cage or a container of any type, of which the one in the drawing is merely illustrative.

Further

REDIRECT EXAMINATION

MR. HERRON: We will ask the Clerk to mark this type of structure for identification.

THE CLERK: That is Exhibit PP.

MR. HERRON: Directing your attention to the testimony you have given concerning the Admore pump, that is the structure of Bradford in which there are four holes in the barrel at a point about an inch and a half or so from the bottom of the barrel, would it make any difference in the operation of that pump if the four holes had been drilled through the puller nut in some such fashion as is illustrated in this exhibit PP?

MR. SOBIESKI: I object to that question, your Honor, on the ground that there is no evidence that any such structure as this was ever constructed, and there is evidence that the Bray structure was construed prior to the application, therefore it is immaterial.

(Testimony of Edward T. Adams)

MR. HERRON: It is not a question of the prior application. I am simply asking whether it would make any difference in the operation of a pump, the location of those holes.

THE COURT: Whether they were a little lower down?

MR. HERRON: Yes; as to whether it would make any difference.

THE COURT: He may answer. He has gone over that field pretty thoroughly.

A   It would make no difference.

MR. HERRON: That is all. We ask that structure be received in evidence as an exhibit.

MR. SOBIESKI: We object to that structure, your Honor, as having no bearing whatever in this case.

THE COURT: It is only illustrative of what he testified to, that is all.

MR. HERRON: I am introducing it just as I would a chart or anything else.

THE COURT: As though he had made a pencil sketch himself and fixed the holes down in the nut itself instead of the side of the barrel, that is all this testimony is.

MR. SOBIESKI: Very well, your Honor.

Exhibit PP for identification is received in evidence as Defendants' Exhibit PP.

(Testimony of Patrick H. Bray)

PATRICK H. BRAY,

the plaintiff herein, recalled as a witness in his own behalf on rebuttal, testified as follows:

DIRECT EXAMINATION

MR. SOBIESKI: Mr. Bray, calling your attention to this Exhibit No. 17-A, which has been described by Mr. Herron as the inverted Bramo, this one here (indicating) has a metal piece directly above the lock. Will you tell us what that metal piece is?

A It looks like a packing sleeve, packed with something. The purpose of that metal sleeve, if packed tight, it would not allow any fluid to go out, would prevent the fluid from getting out of the bottom of the traveling barrel. If any gets in, prevent it from getting out, prevent it from getting in or out. In my opinion, that type of pump would not be operative in a deep well such as we find in this district, where any sand or gas is present in appreciable quantities. It absolutely could not work in a deep well for different reasons. One reason is it hasn't got stroke enough to even pull this lock on the bottom before it would get up. The other reason is, with the ball on the bottom which they generally use in one of these, in this part here, the ball in there—that is, the bottom and this is the top—and another ball up here, makes a strong space for a gas lock; and in all California wells more or less gas; you can't pump one of them that I know of successfully without what they call an anchor below to make a separation of the gas, not have it come through the pump. If you have it come through the pump it holds your balls up all the time, and the idea of this, there would be enough coming through here that would compress this

(Testimony of Patrick H. Bray)

and the balls could not work out or they could not close and you would get no fluid through it. You might get some gas through it.

Q How does this stuffing box affect the operation of the pump?

A I don't think that it would have any effect at all. It is packed pretty tight in there when this is pressed up against that spring in there. It is packed pretty tight. You couldn't move it very handy, but it is packed tight enough so there wouldn't any fluid get by before that packing wore out. How long that packing would wear I don't know. That packing will naturally wear out in there. Now, in speaking about the distance between the balls—by that I mean the valves in this pump, in the Bray and in the Hoeferer pump, the valves are not in the same relative position in the pump as are the valves in this pump.

Q Where are the valves in those pumps?

A One valve here, or the plunger; there is one in the traveling barrel.

Q About where would that be?

A When the pump sets down they would be in the neighborhood of from 6 to 10 inches, depends on how close they would space them. And they would be at the top of the barrel and top of the plunger.

Q How does that spacing of the valves close together at the top of the barrel affect the gas lock you have described?

A Well, it hasn't only that 6 to 10 inches for it to break.

Q What do you mean "for it to break", Mr. Bray?

A In other words, form a pocket of gas in place of— now, if they would run a pump, say, like this, only longer,

(Testimony of Patrick H. Bray)

and have the same hook-up they have in that, they would have more of a chance for a gas lock than they have in this because they would have it longer, because the distance between this ball and the lower ball would keep getting bigger all the time, longer. The greater the distance between the balls the greater the chance of a gas lock.

Q And the closer the distance between the valves, the less chance you have of the gas lock?

A They haven't very much chance for a gas lock with the two balls close together. The top ball opens very freely when the valve drops downward and displaces everything in the tube and directly from there; and, of course, when they are up, when this stroke is up this is supposed to be filled with oil from the top of the plunger here, top of the cage, rather—I don't know what that is on there for—the top of this cage here is all I carry in my pump. Mr. Hoeferer carries on top a cage with a longer seat. From there on up it is open to the other cage which is on the traveling barrel, and the ball seat in there, two valves, you know, and this comes in, I guess, in between these, whatever that stroke is, whether it is 26, 34, 40, 48 or 56. You can get five different strokes in an oil well, and whatever stroke that is, if they have oil in the well it will probably fill that space. If you haven't so much it might fill only part of it, but you might be carrying a longer stroke, but as that space starts down again the top ball opens and discloses everything out in the tubing; in other words, it by-passes all oil, sand, everything else away from your pump proper.

Q Can you state, from your experience in operating a pump in this district, whether in deep well pumps in which

(Testimony of Patrick H. Bray)

gas is encountered the practice is to use pumps of a design having two valves at the top close together?

A Well, I probably was one of the first started this in this country here. There were two other pumps of that kind. One of them was about a year after I started and one was here about the time I started. That was the Ellis pump. They worked the same way, that is, they worked on a traveling barrel, different designed and different made, and I find that most all the people that are pumping wells successfully have come to the upside down pump product. They find them easier on the rods. It is very seldom you have rod trouble with a free traveling plunger. I claim I have got a jet and claim Mr. Hoeferer has not, and the consequence is, if your plunger drops free it goes down to the bottom and it don't whip the ride; in other words, the plunger don't stick; the rods drop down with the weight pretty perfect. They don't buckle. They don't go into any corkscrew. That is, when the barrel drops freely without any undue friction. If the barrel sticks—if the plunger sticks in the barrel, of course, part way down, it causes the rods to bend all the way down, see, and when they come up they come up with a whip. But if you do that often enough and the plunger sticking there, you are going to break off that rod or break off the pin or break off the rod itself, which happens very often in a plunger that sticks.

CROSS EXAMINATION

BY MR. HERRON: As I understand your testimony, Mr. Bray, you say, in effect, that in the event that you are running in a well in which there is considerable gas pressure it is advantageous to have the valves close

(Testimony of Patrick H. Bray)

together as a precaution against what you have termed a gas lock, is that correct?

A   Yes, sir.

Q   In the event a pump is run in a well without an appreciable amount of gas, it makes no difference in that event, does it, whether the valves are near or far apart?

A   It depends on this: Now, you take the smaller wells that they have in Pennsylvania and different places of that kind, they can pump them with pretty near anything. There is no gas. The gas is exhausted, and the consequence is they don't need that precaution. Any well in California, unless it is one of these little ones out here in Los Angeles down a couple of hundred feet, something of that kind—and I am talking about a well that is producing from 50 barrels of oil up—in California, in my experience in Southern California, I actually believe that this is the only system of pump to get away from the gas lock, is to have the two balls close together.

Q   Then, the answer to my question is: That in wells where there is no appreciable amount of gas, irrespective of whether you find the well here or in Pennsylvania, it would make no difference whether the valves were close together or far apart?

A   I don't think it would make any difference where you have no gas in the well.

Q   A pump with the valves far apart would pump oil successfully in that type of well?

A   In that type of well, small wells, shallow wells.

Q   Now, you have said at the time you brought out your structure there were two pumps on the market, one of them Ellis. When did you first see the Ellis pump?

(Testimony of Patrick H. Bray)

A   The Ellis pump, I first saw it, I think, in '27, early part of '27. They had it on exhibition in the National Supply.

Q   Are you sure you did not see it in 1926?

A   I don't think I did, no.

Q   It might have been either 1926 or 1927?

A   Yes. I know it was about that time. I wanted to see if there was any difference in what I was making. I got my pump on a patent and built a pump around it. Understand, I had a patent at that time.

Q   Oh, you built your pump around your patent, not your patent around your pump?

A   I had to have packing means with that patent I had, you understand that. I had a device that was patented as a cone for packing but it didn't call for anything else. I had to have something to protect that packing, and the only concept and idea that I could get at that time was cups. Nobody had any patent on cups. Cups was free.

Q   Are you talking about some patent other than the patent in suit, Mr. Bray?

A   It is a different patent. I am giving a little history of my pump patent.

Q   Never mind about it as long as it is not concerned with the patent in suit. Now, as I understand it, you have testified that in the inverted Bramo, that is the Bradford construction with the tight-fitting gland at the bottom, that that gland, provided as that is, with packing means and spring, would prevent any fluid from passing up between the gland and the lower face of the plunger until the packing became worn, is that correct?

A   Yes.

(Testimony of Patrick H. Bray)

Q Now, I say before the packing became worn and the fluid got in, there would not be any question but the pump would pump under Pennsylvania conditions, would it?

A Oh, yes; it would.

Q What would prevent it from pumping?

A Gas is not always what locks. If you fill that space with oil—

Q Just a moment. You do not understand my question. You have testified that until the packing became worn the packing would prevent oil from passing into that space. Now, directing your attention to the time before the packing became worn, and in consequence, when there was nothing in the space but air, there would not be any question, would there, in your mind, but that pump would pump in a non-gas well such as they have in Pennsylvania?

A Yes; there would, for this reason:—

Q What would stop it?

A For this reason: You can compress air and gas or anything else. You can compress air very easy. It is hard to compress fluid. You have got a solid body of that, understand, and the same pressure all around, and gas and air is different; you can compress them. If that was closed in there, if you got air in there, if you went to work it you naturally compress it and it wouldn't let your plunger drop.

Q Freely?

A It wouldn't freely, no.

Q But at all events, it would pump oil—wait a minute. It would pump oil to the extent that the drop was able to

(Testimony of Patrick H. Bray)

compress the air and, hence, to allow that movement up and down of the traveling barrel, would it not?

A   No; if you couldn't get your plunger to move. Well, you got to get your plunger to move to pump oil, you know.

Q   Right; and you can get it to move against the resistance interposed by air in that pocket?

A   Well, I don't think it would be very long.

Q   I am not asking you about the length now. I am asking you if you would not compress the air and to the extent that the barrel would drop and compress the air you would get a pumping action and production of oil?

A   I think your plunger in going down would compress air the first time if it was in there.

Q   I think it would, too. But that compression would, nevertheless, allow of some change of dimension of the pumping chamber and in consequence result in some pumping of the well, would it not?

A   If that air was compressed very bad in that pump, surely a plunger couldn't drop.

Q   It couldn't drop the full limit, but it would drop to the extent that it was possible to compress the air, wouldn't it, Mr. Bray?

A   No; I don't think it would drop enough to work but very little. You would have such a tight compression there that your rods would bend up above and your plunger wouldn't go any further.

Q   As a matter of fact, the air compresses as you go up and not as you come down, doesn't it?

A   Oh, no. In this here, you mean in this?

(Testimony of Patrick H. Bray)

Q   The Bradford?

A   Well, either way. It don't make no difference which way it compresses if this is a traveling barrel.

Q   You do not compress air, do you, as you enlarge the space between the tight gland and the bottom of the plunger; that is not when you compress the air?

A   You compress it if you have got a 5-foot rod there. That is about 5 foot extends down in there and that works on that. You got a barrel over that, haven't you?

Q   Yes.

A   Well, that barrel, the top of that plunger is supposed to fit tight, ain't it?

Q   Yes.

A   The bottom is already fit tight with a packing?

Q   That is right.

A   And that space in there, if there ain't any oil in it, it must be air; there is something in there.

Q   Correct.

A   When that traveling barrel goes up it compresses air, don't it?

Q   Yes. And when it goes down the space there is enlarged and the compression, if any, released, isn't it?

A   When it goes down it won't work. It would probably release where it comes to but you are doing this with the weight of your rods. Your rods is spaced at the pump on these wells that way and if the plunger don't drop the rods will bend. They can bend very easy in a well. There is a lot of room for them.

Q   The question I am asking is this: Air will compress, won't it?

A   Yes.

(Testimony of Patrick H. Bray)

MR. HERRON: Now, as the barrel moves up the pumping chamber, the space up here, up above the packing, gets larger, doesn't it?

A Yes.

Q Now, as the gland moves downward the effective pumping space, that is, the space between the valves grows less, doesn't it?

A Yes.

Q We will assume that we here have a standing column in a pump equipped at the top with a plunger and a valve. Over that we will place a valved traveling barrel; this is supposed to be tight there and this is tight here, and we will put the other valve there at the top. Now, as the barrel moves up the area of this space which I mark A on this sketch is increased, isn't it? There isn't any question, is there, that as the barrel moves upward on the upstroke the area of that chamber is increased?

A If it could work it would.

Q All right. We are assuming a working pump.

A Yes.

Q And as that area increases oil is drawn down in through your hollow pump up through that valve and fills that area?

A Oh, yes.

Q Now, likewise, we will now assume, Mr. Bray— and if we take the next movement of the pump, when this area is filled with oil, when your barrel moves down the oil is ejected out through your top valve and up this pump?

A Yes.

(Testimony of Patrick H. Bray)

Q All right. We will now turn our attention to the bottom of the barrel, and will you assume in this structure concerning which you have testified that there is a tight fitting gland on the bottom of the barrel, making a tight fit between the bottom of the barrel and the standing column at the point marked B. We now begin the upstroke. As the barrel which is down moves up the area between the bottom of the plunger and the top of the packing gland, which I will call C, that area in there, is restricted, isn't it; it grows less?

A Yes; on a working pump.

Q And that results in compressing whatever is in that area, doesn't it?

A Compression, and if there is an out below it runs it out.

Q But we haven't any out. Now, let us just consider this structure. That results in compressing whatever is there, doesn't it? If it is air, your will compress your air, won't you, and as you compress the air, to the extent that the air is capable of compression you make the effective pumping area, pumping chamber A grow smaller, don't you?

A No. When it goes up it *grow* larger.

THE COURT: That is correct, Mr. Herron.

MR. HERRON: Yes. I am sorry. As you compress it, it makes this grow larger, so, to the extent of the compression, the making of the pumping chamber A larger draws oil up through the bottom of the well into the chamber, doesn't it?

A When it starts up it draws.

(Testimony of Patrick H. Bray)

Q Yes. And when it moves down the oil which has been drawn up will be expelled up to the top of the well, thus pumping the oil, won't it?

A It will be released direct into the tubing if there is any there.

Q Yes. Now, the net result is, that to the extent that the plunger can move up and down as against the tendency of the air to resist compression, to that extent you get a pumping action in the upper or effective pumping chamber of the pump, don't you?

A As I said before, I don't think—

THE COURT: No. Counsel says, to whatever extent it is; just that much.

A Yes. If it is two inches or three inches or four inches, you will get that much, of course.

MR. HERRON: In other words, the fact that you compress air simply restricts the effective pumping stroke of the pump and you get a pump which pumps less oil because of the tendency of the air to resist compression?

A Hang up on them. There will compress some— how much, I don't know. I don't think anybody else would know.

MR. HERRON: Just read the question, please.

(Question read by the reporter)

MR. HERRON: Can you answer that yes or no and then explain it?

A It will compress air.

The sketch to which reference was made is offered and received in evidence as Defendants' Exhibit QQ, and appears at Volume II, page 180 hereof.

THE WITNESS: Can I explain a word about that?

(Testimony of Patrick H. Bray)

THE COURT: Yes.

A   I would like to explain as it looks to me, under my experience. I have had a lot of it. If there ain't any oil in there or ain't any oil can get in there, there ain't very much chance for a pump like this to work successfully because the plunger won't compress very far, anyway. It can't compress all the way up and wouldn't get but a very small stroke. You couldn't pull enough oil in to get it up in four weeks, if it was 3,000 feet.

MR. HERRON: In other words, you feel that from an effective standpoint the stroke would be so restricted as to make not a desirable operation?

A   I figure, as I say right now, that this pump would not work in gas or a deep well. That is my evidence.

Q   Yes. And that is as far as you want to be understood as going in saying it is inoperative, namely, that it won't work in deep wells with gas?

A   That is right.

MR. HERRON: And the pump you are referring to is the inverted Bramo. Mr. Bray, you have spoken of what you call a gas lock. Now, the gas lock occurs between the top standing valve and the valve in the moving part, that is, the valve in the traveling barrel, does it not, and not between two valves both of which are standing, as for example, two valves in a standing column?

A   You have the lower valve in that standing column.

Q   Yes. I am assuming a case in which you have two valves in the standing column and also an upper. Now the gas lock that you talk about is that lock between the valve carried by the traveling barrel and the top valve in the standing column, isn't it?

(Testimony of Patrick H. Bray)

A   There would be just that much difference between the lower ball and the top ball in your standing column to make a gas lock.

Q   Let me see if I understand you.   In other words, you think a gas lock can be formed between two valves which do not move with relation one to the other?

A   They both move.

Q   No, no.   I am now assuming a structure in which there are two valves in the standing column and in consequence those two valves stand still in relation each to the other.

A   They can't stand still.

Q   They do, because they are fixed in the standing column.

A   Wait a minute.   Let me finish and probably you will understand what I mean.   Now, you have two balls in a standing column; you have one down below and one on top, we will say.   That is a standing column.   As your barrel moves up both those open down.   That lets in the gas in between there.   If you get enough in there as you come down again you will compress a little at a time and first thing you know you have got a lock in there.

Q   But that lock is between the top standing valve and the moving valve in the barrel?

A   No; not in the barrel.

Q   You think it is between the two immovable balls, that is, the two fixed valves which do not move in relation one to the other?

A   Yes; it will lock in there.

MR. HERRON: All right.   That is all.

(Testimony of Patrick H. Bray)

A   I know from my own experience, practical experience. The first pump I ever had on the market—the first pump I ever had on the market years ago I had two balls and seats and that was my trouble, and I had to take one of them out.

Q   And you know that, do you, from your experience?

A   Yes.

Further

DIRECT EXAMINATION

BY MR. SOBIESKI: Mr. Bray, calling your attention to this chart Mr. Herron drew, have you ever in your experience encountered a pump with a packing so tight that no oil could get around it from the bottom of the traveling barrel?

MR. HERRON: You exclude, I suppose, the Bradford deposition where he has testified that is the construction of the pump?

MR. SOBIESKI: I am not excluding anything.

A   No oil could get in there, under the circumstances, or anything else, I don't believe.

MR. SOBIESKI: Would, in your opinion, perhaps, a minute amount of oil get in if they put it into the well and started working it?

MR. HERRON: Now, if the Court please, I think we must object on the ground it is apparently an attempt to cross examine his own witness on a statement the witness has made.

THE COURT: I will allow him to answer. He may answer. Could any get in, he says?

MR. SOBIESKI: As a practical proposition, do you believe that that pump which we have here would ac-

(Testimony of Patrick H. Bray)

tually prevent all of the oil from getting in or would a small amount get in?

A I don't think any would get in because I don't think the plunger would even move. It might move an inch or so and it might move far enough, which would be like putting a plunger in the Hoeferer pump or my pump and not raising it up far enough so the barrel worked. Understand, if you can't raise that plunger you can't pump.

MR. SOBIESKI: In your opinion, will a fit on the standing column by the barrel which is so tight as to prevent any oil from getting in there, will that fit be so tight that it will also prevent the barrel from moving up and down?

A Yes. I believe it would compress air hard enough so it would not move. It would not move either way; it would stick.

THE COURT: Do you mean because of the friction effect of the packing on the standing column?

A No. I mean of the pressure that would be in between the bottom of the plunger and this packed off here.

THE COURT: I merely wanted to understand what you meant.

MR. SOBIESKI: These pumps are actually located in the bottoms of the wells, aren't they, Mr. Bray?

A Well, not particularly in the bottom. Some of these wells, you know, you pump some wells 3,000 feet or 4,000 feet; some of them you pump down as low as 7700 feet. It depends on the oil in a well. Wherever you place your shoe that is where your pump will go. The distance that your shoe is run in on the tubing, the shoe is the stopping point for your mandrel that is on your pump proper,

(Testimony of Patrick H. Bray)

and as that goes down and hits that shoe it locks itself at that point. Now, in order to recover any oil from a pumping process it is necessary that the oil be brought to the surface of the ground.

Q And unless the pump would work for sufficient number of strokes to bring the oil to the surface of the ground you would not recover any oil as the result of that pump, would you?

A You can't get any in, never even would get it up. You have got to keep getting it in to get it up on top or the pump won't work.

Q In your opinion, will the inverted Bramo type of pump with the tight fit you have discussed work sufficient to raise oil to the surface of the ground?

A No, sir.

MR. HERRON: Just a moment. We object to that on the ground that the question is unintelligible and that we do not know the type of well, the depth of the well or anything else, the point at which the pump is placed or any of the other factors.

THE COURT: We will let him make his own specifications on that.

A Well, we probably have a few wells here that that pump would be capable of raising the plunger if it was free, if it could run on a free column without any packing in there. We have a few wells here probably 300 feet, some of them less than that, that that pump would probably work in successfully, some little shallow wells from seepage oil.

THE COURT: I was going to ask you as to whether you were attributing anything to the frictional effect of the tight packing on the column?

(Testimony of Patrick H. Bray)

MR. SOBIESKI: Do you understand the Court's question? What effect do you attribute to the frictional effect? What result do you attribute to the frictional effect of the tight packing on the column?

A Well, the tight packing on the column is so tight that it would probably compress itself so hard that in place of the plunger working the rods would bend in place of that, and the consequence is the plunger would stand still and the rods would work up and down.

THE COURT: Do you intend to say that that is because of the compression of the air within that chamber that is there created, or do you say that because the packing is made extremely tight in order to exclude it that that has its effect?

A No; I think—

Q To lock and grip on the standing column?

A I think the packing would probably work up and down on the column all right, but it has no chance for anything to get by it, and the consequence is, if air or anything would be in there, it could not get by it if compressed.

THE COURT: I understand you.

MR. SOBIESKI: In regard to this Ellis pump, your best recollection is that you saw that first in 1927, Mr. Bray?

A I think it was the spring of 1927 I dropped in there one day. I don't know just what date I went in to see how one—they had one on exhibition and I wanted to see in what, if anything, it was like mine.

Q The date was all I was asking about.

A Well, I don't know.

(Testimony of Patrick H. Bray)

Q Now, calling your attention to the Admore liner barrel, Exhibit No. 7, will you describe to the Court the location of the valves in that pump and compare them with the location of the valves in the Bray pump?

A This valve is what we had up before. It is on the bottom. This valve is 16, I think. That is the standing valve, bottom standing valve. The equivalent valve in the Bray pump is this valve marked S.

Q In Exhibit MM. What is the purpose of putting this valve which is at the bottom of the standing column in the Admore in this location, S in the Bray pump?

A Well, to get them close together, and this is all open up to the top of my plunger, has a free access of oil to the top of the plunger, which this ball is set on the standing column, on top of the standing column. That is the only ball in the plunger.

Q I understand this design of having the balls close together is used to aid in prevention of gas locks?

A That is one thing and another thing, it makes a quick action to make a displacement of your oil. As this passes in slow this travels. The load in there between these two balls, as it starts down this cage is fitted with three or four openings to let the oil direct out in the tubing. In doing that you get away from a lot of sand trouble and also your plunger is not working through a column of oil all the time like your traveling plunger. You by-pass it direct into the tubing, which has been a big improvement on the old style pump. This design of pump is the design that is generally used in deep wells in this district at the present time. Most all the upside down pumps are that way. Probably some of them still trying to work the balls below, but I don't know, I haven't

(Testimony of Patrick H. Bray)

seen so many lately. They most all come to two or four balls up there. They can have two or four now. They have a blank cast here with another ball and seat. The idea of that is sometimes as protection, if this ball pounds out and gets leaking, why, the other one will hold up. As this ball comes down it gets an oil cushion. That is in this one but it ain't very many wells it is necessary to have it in. Some wells it is necessary to have that double ball seat in there on account of gas. I guess Mr. Hoeferer uses a good many.

Q. Now, calling your attention to Exhibit 7, does this have what you call the valved traveling barrel?

A Yes. That valved traveling barrel is marked 5 here. There is a working valve on the standing column. That is No. 16. It is not inside of the barrel but it is on the standing column. The standing column starts from the bottom here and goes up to this point right here where that nut is. Oil comes out there (indicating), goes through here. As this barrel moves up it opens this ball and lets in the load, and as it comes down again this ball closes.

THE COURT: At the lower end where does the standing barrel commence?

A At the lower end. That is No. 6. That is the bottom of your barrel.

MR. SOBIESKI: And the lower barrel is on the standing valve column?

A Yes. Now, there may be a ball seat in here. I say, there may be a ball seat in there.

THE COURT: Did you mean to say "lower barrel was on the standing column"? You mean the lower valve?

(Testimony of Patrick H. Bray)

MR. SOBIESKI: Yes. I mean to say the lower valve on the standing column; it is on the standing column there?

A It shows there. Here is the suction here up on top here.

Q Yes. In this model how is the trapped oil ejected from the barrel?

A Well, it is ejected from here. (indicating) Those holes in the bottom of the barrel, as the barrel moves up it can't go past this point if those are perfect without any slippage. It has got to go out through these holes and flows up.

Q What effect with regard to reversibility does the holes in the barrel have?

MR. HERRON: We will stipulate that the holes would obviously make it not reversible unless you plugged up the holes.

MR. SOBIESKI: In your opinion, Mr. Bray.

MR. HERRON: I stipulate because I think it is not really within any of your claims and really not an issue in suit.

MR. SOBIESKI: Is it an expensive proposition to plug up those holes in a barrel, Mr. Bray?

A I don't think, with the amount of work.

Q Does one end of the barrel ordinarily wear more than the other end of the barrel?

A. Yes. In the stroke of a barrel, with my engineering—this shows 33 inches, that plunger 33 inches long. That barrel is close to 10 feet long. Well, the working part of that would be probably ten. It would depend on the stroke, you understand, whether it was 26 inches or 34 inches or 40 inches or 48 inches or 56, that

(Testimony of Patrick H. Bray)

would be the stroke. Now, on the other end of the barrel where the pulling tube is it touches nothing only the guide. That is all that touches, so that keeps the other part of the barrel about the same as it was when it was new. There might be a little rust in there, but those metal guides cuts out afterwards. We try and clean them out when we take them to the top. There ain't very much in them, and generally pretty smooth, about the same as they was when they went out.

Q So, then, by reversing the barrel you get a longer life to the barrel?

A You get a double life.

Further

CROSS EXAMINATION

BY MR. HERRON: There isn't any question in your mind, is there, Mr. Bray, but that the type of pump illustrated and diagrammed on Exhibit 7 is a workable pump?

A Yes; that will work.

Q I think you said in your direct testimony that you consider that you had a free dropping barrel to your pump, is that correct?

A. A free dropping barrel, yes.

Q And that if you did not have a free dropping barrel the sucker rods would tend to "buckle or corkscrew". Is that what you mean?

A Well, if the plunger didn't drop.

Q Freely?

A That is, going down somewhere, the rods are going to bend every place all the way down.

(Testimony of Patrick H. Bray)

Q   You said it would set up a whipback and the rods would buckle or bend?

A   They ain't going to be like the spring because they wouldn't last long. They don't bend much.

Q   You mean they bend to some degree or buckle or corkscrew?

A   They will. Every rod all the way down will bend a little bit, so it makes it a whole lot when you get down. In other words, if you have got probably enough of a base there to pretty near—well, when you go in your well you set your plunger down or your barrel down to the bottom, see, and set it clear down, and then to space the well and give you kind of an idea how the rods work and all, to space a well you pull that up one inch for every hundred feet that you have got of tubing where your pump is setting and then you take about 10 inches more, and taking one inch for every hundred feet you are just taking the stretch out of the rods. If you pull up 10 inches more of it sometimes they hit down when you have to raise it up a little further. After you get that set once, if your plunger drops free there is no reason why the rods would buckle or anything else; but if it don't drop free, if the plunger sticks away up at the top after it comes up there, that will do it.

Q   Exactly. In consequence, in the pump you built you tried to get away from that?

A   Yes. I know from my own experience I have got one that will.

Q   Now, if I correctly understand you, you say that the purpose or one advantage of having valves close to-gether at the top of the pump is to prevent gas lock, and in some cases under extreme gas conditions one valve is

(Testimony of Patrick H. Bray)

not enough and you, therefore, add a second or supplementary valve—

A   I didn't say that was put in to prevent a gas lock. I said it was put in to make a little longer life for the balls. Sometimes a ball with gas pressure will pound and harden, you understand, but if you have an extra ball down here and blind cage, these balls work thoroughly satisfactory.

Q   Pumping under gas conditions you sometimes use two valves at the top instead of one?

A   Now, there are some wells, you understand, where sand will wear. You get silica sand and you get a grinding sound in there, something of that kind; it will chip that seat there and make a leak, and after that forms a leak there you can't pump oil. It goes back in the well.

Q   In other words, the silica sand operating between the ball and its seat will tend to wear away the under surface?

A   Sometimes it will start; sometimes it will only take one little point or something a little hard in there to chip that. Those seats are very hard and the consequence, if you get a start in there the pressure of oil—you have the whole pressure of the tubing from the top down against that, and if it starts through there it won't be over an hour or two until it has cut right straight through there.

Q   Do I understand you to say in the operation of the pump when you are talking of reversing the barrel that the guide takes any wear as between the outside tubing and the guide so that the barrel is not worn and, in consequence, you can reverse it?

(Testimony of Patrick H. Bray)

A   It is not worn on the end where the plunger is working.

Q   Is it worn to any degree down here where it rubs against the tubing at times?

A   It don't rub against the tubing.   This guide protects it from going against the tubing.

Q.   Now, you say in your experience you found in your pump your barrel had to be made to drop.   What type of structure were you building where it did not drop?

A   Well, I will give you an instance.   It ain't a year old—probably a year, about a year old, where I had occasion to put in a pump in the Rio Grande Oil Company. They had worked two different kinds of pumps, probably with as good reputations as anybody's pump in the market. That was an Economy and the fluid pack.   The fluid pack is supposed to be a fine pump.   This well burned down up there at the time of the earthquake and they put in everything new.   They put in a new string of B & B rods; and they have a very funny slick fine sand in that field up there this side of Long Beach.   That sand will get most everywheres, and the consequence is the Economy pump was on a tough proposition and the sand would get in there and swell it up and the plunger was not dropping and they started breaking the rods.   In three months they broke 12 rods.   They thought they had a rotten string that they bought from the B & B, which they didn't have, which I proved there.   They bought one of my pumps and put it in there and I have never had a rod job, never had a rod break.

Q   And that proves to you—

A   (Interrupting) That my plunger was dropping freely.

(Testimony of Patrick H. Bray)

Q   In other words, that your barrel was dropping freely?

A   Yes.

Q   In consequence, you did not break them?

A   No.

Further

DIRECT EXAMINATION

BY MR. SOBIESKI: Now, I call your attention to this exhibit showing the pump shown in the Cummins patent.   In this Cummins patent that is what is known as the traveling plunger type of pump, is that correct?

A   Yes; it looks that way to me.   Hook your rods on here.

Q   In this diagram GG Mr. Adams has shown how the pump would look if the barrel had been made—if the pump had been made in the type of the traveling barrel and moving plunger type of pump.   Now, can you tell us in the type of pump which we would have had if the Cummins patent had been reversed in his way, would there have been any trapped fluid?

A   Is there an opening here?

Q   It shows no opening.

A   There would not be any trapped fluid, I don't believe.   It would leak down here.   It would leak out.   If it couldn't get in here, in and out there, it would be like the other one.   It wouldn't work, I don't believe, not very successfully.   The reason, I believe, would be because there would be no provision made for the escape of trapped fluid.   That is, either to let it in or let it out.   It would never work successfully on that pump.

(Testimony of Patrick H. Bray)

Further

CROSS EXAMINATION

BY MR. HERRON: Mr. Bray, will you be so good as to show us the portion in which the fluid would be trapped, which you refer to?

A   Where it would be trapped?

Q   Yes; where?

A   If it went in there at all, do you mean?

Q   Yes; where would it be trapped?

A   It would be in here or should be in here.

Q   On that drawing where should it be? Where is that trap area?

A   I don't see any.

MR. HERRON: I thought not. That is all.

A   It seems to be all one plunger. I don't know. If that is the standing column there and it was long enough, and it would have to be to pump one of these wells here 7 feet at the least, then there should be a place for this to drop down there. There is room enough in here and between here for oil to come in here, but it can't get up there. It wouldn't work.

MR. HERRON: Oil comes in through here, goes up through the hollow rod?

A   That is pulled in from a well.

Q   Yes; that is what I mean. How are you going to work this down here, which is a space between that and the tubing, which is a recessed space in there? The tubing fits along here and along here.

A   You have got compression there. Remember, you have got the same—hold on. Wait a minute.

(Testimony of Patrick H. Bray)

Q Exactly.

A Your oil that will be pulled, it don't make any difference where it is. We will say from here down to the bottom on the shoe. This is the shoe here.

THE COURT: I notice there is an absence of any packing joints in there at all. It runs on the column, as I read the diagram.

MR. HERRON: I think this is intended to be in the way of metal packing at the end of the piston, is all.

THE COURT: No packing that would compel you to have a limit on your packing which would create that chamber at the bottom.

MR. HERRON: That was the point I was making. We will not go further if the Court sees the point.

MR. SOBIESKI: With this close fit along here there would be a compression in running up and down, is that true, Mr. Bray?

A It would, because your oil is equalized. You understand, there is just as much oil down here and it has got to go somewheres. Of course, when you are going down it pushes the oil up. Of course, you understand, but it would be pretty hard to drop that with that oil down there.

MR. HERRON: In other words, all the compression you mean is the compression caused by the dropping of the barrel down into the well within the tubing, causing a surging out, perhaps, as it drops?

A This drops down.

THE COURT: That is just a common body of fluid that stands in the tube all the way up to the top of the well.

(Testimony of Patrick H. Bray)

MR. HERRON: In other words, there is no compression chamber, as distinguished from a possible what you would term a compression, between the plunger body and the well?

A You have your tube in here and this is all oil down in there and it comes through the weight of the rod.

Q Of course, there is a space of greater or less dimension between the tubing and the pump, isn't there, as Cummins shows here by the dotted line?

A I don't know about that.

THE COURT: There is just the common pressure which is uniform, dependent upon the distance and gravity of the whole column of the fluid on the side. It only has to displace the liquid standing in the tube, so how could there be any compression, as termed such within the patent? Of course, compression against the whole body of fluid standing in the tube.

MR. SOBIESKI: Drawing your attention again to this Admore design in 7, with the holes in the side of the barrel, is it your opinion, Mr. Bray, that ejection of the fluid, while clearly through the side of the barrel, causes as great a disturbance of the fluid underneath the bottom of the traveling barrel and on the top coupler face of the shoe as where, in yours and the Hoeferer pump, the oil is ejected directly down through the nut on the bottom of the traveling barrel?

A No; it hasn't got the same effect at all. In the first place, this barrel is down when it starts up, that is, when this oil starts out. Now, it has got a baffle to hit all the way up. When it hits that baffle—the tubing is the baffle. When it goes out of these holes it hits the tubing, it cracks: in other words, it spreads; it don't go down. There is a

(Testimony of Patrick H. Bray)

body of oil down there and as this goes up it keeps hitting that tubing all the way up as far as it goes. As it comes down it hits just the same way in the lower part. It keeps squirting down straight all the time, right straight down, and it will go through a lot of mud; it will stir a lot of things as it comes through there under pressure.

Q Does the Hoeferer or the Hofco pump eject the liquid down through the bottom of the bottom nut in the traveling barrel?

A Yes.

Q Does, in your opinion, the Hofco pump, by reason of that ejection, get the same result which your pump gets?

MR. HERRON: Do you mean same in point of quantity? I do not understand your question.

A The same results; same results as a jet. Yes; it has the same, the same result that you get with my jet. Either one or the other of the three of them probably handle about the same oil. I haven't never had no trouble with mine, is the one reason. I put in about 500 pumps and I never had any trouble about not dropping on account of anything there.

MR. SOBIESKI: I call your attention to the Bray pump; is the problem of sand forming below the traveling barrel a problem that is frequently met with in oil wells in this district?

A Yes. There is more or less sand trouble in most of the wells, and sometimes where there is a great deal of water—I pumped wells where they were 90 per cent water, and if there is any sand in there they ain't got enough oil to pack that sand to the top. Water won't pack sand. You have to have oil probably 25 per cent up.

(Testimony of Patrick H. Bray)

Q Then this condition is especially aggravated in wells with a considerable amount of water?

A Yes. I generally get oil enough from any wells which I pump to pump. If the water will not stand up I put in a different process.

Q Now, will you explain to the Court on this diagram of the Bray patent just where sand would form if the pump was going to sand up?

A Now, there are different ways for a pump to sand up, your Honor. One is a natural drop down that sand works on down to this point. There is your shoe there and this is your lock on the bottom of this standing column. That lock is stationary but if you get enough sand down here to keep building up, pretty soon this pump won't drop down; it will keep going up here and build up with sand down here. And that is the case a great deal that I find out with the holes in the sides. You haven't got a chance to move this sand after it gets down there. It keeps building up. The consequence is it sands up. That sometimes causes what we call a free-hand stripping job. That means three or four days. You have to pull the tubing and rods together. Sometimes if they don't find the connection, sometimes it will be up 10 feet, sometimes way down below, and it means a double job all the way around and there is no way of getting away from that outside of pulling that when that condition happens. Now, I haven't had any sand trouble of this kind unless it was a heavy well, a well that would heave up sand, would fill everything two or three joints above your pumps.

THE COURT: Do you mean by that that the effect of your jets in that compression chamber—I assume that out

(Testimony of Patrick H. Bray)

of the whole column of the oil the sand will settle out of it down to the bottom?

A Settle down to the bottom and then the jets move it.

Q And is that strong enough to move that so it can be taken out above from that whole column?

A I think so, Judge. I think if we had a glass jar here and put some sand in it and some water and poured it in, you would find out it started to fill.

Q After you got up 10 or 15 feet from the action of the jet?

A Well, it generally starts from the start; starts stirring it.

Q Then you have the sand continually?

A Not so much. It is a bad well where there is even 5 per cent sand, a bad well. Some of these wells won't make probably 1 per cent of sand, so there ain't so much, but there is enough there if it accumulates there and has no way of stirring it up down there to get it out, it is going to stay there and the first thing you know it will sand up, that is, if your plunger is working.

Q What I don't quite appreciate is: After you do stir it up first, how that stirs enough to cause it to go up and come out. Of course, it has got to go out to eventually get rid of it.

A When the barrel comes down again—it has already left a space here; the oil has come down the tubing. The difference of the OD on this pulling tube here and the difference of the OD inside of the barrel, which, in my pump, my pulling tube is an inch and a quarter; my barrel is one and fifteen-sixteenths. Now, that comes down here and comes down, and as it goes up this fills up. As the plunger comes down again you make a displacement here

(Testimony of Patrick H. Bray)

out—and also this oil has got to move up, too, because the barrel has got to get back where it was when it started. Everything moves as the barrel goes down again. Some of the Bray pumps have always been made without holes here, and I find it my experience that I get practically the same effect with the bottom nut having an additional clearance instead of using the ducts, and I get the same results. I have been working them now for quite a while and I don't seem to have any trouble at all. On this pump this barrel here, this bottom nut is of the same dimension as the bottom of the barrel. It is the same outside dimension. Some of these pumps are actually in the well at a slight angle, a little crooked bend. The well is not perfectly straight. Very few straight wells, very few. Now, in my experience this bottom nut down here would tend to guide the barrel in its downward motion into the direction it was going. It won't let it come either one side or the other strong enough to hit the tubing. This guide—now, if you get a very crooked hole you will find out that your guide will wear out in a crooked hole; shows that it is guiding. Where your barrel is not worn at all your guide is worn.

Q In looking at your patent, which is in evidence, on page 1, approximately line 53, you say: "Figure 3 is an axial section of the jet nut or foot bushing of the barrel". Is that the part of the pump which is elsewhere in the claims mentioned and described as a guide?

A Yes; described in one, I believe.

Q Yes. I understand, then, that this bottom nut on the lower end of the traveling barrel is spoken of in your specifications as a "jet nut" or "foot bushing" in addition to being spoken of as "a guide"?

(Testimony of Patrick H. Bray)

A Yes.

Q Mr Bray, is the bottom nut on the lower end of the traveling barrel spoken of in the trade in various names?

A Why, yes; some of them call it "a puller nut".

Q MR. SOBIESKI: Has this nut other names?

A Some of them call it a "guide"; some of them call it "slush nut" and different things of that kind, but I would think that—

Q In your patent you have spoken of it as a "jet nut", "foot bushing", "jet bushing", and "guide"?

A Yes. I tried to get all the names I could.

 Further

CROSS-EXAMINATION

BY MR. HERRON: Now, Mr. Bray, you have testified that you get the same results from your pump whether you made it with a series of ducts, as shown in the litttle cross-sectional sketch of your patent, or whether you made it with an annular opening between the standing column and the inside face of the puller nut, is that correct? ?

A I have never seen no difference in the working of the pump. I get the same results in oil and it does not sand up.

Q Now, as a matter of fact, would it make the pump inoperative if the puller nut were entirely eliminated so all that happened would be the barrel would churn up and down in the oil?

A Yes; it would.

Q Did you ever do that?

A No time.

Q You don't know that, do you?

(Testimony of Patrick H. Bray)

A   Yes; I know it, for this reason: I know that a barrel could not work that way unless they had something to guide to keep it from rubbing up and down the tubing. You are wearing a hole every time you do that.

Q   Let us forget about wearing and assume it would rub against the tubing and wear it, nevertheless would it become inoperative because it would sand up; you wouldn't say that, would you?

A   Ask that again.

Q   You would not say it would become inoperative and sand up, would you?

A   Yes.

Q   You would say that?   Why?

A   Because you have no chance to move it.

Q   In other words, you think the only thing that operates to move that sand is your jet action?

A   Is the pressure.

Q   And not any churning up and down of the barrel in the liquid in the bottom of the well, nor any tendency of the fluid to move downward and across the face of the clutch and up through and past the seal?

A   It don't do that.

Q   You don't recognize any such thing?

A   It don't do that.   It don't go that way.

Q   How are you going to get fluid past this point here? In other words, you say there isn't any motion of the fluid down past any tubing around over the face of the clutch and up, escaping as a leakage past your packing; you don't recognize any such force?

A   No.   I don't think anything could pass this packing as long as there is any packing in this pump.   This is a cone the same as your pump.   If that packing is put

(Testimony of Patrick H. Bray)

on there—give me my model there. I can explain it probably a little better.

Q I get your thought. I simply say, you think there is no passage of fluid between the packing and the barrel in your pump?

A As long as that packing is in there there is nothing that can escape either in or out.

Q All right. It is perfectly clear to me what you say.

THE WITNESS: Now, if it please your Honor, I would like to show this.

MR. HERRON: All right; go on.

A It won't take me but just a second. Take this off, pull it out and show you here that packing is made and runs in a well and it don't wear out only the top at one time. Now you notice, your Honor, that there is a cone there. That cone or packing is bound spirally around here. As the pump moves up this packing is compressed up against that swedge or cone and it forms it very tight where it goes up. Now, this packing in my two and one-half pump is 3/8, and when it is up here it is compressed to about 5/16. Now, the only thing, as this moves down again where there is no wear still it is packed off. Now, these grooves in here, you understand, is cut in that way; that one is up to keep any pressure away from the packing for protection, and also any sand that would come down. Here is a hard ring; that ring there is hard steel. It is smaller than this. That is another protection to keep sand out. Now, these in the pumps here are built the other way. They are for suction. As the pump goes up the oil fills in across here and forms an oil seal. Each one of these is an oil seal, and this is only half of the size, but the idea of a close fitting pump. Now, I have worked

(Testimony of Patrick H. Bray)

this pump in wells after pumps of all metal was emulsifying oil, cutting that. I have taken and kept records of it where the pump was cutting 80 per cent, where it was like a molasses, thick and red, and fetched it down to 9 because there was no slippage in it. If you can get slippage you can cut oil, and if your oil has slippage down here you are going to cut the oil.

MR. HERRON: Do I understand by that, Mr. Bray, that you mean you have a pump which is 100 per cent efficient in sealing off any escape of liquid between the pumping chamber past the packing of your pump, or in preventing leakage passing in the opposite direction?

A As long as there is one ring of packing in there it will do that. When it goes—now, here is the way I find out: When it goes the well will immediately fall off 20 to 30 barrels.

Q In other words, until it is completely gone you have 100 per cent seal, in your opinion?

A That is, I figure none getting by it.

Q Like the One Horse Shay; when it goes, it goes all at once and up until that time it is perfect. Now, you have said that your pump never sticks. Never sands up unless there is a heave in the well of sand sufficient so the sand mounts up—did you say to about the third joint of casing from the pump?

A Oh, it sometimes goes up quite a ways. It depends on the well.

Q In other words, your jet will keep free from sand in any condition except a heave of such dimension that the sand will go up a joint or two in the casing?

A There is another condition, say, for instance, a man hasn't got very much oil in his well.

(Testimony of Patrick H. Bray)

Q   Let us take a typical one.

A   That is not a heave. Now, I am giving you two conditions.

Q   I am talking about the heave. Let us do one at a time.

A   All right; I will explain.

Q   Now, do I understand your testimony to be that the jet action of the structure of the pump you are making is so efficient as that the pump never sticks unless there is a heave of such proportion as to cause the sand to pass up level with the second or third joint of the casing?

A   Yes.

Q   And you ascribe the fact that your pump does not stick except under those conditions to that jet action only and nothing else?

A   That is the only thing I can place it by.

Q   Yes. And you testified you never have set down a pump with the puller nut off, or in any other fashion left entirely open at the bottom end to test whether that pump would sand up or not under the same conditions, have you?

A   I never put one in but I know it will. There is no action there.

Q   You, yourself, know that but you never experimentally proved it?

A   You know, you have got to know with your experience. My experience is I never put in a pump that way, because I think I would be crazy if I did.

Q   All right. Now, have you any experimental data or have you anything else upon which you gauge the effectiveness of your jet action except the one fact that in

(Testimony of Patrick H. Bray)

your experience your pumps have not stuck except in case of a heave?

A Yes; I figure this: I figure on the results of the oil that you get on top and the length of time you keep going with a pump means a lot. If you are pulling the pump every few days or something like that on account of some sand going in there, you ain't getting the results, are you?

MR. HERRON:

Will you read the question?

(Question read by the reporter)

A I have no data outside of practical experience with it and the pumps that I have put in. I found other pumps under the same conditions did sand up and mine did not.

Q Now, what sort of pumps were they?

A Well, they were different kinds of pumps. We will say Neilsen is one and O'Bannon pump was another one.

Q Did you find that Neilsen and O'Bannon pumps sanded up under conditions that yours did not?

A Yes; under the same conditions.

Q Now, what were the conditions under which they did sand up? Let us get concrete and tell what the conditions were.

A The sand did build up below where the oil came out and stay there, and the consequence was they kept building up a little bit. If your pump goes down to that sand it can't go any further and you have to work from there up.

Q How do you know that was true in the case you mentioned?

A Had to pull the well.

Q In other words, that pump was stuck?

A Was stuck.

(Testimony of Patrick H. Bray)

Q Now, you don't know what stuck it, do you?

A Yes. If the puller don't drop, you know, you have got something.

Q Something that stuck it?

A Yes.

Q You don't know what it is?

A Mostly sand, you can count on that. There is something. Now, I will put it this way so you will understand probably better. All oil men—

Q Now, let us keep on the subject I asked.

A This is on the subject. I want to explain that. You asked me a question and I was trying to explain it here. When this sand, as I say, builds up below that oil outlet or hole in the side it keeps building up. Your plunger is going up here; you are going all the time. And a man that understands his business will know it is hitting something, from the top. You can feel it. It must be hitting something solid or it would have gone down. Anybody who will take hold of a polish rod, he can feel that from the top. He don't have to guess at that. He knows it is sand.

Q Now, where did the sand come from?

A Where it comes from? It settles to the bottom.

Q From where? Just illustrate on the diagram where it settles.

A Take it all the way up; take, for instance, you are making 5 per cent sand and your pump is working very nice and you have it shut down for 4 or 5 hours, or maybe half a day or a day, and sand and that is going to settle all back down there to the bottom as long as it ain't moving. Even with oil it will do that.

(Testimony of Patrick H. Bray)

Q   In the pumps you mention had the wells been shut down?

A   Had they?

Q   Yes; in the cases where there was a sticking that you mentioned.

A   No. Very often—

Q   I am talking about the particular instance you refer to, where you say an O'Bannon pump stuck, had the well been shut down before that sticking occurred?

A   I don't know. I didn't drill them. I don't know that.

Q   You don't know?

A   No; I don't know that.

Q   So, you don't know whether the sticking was caused by the fact that the pump had been shut down and, hence, all the sand in this upper column had slowly settled down and packed the thing in full, do you?

A   Well, I will probably put a witness on that will know. I don't know that.

Q   You don't know yourself?

A   No.

Q   Now, in these other cases where you say other pumps stuck, do you know whether in those cases the pump had been stopped so that you had that same gradual settling of sand with the consequent building up of sand all the way up here?

A   I don't know what pumps you are talking about.

Q   I think you testified there were some other pumps.
A   I mentioned two.

Q   All right. What was the other one?
A   The O'Bannon and the Neilsen.

(Testimony of Patrick H. Bray)

Q Are you testifying now about an actual case of the Neilsen stopping due to sanding up?

A I know where they pulled them out and I replaced them with my pump.

Q Now, please, I am asking if you are testifying about an actual case that you know of yourself, as distinguished from heard about, in which Neilsen pump sanded up?

A Yes.

Q All right. Now, in that particular case do you know whether the well had been shut down before that sanding occurred or what it was, do you know that?

A It wasn't shut down.

Q It was not. How long had it been before the first shutdown had been?

A It just stopped pumping.

Q I asked you when the first shutdown had been how long before the time of the sanding up had the last shutdown been, do you know?

A Well, they worked it for quite a while. They worked it until I got down there. It was my pump. They were working it but it wasn't working there.

Q In other words, the pump was actually working and they pulled that out when you got there, but it simply didn't pump oil?

A You could work that up on top. It was not pumping oil. You could work it if it was only 6 inches. The rods would do the balance. If it was all plugged with sand up to seven feet or six feet and a half, and you had seven feet of rod you could still work that as far as it would go, but the rods would bend.

(Testimony of Patrick H. Bray)

Q In other words, they were pumping the pump and you had no trouble in pulling it?

A Yes, they did have trouble with pulling it. They couldn't pull the sand valve. They pulled a wet job. They couldn't pull that pump out with that sand in there.

Q Well, did they?

A No; they didn't. They pulled tubing.

Q They pulled the tubing?

A Yes.

Q How far had that sand packed? How far was the sand extending up in the casing, do you know?

A Oh, now I don't know. It was stuck in the tubing when it come out, and I was waiting there for them to burn it out or pound it out.

Q And you don't know how far the sand went up into the tubing and the pump?

A. No. I wasn't interested enough really to stay there that long to see.

Q So, in point of fact, you don't know how far the sand went up?

A I know it was sand.

Q Or you don't know what had caused the sand to pack, do you?

A Yes.

Q The answer is "No"?

A I say yes; I know it was sand.

Q Or how much there was or what caused it to sand, you don't know?

A I didn't measure it.

Q Do you know whether there had been any heave, any slight heave in that well?

A No.

(Testimony of Patrick H. Bray)

Q   How do you know there had not been?

A   Because I pumped it myself.

Q   I mean, how do you know there hadn't been a slight heave before it stuck?

A   Do you know what causes a heave?

Q   I know.   How do you know there hadn't been?

A   Some wells can't heave because they ain't got very much there to heave.   Other wells have quite a lot of floating sand.   They handle that pretty well but they don't get a heave.

Q   As a matter of fact, wasn't the presence of sand between the plunger and the barrel the cause of the failure of the pump?

A   Yes; sure, that is what I say, between the plunger and the barrel.   It had to be there.

Q   Sand in the pump, sand in the pump between the—

A   No; the sand come down here.

Q   Wasn't there also sand in the pump between the plunger and the barrel inside the pump; sand inside the pump, wasn't there?

A   No; there was no sand inside the pump.

Q   Did you examine it to see if there was?

A   The pump would work.   You could pull it to see, a little ways back and forth.   It wasn't that and here is where it was stuck.   The barrel was stuck between the tubing and the barrel.

Q   Did you examine the pump to see whether there was not sand between the plunger and the barrel?

A   They couldn't pull it out.

Q   Can you answer my question, please, that I asked you?

A   That is what I am trying to do.

(Testimony of Patrick H. Bray)

Q Just listen to it.

THE COURT: You looked at it, of course?

A Yes; I looked at it. But you could work the barrel but you could not pull it out because the sand valve in there was stuck.

MR. HERRON: Oh, exactly; the standing barrel inside the barrel was stuck.

A This column was stuck. You couldn't pull it out, the mandrel.

Q In other words, when you had the pump on the ground you couldn't pull this out of the pump because this part was stuck in the barrel, is that correct?

A Yes, about here. This part down here was stuck. Your sand drops down here.

Q Do you say that after you got the pump on the ground you could not pull the barrel with relation to the plunger?

A Now, you get this right this time. You have got 7 foot of a rod here; you have got 7 foot of a pulling tube in these ordinary pumps. Some of them has got more. If this builds up with sand down here, this is going to fill in all with sand and your plunger won't drop down only to that sand, and finally you won't work over 6 inches, but that still is working, but it can't pump because it can't go down there so far because it has got to work back up.

MR. HERRON: Please read the question, Mr. Reporter.

A That is the nearest explanation I can give.

MR. HERRON: Just listen to the question. I think you can answer.

(Question read by the reporter)

(Testimony of Patrick H. Bray)

MR. HERRON: That is, after you got it up on the ground at the top.

A No; you couldn't pull the barrel or you couldn't pull the plunger, either one. They go together. If you could pull the plunger you could pull the barrel.

Q By the way, did you ever know of a fluid-packed pump to sand up?

A Yes, sir; plenty of them.

Q Many. Of your own personal knowledge, yourself?

A Yes, sir; where I replaced them.

Further

DIRECT EXAMINATION

BY MR. SOBIESKI: Mr. Bray, when this pump came out of the well which you had described, will you tell us where the sand was on the pump that you saw?

A Got it out of the tubing. The tubing was pulled out altogether, laid there, and probably was taken away that way to the shop in the tube. But they couldn't pull it out—they couldn't pull this out and, in consequence, it was sanded up. That is what caused it.

Q You mean they couldn't pull the entire pump out of the tubing?

A Yes. When everything is clear everything comes out. That is right. This mandrel here that fastens in her, and also fastens into the lock in the shoe, if they can't pull that of course they have got a sanded-up job.

Q And the pump was stuck in the tubing which goes around the pump?

A Was all stuck down here and was filled up probably to the bottom of the plunger.

(Testimony of Patrick H. Bray)

MR. HERRON: You say "probably". Do you know how far it was filled?

A No. I know it was stuck, anyway.

MR. SOBIESKI: Well, in order for the sand to prevent you from pulling the pump out of the tubing the sand must have been packed—was the sand packed tightly between the pump and the tubing?

MR. HERRON: He has just testified he did not examine it, Mr. Sobieski.

A I didn't examine this, you understand, because I couldn't get it out.

THE COURT: I think we understand that when you got there and looked at it this pump was in the joint of the tubing, wasn't it?

A Yes.

Q And you tried to pull it out and it would not come?

A Yes; and also, they couldn't pull it before they started out. You see, they always try to pull their barrel out before they start up because if they don't they get a wet job, your Honor.

Further

CROSS EXAMINATION

BY MR. HERRON: Was the Neilsen pump you have mentioned similar to your pump in the arrangements of the pump for jetting action, or similar to the Hofco or both?

A Holes in the side.

Q Was there any clearance between the standing guide column and the puller nut in that pump, the Neilsen pump?

A I will swear, I don't know. I never looked at one of them very close.

(Testimony of Patrick H. Bray)

Q   You don't know.

A   I don't monkey around people's shops.

Q   So you don't know whether the clearance was great or little?

A   You would have to take that off and measure it and they won't let you do that, you know.

Q   Where was that well we are speaking about?

A   It was in Inglewood.

Q   Could you give us the number?

A   Bell Oil. Yes; Bell Oil Company. I don't know what the number was. They are still in business.

Q   And when did this thing happen?

A   Well, I would have to go up home to get you my records.

Q   Do you know approximately the year?

A   Oh, I don't know. Six or seven months ago.

Q   Oh, just recent history?

A   Yes.

Q   I believe you testified that in your pump the packing is of such a character that you will get 100 per cent pack of the pump as long as there is a ring of packing left; is that, in effect, your testimony?

A   I said there couldn't anything slip about it.

Q   In consequence, if nothing could slip about it, it would mean 100 per cent seal, wouldn't it?

A   Well, you might place it that way.

Q   Isn't that a fair construction? I am asking you.

A   I think it will.

Q   Yes, all right. Now, as I understand it, you say that as soon as the last ring goes you will get an immediate drop of the pressure and then the whole packing goes and the pump ceases to exert any pressure at all, is that it?

A   Oh, no. The pump will pump, but I say—

(Testimony of Patrick H. Bray)

Q Oh, the pump will pump without any packing in it?

A Yes; and it will drop off from 20 to 50 barrels on account of the slippage.

Q You mean, do you, that after your packing has gone you get still an 80 per cent efficient pumping pump?

A I said from 20 to 50.

THE COURT: The question is: Did you mean to say that when your packing wore, that your whole packing, as counsel incorporated that in his question, that immediately the whole packing would go, is that true?

A If the whole packing is worn out immediately the fluids—

Q Well, what I am trying to get at is: Would the packing all wear out at once after it began to wear?

A No.

Q Would it be a gradual process or not?

A No. It wears out—them top rings, as it is pressed up against the spring, the top ring is on a cone—yes, I showed you that—and that is the only part that wears, and it all wears there until the last one is pushed up in that way.

THE COURT: I see.

MR. HERRON: So, then, there is a complete release of the pressure of your pump, is that what you mean?

A I say, when that packing is gone I get a certain slippage there, enough to lose that much fluid. Now, I have oil seal grooves in this that do—here they are here (indicating)—each one of them is an oil seal groove and they still leave that oil get a suction.

Q. Now, what percentage of the effective action of the pump would you have after the cone-shaped element

(Testimony of Patrick H. Bray)

you referred to has completely failed? What percent of the effective operation of the pump would you get?

A   Well, probably from 50 to 80.

Q   Still pump 50 to 80 per cent?

A   We always know in working this kind of pump when that packing is gone. We know when to pull it. The packing lasts in here all the way from four weeks to as high as six months.

Q   And then you will observe when that packing goes that drop in effective action in the pump.

A   And if we pull it out we find no packing. Once in a while a little piece about that big (illustrating).

Q   After you observe that drop in productivity of your pump and you therefore know that the upper packing is worn out, does the pump stick?

A   No.

Q   The pump, I suppose, is continued in operation from the time that that drop off in the production is found through the tour in which it occurs until it is located and until you get ready to pull the pump and repack it, is that right?

A   Yes; and they try and do that as quick as possible, for this reason: They are losing this oil that they should be getting when that packing is completely gone. We haven't got as tight a barrel and plunger as we should have and you get a certain slippage by those metal guides and they cut the oil into emulsion. We see in here no oil can emulsify because it can't go by.

(Testimony of Patrick H. Bray)

Q In other words, when you get the emulsion you know that the oil is going up past the seal provided by your metal sealing devices?

A Yes.

Q You know it is slipping. Now, how long—and if I understand it, in operation the pump is kept continually in motion notwithstanding the fact of the decreased production, until you get ready to pull it; that is, in order to avoid stoppage of the pump with the consequent settling of sand down between the barrel of the pump and the tubing, is it not, so it won't stick?

A The pump will therefore act below the plunger just the same as it would if the packing was in, no difference about that.

MR. HERRON: Just read him my question, please.

A That is what I am trying to get at.

MR. HERRON: I think you do not understand, so we will just read it.

(Question read by the reporter)

A My answer to that would be that sometimes they don't have time to pull it right at the time and they keep pumping and they get what they can get.

Q And that pumping sometimes goes on a matter of many hours or many days?

A Oh, many weeks sometimes.

Q And you find that when a pump of that sort such as yours has gone on operating in that condition for many weeks and you get on the ground with your new pump, you can pull the pump out through the operation of the

(Testimony of Patrick H. Bray)

puller nut and its corresponding upper part and put in your new pump without difficulty?

A   I don't put in no new pump.

Q   And repair the old one, as the case may be?

A   Right on the floor, yes.

Q   And even after operating many weeks, as you say, you do not find that under that condition your pump sticks?

A   No; it don't stick.

Q   Now, you say the operation below the plunger is not changed. Is not changed by the fact that the packing up here, I suppose, which would correspond to that packing represented in that case—that packing has worn, is that true?

A   I din't just quite get the first part.

Q   I understand you to say that the operation below the plunger here, that is, between the plunger—in other words, the operation of this annular space II, is not changed by the fact that there has been a failure of the packing at the point CC?

A   No change as far as jetting of the oil or a flow of the oil coming down through this nut is concerned. It fills up when the pump goes down and releases when it goes up, just the same as if there never was any packing.

Q   In other words, you feel that your pump would just as effectively jet and remove sand from off the face of the puller nut if you did not have this tight fitting packing, as though you had a tight fitting packing?

A   Yes, sir.

(Testimony of Fred H. Hayn)

FRED H. HAYN

called as witness on behalf of the plaintiff, in rebuttal, testified as follows:

DIRECT EXAMINATION

My name is Fred H. Hayn. I am a patent attorney, graduate of Cornell University, mechanical engineer, electrical engineer, admitted to the Bar, a member of the Federal Courts, generally, the Supreme Court of the United States, and been an assistant examiner in the Patent Office for over 21 years. United States Patent Office; and reached the grade of first assistant examiner and have been in charge as acting examiner in several divisions of the office. My duties in the Patent Office entailed the reading and interpretation of patents. Sometimes at the rate of 50,000 a year. Of course, those vary widely. I have examined small condensers, steam engines, steam and vacuum pumps, carburetors, in the office generally.

I am familiar with the Bray patent No. 1,840,432, involved in this suit.

Q For your convenience I hand you a copy of it. I also hand you a copy of the Northrup patent which has been introduced in evidence by the defendants, No. 1,378,268. Are you familiar with the Northrup patent?

A Oh, yes; pretty clear what is shown here.

Q In a general way?

A Yes.

Q Does that patent have on it a lock on the bottom of the traveling barrel?

A I am unable to find any.

(Testimony of Fred H. Hayn)

Q Does that, in your opinion, contain a bottom guide apertured for the ejection of liquid from the traveling barrel in a downward direction?

A The reference numeral 24 on page 2 of the patent, lines 12 to 18, there is a reference to a threaded bushing or guide 24. Now, that bushing or guide discloses a sort of clearance. I would say it does not touch the barrel inside.

MR. HERRON: You mean it does not touch the standing column inside?

A It does not seem to me to be—there is no reference character denoting it that I can see.

MR. SOBIESKI: What figure are you pointing to?

A Pointing to Figure 1. Now, let's see; 32—no; there is not one word mentioned as to what that is there. We cannot guess as to what a drawing shows. There must be some reference in the description for a disclosure. You can't read into a patent that which—I answer I can find nothing whatever of any of the oil feature counsel referred to.

Q Do you find anything in that patent which contains a device similar to the claims described in the Bray patent?

A You must remember that there is nothing new under the sun. Every element that is found in the Bray patent here has been used time and time and again. Even the Edison phonograph, Every element was old. These are a separate—what we call combination claims. Each element itself may be old. There must be some new advantage or result, or an old result in a new and materially better way to be patented. I find in the Northrup patent the valves 28 and 19. I find some packing elements there

(Testimony of Fred H. Hayn)

but I do not find the lock. I do not find—that is to say, meaning, of course, in the claim 1 here, this coupler and the anchor and the non-rotatively interlocked with the lower guide coupler—I do not find that at all. There are some holes there. They are not even mentioned by a reference character. They are, I think, referred to in a general way on page 2 of the Northrup patent, lines 33 to 37. Now, that seems to be more or less guess work, and certainly those holes in there are not the equivalent of these apertures 5 in the Bray or the—

MR. HERRON: If the Court please, I think I must object to the witness testifying as to whether they are equivalent or not. That, again, depends, as I see it, on a construction of the claims of the Bray patent and on its scope, a matter properly for the Court.

THE WITNESS: I will change that. I will change that. I will say it does the same thing.

MR. SOBIESKI: This witness is an expert witness.

MR. HERRON: He is not the Court.

A You don't have to use the word "equivalent". I will say it does the same thing, see. In other words, to anticipate, you must anticipate the structured means, element by element, a literal meaning function and result, what they will do—

MR. HERRON: In your opinion?

A In my opinion.

MR. SOBIESKI: In your opinion, Mr. Hayn, would it require inventive skill as distinguished from simple mechanical ability to change the structure which you find in the Northrup patent to a structure such as you find in the Bray patent?

A I consider that it would require an entire reorganization of the structure of the Northrup patent to enter

(Testimony of Fred H. Hayn)

the field of what is disclosed in the Bray patent. In other words, I am unable to reconcile the bearing of this patent on the patent in suit.

Q. Then, in your opinion, Mr. Hayn, a change from the structure shown in the Northrup patent to the structure shown in the Bray patent would require inventive ability as distinguished from mechanical skill?

A Absolutely.

CROSS EXAMINATION

BY MR. HERRON: Mr. Hayn, it is a fact, is it not, that in Figure 1 of the Northrup patent there is shown an aperture as between the standing column—

A 1?

Q —1; and the puller nut 24?

A I could not undertake to testify that. There seems to be a white space there but whether that is an aperture or not I could not testify.

Q As a man who has examined, perhaps, 50,000 drawings a year in the Patent Office, would you say that that which appears in Figure 1 of that patent is the customary way of indicating a space or opening?

A It might be.

Q That would be how you would indicate it, would it not, if you desired to indicate it in a half-section such as that?

A If I wanted to know what that space was I would ask the attorney to tell what it was for and indicate it by means of a reference character.

Q But you would not have any trouble in realizing, by looking at it, that it indicated a space, would you?

A It might and it might not. I could not tell about it.

(Testimony of Fred H. Hayn)

Q What would you say it did indicate if it did not indicate space?

A It might be a draftsman's error.

Q Do you think that in the case of a draftsman's error it would be on both sides an equal clearance shown like that?

A Oh, yes.

Q In other words, now, irrespective of whether the draftsman intended to indicate a space or not, that which is actually indicated there is space, isn't it?

A I would not even say that. They would have to tell me what that is there before I would say that was a space.

Q Suppose we just hand you this drawing now, just Figure *I,* and we ask you: Is a space there indicated; what would your answer be? What is your answer?

A Well, I would have to say that there is a white indication between there. It might be a space. Whether it was intended to be one I would not say.

Q But at all events, you would say that there was a white line there which indicated a space?

A Exactly.

Q Unless an error has been made?

A Yes.

Q Yes; thank you. Now, would you say that it would involve any exercise of the inventive acumen as distinguished from mere mechanical skill to provide on the top face of part 6—Figure 2, part 6, or Figure 1, part 6?

A I got that number but I want to see what the statement is. Yes. Now, go on, please.

Q All right; let us get the clutch first. Calling your attention now to the patent to Cummins and to the clutch

(Testimony of Fred H. Hayn)

of that patent—the jaw clutch which appears here in part 21 and part 22.

A   What did you call the part 21?

Q   The two cooperating parts of a clutch.

A   Well, the specification in the Cummins patent refers to it as "notches" 21, designed to engage lugs 22.

Q   All right. In layman's language that would be a clutch, wouldn't it? Wouldn't you say so, with the background of your long experience, that makes a clutch?

A   I would not undertake to say just what a clutch is. Nor would I undertake to say that that is a clutch. It might be a fastener, two engaging members, one a female member and one a male member. I would go that far.

Q   Mr. Hayn, if you will please examine that patent and after you have examined it, in the light of your experience, will you tell me whether you have any substantial doubt as to whether those two parts are a clutch?

A   It depends what the patentee means here, if anything. If he means by a clutch—

Q   Read your patent and see what you think it means, in the light of your experience.

A   I have read here on page 1, 78 to 89, and I do not see any reference to any clutch. I see "notches" and "lugs 22". I can't call that a clutch.

Q   You can't tell whether to call that a clutch or not?

A   I can't say that that is a clutch.

Q   You just don't know?

A   No. I say they are two engaging members, one a female member and one a male member. If that be called a clutch, then it is a clutch.

Q   All right. Now, we will use the terminology that you prefer. Would you say it would amount to the exer-

(Testimony of Fred H. Hayn)

cise of inventive art now to apply to the top face of that part of Northrup labeled 6 one of those cooperating parts, and to the bottom of part 24 of Northrup, the other co-operating part, so as to engage those parts together by the dropping of the barrel?

A   I would say it would take invention of a high order that would amount to a reorganization of this Cummins construction.

Q   Now, let us see if we understand each other. Now, I am asking you to simply add to the top face of the part 6 of Northrup the male part of that fitting part arranged on the bottom of the puller nut 24 of Northrup the female part of the fitting shown in Cummins. Would that simple change come to the order of reorganization of the parts, now, just that?

A   Just that.

Q   Would that involve an invention of a high or any other order?

A   I would say it involved invention of a high order; yes.

Q   As distinguished from the exercise of mere mechanical skill?

A   I would say so, decidedly.

Q   You would not have felt, then, as an examiner, it would have been obvious to you that the one part could have been interposed, that the clutch parts, as I call them, or the cooperating parts, as you call them, could be interposed on the Northrup structure in the parts I have indicated by any ordinary mechanic who would be handed those two drawings and told to put the cooperating parts of Cummins on the pump of Northrup?

A   There is nothing in the Cummins patent that suggests anything like that.

(Testimony of Fred H. Hayn)

Q   That is not what I asked you.

A   I am telling you why I consider it of an invention—

MR. HERRON: Read the question, Mr. Reporter.

(Question read by the reporter)

MR. SOBIESKI: I object to that question, your Honor.   It assumes that the mechanic is being told by someone what changes to make in a certain structure. Of course, if someone else had conceived the idea, I suppose the mechanic could execute it any way he was directed to execute it.

MR. HERRON: I submit the question.

A   I still maintain—

THE COURT: Let him answer, and exception noted. Go ahead.

A   —it would take invention of a high order.   These simple changes are frequently those of tremendous importance.

MR. HERRON: Yes.   And you consider that that simple change would have been one of those simple changes of tremendous importance?

A   I would look at it as invention.

Further

DIRECT EXAMINATION

Q   BY MR. SOBIESKI: Mr. Hayn, directing your attention to patent No. 1,513,699, the Ellis patent which has been introduced in evidence by the defendant in this action, are you familiar with that patent, which I now hand you?

A   Oh, yes.

(Testimony of Fred H. Hayn)

Q Do you find in that patent a lock on the bottom of the traveling barrel?

A There is nothing that I can see in here that is a lock, either in the specification or on the drawings here. In order to determine what a certain element is there has got to be some sort of description, some reference to it; and I see no reference to any lock in here.

MR. HERRON: Well, we do not assert there is one, so I guess we can—

MR. SOBIESKI: Do you find in that patent a traveling barrel with a lock on the bottom being apertured for the ejection of liquid downward out of the barrel?

A Well, that is a little bit too much to swallow, but I think I can get what you mean. On Figure 5 you have got a cap nut. I don't see any—there are some apertures there. For instance—let's see; oil passages 22, yes, but they open laterally. I don't see what bearing that has on the patent in suit here.

MR. HERRON: We ask that the last be stricken. I have no objection to him telling what is there.

THE COURT: It may be stricken.

MR. SOBIESKI: In your opinion, Mr. Hayn, would it require the exercise of inventive ability as compared with mere mechanical skill for the structure described in the Ellis patent to be changed to the type of structure described in claims 1, 2 and 3 of the Bray patent?

A I would say so.

Q It would require inventive ability?

A It would.

(Testimony of Fred H. Hayn)

Further

CROSS EXAMINATION

BY MR. HERRON: Now, Mr. Hayn, what would have to be done to Ellis to make it the structure described in 1, 2 and 3 of the Bray patent?

A What would you have to read in?

MR. HERRON: Yes; go ahead and do it. I would like to know what you consider would be those changes. Claims 1, 2 and 3. You may take them separately, if you find different structures involved.

A You would have to rebuild the Ellis structure.

Q Please explain in detail how.

A Well, you would have to throw away the valve in the Fig. 5.

Q You mean the cage and the cooperating valve?

A Yes; throw it away.

Q And the throwing away is taught in the patent, of course?

A Oh, no.

Q No?

A No.

Q Doesn't the patent teach that under some circumstances you may have that valve with Fig. 5?

A "Under some conditions the valve ring may be omitted"—

Q The word is "may", is it not?

A Yes. "as the passages in the cap nut permit the free flow of liquid and prevent the formation of a dead chamber containing trapped liquid which will resist the free operation of the pump." I consider that an indefinite statement and I would have to be told or explained by

(Testimony of Fred H. Hayn)

the patentee how he would have an operative structure and still omit that valve ring.

Q In other words, you believe, do you, that if the valve ring were omitted that the structure would not be operative; is that your belief?

A Well, it might make a couple of strokes and might squirt a little oil there, possibly.

Q By that you mean that you think possibly it would be operative?

A It might be—well, it depends what you mean by "operative". If you mean whether the parts would work—

Q I mean that they would work, pump oil with probably some degree of success.

A Within limits, doubtless.

Q How would the ability of that pump to work in pumping oil with some measure of success be changed by the removal of ring 25 and the cage in which it is confined? Please explain it in detail.

A Well, there would be—so far as I could see, there would be no control of the lifting of the fluid. Now, whether that—

Q Do you think the control of the lifting of the fluid in that pump is that the lift of that pump is regulated by this ring 25, this fibre ring 25: is that what gives this pump its lift?

A No. I would say it was interfered with, the control by the valves. On page—

Q You do not mean this ring 25 and the cage in which it moves, 27—26-27, would interfere with the operation of the valves, do you?

A Oh, no. Oh, no.

(Testimony of Fred H. Hayn)

Q Then, what do you mean? What would happen if you removed your ring. If you remove the ring 25 and the cage 26 and 27—that fibre ring 25; I am calling your attention now to Figure 5, and the cage in which it is designed to move, 26 and 27, what would be the effect of that removal in the operation of the pump?

A Probably improve it.

Q All right. So, I take it the first thing you would do, if you were going to rebuild it to conform would be to remove the ring and the removal of which would help the pump?

A Yes.

Q What next would you do?

A I think I would bore a few holes 5 somewhere.

Q Haven't you got the ducts 22?

A No, but those are positioned laterally.

Q Laterally to what?

A Laterally to the length of the pump—vertical to the length of the pump.

Q Do I understand you as reading this drawing as saying that the ducts 22 are so placed as to open laterally with relation to the movement of the pump? Directing your attention to Figure 3, ducts 22?

A Those seem to be vertical ducts but they connect with a lateral opening there in Fig. 5.

Q What is the lateral opening with which they connect?

A There don't seem to be any reference character to indicate it, but Fig. 5—I am pointing to the heavy lines that are crossed by the leader from the numeral 23; the heavy lines, parallel curved lines that go above and below

(Testimony of Fred H. Hayn)

that evidently open into the ducts 23. That is what I mean by the lateral opening.

Q Would you be so good as, on the Court's copy, to just draw a line to what you call the lateral opening? I understand you to say the ducts connect with some lateral openings, but I would like to see it.

A Here is the cut out portion. Now, this is the lateral opening. Do you want me to mark it or write it?

Q Just call it AA.

A Lateral opening AA there.

Q Now, Mr. Hayn, what you term your lateral opening is the space between the bottom face of the puller nut 21 and the top face of the ring 26, in which space it is designed that the ring 25 should move up and down, is that it?

A I can't answer that question at all because I don't know what you mean by "puller nut". There is no evidence in here at all of the puller nut.

Q All right; I will tell you what I mean by "puller nut". When I say "puller nut" I mean member 21 of Fig. 5.

A If you will repeat your question, using that as "member 21" possibly I can arrive at what you are driving at.

MR. HERRON: Would you be so kind as to read the question?

(Question read by the reporter)

MR. HERRON: I am sure he can answer that question. Read it to him again.

A It is ambiguous. I don't know what you are talking about.

MR. HERRON: In what respect, Mr. Hayn, is it?

(Testimony of Fred H. Hayn)

A   You mention a ring and then you mention a space and then you mention a puller nut and then you change that to a member.

Q   What you call the lateral space, the lateral opening, Mr. Hayn, is, is it not, the space between the lower surface of part 21 of Figure 5 and the top surface of part 26 of Figure 5?

A   I understand it now.   That is not strictly true because, you will notice at the left where the leader from the numeral 27 indicates a solid material, hence, the opening does not go all the way around it.

Q   Well, doesn't that indicate to your mind simply that two pieces of metal which come down to hold ring 26 in position with relation to the lower part of part 21?

A   Yes; it does.

Q   And what you have termed the lateral openings is the space between the part 26 and the lower part of 21, so held from coming in contact with each other?

A   Oh, yes.   And that is what I call the lateral space.

Q   Now, with that explanation, will you explain to me what you meant when you said that the ducts 22 connected with the lateral space?   Is that what you said they did?

A   Yes; as far as I can see now.

Q   All right.   Now, just what is that action?

A   Well, if the oil would go up there—up through the ducts 23 it would go out laterally.

Q   If it would go up through the ducts it would go out laterally by which way?   Do you call "up" toward the top of the patent?

A   Yes.   I don't know whether I have answered the question or not.

(Testimony of Fred H. Hayn)

Q I asked you your use of the word "up"—which direction?

A "Up" means toward the top of the patent. Toward the top of the device.

Q And, hence, you say if the oil would go up the ducts 22 of part 5—the oil would go out at the top of that laterally, is that what you mean?

A Since there is a space there, I do not see any reason why the oil should not go out unless it were held from going out by means of that ring 25.

Q I guess I will have to ask you to draw a picture, showing me where you mean the oil would go out. I am afraid I do not get it.

A Through the same lateral opening—I will repeat. Here are the ducts, pointing to 22, by the reference numeral 22. The lateral opening is, as I have explained—

Q Yes; you defined that.

A Then, since there is nothing to prevent it unless the ring 25 were positioned there, I do not see why the oil should not go out laterally up here from it that way; if that is what you are driving at. I don't know.

Q I am driving at your construction, of course. Is there anything to prevent—I assume that the oil would go in this direction as well as the other way. Now, we are assuming that the oil is moving downward; is there anything in that structure which would prevent it from moving down through those ducts and against anything that might be down at the level of the line indicated in the patent?

THE COURT: Your question is assuming the taking off of the cage?

(Testimony of Fred H. Hayn)

MR. HERRON: No, your Honor. My question first assumes with the cage on and then with the cage off. I am taking it both ways in order that I may get Mr. Hayn's reaction. The question now contemplates the cage on.

MR. SOBIESKI: I understand your question now contemplates the cage on but the fibre ring out?

MR. HERRON: Yes; with the fibre ring out, and then I will place the fibre ring in and see what he thinks about it.

THE COURT: Is there any occasion for much argument on that? Of course, with the ring out the first direction of the oil on being expelled would be perpendicular unless it is interfered with. It is bound to be perpendicular.

A I agree with that, but I don't know why the oil should not go out laterally, as I explained there. There is an opening there and if the traveling barrel moves up and down it is bound to come out on the side there, and I don't doubt but what there will be an upward and downward movement of the oil.

MR. HERRON: In other words, you feel that if the oil came out under any pressure at all that it would have the tendency to persist somewhat in its motion and pass straight down in the direction of the jets, is that not right?

A I can't read any jets in here.

Q Strike that unfortunate word and let me put it this way: Mr. Hayn, if the oil came out with any velocity at all or with any tendency to take a forward direction and a downward direction, that tendency would persist, would it not, and cause it to be discharged downwardly?

(Testimony of Fred H. Hayn)

A I am afraid that I will have to say I don't know. I could not testify to any theory as to the operation of that device or as to any theory in suit here.

Q In other words, you feel that you do not know anything as to the theory of how that device operates?

A No; I would not put it that strong. I would say that I can't follow that theory that you are trying to initiate there. I can't follow it at all. I don't understand it.

Q I see. Now, let us see if I can make you understand it. My question, is whether if liquid is ejected through those jets with a downward direction, that direction would not continue in the direction in which it was originally given and pass within the ring 26 with a downward motion?

A I am afraid I can't read anything like that into this structure here.

Q I am not asking you to read anything into the structure. I am asking you a physical fact; would that occur, as a fact, or wouldn't it?

A It might and it might not. I am not prepared to say whether it would or would not.

Q Now, if the cage were entirely removed, that is, part 26 and 27, then would there be any question in your mind that the direction of the flow through the ducts 22 would be downwardly?

MR. SOBIESKI: I object to that question on the ground that the patent does not in any way indicate the cage would be removed.

THE COURT: Well, of course, you examined on the question of mechanical contrivances and ordinary mechanical knowledge.

(Testimony of Fred H. Hayn)

MR. HERRON: If that were removed what would be the effect of it?

MR. SOBIESKI: You are now saying that if the entire cage were removed—very well. By that, you mean the structure described in Figure 5 of the patent?

MR. HERRON: I do not. I mean that portion of the structure exhibited as Figure 5, consisting of parts 26 and 27.

THE COURT: To get it all clear, the cage extension on 21.

MR. HERRON: Yes. Thank you, your Honor. That is true. I mean the removal, as the Court has suggested, of the cage extension of part 21.

MR. SOBIESKI: We object to that on the ground there is nothing in the patent to show that is removable. It is a solid block.

THE COURT: The witness has answered your question that it would require more than mechanical skill to convert that into the thing that Mr. Bray has. Now, he may cross examine on that.

MR. SOBIESKI: Very well.

THE WITNESS: Let us have the question.

MR. HERRON: If that portion of Figure 5 which consists of the cage extension on part 21 were removed would the direction of flow through the ducts 22 be downwardly on one stroke of the pump and, presumptively, upwardly on the other?

THE COURT: That means if you had no cage at all and had those grooves, long grooves with the column fitting inside and the fluid being expelled, what would be its direction?

(Testimony of Fred H. Hayn)

A   Oh, I suppose, if the structure be reorganized, just as pointed out there, there would be an upward and downward movement of the oil.

MR. HERRON: And that movement would persist in a downward direction?

A   I think so.

Q   Now, if you will examine Figure 4 you will notice, I think, that part 25 is shown as not entirely filling the annular opening between 12 and 26.

A   There is, like in the other case, there is a white space there but the specification of the patent does not tell what it is, so I can't tell what it is.

Q   And it would not occur to you that it was intended to indicate space?

A   Not at all.

Q   You would assume, would you, that the draftsman, by error, had made that dotted line?

A   I would not assume anything of the sort. It would have to be explained to me. There is no reference made in the specification and the thing must be certain and not guess work.

Q   And, accordingly, that matter of the drawing means to you nothing?

A   Nothing.

Q   All right.   Now assuming that the dotted line was intended to represent space, not an error, in that event there would be a downward direction of the fluid along the side of the standing column and through that space, would there not?

A   With your assumption, probably.

Q   Yes. You think, actually?

A   Yes.

(Testimony of Fred H. Hayn)

Q Now, what else would you have to do in order to reconstruct the Ellis pump to make it into a Bray pump, except to remove that downward cage extension of Figure 5 of the patent?

A I would have to put a lock in there, or at least whatever he calls it in here. Bottom guide on the barrel running on the column and having a bottom face forming a coupler, and an anchor for the column presenting a top face forming a complementary coupling to be non-rotatively interlocked with the lower coupler guide. That part is not there.

Q By "that part", how much of that part is not there —all of it?

A All of it.

Q Is there any anchor here?

A I don't know anything about any anchor. You haven't asked me about it.

Q Don't you know anything about this patent except what I ask you about?

MR. SOBIESKI: I object to that question, your Honor, as being immaterial; already asked and answered.

MR. HERRON: Well, I will not insist on it, on the ground it is embarrassing, perhaps.

A Oh, no. No. I will answer whatever you want.

THE COURT: You may ask him whether he made a study of this patent beforehand. That is proper.

A Oh, I have made a study of it. I haven't read every word of it. I don't think it is necessary.

MR. HERRON: In other words, you have testified it would be invention to change this over to Bray with-

(Testimony of Fred H. Hayn)

out having read every word of the patent and knowing what it teaches, is that what you mean?

A   I mean exactly you can tell by taking out the pieces in the drawing, taking out all the specification, taking the function and result, you can get that.   Patent examiners do that every day of the year.

Q   I presume that is why the Court frequently knocks out *out* the patents in passing on them.

MR. SOBIESKI: I object to that last remark and move it be stricken.

MR. HERRON: Now, please read the patent so you know what you are talking about, and then let us see if we can expedite the matter.

THE COURT: I am sure the courts would be glad if the examiners had the final word.

THE WITNESS: The patent lawyers request that we, not the agency of the courts, construe the patents.

A   Oh, there is unquestionably a statement here about the anchoring feature but I do not find any similarity in this structure 4 and 3 in the Bray patent.

MR. HERRON: In other words, you draw no distinction between the two so far as either anchoring feature is concerned?

A   Well, I would not go quite that far.   The Ellis patent may anchor one way and the Bray patent anchor another way.

Q   That is right; but they both anchor, don't they, and hold the standing column?

A   I suppose I will have to admit that.

Q   Well, do you admit it?

A   Yes.

(Testimony of Fred H. Hayn)

Q All right. Now, there being no difference in their both anchoring, we will now pass from the anchoring and ask what other changes you would have to make other than the removal of the cage to make the Ellis into Bray?

A Well, of course, there is quite a different packing means.

Q Do you consider that so far as the question of making this pump over into a structure which is similar to the structure taught in Bray is concerned that the packing means are material?

There is no claim as to packing means.

A No; I don't think so.

Q Now, let us just disregard the things that are not material and are not covered by the claims and get down to the important part of the structure.

A The structure, I am unable to find any Ellis structure shown in Figs. 3 and 4 of Bray, nor do I find any— I do not find the structure shown in Figs. 3 and 4 in Bray in Ellis.

Q All right. Let us just take that now. How would you have to modify Figure 3 shown in Ellis in order to make the Bray structure? How do you have to modify Fig. 3?

A What part of Fig. 3?

Q The part of Fig. 3 that you say is different, sir.

A I was referring to Bray. I say I am unable to locate in Ellis the structure shown in Figs. 3 and 4 of Bray—Figs. 3, 4 and 5 of Bray. That is not found in Ellis.

Q That is not what I am asking you. I am asking you what you have to do in Ellis to make it into a Bray.

(Testimony of Fred H. Hayn)

You say it would involve invention of high order. I am asking you what involves invention of a high order to make the two the same.

A You would have to reorganize the Ellis construction there. You would have to change the entire lower part of 21—I expect that is what you want me to answer —or the lower part of 12.—

Q Wait one minute. You say you would have to change the entire lower part of 21; you mean take the cage off, as you heretofore testified?

A Take the cage off.

Q All right. Let us pass that and come to something else. Now, what else would you have to do?

A Then he would have to screw on that part—first, he would have to provide it with internal threads. I don't know whether it has got any there or not.

Q Anyhow, there is nothing of a high inventive order in that?

A It might be, to get a new result, new advantage, certainly.

Q What part are you going to apply the internal threads to?

A Now, you see in Fig. 3 of Bray you have the part B. Now, if you took as equivalent to that the part 21 you would have to provide—well, you are throwing away that part 21 now. All right.

Q What for? What would you throw that away for? Doesn't the Bray patent call for an apertured guide?

A Oh, yes.

Q And the part 21 with cage off, or with it on, for that matter, is an apertured guide, isn't it, as shown in Figure 3?

(Testimony of Fred H. Hayn)

A   Well, I don't think I can find any apertured guide in there.   It is not mentioned in the specification here.

Q   Looking at the structure, is it or isn't it an apertured guide?

A   I cannot read into a patent that which is not clearly shown and described.

Q   Assuming now that part 21, as shown in Figure 5 of the Ellis patent, runs on, in the sense that the ribs formed by the making of the ducts in part 21, runs on the standing column 12.   In that event, is it not an apertured guide?

A   I don't know.   Because I would not attempt to say what the patentee had in mind unless he put it in the specification.

Q   I am asking you what the patent says.   I am not asking you to guess at what the patentee had in mind. Reading the patent, the teaching of the patent, isn't the structure I have just put to you, operating in that fashion in that Ellis structure, an apertured guide?

A   Well, I expect I have to answer that it could be considered so.

Q   Well, you do so consider it and it is; that is the fact, isn't it?

A   I do not.

Q   Oh, you do not so consider?

A   I say whatever is disclosed in the specification.   I know there must be the disclosure.   I know what you read into it.

Q   Do you mean it is or is not an apertured guide, in answer to the question I have just heretofore asked you?

A   You mean, irrespective of what the patent discloses?

(Testimony of Fred H. Hayn)

MR. HERRON: Will you please read the preceding question?

(Question read by the reporter)

A With that premise there, that you assume those things, I will say yes.

Q Now, isn't it a fact that the patent discloses it as being precisely that, that is, in Figure 3?

A I can't find it.

Q Doesn't Figure 3 so show it?

A What is the antecedent of it?

Q Let me read you this part of the specification, page 1, line 90: "At the lower end of the working barrel 19 a cap nut 21 is threaded therein and has a bearing upon the standard 12. The interior surface of this nut is formed with a series of oil passages 22 intermediate of the bearing surface 23 so as to permit the free discharge of oil from the dead chamber 24 which exists between the nut and the valve at the upper end of the standard." You find that in the specification, do you not?

A Oh, yes. Yes.

Q Now, applying that specification to Figure 3, as the Figure is shown in the patent, there isn't any question in your mind, is there, that you there have an apertured guide?

A There is considerable doubt in my mind that that is an apertured guide. You have got an outer member; you have got an inner member, the outer member traveling on the inner member and some grooves cut away. Now, whether you would call that a guide I would not be in a position to say.

(Testimony of Fred H. Hayn)

Q   What, to your mind, is a guide?   What is a guide?

A   Well, something that runs on something else, I suppose.

Q   And imparts direction to it, perhaps?

A   What is that?

Q   And imparts direction to that thing which moves, don't it?

A   Not necessarily.

Q   In other words, a guide is something that runs on something else, would you say?

A   Possibly.

Q   Now, let us take a locomotive guide; you know what that is, don't you?   The crosshead?

A   I don't hardly call that a guide.

Q   You would not?

A   That is a good and full mechanical structure.

Q   I will find out from a railroad man what I do mean.   A crosshead, is that a guide?

A   Crosshead?

Q   Does the crosshead run in a guide on a locomotive? Do you know?   If not, don't hesitate to say so.

A   I would certainly hesitate to say what was and what was not a guide.   I would want to consult Webster's dictionary or some standard treatise on what a guide is or is not before I would testify.

Q   You did not find it necessary to consult Webster's dictionary before you testified that this Ellis structure was not the same as Bray, did you?

A   Oh, no.

Q   What did you think Bray meant when he used the word "guide" in giving that testimony?

(Testimony of Fred H. Hayn)

A   Well, one part moving on another there. You know, these terms are used very loosely in practice.

Q   How did you use it in giving the answer you have given when you testified you did not find in Ellis the same structure as you found in Bray?

A   Well, I did not mean to testify that there was no similarity between the two. I didn't mean that.

Q   How did you interpret the word "guide" when giving that answer?

A   Well, in that case there is the one part moving along on the other, being guided by the other.

Q   That is the sense in which you understand the Bray patent and the sense in which you used the term "guide"?

A   Yes.

Q   Thank you. In that case, just using this for another purpose—I think you were not here at the time. It makes no difference—would you not say that the two wooden pieces here impart a guiding—that is, the wooden parts marked C and the corresponding—oh, yes. That the part B imparts a guiding motion to the part D in Exhibit OO? Want to take a look at it?

A   Never mind. That, as I say, is an entirely unenclosed—

Q   I am not interested in a dissertation or a lecture. Will you please answer the question as to whether that part B imparts a guiding motion to this part D?

A   I cannot read that into this structure here.

Q   I don't ask you to. I simply ask you for the simple mechanical fact.

A   Do you mean one part moving on another?

(Testimony of Fred H. Hayn)

Q Whether the part B imparts a guiding motion to the part D in this model; now, does it or does it not?

A Oh, it is possible. I suppose I will have to answer yes.

Q Well, do you so answer?

A Yes.

Q All right. Now then, since you find in both the Bray and the Ellis construction, Figure 2, with the removal of the cage we have mentioned and the cooperating valve an apertured guide running on the column, what else would you have to do in order to make this into the Bray structure?

A I will answer that I don't admit it. I don't admit it at all.

Q I didn't ask you to admit anything, if you will just listen to the question.

A You said "since you find". Now, I don't admit that premise. Give me a clear question.

Q Oh, then you don't admit that there is in both Bray and Ellis, if the cage is removed, a guide; I thought you just got through saying you thought that you would have to. I thought you did.

A I think you are mixing my language up there. I decline to read into Ellis that which is not there.

Q Mr. Hayn, I am not asking you to read anything into Ellis. I have asked you repeatedly, and I ask you again, whether, bearing in mind lines 90 to 96 of the specification of Ellis, page 1, and bearing in mind the drawing illustrated in Figure 2 of Ellis, and likewise Figure 5 of Ellis, whether Figure 5 is not—and also

(Testimony of Fred H. Hayn)

Figure 3, whether you do not find in Ellis an apertured guide running on a standing column 12?

A   Oh, I suppose I will have to admit that, yes.

Q   Now, do you admit it?

A   I do admit it.

Q   There isn't any question about it and you are not going to back up on it?

A   Not unless you put in more presumption.

MR. HERRON: Now, then, read him my previous question.

(Question read by the reporter)

A   I would have to reorganize the lower part of that which we have now called the guide.

Q   By doing what to it?

A   Well, replacing the lower structure, that structure 4 and 3.

Q   Wait a moment.   I am asking you what you have to do to Ellis to make it into Bray.   Now, look at Ellis and tell us what you have to put into Ellis and then we will have it straight.

A   I would have to add a substitute for that bearing member there.

Q   "Bearing member", you mean what?

A   The bearing described on line 92, page 1 of the Ellis patent; I would have to take that bearing member and I would have to cut it away to provide the female member 3.   That is the male member, and 5—3 and 4 there.   In other words, I would have to reorganize the

(Testimony of Fred H. Hayn)

Q Or 23—you are right. Now, when you say "re-organize" you mean that you would have to cut a slot in it to provide *of* a female member of a jaw type clutch, is that right?

A I would have to do more than that.

Q What more than that would you have to do?

A Well, we would have to cut off the pipe 12 there.

Q. No. You misunderstood me. I mean now with reference to part 3 in this reorganization of the—Figure 5, part 21 and the reorganization of it which you would have to do, what would you have to do except to cut a slot through there so as to provide a female portion of an engaging clutch?

A We haven't that lower member there in Figure 4 there. That will have to be provided, in Figure 4 of Bray.

Q What lower member?

A Figure 4 of Bray. Don't you see that? You would have to put that in there.

Q Oh, you would. In other words, you feel that the ducts of Ellis—ducts 22 of Ellis are not the equivalent of the ducts 5 of the Bray, Figure 4?

A I would not quite say so, because you have this lateral opening there.

Q But the lateral opening has been done away with by the removal of the guide.

A Then, aren't the ducts done away with, too?

Q Certainly not. Just look at the drawing.

A Then I don't know what you are talking about.

Q All right. Turn your attention to Figure 5 of the Ellis patent; if you remove the cage-like extension of the

(Testimony of Fred H. Hayn)

part 21, the ducts will still be left in the part 21, won't they?

A   I see what you are driving at.   I didn't understand you.   Yes.

Q   Yes; they will.   All right.   Now, then, I ask you whether those ducts which are still left are or are not the equivalent of the ducts shown in Figure 4 of the Bray as part 5?

A   I would not say so.   In order to be an equivalent you have got to have a substantial identity of means, a substantial identity of function, a substantial identity of result; and you haven't got it by mentioning those things you mention.

Q   All right.   I am asking you now whether the ducts are equivalent—whether the ducts shown in 22 in Ellis are the equivalent of the ducts as shown in Figure 4 of Bray marked part 5?

A   I would not say so, no.

Q   Why not?

A   Because there is not a substantial identity of means, a substantial identity of function, and a substantial identity of results.

Q   In other words, you think in order to have a substantial identity in all those respects Bray has to have a series of holes bored through the nut, such as are shown in that drawing?

A   Not necessarily.

Q   Well, what does he have to have?

A   He has to have those vertically arranged ducts there without any transference such as you have in Ellis here.

(Testimony of Fred H. Hayn)

Q You have no transference, because you have taken the cage off.

A In other words, here you are reorganizing?

Q Certainly, sir, within the limits of my assumption. Let us keep the cage off and it is off in all the questions I ask you.

A With those assumptions I will have, of course, to admit yes.

Q Yes; and you do admit "Yes", do you?

A I admit, with your assumptions, Yes.

Q. Now, that we have finally found that we do have in Ellis an apertured guide traveling on the standing column, what else would you have to do in order to change Ellis into Bray?

A I haven't admitted eventually according to your assumption. I would like to have that made clear. You are reorganizing the structure—

Q Just a moment. Have you told me or haven't you told me that, considering the specification of Ellis and the drawings of Ellis you do find in Ellis, Figure 2, an apertured guide traveling on the standing column?

A I do.

Q Now, that is not my assumption but that is what you find in the drawing, is that right?

A Yes.

MR. HERRON: All right. Read the question to him.

A You have been questioning as to the equivalency of one and the other, and I won't admit that.

MR. HERRON: Read the preceding question. I don't care what you admit. I want you to tell me facts;

(Testimony of Fred H. Hayn)

the Court will do the admitting. Will you be so good now as to read the preceding question?

(Question read by the reporter)

A You have to provide Ellis with a structure disclosed at 3 and 4; you would have to provide Ellis with a structure disclosed in Figs. 3 and 4 of Bray. In other words, that specific form of coupler. I said that a number of times.

Q Now, what is the specific form of coupler shown in Bray?

A It is a notched member.

Q In other words, there is a female notch—that is, a notch making a female member?

A Yes.

Q Of the coupler nut of the guide?

A I won't admit any coupler nut.

Q Oh, all right. Well, on what we have called part 21 of Figure 5 of Ellis?

A Yes.

Q Now, would it involve invention to so groove part 21, Figure 5 of Ellis, as to make a notch which would provide the female part for the reception of a male part located on the anchor of Ellis?

A I would consider it so, decidedly. You must remember—

Q In other words, Mr. Hayn, if you were handed a patent of Mr. Bray, or let us take one of the preceding art patents; if you were handed a Cummins patent—by the way, have you ever been a mechanic?

A Oh, yes. I have—

Q I mean, are you a skilled mechanic?

A I think so, in certain effects.

(Testimony of Fred H. Hayn)

Q Have you ever done work in a shop?

A Oh, yes. I have made machine tools; I have worked in the machine shop.

Q And served your time in the machine shop, as I think they put it?

A Yes.

Q Well, now, if you had been in a machine shop and someone had come in and they had handed you the patent to Ellis with this cage cut off and they had likewise handed you Cummins and they said, "I want you to put onto the Ellis pump the clutch shown in the top part of Cummins as the bottom of Cummins as part 14 with the cooperating part 19", would you have said, "I can't do it; it would take an inventor to do that" or would you have walked over to the lathe and put the female part on one and the male projection on the other? What would you have done?

A I believe I would ask him what he would use it for.

Q I think you probably would have. Could you, as a mechanic, have done it.

A Under the man's direction, I certainly could have put in a slot somewhere and ground off a projection. I certainly could do that, certainly. If I was told what to do, of course.

Q And there would have been nothing inventive about that?

A That is another proposition.

Q About making those slots, the slot in the tongue, mechanically?

A Nothing inventive about that.

Q That is right. Now, if you had been handed the Ellis patent and Mr. Bray had said to you "I would like

(Testimony of Fred H. Hayn)

you, Mr. Hayn, to add a clutch at the bottom of that Ellis pump so as to prevent the barrel from rotating" and said "Here is a bunch of patents in the prior art, including Cummins", would you have been able to have done in with that direction?

A   I certainly would not unless I was told exactly to add what was wanted.

Q   In other words, you couldn't have looked at the Cummins drawing and seen that clutch there and have placed that clutch upon Ellis unless somebody told you just how to do it and where to do it, is that right?

A   I am afraid I would have to be told what to do.

Q   And if you had, by any wild chance, found out how to do it without aid, you would have thought you had invented something, is that right, of a high order?

A   Well, as a mechanic, I don't believe that would enter into my consideration at all, whether it was an inventive act or what it was.   I would do that as a mechanic would do, what he was told to do.   And I don't think that any—in my opinion, no mechanic would take two patents here, take the Ellis patent and the Bray patent, and so rearrange and reorganize as to make Ellis just so you have got here the same thing you have got here in Bray.   It couldn't be done.

Q   In other words, you have never known a mechanic, in your experience, to take two patents and combine their parts, as a simple mechanical operation?

A   Never, never.

Q   And where did you do your time?   I mean, where you served your time as an apprentice?

A   Oh, I worked around Buffalo, New York, you know.

(Testimony of Fred H. Hayn)

Q Now, what kind of a shop?

A Oh, we worked in a shop of a friend of mine, neighbor. He is quite an inventor himself there. He had a lathe and I worked around him there.

Q Sort of a little home shop?

A Yes; and we worked, oh, in other places.

Q In any large mechanical shops, where you know whether or not the average mechanic would do that?

A Oh, yes, yes.

Q What were they?

A Well, I can't tell. This must be about 40 years ago.

Q And you don't recall where you were employed?

A It would be pretty hard to say.

Q Yes. Now, would it be necessary to make any other changes than the ones you have mentioned in order to transform Ellis into Bray, Figure 2 of Ellis into Bray?

A Well, so far as claims are concerned, I would say probably not unless—

Q Let us not bother about the word "probably". Now, would it or wouldn't it? You are the expert and I am asking you.

A Well, you would have to bore those holes 5 or we would have to have the—

Q You have already talked about those holes 5.

A Then there is nothing more.

Q We understand each other, that if you made the changes in Ellis which you have described the resulting structure would be Bray, as you understand the Bray patent?

A I would say so.

(Testimony of Fred H. Hayn)

Further

DIRECT EXAMINATION

BY MR. SOBIESKI: Mr. Hayn, with respect to this Ellis patent here, this Figure 6—no, Figure 5, Figure 5, from your knowledge of fluids, assume that the cage is on the bottom of the traveling barrel and the fibre ring is out, would more of the fluid be likely to go in a lateral direction through the bottom nut shown in Figure 5 than the amount of fluid that went in a downward direction out of the nut?

A Oh, obviously, I think, that the larger area and lateral size here, this being admitted that this is a bearing and close to the standing column there, unquestionably there would be much more liquid to go out laterally than down below there. The amount that would go down would be very small.

Q I think that is all on this question. I show you, Mr. Hayn, a copy of the O'Bannon patent which has been introduced in evidence here, No. 1,454,400, and that is a patent for a pump, is it not?

A Well, there is a part of a pump disclosed, but the patent is for packing.

Q Yes. And does that show a pump of the type of a traveling barrel pump, Mr. Hayn?

A Oh, yes. The object of the invention is to provide a packing adapted to use between the plunger and traveling barrel or cylinder and sealing the same in a highly efficient manner, so we will have to admit that.

Q. Does the traveling barrel have on it a bottom nut?

A I don't see any nut. No.

THE COURT: He calls it the head member 12.

(Testimony of Fred H. Hayn)

MR. SOBIESKI: I guess I didn't use the right term.

A   Head member 12.

Q   Yes.

A   Oh, yes. "The plunger consists in a hollow tubular member as shown and the construction of which is well known and which passes through the head member 12 in the lower end of the traveling barrel B."

Q   Do you see in that patent, Mr. Hayn, a device for locking the bottom of the traveling barrel?

MR.HERRON: We will stipulate there isn't any. We have not asserted there is any and we are not going to assert there is any.

MR. SOBIESKI: Yes; there is no locking.

Q   Does this patent disclose a method of ejecting the liquid from the traveling barrel in a downward direction out of the bottom nut?

A   Well, there seems to be some holes 16' there, and I don't see how you can eject any fluid downwardly there. I should say the ejection is laterally. That is what you asked me.

Q   And those holes are put in what part of the pump structure?

A   They are in the outside—I am trying to find out what that outside stationary member is designated and I can't see. By 13, I believe.

Q   Can't you read the number?

A   No.   But he calls it upstanding valve there.

Q   This part here?

A   15; that is, the inner side is called plunger. Well, it is the outer part, anyway.

MR. HERRON: The outer part of what?

A   The outer part of the mechanism of the pump.

(Testimony of Fred H. Hayn)

Q   You mean what we call the traveling barrel?

A   I don't know what you call the traveling barrel.

Q   Well, what would you call the traveling barrel?

A   B is the traveling barrel, B.   B is the traveling barrel.   That is where, obviously, the holes are, in the traveling barrel B in the outside.

Q   You find them without any trouble and without referring to the specification as to—

A   I refer to the specification to be sure, because I am never sure what the specification means.

Q   Do you refer to the location of just those holes?

A   Yes.

Q   When?

A   Just now.

Q   Would you call my attention to the line?   Would you give me the line?

A   What particular line?

Q   To which you just referred?

A   Well, "with the running barrel B", line 70.

Q   I am talking about the holes.   You said you just referred to the specification.   What line?

THE COURT: Page 1, column 2, lines 96 and 97.

A   Yes.

Q   BY MR. HERRO*M*:   That was what you referred to?

A   Yes.

MR. HERRON: Yes; all right.

A   Yes.

MR. HERRON: Yes; all right.

MR. SOBIESKI: Mr. Hayn, with that type of structure, in that type of pump disclosed in the invention, with the absence of the lock and with the holes for the ejection

(Testimony of Fred H. Hayn)

of the liquid being made in the side of the traveling barrel, would you consider that the change from that structure to the structure described in the Bray patent involved inventive skill?

A Oh, yes. This is nothing but a piece of junk; and with those holes in the side there the means function and result here entirely different, the way I look at it.

Further

CROSS EXAMINATION

BY MR. HERRON: What is there about this structure that leads you to designate it as a piece of junk? And operative pump is shown there, isn't there?

A Junk, you know, has its uses and there is no disparaging about that.

Q You do not intend that this remark that this pump is nothing but a piece of junk to be disparaging at all. Do you feel it is an operative pump, or isn't it, so far as you can tell?

A Oh, it is probably operative.

Q Well, is it or is it not?

A Oh, yes. It wouldn't have passed the Patent Office if it was not.

Q In your opinion, anything that is passed by the Patent Office is, therefore, operable?

MR. SOBIESKI: We will object to that, your Honor.

A Not always.

MR. HERRON: I thought not. Mr. Hayn, as you have said, this is an operable pump?

A Oh, yes.

(Testimony of Fred H. Hayn)

Further

DIRECT EXAMINATION

BY MR. SOBIESKI: Mr. Hayn, calling your attention now to the Neilsen patent which has been introduced into evidence here, No. 1,717,619, which is described as a multiple tube inserted pump—Calling your attention now to the Figure 2 in this pump, do you find a structure there that could be called a traveling barrel? First, I will ask you this: Do you find in that patent a structure which is locked on the bottom of the traveling barrel?

MR. HERRON: We can shorten it. We do not contend there is.

MR. SOBIESKI: You stipulate that there is no lock on the bottom of the traveling barrel?

MR. HERRON: Why, certainly, yes.

MR. SOBIESKI: Now, calling your attention to this white line between Figure 39—the part marked 39, rather, and the standing guide column, does that, to your mind, indicate a—

A (Interrupting) That indicates nothing at all, the same as the other structure there. I would have to find out in the specification what those white lines are.

Q All right. Now, calling your attention here to the specification on page 2, and reading "During the downstroke of the plunger the liquid in the chambers 15, 17 is forced through the ports 18 into the cage 6, thus washing over the ball 7"— Providing for the downward motion. Do you find, therefore, in that patent a bottom nut on the traveling barrel which is apertured for the ejection of liquid downward through the traveling barrel?

(Testimony of Fred H. Hayn)

A I don't see how you could have anything like that. These apertures 18 are lateral.

Q I am not referring to 18 now. I am referring to the aperture between 39, or, you might say in the nut which is 39.

A Let us find out what that 39 is and be sure. "A foot nipple 39 screws onto the bottom of the plunger 37, said nipple being adapted to engage the coupling 33 to affect the removal of the pump from the well." I must read to get anything like that in there. There is probably shown that there may be a clearance there, but I certainly would not accept or undertake to testify as to what those white spaces were.

Q You see no lock in that patent?

A No.

Q But it is possible, of course, that this white is a space?

A Oh, it is possible that it is to indicate that that is a clearance space there. There might be some function for it but I don't find any mention of it in the specification.

Q In your opinion, would a change in the structure, then, from this structure described in the Neilsen patent to the structure described in the Bray patent, require exercise of inventive skill?

A I should say, invention of a high order.

Further

CROSS EXAMINATION

BY MR. HERRON: Now, Mr. Hayn, you have read the specification in its entirety, have you?

A Not every word, no.

(Testimony of Fred H. Hayn)

Q Well, as a matter of fact, don't you find in that patent the teaching of an ejection of liquid between part 39 in Figure 2 down around the space between part 30 and part 40, in through the opening 41, in order to wash off 31?

MR. SOBIESKI: What part are you reading right there?

MR. HERRON: Pardon me. I should have told. I am reading—wait a minute. Just strike it out and we will get it right. Now, calling your attention to the annular space 38—

A That is on a chamber; chamber 38.

Q All right. You understand what I am talking about, then, chamber 38?

A Yes.

Q And, that chamber is lessened or increased in area, capacity, as the barrel of the pump moves up and down, is it not?

A Yes.

Q Now, as the barrel of the pump moves up with relation to the stationary plunger 33 that space will be decreased in area, will it not?

A As it moves up?

Q As the bottom of the barrel moves up in relation to the fixed plunger?

A Yes.

Q The effect of decreasing that area will be to pass the liquid in that chamber, to the extent of that decrease of area, down through what is shown as the white passage existing between part 39 and the standing column; that is true, isn't it?

A I can't follow your theory at all.

(Testimony of Fred H. Hayn)

MR. SOBIESKI: What part are you referring to there?

MR. HERRON: Beg pardon? Is my question obscure?

A In a way it is.

MR. HERRON: Please read it.

A But you are asking me to adopt a certain theory of operation of this pump and I can't do it.

MR. HERRON: Just please read the question. I am not asking you to adopt anything, Mr. Hayn, if you will just listen to the question. If it is obscure, and you will tell me wherever it is obscure, I will try to give you my thought.

A Don't you mean volume there? I don't know what area you are talking about.

Q Oh, call it volume, area, or cubic content of the chamber; suit yourself. You know what I mean, and if you will just be good enough to answer the question.

A I will try to.

MR. HERRON: Answer the question. Read it to him, please.

(Question read by reporter)

MR. HERRON: (continuing): Or the cubic content.

A I find no evidence of any such statement in the patent.

Q I didn't ask you that. I asked you if that would not be the effect of that mechanical operation?

A Might be.

Q Well, would it be?

A I would not undertake to say.

(Testimony of Fred H. Hayn)

Q If the packing of the plunger 33 were tight against the barrel would it be the effect of that mechanical operation, would that be the effect of that mechanical operation?

A I am sorry, but it would have to be sheer guess work to say "Yes".

Q In other words, is it sheer guess work for you to say that if the cubic content of the chamber 38 is, through the compression caused by the moving upward of the barrel carrying the part 39, lessened, that that fluid would naturally flow out through the duct shown in white between the part 39 and the blue, if you consider that as a duct?

A Yes. I will have to answer "Yes" to that.

Q And you do so answer, do you?

A I do so answer, yes.

Q All right. The effect of such action would be, would it not, to wash over the upper or top surface of part 30?

A I can't read that in there. You have got a surface there, whether it does or not. If you want to assume those statements that you are assuming, I will have to say "Yes", but I can't say that it is shown here.

Q Can't you say that this drawing shows anything at all?

A Oh, yes; but not what you are trying to get me to say.

Q What is there about this drawing—what is there about it that does not indicate that, in point of fact, the operation which I have described would be the mechanical result of the movement of that barrel up against that plunger?

(Testimony of Fred H. Hayn)

A   I can't accept that at all. The disclosure must be certain here. You must have the description in the specification here to tell what it does on here, and if you are trying to read some of those fanciful—pardon me for saying so—those statements you make here, I cannot do it.

MR. HERRON: Pardon me, I think you misunderstand my question. Read it to him, please.

(Question read by the reporter)

MR. HERRON: (continuing) That is, the movement of the barrel relative to the plunger, to be exact.

A   With your assumptions there, of course, I will have to say yes.

Q   I am not assuming anything. I am asking you a question and asking you to answer it. Now, will you please do it? Read it to him.

(Question again read by the reporter)

A   I cannot answer that at all. You want to know—

Q   I want to know precisely and only what is conveyed by the words of the question; just that and nothing more.

A   *I* don't convey anything to me definite. The showing here, it doesn't convey anything to me.

Q   Just a moment. If you please, tell me what is obscure and I will clear up that obscurity. Read it to him, and please tell me.

(Question again read by the reporter)

Q   What is obscure about the question? I will be glad to explain, Mr. Hayn.

A   I would like to answer the question if I possibly can. Now, you want to know, as I understand that question—

(Testimony of Fred H. Hayn)

Q I want you to just answer the question.

A I don't know. I can't answer it.

Q Why not?

A Because I don't know what you are talking about.

Q Please tell me what you don't understand about the question and I will clear it up for you.

A I am trying to tell you. You are asking me to explain what there is about that showing here that will have the result in the flowing of that fluid downward. Now, I have testified, I believe, that I don't know what those white spaces are and I can't read those things into that structure the way you—

Q Read them as a space, sir, and then answer my question. You don't think they are anything but a space, do you, honestly, now? That is what you think they are, don't you?

A Well, the natural assumption would be they are a space but you can't be certain.

Q That is your best opinion, isn't it?

A That is my best opinion.

Q Please act on your best opinion and then so answer the question.

A You mean if those were spaces would the oil flow downwardly?

Q I am sure, since you say, in your best opinion, they are spaces; so, reading the drawing what is the effect of moving the barrel up with relation to the standing plunger?

A Well, very likely the fluid would flow down below there, with your assumptions.

Q Not "very likely". I want your best opinion.

A And, with my best opinion.

(Testimony of Fred H. Hayn)

Q Now, you mean it very likely would, or that, from a mechanical standpoint it would of necessity do so?

A Yes. Remember, the structure is so indefinite here, the showing is indefinite and I have got to answer in respect to what I see here.

Q Mr. Hayn, I don't want to be captious, but can you tell me what you think, not what you would probably think if I pressed you to express that? Is it your answer, honestly intending to convey the thought that your best opinion is that the fluid would move downward, as you have indicated, through that space? Now, is that your best opinion?

A The best opinion, of course, it would have to be with this reservation; if the structure is as assumed, I will have to say yes, but I can't go any further than that.

Q Well, I regret your inability, but let it pass. Now, assuming that that did occur and that the fluid did pass downward through the space which you say it is your best opinion there exists, would that downward flow not wash off the top surface of the part 30 to the extent that it exerted any action?

A I can't read that in there again.

Q In other words, you don't know whether a flow down there would tend to wash that part off or whether it would not?

A If there is a flow there it certainly would wash it off there, but there is no indication there is such a flow there at all.

Q Now, will you turn to the Bray patent and show me what in the drawing there indicates that there is a flow washing off the clutch surface?

(Testimony of Fred H. Hayn)

A Well, there is on page 1, lines 24 to 29, that reference to the liquid. That is one place. Of course, that is a general statement.

Q I take it it is your opinion that the specification of the Bray patent teaches the creation of jets washing off that surface down below?

A It would have to do that? "Further, the bushing is provided with a series of ducts 5 from top to bottom so that as the barrel slides up from bottom position, liquid will be forced down in strong jets and wash sand from the top of the pin P so that this can be readily pulled from its seat in the lock box L."

Q So there isn't any question in your mind but that the Bray patent contemplates, in your opinion, the washing off of those surfaces?

A Well, I am not going to construe the patent.

Q You do not care to construe that patent?

A No.

Q Now, comparing the structure shown in the Bray drawing—that is, the provision made there for washing of subjacent parts with the structure of Neilsen we have just been discussing, to-wit, that chamber and that space, would you say the same result would occur from the operation of each of the structures shown?

A I can't find anything in Neilsen whatsoever that is anticipatory.

Q I did not ask you that. I ask you to answer my question. Read it to him.

THE COURT: Read the question, please.

(Question read by the reporter)

A Well, in the first place, I don't find that locking means. No.

(Testimony of Fred H. Hayn)

Q Why not?

A That is what I tried to explain to you.

Q Well, now, do it.

A Because I don't find the structure in Figs. 3 and 4 in Bray in Neilsen, nor do I find any vertical ducts.

Q I didn't ask you that. I ask that you answer the question. You have said that the drawing, in your best opinion, indicates an opening; you have said as that chamber is restricted in area fluid would flow through the opening. I have asked you if that fluid would flow across the shoulder of the top of 30 of Neilsen and you have said it would.

A Just a minute. Just a minute, now. You are getting concrete. Shoulder 30 of the Neilsen.

Q The shoulder on part 30 of Neilsen, I mean the shoulder, the upper, top shoulder of part 30 of the Neilsen.

A Oh, I don't find any shoulder there. I find a line there but that is all. I can't see what you want.

Q Doesn't that indicate a shoulder?

A Not at all.

Q What would you call it?

A I don't know what it is called. It is the top of a certain structure.

Q You know what I am talking about, don't you?

A I·do. Right here (indicating).

Q Now, then, I will define my terms. I call it a shoulder. Now, then, you know what I am talking about. Now read the question.

(Question read by the reporter)

Q Now, that is all true, isn't it? That is all true, isn't it?

(question read by the reporter)

(Testimony of Fred H. Hayn)

A   It is true with that same reorganization that comes before, if that is an opening there for the oil to go down it certainly would.

Q   That is right.

A   But there is an "if" there.

Q   And you state, in your best opinion, you think it is?

A   Yes.

Q   And I am content to accept it. Now, then, with that as a basis, please answer the rest of the question. I ask you now to compare that action with the action in the Bray through the ducts 5 and tell me why they don't work the same, since you have said they don't.

A   I am not trying to evade counsel's question. I am trying honestly to reply. As a matter of fact, I have tried all the way through.

Q   I am throwing no brickbats.

A   If I have seemed obstreperous, I did it utterly unintentionally. Now, you have got these ducts 5. Now, you want to know whether this hypothetical space in here—

Q   My dear fellow, I want you to answer the question. Now, please read it to him.

A   I am trying to.

Q   Don't interpret it to make it something different. I want you to answer my question, not yours. I will ask that you answer that. Now, please read the question.

(Question read by the reporter)

A   My answer must be to accord with the assumptions. If you agree with those assumptions that you put there in the beginning, that my answer would follow those, I would say yes.

(Testimony of Fred H. Hayn)

Q Well, do you say yes to the question I asked?

A Not exactly as put, no. I can't admit something in here that you see in here and say this does the same thing. Well, I can't do it.

Q. You can't admit anything, despite the fact that it is your best opinion that is what it is, is that what you mean?

A Well, no; just a little more than best opinion. I have got to be shown.

Q Now, I am going to ask the reporter once more to read you the question and I want you to tell me honestly whether you can answer it yes or no. If you can't, just say so and we will just drop the discussion right here.

A Yes, read it again. I will try my best.

(Question again read by the reporter)

Q Now, do they work the same or don't they?

A Since you start with that as a basis, I will have to say yes.

Q They do work the same?

A Yes.

Further

DIRECT EXAMINATION

BY MR. SOBIESKI: Mr. Hayn, calling your attention to Exhibit No. 7, in the files herein, which is a catalog of the Admore liner barrel pump, containing on the left side a picture or diagram of that pump, assuming that you are familiar with that drawing?

A In a general way, yes. Of course, it is just a structure there and I will admit I know it pretty well.

(Testimony of Fred H. Hayn)

Q   Yes. Do you find, calling your attention to the structure marked I in red on that there—that is the barrel?

A   It must be.

A   Now, I cannot be certain on these things, and when you hand me a sketch, practically nothing but a sketch, why, it is pretty hard to testify in respect to it. But it has a screw thread at the top there and, naturally, that would follow that one would be the traveling barrel.

MR. SOBIESKI: Now, then, near the lower part of this traveling barrel there is a dark spot. And there has been testimony that that represents certain holes in the traveling barrel—two dark spots, I believe, to allow the liquid to escape.

A   Yes.

Q   That is the testimony. Now, with respect to the bottom of this traveling barrel, you see there a nut with a female coupling device?

A   Oh, yes.

Q   And the male coupling device?

A   Yes.

Q   Now, it has also been testified that this device with the bottom nut on this—bottom nut 6, has a fairly close fit to the standing column but that there is room for some liquid to escape alongside of 6. Now, assuming a structure such as I have described, such as we find here, such as a structure with a traveling barrel with the bottom nut of a coupler face and holes in the side of the traveling barrel and the fit of the bottom nut being fairly tight to the traveling column; assuming that situation, is it your opinion that it would involve inventive skill to make any change of that structure to the structure described in the Bray patent?

(Testimony of Fred H. Hayn)

MR. HERRON: Now, we would ask that counsel define his term "fairly tight", in deference to the witness' inability to understand anything that is not definite, as well as to make a better record.

MR. SOBIESKI: By that—

MR. HERRON: You assume, I suppose, the clearance which has been testified exists in the structure represented by 13 and 14, is that correct; and that is—

MR SOBIESKI: I believe that is correct; yes.

MR. HERRON: I will get that from Mr. Adams here so we will know what we are talking about. What is the clearance between the puller nut on pump 13 and 14, the Admore?

MR. ADAMS: That is hard to estimate. That is slightly over a sixteenth of an inch.

MR. HERRON: Over all or on one side?

MR. ADAMS: Over all.

MR. SOBIESKI: Assuming such a clearance and striking out the words "fairly tight"—

MR. HERRON: I beg your pardon?

MR. SOBIESKI: Striking out the words "fairly tight".

MR. HERRON: And inserting—let us put it this way: Having part 6 with an interior diameter of one-sixteenth of an inch larger than the column over which it moves.

MR. SOBIESKI: Do you understand the question?

A I think I do. I would testify that, in my opinion, it would certainly be invention, assuming, of course, which I cannot do, this is clearance space and that those are holes, but granting that they are, they are laterally positioned while in Bray they are vertically positioned,

(Testimony of Fred H. Hayn)

and I certainly cannot find any teaching of this here of the Bray structure.

MR. HERRON: In other words, if I understand you correctly, you say it would be invention to change over the Admore liner barrel made as counsel has just now described it to you, over to the teaching of the Bray patent?

A   Absolutely.

Q   And the invention is of a high order, I assume?

A   Yes.   Yes.

Further

CROSS EXAMINATION

BY MR. HERRON: Now, would it involve invention to close the holes shown in that structure?

A   It might under certain conditions.

Q   Would it?   Just assume—

A   I would say yes.

Q   Take the clearance you have got—

A   Even if that is a small item there, there is no measure of invention, you know.   It might be slight, it might be medium, it might be more.

Q   Just a moment.   Let us take the structure.   You have testified you are reasonably familiar with it and have assumed to say what it does; would it involve invention to simply close those holes and leave that pump in all other respects the same?

A   I would say so.

Q   And in what would that invention, in your opinion, consist?   What would be the change in operation that would arise or in result that would rise to the dignity of invention?

(Testimony of Fred H. Hayn)

A Now, you are asking me to testify on something that I don't know anything about. You have got a vertical—

Q I am asking you to testify on something you have been glad enough to testify to your counsel's questions about. Now, please just answer my question.

A Just read the question, please.

(Question read by the reporter)

MR. HERRON: Describe to us the inventive act involved.

A Well, the invention would be the provision of a series of ducts.

Q No, no, no, no. My question says to simply close the holes and leave the pump in all other respects the same, close these lateral side holes, the four holes, exhibiting you the concrete structure.

A I see what you mean. You are plugging up these.

Q I am plugging up these holes, those four holes, yes; and I am doing nothing else. Literally nothing else; would that amount to invention?

A I would say so.

Q And what would be the inventive act; what is it? In what does the invention lie?

A The inventive act in closing those holes there would be to produce a new advantage, new result, or an old result in a better way.

Q Taking each of those things it would do, just tell us what it is.

A Why, the flow of the oil, of the fluid in the barrel, in other words, for the purpose of getting rid of the sand.

(Testimony of Fred H. Hayn)

Q In other words, you feel that it would involve invention to close those holes and therefore produce a greater flow of oil down through the annular space?

A You are bringing in theory again. I can't accept anything like that.

Q I am not going to argue the question whether you accept it or not. I will ask that you read the question, please.

(Question read by the reporter)

Q Would closing the holes produce any result?

A Yes.

Q What would be the result?

A Prevent the oil from flowing out the closed holes.

Q Yes; and what else, if anything? In other words, you think it would substantially change the operation of the pump simply to close the holes and prevent the oil from flowing out?

A Not necessarily the operation of the pump; a pump is a pump, whether it pumps oil or—

Q Oh, well, forget it.

A Give us some concrete questions.

Q All right; I will give you some concrete questions. If you close those four holes you have said that would amount to invention?

A Yes.

Q Now, I have asked you what the invention is, in what does it lie.

A Closing the holes.

Q Certainly, but what new or useful result does the closing of the holes bring about that you have to have in order to constitute invention?

(Testimony of Fred H. Hayn)

A   A better operation of the pump, as provided in the Bray structure here.

Q   And in what does that better operation lie?

A   In the jet action. That is probably what you wanted me to answer.

Q   It was, precisely.

A   All right. Why didn't you say so in the beginning?

Q   It is true, isn't it?

A   Yes.

Q   If you closed those four holes and thereby produced a stronger flow or a more copious, according to your theory, as I take it—

A   I am not advancing any theory.

Q   All right. According to your understanding of the operation of the structure, a more copious flow of the fluid downward into that amounts to invention?

A   Yes.

Q   In other words, it amounts to invention to increase or decrease the amount of that jet, is that right?

A   I wouldn't say that. That is going a little bit too far and putting in too much theory for me to accept.

Q   If the invention does not lie in the increasing of the force of the jet, in what does it lie?

A   I am not here to define "invention", if you please. I don't know what it is. If you can do it, go ahead. I can't.

Q   You have just said it would involve the employment of the inventive faculty to plug the holes and produce the increased jet.

A   Yes; but I can't define that.

(Testimony of Fred H. Hayn)

Q   Now, I ask you if the production of the increased jet is invention?

A   It might be.

Q   Well, is it in this particular case?

A   I couldn't say.

Q   Well, haven't you just said it was?

A   I am afraid I consider what you are trying to get out of me is an equivalent. I am willing to answer the question, but it seems to me you are nothing but quibbling.

Q   I just want to understand you. Have you or have you not testified that there would be the exercise of the inventive faculty in closing the four holes simply?

A   Yes.

Q   Have you not testified that that invention or the new and useful result which causes the closing of the holes to amount to invention is that you get an increased jet?

A   Yes.

Q   It follows, therefore, does it not, that you consider to get an increase in the velocity of a jet amounts to invention?

A   Not always.

Q   Well, it does in this particular structure?

A   Possibly.

Q   In other words, you consider—well, does it?

A   Possibly.

Q   Well, does it?

A   He is trying to pin me down to an opinion as to what is an invention and I cannot give it.

Q   I certainly am.

A   I am not in any position to define what invention is.

(Testimony of Fred H. Hayn)

Q I certainly am, since you have given your opinion. I want to know if the increase in the amount of jet action in this particular structure concerning which you are testifying, produced by the closing of those four holes— now, I want to know if the increase in the amount of the jet action occasioned by the closing of those holes is, in your opinion, such a change as amounts to invention?

A I will have to say yes, of course.

Q And you do so say?

A I do so say.

Q Do you so believe?

A I do so believe.

Q. Yes. How much increase—now, considering this particular structure, this structure, not generally—how much of an increase in the effective force of the jet, or let us say in the velocity of the jet, would there have to be produced by the closing of those holes in order for the change so produced, in your opinion, to amount to invention?

A I can't answer any question like that. That is entirely beyond me as to any theory that you are trying to get me to deduce from what you asked me.

Q Do you mean that you are utterly without any opinion as to what the answer is to the question I have just asked?

A That is precisely what I mean.

Q Then how are you able to say that the increase in jet action in this particular structure which would be occasioned by the closing of the four holes would result in any change in the jet action of such an order as that the closing of the holes would amount to invention?

(Testimony of Fred H. Hayn)

A   Of course, you are asking me a conclusion there and I am leading off to reply. Invention is such a mythical thing that all you can give is whether or not it is invention. Invention cannot be defined, and the best answer I can give you as to my ability is that it would amount to invention. Now, to go and take those various details in there and say what is invention, I would not say.

Q   Do you know how much the increase of that jet is?

A   No; I don't.

Q   Do you mean that any increase whatsoever, no matter how slight—

A   Might, yes.

Q   Not "might". —would result in invention?

A   It might and might not.

Q   Would it?

A   I can't answer that definitely. I would like to answer your questions but you are treading on ground that cannot be answered.

Q   Well, I am going to insist on your answering it, and the question is this: Would any increase in the velocity of the jet in this particular structure amount to invention?

A   I can't answer it.

Q   Would it or wouldn't it?

A   I don't know.

Q   Have you any opinion?

A   I have none.

Q   You have none at all?

A   None.

(Testimony of Fred H. Hayn)

Q Then, would a great increase in the amount of velocity amount to invention?

A I can't answer that, either.

Q Then, would any amount between a slight increase and a great increase amount to invention?

A Might.

Q And might not?

A And might not.

Q You can't answer?

A The same way for all of them.

Q You don't know?

A I don't know.

Q Then, why did you say that the increase in velocity produced by the closing of these four holes in this structure would amount to invention, as you have sworn under oath you believe?

A Because, in my opinion, it would. I am unable to define invention. If you can do it, go ahead.

Q Is your opinion founded upon reason, anything in reason?

A On reason and by my judgment of that structure that is there.

Q All right. Give us precisely the process of reasoning whereby you deduce that to close those holes would result in such a change of jet action as would rise to the level of invention?

A Well, I will tell you that. I think I can answer that. By plugging up those holes there you are not having this lateral action and you close them up and you get an advantage—it might be a slight one—but you get an advantage there by having the vertical action.

(Testimony of Fred H. Hayn)

Q And you think any advantage, no matter how slight, would amount to invention?

A Yes, yes, yes; decidedly.

Q Any advantage?

A In any advantage.

Q Whether it amounts to advantage or performed useful labor or did not, it would still amount to an invention, is that right?

A Of course, you can't have a frivolous result. Naturally, you have got to have something that is advantageous.

Q Precisely.

A If it is frivolous and you get no useful result, why, of course it is no invention, that is all.

Q All right. In your opinion, now, to what extent does the velocity of jet have to be increased in order to get a useful result?

A I can't answer that.

Q Let me ask the question—

A You are repeating what you said before.

Q (Continuing) In this particular structure concerning which you are now testifying?

A That is merely a matter of degree, and ordinarily, a matter of degree does not amount to invention.

Q Exactly.

A Is that what you want me to say?

Q And you don't know?

A I don't know.

Q And you don't know to what degree and point that fact of closing of the holes increases the jet?

A No.

(Testimony of M. Rennebaum)

M. RENNEBAUM,

called as a witness on behalf of plaintiff, in rebuttal, testified as follows:

THE CLERK: Will you state your name?

A   M. Rennebaum.

DIRECT EXAMINATION

BY MR. SOBIESKI: What is your business or occupation, Mr. Rennebaum?

A   I am a production man; have been in the production business for the last 15 years.

Q   Have you had any experience in relation to oil well pumps?

A   Yes; that is practically all I have done in the last 15 years.

Q   Will you give us a short statement of your experience along that line?

A   I have been with the Associated Oil Company for 8 years and I had charge of the Cole Oil Company production for them for 3 years, and the last couple of years I have been out pulling wells for different small companies.

Q   When you had charge of your production—you say you had charge of production: in a general way, what were your duties for these companies?

A   Well, it was looking after the wells, looking after production. If the production was off, why, we would pull out a pump and see what was wrong with it. If anything was wrong, why we remounted it to draw it back on.

Q   You had charge, then, of pumping operations for these various oil companies you have mentioned you were previously employed by?

A   Yes.

(Testimony of M. Rennebaum)

Q Have you operated Bray pumps and the Hofco pumps?

A Yes; I have.

Q You have. Calling your attention to this Exhibit II, do you recognize that as illustrative of the design of the Hofco pump?

A Yes.

Q On Figure 2 of this I see a structure designated as D. Can you tell us what that structure is called?

A Well, that is called the lock nut. Some people call it the sealer nut. It has various names.

Q Now, you have also operated, you say, a number of Bray pumps?

A Yes.

Q In the operation of the Hofco pump, in your opinion, liquid which is trapped in the barrel here is ejected out of the barrel, is that correct?

A. Yes.

Q In what way? How is it ejected out?

A Well, when the pump goes up it traps it in and when it goes down it forces it out.

Q Have you also operated pumps which allow for the ejection of the liquid through the side of the barrel of the pump?

A I have. The Neilsen pump is one of the old established pumps with holes in the side. I found it wasn't very good in some wells; in some wells it operated all right. In other wells it didn't.

Q In your opinion, as an oil man, is the operation of the pump with the oil going in and out of the bottom nut on the traveling barrel more efficient in the aid of the prevention of sand and emulsion over the barrel than where

(Testimony of M. Rennebaum)

the oil goes in and out through the holes in the side of the barrel?

A   And for this reason: When you put holes on the side here you are so much farther away from the sand. It hits up against your tubing here. You haven't the same force that you have if the oil comes straight down here, because if it was washing over that standing valve, every time it made a stroke it forced the oil down here and lets all of this up and oil would squirt out here on the side. Of course, it would drop down here. It would not have the same force if it went straight down.

Q   In your opinion, which method is more efficient in the aid of the prevention of sand?

A   The Hofco or the Bray.

Q   You have operated both the Hofco and the Bray pumps; have you noticed any difference in whether those pumps sand up or the way they operate with respect to sand accumulation below the traveling barrel?

A   No; there is practically no difference.

Q   You have known practically no difference?

A   No; both give the same.

Q   In your experience, you find they give the same results?

A   Both work free and give the same results.

Q   And this bottom nut down here, you say that is called by various names?

A   Yes.

Q.   Will you tell us some of the names that is called by in the industry?

A   Some of us call it a lock nut and some of us call it a sealer nut. It is a puller nut. It holds this bottom standing valve here. If you don't have this nut on here

(Testimony of M. Rennebaum)

you couldn't pull this standing valve. It is also used in case you had a rod job. If you don't have this nut here straight across you couldn't screw up your rods, or if you had a sand job you couldn't unscrew your rods because the barrel would just rotate around.

Q   You say this operates as a pulling nut?

A   Yes; it is a pulling nut. Also, it is hitting down. When we put this well on, as a rule we also hook it on the bottom and let it hit down so as to lock it in the shoe.

Q   Yes. And, as I understand, by pulling nut, how does it act as a pulling nut?

A   Well, when you get ready to pull the well, when you hook on your rods and you pull up your plunger here, this here inside here is very much smaller than this tube in here. When she comes against it you pull your standing valve out. Otherwise, you couldn't pull your standing valve.

Q   This same nut, that is sometimes called a guide?

A   Absolutely; it acts as a guide also.

CROSS EXAMINATION

BY MR. HERRON: What do you mean by that?

A   Well, it has a hollow in here and this tube in here, this plunger, travels up and down on that and keeps it in place.

Q   As a matter of fact, doesn't the packing hold it in place?

A   No; there is no packing down here.

Q   I meant up above, up above the packing.

A   There is packing up above; yes.

(Testimony of M. Rennebaum)

Q And that would hold it under normal conditions so the pump would run up and down straight?

A No; you couldn't pump if you didn't have that in here?

Q If you didn't have what?

A The lock nut in here.

Q In other words, you would get no pumping action whatsoever if you did not have part D of that picture, is that it?

A You would have a little action there, yes.

Q How greatly would the volume of the production of your pump be decreased by the elimination of the part D?

A Well, if you took that nut out entirely you would pull this plunger entirely off of your standing valve.

Q Now, I am not talking about that. I am talking about the fact that you say if you took the nut off it would decrease your production. Why?

A Well, I will say that—well, I will take that statement back.

Q I thought so. Now, in point of fact, there is a space between the column, the standing guide I and that part D, isn't there, so that that ring really runs over the standing column A and the guide D in the sense that my hand or the ring from my fingers runs over the pencil (illustrating)?

A Yes.

Q It may be at times that if it ever should encounter it the sides of the ring would act as a stop and stop further motion?

A Yes; guide that there.

(Testimony of M. Rennebaum)

Q Yes. In other words, stop it against further deflection against the side?

A Yes.

Q I presume, if it went over, by the entering of the two metal pieces into juxtaposition, that is, if it ever got froze—

A (Interrupting) Yes; depends on whether the hole is crooked.

Q —it would not run as I have illustrated with my hand over the pencil out of the packing which keeps it straight? The answer to that is "Yes", isn't it? That is right, Isn't it? I say, unless the hole is crooked it would keep in position up above?

A The packing helps it better.

Q What you mean to imply, I take it, about the performance of both Hofco and the Bray pump is that they are both good pumps?

A Yes; absolutely.

Q You would not say that they never sand up, would you?

A The only time that I have ever seen either of them sand up is when they pumped all the fluid that they could pump and had the sand in the barrel; couldn't get no more fluid.

Q That has been your experience when they have something called a heave?

A No; not a heave. Just a case where they didn't get enough fluid.

Q I take it they both would stop where you really would stop the pump and let her set for some days, when this sand would settle down and pack around those parts; you would expect that, wouldn't you?

(Testimony of M. Rennebaum)

A   It all depends upon the condition of the well, how much sand.

Q   You would expect a settling of the sand, and even though you would pump up and down your barrel in that event, the jet would not be strong enough to dig out the sand, would it?

A   Yes; it would.   I have done that.

Q   What you have actually done is to set down and hammer?

A   Well, we call it drilling out.

Q   That is it exactly.   In other words, you come down on that packed sand with the barrel and drill it out just as you would drill?

A   You don't come down quite to the sand.   As long as there is a little fluid coming in there the motion of the pump in the oil coming through these holes here works as a jet and forces it up.

Q   In other words, it is coming down and as long as you can come down close to that sand you, in effect, drill it out?

A   Yes; Hofco.

Q   As a matter of fact, or any other pump; that is true of any pump, isn't it?

A   No; I wouldn't say that.

Q   Of what pump isn't it true?   Let us put it the other way: Of what other pumps is it true?

A   Well, you take these two pumps of the Bray and the Hofco.   I have used it as an upside down pump. There was the Nielsen pump; I have never had such good success with it.

(Testimony of Dan W. Hoferer)

Q   Well, what other pumps have you used?

A   I have used—not the Republic, but the B & B.

Q   They were all pretty good pumps, weren't they?

A   Yes; they are all good pumps.

Q   Those pumps will likewise do the same thing Hofco and Bray will do, so far as drilling out the sand?

A   Yes; they will if there isn't too much sand in it.

MR. SOBIESKI: That is all.   We rest.

DAN W. HOFERER,

a defendant herein, called as a witness in his own behalf, in surrebuttal, testified as follows:

DIRECT EXAMINATION

My name is Dan. W. Hoferer.   I am a manufacturer of oil well pumps, and I am manufacturing the Hofco pumps concerning which stipulation has here been entered into.   I have been engaged in the pump manufacturing business and in the production of oil since 1923.   I have superintended the installation of many Hofco pumps in wells.   I have observed their operation.   I think I am familiar with it thoroughly.

Q   Now, the last witness talked somewhat of the sanded conditions in which, we will say, that the fluid in a well has been more or less drawn out and considerable amounts of sand have gone in, where he has gone out and described what he calls a drilling operation in which he drills out the sand.

A   I heard that, yes.   I know the condition to which he refers.   That is a natural condition which occcurs; that is, pumps sometimes sand up under certain conditions.

(Testimony of Dan W. Hoferer)

Q And when they do, describe briefly to the Court what is done to constitute the operation which the last witness testified concerning.

MR. SOBIESKI: Has the witness testified of any experience he has had in operating these pumps, other than installing?

MR. HERRON: Oh, yes.

Q You have installed many pumps, haven't you?

A I have superintended the installation of pumps since 1923.

MR. SOBIESKI: Installation only?

A The installation and operation under my supervision.

MR. HERRON: In other words, you service your pumps and take care of them and all that sort of thing?

A We direct the method in which they should be operated. I say "we"—I mean myself, because I was in charge of the particular department. In referring to the sanded condition which you speak of now, I wish to follow what this witness just testified to prior to me.

Q I just wish you would testify, first, assuming that the pump is sanded up. Just explain to us what you do in order to withdraw the pump from the hole if you are wanting to withdraw a sanded pump.

A If sand has packed up around the standing column to the extent of, say 12 or 14 inches during ordinary operation, it is impossible to withdraw that pump without lowering the traveling barrel and letting it or permitting it to actually drill down and hit bottom to stir up the sand and to permit the removal of same.

(Testimony of Dan W. Hoferer)

Q   That is to say, if you have a sanded well where the sand is up in your pump 13 inches or more from the bottom you then set the barrel down?

A   Lower the rods, bringing the barrel up and down until it actually impinges upon and strikes against the sand which is there below.

Q   And then drilling it out with the bottom nut?

A   You drill it out with the actual contact, not with any motion of fluid.   I carry that out a little farther with my reasons, that if we do not lower the pump and actually drill the sand out by direct contact it is impossible to remove it and we find, when we have a stripping job, that is, removing the tubing and the pump all inact, then upon examination after the tubing with the pump locked therein, that it is firmly held with sand; and that only takes 12 inches of sand and you cannot pull your pump out.

Q   And you do run into that condition?

A   We do; very common.   And the way we meet it is by actually lowering the barrel down until it pounds against the sand and wash it away.

Q   And churn it up and then finally—

A   Continuous contact and jarring action of the pump against the sand, and finally stir it to the extent it will reach its maximum pump stroke.   When it does, you can remove it.

Q   But in any event, your pump is sanded up, as you have testified.   Can you remove by reciprocating the barrel up and down without having the barrel actually coming in contact with the sand below?

A   No; you cannot.

(Testimony of Dan W. Hoferer)

CROSS EXAMINATION

BY MR. SOBIESKI: Mr. Hoeferer, I call your attention to this catalog. I believe there is a catalog in evidence of the Hofco Pump Company?

A Yes.

Q And you recognize that as being one of the catalogs put out by you?

A It was up to 1932. This is an old folder, the original that we put out when we first came out with my pump —I beg your pardon. We had one prior to that. We don't use that folder and haven't since 1932. I have a new catalog.

MR. SOBIESKI: Do you have one of those catalogs here?

A Yes; I have one of the catalogs with the red folder but I left it in my car. I am sorry.

MR. HERRON: I assume it will be stipulated that that folder concerning which the witness has just testified is identical with Exhibit No. 4?

MR. SOBIESKI: Yes.

MR. SOBIESKI: Do you mean to say you had none of those printed or none of them distributed since 1932?

A We have not distributed any, to my knowledge, since 1932. I fix that date on account of Mr. Bray's notification to me of infringement.

Q You have not distributed any of these since that time?

A Not to my knowledge, no, sir; and my instructions are issued specifically against that practice.

Q Calling your attention to the back page, where it says: "Likewise, the surge of the fluids in and out of the

(Testimony of Dan W. Hoferer)

lower end of the reciprocating barrel keeps all sand and foreign sediment stirred and permits. pulling of pump at any time without undue trouble." Is it your contention that the surge of fluids in and out of the lower end of the reciprocating barrel does keep the sand and foreign sediment stirred?

A   Only to the extent of the area inside of the annular space and above this so-called puller nut. It was not my intention to claim any jetting action or any forceful flow of the fluid which would wash the sand free because your fluid—

Q   Well, the question that I was asking is: It says "permits pulling of pump at any time without undue trouble." Now, as I understand it, the sand which prevents the pulling of the pump is the sand which would accumulate below this traveling barrel; is that not correct?

A   That is partly correct but that is not what I had reference to.

Q   This statement, then, that "the surge of the fluids in and out of the lower end of the reciprocating barrel keeps all sand and foreign sediment stirred and permits pulling of pump at any time without undue trouble", you say that that does not refer to the sand below the bottom of the traveling barrel?

A   That is right.

Q   It does not refer to that?

A   It does not refer to that.

Q   And yet, the sand which does prevent the pulling of the pump is the sand which accumulates below the traveling barrel and by what you have described—

A   As 12 inches of sand.

(Testimony of Dan W. Hoferer)

Q —as 12 inches of sand—and what was the word you used?

MR. HERRON: "Drilled out."

MR. SOBIESKI: —by drilling operations this traveling barrel can clean out that sand?

A Yes; that is right. Wait a minute now.

Q The question is kind of involved. I will restate it.

MR. HERRON: Perhaps you can get it by just asking him what he does mean by that, to explain the action.

MR. SOBIESKI: In the drilling operation you have described when the well is sanded the traveling barrel is moved up and down to drill out the sand below the traveling barrel, is that correct?

A That is correct, providing you lower it so that its maximum stroke will be below normal usage. Now, what I have reference to with the surging action of the fluid keeping sand away so it prevents sticking, to permit its removal, is that sand must not accumulate in here to prevent the removing of the barrel at all; in other words, we must move the barrel in order to get a chance to drill out if there is sand present.

Q But the sand that prevents the pulling of the pump is the sand which accumulates below the traveling barrel, is that correct?

A That is directly correct.

Q Then, in other words, if you had a barrel which had a device to prevent the sand accumulating so that you could pull the pump at any time, you would have a device so you could keep the space below the traveling barrel free from sand, is that correct, or is it too involved?

A I think I can answer it, but let's have the question again.

(Testimony of Dan W. Hoferer)

MR. HERRON: Do you understand him?

A   I think that is correct, providing the device is interpreted a little clearer there. I am not sure what he means by the device here unless he means a combination of the puller nut or an extension downward of some form in which he gets motion of fluid at the face of the puller nut or in this area around the lower part of the standing column, such as National Supply use right now. Now, they have a pump with a wash feature incorporated in their pump. They structurally have the extension of the device you mention. Mr. Sobieski is not quite clear as to how that will be accomplished.

MR. HERRON: Well, he does not understand your question.

MR. SOBIESKI: I will ask another question.

Q   I understand when you pull your pump the whole thing comes out, including the barrel and the standing column both, is that correct?

A   That is right.

Q   So that when you are discussing the type of sand that you had in mind when you want to be able to pull the pump at any time, you are not concerned with whether the barrel can move up and down on the pump, are you?

A   Yes; you are, very much so.

Q   Why is that?

A   The barrel must be movable so you can get a jar action or you cannot move it. May I cite a reference?

Q   No.

A   All right.

(Testimony of Dan W. Hoferer)

REDIRECT EXAMINATION

BY MR. HERRON: Now, give us the illustration which you had in mind as the reason for the answer you gave counsel.

A I designed the Fluid Packed Pump Company's removal feature in 1929 in which I had a key type hooking the device, where there was no chance for a jar action of any type. We simply set down with the rod and turned a 90 degree turn and pulled the rods up. Very seldom could we remove that pump. No time did we have a chance to take a run and thereby jar it out. Your strain is too easy. It is too far away and you are on the stretch of the rods and you must have a jar action to successfully remove your pump.

Q. Do I understand it is the fact, Mr. Hoeferer, that as you have used this pump in drilling out where the sand is not up to great depths, you can put your hand on the polish rod and you can actually detect the fact there is a particular action?

A Yes.

Q And a driller who is an experienced driller can tell by the feel at the top of the earth very well whether the tool is drilling?

A Yes; he can tell whether it is striking solid matter at the top or the bottom of it, sir.

Q In order to get your pumps out when they become sanded, as you have testified they do, despite the fact in your literature in there they do not, have you made use of simply reciprocating the barrel?

A You have a specific—I will say, on almost each case I have reference to keeping a part movable so I can get the jar action which is necessary to remove a pump.

(Testimony of John H. Bray)

Q   That is all.

A   That is true in each one of the some 70 or 80 different types of inverted pumps.

JOHN H. BRAY

called as a witness on behalf of plaintiff, in rebuttal, testified as follows:

THE CLERK: Will you state your name?

A   John H. Bray.

DIRECT EXAMINATION

BY MR. SOBIESKI: What is your business or occupation, Mr. Bray?

A   Petroleum engineer.

Q   I hand you here a catalog of the Hoeferer Pump & Mchine Company—

MR. HERRON: Similar to Exhibit 4, I think.

MR. SOBIESKI: —similar to Exhibit 4, and ask you if you have ever seen that one before.

A   Yes.

Q   When did you first see it?

A   On the 3rd of June of this year.

Q   At what place?

A   At the place where I procured it.

Q   Where was that?

A   In Mr. Hoeferer's—the Hofco Pump Company's plant in Long Beach.

Q   Whereabouts in Long Beach?

A   I don't know the exact number but it is on Cherry Avenue.

(Testimony of John H. Bray)

Q   How did you get it?

A   Went in and asked for it.

Q   What did you say?

A   Asked if I could have a catalog of Mr. Hoeferer's pump.

Q   Did they give you two of them?

A   Gave me two of them.

Q   What did you do with them?

A   Gave you one and kept this copy myself.

Q   Has it been continuously in your possession ever since that time?

A   Yes, sir.

CROSS EXAMINATION

BY MR. HERRON: You asked for a folder like they used to have of the Hofco pump, didn't you?

A   No, sir.

Q   And that is what you got?

A   I asked for a catalog.

Q   You asked for a folder such as they used to have, specifically describing it, and the girl went and looked them up?

A   The girl was not there.   Some young man in the plant.

Q   Some kid gave it to you on Sunday?

A   He took it out of a box with a great many.

MR. HERRON: We will stipulate he testified concerning one identical with the one the Court has.

(Testimony of John H. Bray)

MR. SOBIESKI: All right.

It is ordered that the foregoing statement of evidence as prepared by solicitors for the respective parties, together with the Exhibits therein referred to and embodied in the book of Exhibits and the following Physical Exhibits, heretofore ordered to be transmitted to the Clerk of the United States Circuit Court of Appeals, for the Ninth Circuit, to-wit:

Plaintiffs' Ex. 2 and 3, Defendants' Exhibits A, 17A, 17B, JJ, KK, LL. OO, PP;

That part of defendants' Exhibit C, referred to as Exhibit No. 2, Ex. No. 7, Ex. No. 10, Ex. No. 11, Ex. No. 12, Ex. No. 13, and Ex. No. 14;

That part of defendants' Exhibit No. E, referred to as Ex. No. 17; and

That part of Defendants' Ex. G, referred to as Exhibit No. 18; is a true, full statement of the evidence and is hereby duly allowed and settled and approved as the statement of evidence and may be filed as a part of the record on appeal.

March 7—1936

Wm P James
U. S. DISTRICT JUDGE.

[Endorsed]: Received copy of the within this 18th day of February 1936 Burke and Herron By Mark L Herron Attorneys for Defendants and Respondents. Filed Mar 7-1936 R. S. Zimmerman, Clerk By Edmund L. Smith Deputy Clerk.

[TITLE OF COURT AND CAUSE.]

MEMORANDUM OF CONCLUSIONS AND ORDER FOR JUDGMENT.

Plaintiff sued Defendant Corporation to recover for alleged infringment of Letters Patent No. 1,840,432, application for which was filed in the Patent Office February 4, 1929. An injunction is asked for.

The alleged invention relates to pump structure, particularly those used in deep wells. Such pumps are of particular use in the production of crude oil.

The inventive art, in the direction of which plaintiff exercised his originative faculties, was crowded with pump devices. Many patents have been issued to inventors in this field, and when plaintiff sought his patent he had little scope in which to extend improvement over the prior art. It is so admitted, as indeed it must be.

When it is desired to raise fluid from the bottom of a deep well, sometimes from thousands of feet below the surface of the ground, there is first the well casing. That is the casing which is inserted at the time of drilling to protect the sides of the hole and seal off water strata. Then a pump string is put down, inside which the pump is to work to force the column of liquid to the surface. A pump base is screwed into the pumping string at the bottom of the well, through the center opening of which the oil runs in to be taken care of by the lifting apparatus. Attached to this opening is a standing tube of length proportionate to the pump barrel used. The pump barrel is fitted to this tube and operates up and down with the impulse communicated to it by power means on the surface of the ground through pump rods leading therefrom.

Oil or other fluid is sucked into the barrel on the downward stroke of the pump barrel, and, through an arrangement of valves, on the upward stroke is expelled
into the pump casing. The latter overflows at the surface
as the pumping action is continued. This is the conventional method of operation which was practiced for years
before plaintiff sought to make an improvement in pump
barrels. It was ordinarily provided that there should be
means of escape for oil and sand that might interfere
with the desirable vacuum action of the barrel, and it had
been customary to provide holes on the barrel to eject the
fluid and sand laterally on the upward stroke of the pump
barrel. It was desirable to have a locking means between
the pump barrel and the standing colunm base, whereby
the barrel could be lowered until its base was in contact
with the column fixture, and when locked, the base fixture
could be unscrewed through the turning of the pumping
rods, allowing the entire pumping apparatus to be lifted
from the hole.

Plaintiff's device included two features: An arrangement for expelling sand and fluid from the barrel,
and an arrangement for locking the column to the barrel.
For the latter he claimed originality because of "interlocking" features, and for the former he claimed originality
because, as he declared, his method secured the result of
washing clean the interlocking part attached to the base,
which base would otherwise be covered with sand and
prevent proper interlocking contact when the barrel was
lowered with its cooperating notches. His claims present
no other features of asserted novelty.

In Claim 1, the combination described included nothing
that was not common to deep well pumps, except the

coupler contrivance. And the practice of using notched contact means between pump barrels and the fixture base operated in connection therewith, was well known. Moreover, it would require but the mechanical knowledge of any driller or pump operator to adapt such a contrivance to the designed use. It had been so adapted by others before the plaintiff sought to use it. Claim 1 is invalid as presenting no novelty.

Claims 2 and 3 (the remaining claims of the Patent) I have analyzed and divided as follows:

Claim 2: A combination, with a pump of the class having a standing guide column with a working valve and a valve traveling barrel running on said column:

(a) Of a bottom guide, on said barrel, running on said column and having a bottom face forming a coupler; and an anchor for the column presenting a top face forming a complementary coupling to be non-rotatively interlocked with the lower guide coupler.

(b) Said guide having a series of venting ducts which open directly toward the anchor coupler face so that during the strokes of the barrel, ejected liquid is impelled toward the anchor top face to aid in prevention of said accumulation thereon.

Claim 3 contains nothing not included in Claim 2 except the words "said guide being aperatured for the ejection of liquid from the barrel directly toward the effective face of the lower, standing coupler."

Venting ducts described in Claim 2 are the same as "aperatures", referred to in Claim 3.

It is very plain that the patentee claims nothing for any particular form of standing column or working barrel, except as to the two features: coupling means and means for the ejection of fluid directly upon that coupling. As I have indicated, the coupling means presents no features of inventive novelty over the prior art or public use. Mechanical skill only is involved in rearranging the coupling means; its use was well known.

Only one feature remains for discussion: Was the arrangement of venting ducts at the lower end of the barrel, and on a "bottom guide" affixed thereto, a novelty in the art considered, and if so, is the invention of broad or pioneer character, so as to include a wide range of equivalents. My conclusion on the several features of the problem are, first: The patentee contemplated that the guide used should have a close or ordinary mechanical fit; he intended to obtain the maximum of jet action to accomplish the desired effect in washing off the coupler face. His invention is brought within a narrow field and is not entitled to a wide range of equivalents. The teachings of the patent are specific in terms as the patentee states: "The bushing is provided with a series of ducts 5 from top to bottom so that as the barrel slides up from bottom position, liquid will be forced down in strong jets and wash sand from the top of pin P so that this can be readily pulled from its seat in the lock box L." If perchance an operator does not employ a guide with a running fit, and so allow some discharge of fluid to get by between the guide and the circumference of the standing column, he is not perforce following the specifications of the claims, hence is not infringing the plaintiff's patent right. That is the most that can be asserted against the defendant in this case.

There was strong evidence offered as tending to show that the effect claimed for the jet action would not result in operation—this to the point of lack of utility. I am not disposed to conclude that the patent is void for that reason.

It might, of course, be interesting to follow by description the various patent forms that have been introduced in evidence. While I have given much study to the argument made in the briefs, and to the evidence introduced, it would take many, many pages to set forth any adequate analyses of the evidence. The facts determine the decision. The conclusions arrived at have been stated.

Therefore: I hold that Claim 1 of the patent involved, is void for want of invention. I hold that Claims 2 and 3 are valid, limited closely to the feature of having a bottom guide running on the standing column, with a running fit, and the guide being provided with vents or aperatures for the ejection of liquid directly toward the effective face of the lower coupler member; that the invention is to be strictly limited to the form described; that no infringement is proved.

Findings and decree will be entered accordingly. An exception is noted in favor of plaintiff.

Dated this 16th day of April, 1935.

Wm P. James
U. S. District Judge

. [Endorsed]: Filed Apr. 16, 1935.

[TITLE OF COURT AND CAUSE.]

FINDINGS OF FACT AND CONCLUSIONS OF LAW

The above entitled cause having been brought on for final hearing upon pleadings and proofs and the cause having been duly argued by counsel both orally and upon written briefs subsequently filed, and the Court having duly considered the evidence and arguments of counsel and being fully advised in the premises, now finds the following:

FINDINGS OF FACT

I.

That this is a cause in equity arising under the patent laws of the United States over which this Court has jurisdiction.

II.

That United States Letters Patent No. 1,840,432 was granted to plaintiff, Patrick H. Bray, January 12, 1932, on an application filed February 4, 1929, for "PUMP".

III.

That the title to said Letters Patent No. 1,840,432, is in plaintiffs, Patrick H. Bray and Catherine Patricia Marquette.

IV.

That the claims in issue are as follows:

1. The combination, with a pump of the class having a standing guide column with a working valve and a valved travelling barrel, running on said column; of a bottom guide, on said barrel, running on said column and

having a bottom face forming a coupler, and an anchor for the column presenting directly to said coupler face a top face forming a complementary coupling to be non-rotatively interlocked with the lower guide coupler.

2. The combination, with a pump of the class having a standing guide column with a working valve and a valved travelling barrel, running on said column; of a bottom guide, on said barrel, running on said column and having a bottom face forming a coupler, and an anchor for the column presenting a top face forming a complementary coupling to be non-rotatively interlocked with the lower guide coupler; said guide having a series of venting ducts which open directly toward the anchor coupler face so that during the strokes of the barrel, ejected liquid is impelled toward the anchor top face to aid in prevention of sand accumulation thereon.

3. The combination, with a pump of the class having a standing guide column with a working valve and a valved travelling barrel, running on said column; of a bottom guide, on said barrel, operating on said column and having a coupler part on its bottom end, and an anchor for said column having a top coupler part complementary to the coupler part of the said guide, whereby the engaged parts may be meshed in non-rotative interlock, and said guide being apertured for the ejection of liquid from the barrel directly toward the effective face of the lower, standing coupler part to aid in preventing sand settlement thereon.

V.

That defendants' pumps charged to infringe the claims in issue and as manufactured and sold by defendants are shown by plaintiffs' Exhibit 4 and defendants' Exhibit II;

that they are of two kinds, one employing a "jaw clutch" coupler part illustrated by defendants' Exhibit II, the other a "notched ring" coupler part illustrated by defendants' Exhibit II and by physical Exhibits JJ, KK and LL; that defendants' pumps are of the standing column type with a reciprocating barrel having a puller nut at the lower end surrounding the column and allowing some discharge of fluid between the puller nut and the circumference of the column as distinguished from a close, ordinary mechanical or running fit of the puller nut on the column; that the puller nut in one type of defendants' pumps is notched to cooperate with a complementary notched part at the anchor of the standing column, while in the other type of defendants' pumps a notched ring surrounds the column and is interposed between the lower plain face of the puller nut and the upper plain face of the anchor for the column; that the puller nuts of both types of defendants' pumps are not provided with a series of ducts or any duct, or with vents or apertures, or any duct or any aperture for ejection of liquid toward the effective face of the notched part at the anchor.

VI.

The Court finds all of the elements of claim 1 of said patent No. 1,840,432 in issue, in the same or equivalent combinations in the prior art in evidence, both patented and unpatented.

VII.

The Court finds all of the elements of defendants' pumps in the same or equivalent combinations in the prior art in evidence, both patented and unpatented. [M.L.H. J.G.S.]

VIII.

The Court finds that the locking means between the pump barrel and the standing column base whereby the barrel could be lowered until its lower end was in contact with the column fixture and be thereby locked, and which is used in defendants' pumps, does not either in plaintiffs' or defendants' pumps effect any functional relationship or mode of operation of the parts different from the pumps found in the patented and unpatented prior art in evidence, and further finds that the practice of using notched contact means between a pump barrel and the anchor for the column or the column fixture was well known in the patented and unpatented prior art in evidence. The Court further finds that it would require merely the mechanical knowledge of any driller or pump operator skilled in the art to adapt such notched contact means to the designated use, and further finds that it had been so adapted by others before the plaintiff sought to use it.

IX.

The Court finds that the essence of the invention claimed by claim 1 of patent No. 1,840,432 lies in the interlocking feature described in finding VIII hereof and that said interlocking feature claimed in said claim 1 does not present any novelty or amount to any invention over the prior art in evidence both patented and unpatented, and further finds that said claim presents no other feature of asserted novelty.

X.

The Court finds that the essence of the invention claimed in claim 2 and in claim 3 of said patent No. 1,840,432 resides in an arrangement for expelling fluid from the barrel, for which the patentee claimed originality in this, that he asserted his method secured the result of washing clean the interlocking part at the column fixture, which part would otherwise be covered with sand and prevent proper interlocking contact when the barrel was lowered with its cooperating notches, and the Court further finds that said claims present no other feature of asserted novelty except the interlocking feature which is the subject of findings VIII and IX hereof.

XI.

The Court finds that the patentee contemplated that the guide used should have a close or ordinary mechanical fit with the standing column and that he intended to obtain the maximum of jet action to accomplish the desired effect of washing off the coupler face of the column fixture, and further finds that the teachings of the patent are specific in terms, as the patentee states "the bushing is provided with a series of ducts 5 from top to bottom so that as the barrel slides up from bottom position liquid will be forced down in strong jets and wash sand from the top of pin P so that this can be readily pulled from its seat in lock box L."

XII.

The Court finds that neither of the said defendants, as shown by the evidence, employs in his, or its, pump a guide with a close, ordinary mechanical or running fit; that they, and each of them, thus allow some discharge of fluid to get by between the guide and the circumference of the standing column, and further finds that neither of said defendants is following the specification of claims 2 and 3 of patent No. 1,840,432.

XIII.

The Court finds that claims 2 and 3 are limited closely to the feature of having a bottom guide running on the standing column, with a running fit, the said guide being provided with vents or apertures for the ejection of liquid directly toward the effective face of the lower coupling member or column fixture and that the invention is to be strictly limited to the form as so described.

XIV.

The Court further finds that none of the pumps in evidence, manufactured, used or sold by either of said defendants employs a guide which is provided with vents or apertures for the ejection of liquid directly toward the effective face of the lower coupler member or column fixture.

XV.

The Court finds that none of the pumps of defendants made like those in evidence infringe upon either claim 2 or claim 3 as so limited.

CONCLUSIONS OF LAW

I.

That claim 1 in issue is null and void for want of invention over the prior art in evidence.

II.

That claims 2 and 3 in issue are, and each of them is limited closely to the feature of having a bottom guide running on the standing column, with a running fit, the said guide being provided with vents or apertures for the ejection of liquid directly toward the effective face of the lower coupling members and that the invention is to be strictly limited to the form described.

III.

That said claim 2 and said claim 3 in issue are, and each of them is, as herein limited, good and valid in law.

IV.

That the defendants, and each of them, have not infringed any of the claims of said Letters Patent and have not infringed said Letters Patent.

An exception is noted in favor of the plaintiff

April 23, 1935

Wm P. James
United States District Judge.

Approved as to form:

McAdoo & Neblett
John G. Sobieski
4/23/35
Counsel for Plaintiffs.

BURKE & HERRON and RUSSEL GRAHAM,
By Mark L. Herron
Counsel for Defendants.

[Endorsed]: Filed Apr. 23, 1935.

IN THE DISTRICT COURT OF THE UNITED
STATES IN AND FOR THE SOUTHERN
DISTRICT OF CALIFORNIA
CENTRAL DIVISION

****

PATRICK H. BRAY, and)
CATHERINE PATRICIA
MARQUETTE,

Plaintiffs,) NO. X 50 J IN EQUITY

– Vs –) FINAL DECREE

HOFCO PUMP, LTD.,
D. W. HOFERER,

)

Defendants.

— — — — — — — — —)

The above entitled cause having come on regularly to
be heard before the Court, Honorable William P. James,
presiding, on pleadings and proofs, and the Court having
heard the evidence of the parties and having considered
the briefs filed by the parties and taken due course for
consideration, and being fully advised in the premises,
and having prepared Findings of Fact and Conclusions
of Law pursuant to said Findings of Fact and Conclusions
of Law;

IT IS NOW ORDERED, ADJUDGED AND DE-
CREED:

1. That claim 1 of Patent No. 1840432 is null and
void.

2. That claims 2 and 3 in issue are, and each of them
is limited closely to the feature of having a bottom guide

running on the standing column, with a running fit, the said guide being provided with vents or apertures for the ejection of liquid directly toward the effective face of the lower coupling members and that the invention is to be strictly limited to the form described.

3. That said claim 2 and said claim 3 in issue are, and each of them is, as herein limited, good and valid in law.

4. That the defendants, and each of them, have not infringed any of the claims of said Letters .Patent and have not infringed said Letters Patent.

5. That the Bill of Complaint is dismissed; and

6. That defendants recover from plaintiff herein their costs and disbursements herein expended, said costs to be taxed by the Clerk of this Court in the sum of $122.50.

Done and Ordered this 23rd day of APRIL, 1935.

An exception is noted in favor of plaintiff

Wm P. James
DISTRICT JUDGE

Approved as to form:

McAdoo & Neblett

John G. Sobieski

   Counsel for Plaintiffs 4/23/35

BURKE & HERRON AND RUSSELL GRAHAM

By Mark L. Herron

   Counsel for Defendants.

Decree entered and recorded Apr. 23, 1935

R. S. ZIMMERMAN Clerk,
By Murray E. Wire, Deputy Clerk.

[Endorsed]: Filed Apr. 23, 1935.

[TITLE OF COURT AND CAUSE.]

PETITION FOR APPEAL

TO THE HONORABLE WILLIAM P. JAMES, District Judge:

The above-named plaintiffs, feeling aggrieved by the decree rendered and entered in the above-entitled cause on the 23rd day of April, 1935, do hereby appeal from said decree to the United States Circuit Court of Appeals for the Ninth Circuit, for the reasons set forth in the assignment of errors filed herewith and pray that said appeal be allowed and that citation be issued as provided by law, and that a transcript of the record, proceedings, and papers and documents upon which said decree was based, duly authenticated be sent to the United States Circuit Court of Appeals for the Ninth Circuit under the Rules of such court in such cases made and provided; and your petitioners further pray that the proper order relating to the required security to be required of him be made.

> PATRICK H BRAY and
> CATHERINE PATRICIA MARQUETTE
> By Joseph F Westall
>
> Their Attorney.

McADOO & NEBLETT
JOHN G. SOBIESKI
JOSEPH F. WESTALL
   Solicitors and of counsel for plaintiffs.

[Endorsed]: Filed Jul. 23, 1935.

IN THE DISTRICT COURT OF THE UNITED
STATES SOUTHERN DISTRICT OF CALI-
FORNIA CENTRAL DIVISION

PATRICK H. BRAY, and CATH-)
ERINE PATRICIA MARQUETTE,)
)
               Plaintiffs,)
) IN EQUITY
         vs.   ) NO. X-50-J.
)

HOFCO PUMP, LTD., ET AL

              Defendants.)

———————

ASSIGNMENTS OF ERROR

Now come the Plaintiffs in the above-entitled cause and
file the following assignments of error upon which they
will rely upon their prosecution of an appeal in the above-
entitled cause, from the decree entered in said cause on
the 23rd day of April, 1935.

The United States District Court for the Southern
District of California, Central Division, in entering the
above-mentioned decree, erred,—

I.

In finding in its Memorandum of Conclusions and Order for Judgement that the inventive art, in the direction
of which Plaintiff exercised his originative faculties, in
the production of the subject-matter of the patent in suit,
was crowded, and, on the contrary, in failing to find with
relation to the claimed subject-matter, that the art was

not crowded and that said subject-matter within a proper scope of the claims in suit was novel.

II.

In finding in its Memorandum of Conclusions and Order for Judgement that in the conventional method of operation prior to the invention of the patent in suit that in pumping wells oil or other fluid is sucked into the barrel on the downward stroke of the pump barrel and in failing to find, on the contrary, that it is on the upward stroke that the oil or other fluid is pumped into the barrel.

III.

In finding in its Memorandum of Conclusions and Order for Judgement that in the conventional method prior to the invention of the patent in suit that in pumping wells oil is expelled into the pump casing on the upward stroke of the pump barrel and in failing to find, on the contrary, that it is on the downward stroke that oil is expelled into the pump casing.

IV.

In finding in said Memorandum of Conclusions and Order for Judgement that "many patents have been issued to inventors in this field, and when Plaintiff sought his patent he had little scope in which to extend improvement over the art, and on the contrary, in failing to find that within such scope the claims define a great and very valuable improvement over the art.

V.

In finding in said Memorandum of Conclusions and Order for Judgement as to the state of the prior art "It was ordinarily provided that there should be means of escape for oil and sand that might interfere with the desirable vacuum action of the barrel, and it had been customary to provide holes on the barrel to eject the fluid and sand laterally on the upward stroke of the pump barrel," and in implying in said finding that validity or scope of the claims in suit was under the law, in any way affected thereby.

VI.

In finding in its Memorandum of Conclusions and Order for Judgement "It was desirable to have a locking means between the pump barrel and the standing column base, whereby the barrel could be lowered until its base was in contact with the column fixture, and when locked, the base fixture could be unscrewed through the turning of the pumping rods, allowing the entire pumping apparatus to be lifted from the hole," contrary to the great weight of the evidence.

VII.

In finding in its Memorandum of Conclusions and Order for Judgement that "Plaintiff's device included two features: An arrangement for expelling sand and fluid from the barrel, and an arrangement for locking the column to the barrel. For the latter he claimed originality because of 'interlocking' features, and for the former he claimed originality because, as he declared his method secured the result of washing clean the interlocking part attached to the base, which base would otherwise be covered with sand and prevent proper interlocking contact when the barrel was lowered with its cooperating notches. His claims present no other features of asserted novelty", and thus in ignoring the combinations of elements as set forth in each of the claims of said patent in suit as constituting the patented subject-matter.

VIII.

In the finding immediately above assigned as error in segregating from the claims in suit alleged features of asserted novelty and in ignoring the claims as written by substituting, in effect, such alleged features of asserted novelty in place of the claims.

IX.

In finding in its Memorandum of Conclusions and Order for Judgement that in claim 1 of the patent in suit the combination described included nothing that was not common to deep well pumps except the coupler contrivance, and, on the contrary, in failing to find that the combination of Claim 1 was notwithstanding antiquity of certain elements, novel as a combination.

X.

In finding in such Memorandum of Conclusions and Order for Judgement that it would require but the mechanical knowledge of any driller or pump operator to adapt what is designated in said Memorandum as "coupler contrivance" to the design described and in the combination covered by Claim 1 of said patent in suit.

XI.

In finding in said Memorandum of Conclusions and Order for Judgement that what the Court has referred to as the "coupler contrivance" had been adapted by others before patentee of the patent in suit, to the use in combination with remaining elements of each of the claims in suit as in said patent desclosed.

XII.

In finding in said Memorandum of Conclusions and Order for Judgement that Claim 1 of the patent in suit is invalid as presenting no novelty; and, on the contrary, in failing to find that according to the great weight of the evidence that the combination of said Claim 1 was a novel combination involving patentable invention.

XIII.

In finding in said Memorandum of Conclusions and Order for Judgement that the coupling means presents no features of inventive novelty over the prior art or public use; and in failing to find, on the contrary, that the combination of each of the claims in suit including such coupling means does in view of the great weight of the evidence, present inventive novelty over the prior art, patented or otherwise.

XIV.

In finding in said Memorandum of Conclusions and Order for Judgement that mechanical skill only is involved in the arrangement of coupling means included as features of any of the claims of the patent in suit and that any such re-arrangement of coupling means was well known in the art prior to the invention of the patent in suit as contrary to the great weight of the evidence.

XV.

In finding in said Memorandum of Conclusions and Order for Judgement that the patentee contemplated that the guide used should have a close or ordinary mechanical fit, and, on the contrary, in failing to find that as a matter of law the specification and drawings of the patent in suit must be construed in the light of mechanical skill and that the drawings are illustrative only and not working drawings and that, as a matter of law, the patentee is not limited to details illustrated where language of his claims properly interpreted so as to cover the combination forming the essence of his invention requires otherwise.

XVI.

In finding in said Memorandum of Conclusions and Order for Judgement that the patentee in suit intended to obtain the maximum of jet action and that such contemplated that the guide used should have a close fit and that such close fit was necessary to accomplish the desired effect of washing off the coupler face, as contrary to the weight of evidence and as implying that intentions of the patentee are to be read into claims.

XVII.

In finding in said Memorandum of Conclusions and Order for Judgement that the invention of the patent in suit is brought within a narrow field and is not entitled to a wide range of equivalents; and in failing to find, on the contrary, that each of the claims of the patent in suit is entitled upon due and full consideration of the prior art as properly put in evidence to a sufficient range of equivalents to include defendants' devices charged as infringements in this suit.

XVIII.

In failing to find that, having illustrated and described one practicable form in which the invention claimed may be embodied, that the patent in suit, in contemplation of law, covers all equivalent forms whether described in the specification or not, and that the forms charged to infringe are such equivalent forms.

XIX.

In finding in said Memorandum of Conclusions and Order for Judgement that the statement in the specification of the patent in suit (page 1, line 75, et seq) reading "The bushing is provided with a series of ducts 5 from top to bottom so that as the barrel slides up from bottom position, liquid will be forced down in strong jets and wash sand from the top of pin P so that this can be readily pulled from its seat in the lock box L," must be literally and narrowly construed and read or implied as a limitation into any of the claims in suit; and in not finding that broad language of the claims, and particularly of Claim 1, is entitled under the law to be construed in accordance with its terms and without any such implied limitation from the specification or drawing.

XX.

In failing, contrary to the law, to construe the claims in suit and each of them in accordance with their respective terms and in implying into said claims details and incidents of the specification which specification described and illustrated but one particular embodiment.

XXI.

In implying into the claims or any of them "a guide with a running fit" and in holding that if an operator does not employ such a fit but allows some discharge of fluid to get by between the guide and the circumference of the standing column he is not utilizing the combination of the claims in suit or any of them and is not infringing any of said claims.

XXII.

In finding and holding in said Memorandum of Conclusions and Order for Judgement that Claim 1 of the patent in suit is void for want of invention; and failing to find on the contrary, that said claim is in all respects valid and of sufficient scope to cover and include any of the devices contended on the trial of this cause to be an infringement or infringements.

XXIII.

In finding or holding in said Memorandum of Conclusions and Order for Judgement that Claims 2 and 3 of the patent in suit must be limited closely to the feature of having a bottom guide running on a standing collar with a running fit and the guide being provided with vents or aperatures for the ejection of liquid directly toward the effective face of the lower member and that the invention is to be strictly limited to the form so described; and that no infringement is proven; and in failing, on the contrary, to find and hold that neither or none of said claims should be limited to such features described in the specification but that each of said claims should be construed according to their terms and with sufficient breadth to include defendants' devices charged to infringe.

XXIV.

In finding in said Memorandum of Conclusions and Order for Judgement that Claims 2 and 3 of the patent in suit are properly to be strictly limited to details of form described in the specification, and, on the contrary, in not construing said claims as they were worded and granted by the patent office.

XXV.

In ordering and directing findings and decree prepared and entered in accordance with the Court's Memorandum of Conclusions and Order of Judgement.

XXVI.

In finding (paragraph V of the Findings of Fact) that both devices of Defendants' pumps are not provided with a series of ducts or any duct or with vents or aperatures, or any duct and any aperature for ejection of liquid toward the effective face of the notched part at the anchor and in failing to find that said types of Defendants' pumps are provided with aperatures within the meaning of the claims in suit.

XXVII.

In finding that in paragraph VI of the Findings of Fact that all of the elements in Claim 1 of said patent No. 1,840,432, in issue in the same or equivalent combinations are found in the prior art in evidence, both patented and unpatented, contrary to the evidence.

XXVIII.

In finding or implying in paragraph VI of the Findings of Fact that the combination of Claim 1 can properly be anticipated by alleged equivalents of the prior art.

XXIX.

In finding in paragraph VIII of the Findings of Fact that the locking means between the pump barrel and the standing column base whereby the barrel could be lowered until its lower end was in contact with the column fixture and be thereby locked, and which is used in defendants' pumps, does not either in Plaintiffs' or Defendants' pumps effect any functional relationship or mode of operation of the parts different from the pumps found in the patented and unpatented prior art in evidence, contrary to the weight of the evidence.

XXX.

In finding in paragraph VIII of the Findings of Fact "that the practice of using notched contact means between a pump barrel and the anchor for the column or the column fixture was well known in the patented and unpatented prior art in evidence," contrary to the weight of the evidence.

XXXI.

In finding in Paragraph VIII of the Findings of Fact "the Court further finds that it would require merely the mechanical knowledge of any driller or pump operator skilled in the art to adapt such notched contact means to the designated use, contrary to the weight of the evidence; and in failing, on the contrary, to find that it required the invention covered by the claims in suit to make such adaptation.

XXXII.

In finding in Paragraph VIII of the Findings of Fact that the notched contact means described in said finding had been adapted by others to the designated use before patentee's invention of the subject matter in suit, contrary to the great weight of the evidence.

XXXIII.

In finding in the Findings of Fact, Paragraph IX that the essence of the invention claimed by claim 1 of patent No. 1,840,432 lies in the interlocking feature described in Finding VIII of the Findings of Fact, and in substituting such alleged essence of invention for the combination of the claim as written.

XXXIV.

In construing said claim 1 piecemeal instead of in accordance with its terms as a combination, and in attempting to substitute an alleged essence of invention for the claim as written, contrary to law.

XXXV.

In failing to find that said claim 1 of the patent in suit is not met in terms by any prior art structure and that it cannot be properly anticipated by finding one similarity in one structure and another in another and still another in a third to build up the combination of the claims or any of them.

XXXVI.

In failing in Findings of Fact, Paragraph IX to heed or apply the law to the effect that it is conclusively presumed that each element of the claimed combination is old whether in fact or not and that the only novelty in invention, under the law resides in the combination as claimed.

XXXVII.

In effect rewriting claim 1 for the purpose of passing upon issues of alleged anticipation and of infringement.

XXXVIII.

In finding in Findings of Fact, Paragraph X an alleged essence of the invention, and in substance in substituting such alleged essence of invention for the claim as written.

XXXIX.

In finding in Findings of Fact, Paragraph X that the essence of invention claimed in claims 2 and 3 of the patent in suit resides in the arrangement for expelling fluid from the barrel, and in effect in substituting this alleged essence of invention for the combination of the claims as written.

XL.

In failing in Findings of Fact, Paragraph X to recognize and apply the law that each element of a claim is conclusively presumed to be old and that there is no such thing as a principal element, and that the issue is not whether any one or more or any principal element or feature is new or old but whether the combination of the claim as written is novel.

XLI.

In assuming in such Findings of Fact as those of Paragraphs IX and X that a proper or permissable interpretation of a patent requires a combing through the art to discover what if any feature or element is novel, or asserted to be novel, and then substituting such alleged feature of novelty in place of the claim as written for the purpose of passing upon questions of validity and infringement; and in failing to treat the claim as an indivisable unit in which each element is equally important for the purpose of interpretation.

XLII.

In finding in Paragraph XI of the Findings of Fact that patentee contemplated that the guide used should have a close or ordinary mechanical fit with the standing column, and that he intended to obtain the maximum of jet action to accomplish the desired effect of washing off the coupler face of the column fixture; and in implying such assumed contemplation or intent as a limitation into any of the claims in suit.

XLIII.

In finding in Paragraph XI of the Findings of Fact that the teachings of the patent are specific in terms (meaning the statements of the specification) and in reading such alleged teachings (portions of the specification or illustrations from the drawings) as limitations into the claims as granted, thus in effect rewriting the claims to cover narrow details of structure of a single embodiment when the grant as contained in the claims is of a much wider exclusive right.

XLIV.

In misapplying the familiar rule that claims are to be construed in the light of the specification as meaning that claims may be rewritten in the light of the specification to include something more or different from what their terms express.

XLV.

In such findings as that of Paragraph XI of the Findings of Fact in effect substituting parts of the specification of the patent in suit for the claims as granted.

XLVI.

In the Findings of Fact Paragraph XII reading into the claims 2 and 3 a close mechanical or running fit of guide and with the claims so altered, in finding that neither of said defendants has utilized the subject matter of claims 2 and 3 of the patent in suit.

XLVII.

In finding in Paragraph XIII of the Findings of Fact that claims 2 and 3 are limited closely to the feature of having a bottom guide running on the standing column, with a running fit, the said guide being provided with vents or apertures for the ejection of liquid directly toward the effective face of the lower coupling member or column fixture, contrary to the fact that neither of said claims contain any such limitation.

XLVIII.

In failing to give to the language of said claims 2 and 3 a sufficient breadth of interpretation to include defendants' devices charged to infringe.

XLIX.

In finding (Paragraph XIV of the Findings of Fact) that none of the pumps in evidence, manufactured, used or sold by either of said defendants employs a guide which is provided with vents or apertures for the ejection of liquid directly toward the effective face of the lower coupler member or column fixture, contrary to the evidence and in assuming that the fact so found avoids the charge of infringement.

L.

In finding, Paragraph XV, of the Findings of Fact that none of the pumps of defendants made like those in evidence infringe upon either claim 2 or claim 3 as so limited; and in implying that any limitations as imposed by the Court are proper under the law; and, on the contrary, in failing to find that both claims 2 and 3 are valid and infringed.

LI.

In failing to find that claim 1 of the patent in suit is valid and that it is infringed by each of the devices contended on the trial of this cause to be infringements.

LII.

In finding as a conclusion of law that claim 1 of the patent in suit is null and void for want of invention over the prior art in evidence.

LIII.

In finding as a conclusion of law that claims 2 and 3 in issue are and each of them is limited closely to the feature of having a bottom guide running on the standing column, with a running fit, and said guide being provided with vents or apertures for the ejection of liquid directly toward the effective face of the lower coupling members and that the invention is to be strictly limited to the form described; and in failing to find that there is no warrant in law or fact in reading any such limitations into claims 2 or 3 or either of them.

LIV.

In finding as a conclusion of law that defendants and each of them have not infringed any of the claims of said letters patent and have not infringed said letters patent; and in failing to find that each and every of said claims have been infringed by defendants in the manner and by the means manufactured and sold as charged in the complaint.

LV.

In decreeing that claim 1 of patent No. 1,840,432 is null and void, and on the contrary in failing to decree that said claim is good and valid in law.

LVI.

In decreeing that claims 2 and 3 in issue are, and each of them is limited closely to the feature of having a bottom guide running on the standing column, with a running fit, the said guide being provided with vents or apertures for the ejection of liquid directly toward the effective face of the lower coupling members and that the invention is to be strictly limited to the form described; and on the contrary in failing to decree that said claims in suit must be taken as they are written and that the Court is not authorized to read limitations into them for the purpose of avoiding a charge of infringement.

LVII.

In decreeing that defendants and each of them have not infringed any of the claims of said letters patent and have not infringed said letters patent; and in failing to find and decree on the contrary, that said letters patent and each of the claims thereof have been infringed by each of the devices charged by plaintiffs to be infringements.

LVIII.

In decreeing that the Bill of Complaint be dismissed; and in failing to decree that this cause be referred to a proper officer of the Court to take an accounting of damages and profits as prayed; and in likewise failing to direct the issuance of an injunction against any and all of the acts of defendants and each of them complained of as infringements.

LIX.

In decreeing that defendants recover from plaintiffs their costs and disbursements in this cause expended, and, on the contrary, in failing to decree that plaintiffs recover from defendants and each of them plaintiffs' costs expended in this cause.

LX.

In failing to find and decree that any of the alleged prior uses have not been established by evidence beyond a reasonable doubt, and that none of such defenses can be sustained.

LXI.

In failing to find and decree that no defense of novelty or want of invention over the art has been sufficiently proven.

WHEREFORE, applicants pray that said decree be reversed, and that the District Court for the Southern District of California, Central Division, be directed to enter an order or decree reversing the final decree of said District Court and to enter an interlocutory decree finding the patent in suit valid and infringed and referring this cause to a Master to be appointed by said court to determine and report to said court the damages and profits resulting from such infringement.

> PATRICK H. BRAY and
> CATHERINE PATRICIA MARQUETTE
> By Joseph F Westall
>> Solicitor and of Counsel for Plaintiff.

McADOO & NEBLETT
JOHN G. SOBIESKI
JOSEPH F. WESTALL
> Solicitors and of Counsel for Plaintiffs.

[Endorsed]: Filed Jul. 23, 1935.

[TITLE OF COURT AND CAUSE.]

ORDER ALLOWING APPEAL

On motion of Joseph F. Westall, Esq., solicitor and counsel for plaintiffs, it is hereby ordered that an appeal to the United States Circuit Court of Appeals for the Ninth Circuit, from the final decree heretofore on the 23rd day of April, 1935, filed and entered herein be, and the same is hereby allowed, and that a certified transcript of the record testimony, exhibits, stipulations, and all proceedings be forthwith transmitted to said United States Circuit Court of Appeals for the Ninth Circuit. It is further ordered that the bond on appeal be fixed at the sum of two hundred and fifty dollars ($250.00).

Dated this 23 day of July, 1935.

Wm P. James
United States District Judge.

[Endorsed]: Filed Jul. 23, 1935.

[Title of Court and Cause.]

BOND ON APPEAL

KNOW ALL MEN BY THESE PRESENTS:

That we, PATRICK H. BRAY and CATHERINE PATRICIA MARQUETTE, as principals, and CHAS. J. MILLS as surety, are jointly and severally held and firmly bound unto Hofco Pump, Ltd., and D. W. Hoferer, in the penal sum of Two Hundred and Fifty Dollars ($250.00) to be paid to them and their successors, executors, administrators and assigns; to which payment, well and truly to be made, we bind ourselves and each of us, jointly and severally, and each of our heirs, executors and administrators, by these presents.

Sealed with our seals and dated this 17th day of August, 1935.

WHEREAS, the above named PATRICK H. BRAY and CATHERINE PATRICIA MARQUETTE have taken an appeal to the United States Circuit Court of Appeals for the Ninth Circuit to reverse the decree made, rendered and entered on the 23rd day of April, 1935, in the District Court of the United States for the Southern District of California, Central Division, in the above entitled cause;

AND WHEREAS, said District Court of the United States for the Southern District of California, has fixed the amount of Plaintiffs' bond on said appeal in the sum of Two Hundred and Fifty ($250.00) Dollars;

NOW THEREFORE, the condition of this obligation is such that if the above-named PATRICK H. BRAY and CATHERINE PATRICIA MARQUETTE shall prosecute their said appeal, and any appeal allowed to be taken to the Supreme Court of the United States to effect, and answer all costs which may be adjudged against them, if they fail to make good said appeal, then this obligation shall be void; otherwise to remain in full force and effect.

Patrick H. Bray

Catherine Patricia Marquette

Principals.

Chas. J. Mills

Surety.

STATE OF CALIFORNIA)

                         : ss:

County of Los Angeles   )

On this 17 day of August, 1935, before me J. F. BISHOP, a notary public, personally appeared CHAS. J. MILLS known to me to be the person described in and who duly executed the foregoing instrument, and acknowledged that he executed the same.

And the said CHAS. J. MILLS being by me duly sworn, says that he is a resident and householder of the said county of Orange and that he is worth the sum of $1000 over and above his just debts and legal liabilities and property exempt from execution.

Chas. J. Mills

Subscribed and sworn to before me this 17th day of August, 1935.

[Seal]   J. F. Bishop
    Notary Public in and for the County of Los Angeles,
    State of California.

My Commission Expires Aug. 14, 1936.

Examined and recommended for approval, as provided in Rule 28.

                    Joseph F. Westall
                    Attorney for Plaintiff-Appellants.

I hereby approve the foregoing bond this 19 day of August, 1935.

                    Wm P. James
                    United States District Judge.

[Endorsed]: Filed Aug. 19, 1935.

[TITLE OF COURT AND CAUSE.]

STIPULATION AND ORDER OMIT*T*NG TITLES, ETC. FROM PAPERS

IT IS HEREBY STIPULATED by and between the respective parties in the above-entitled cause, through their respective counsel, that the title and caption of the cause on the respective papers appearing in the printed transcript of record may be eliminated and the following substituted therefor: "Title of Cause".

IT IS FURTHER STIPULATED that the complete filing reference by the Clerk on the respective papers appearing in the printed transcript of record may be eliminated and substituted therefor the word "Filed" followed by the date on which the paper was filed.

IT IS FURTHER STIPULATED that all acknowledgements of service on the respective papers appearing in the printed transcript of record may be eliminated.

DATED this 26th day of February, 1936.

Burke & Herron
By Mark L. Herron
Attorney for Defendants.

Alan Franklin
Attorney for Plaintiffs.

It is so ordered, Feb. 26, 1936.

Wm. P. James
Judge.

[Endorsed]: Filed Feb. 26, 1936.

[TITLE OF COURT AND CAUSE.]

ORDER FOR TRANSMISSION OF ORIGINAL EXHIBITS TO CLERK OF THE UNITED STATES CIRCUIT COURT OF APPEALS FOR THE NINTH CIRCUIT.

And, now, to-wit, March 7, 1936, IT IS HEREBY ORDERED: That the following original Exhibits offered in evidence at the trial of this cause, being Plaintiffs' Exhibits 2 and 3, and Defendants' Exhibits A and B; Defendants' Exhibits C, D, DI, E, F, G, H, I, J, K, L and M, together with the exhibits attached to each thereof, to-wit, the exhibits referred to in said depositions as Exhibits 1, 2, 3, 4, 5, 6, 7, 8, 9, 10, 11, 12, 13, 13A, 14, 15, 16, 17, 18, 19, 20, 21 and 22; Exhibits EE, FF, GG, HH, II, JJ, KK, LL, MM, NN, OO, PP, QQ, 17A and 17B, be transmitted to the United States Circuit Court of Appeals for the Ninth Circuit, and filed with the Clerk thereof, for use upon the hearing of the appeal herein.

Wm P. James
UNITED STATES DISTRICT JUDGE

[Endorsed]: Filed Mar. 7-1936.

[Title of Court and Cause.]

PRAECIPE FOR TRANSCRIPT OF RECORD ON APPEAL

To the Clerk of the above-entitled Court:

Please issue, in accordance with and in response to appeal allowed, a certified transcript of record on appeal from the decree entered in the above-entitled cause, including the following:

1. Bill of Complaint.

2. Amended Answer.

3. Memorandum of Conclusions and Order for Judgment, dated April 16, 1935.

4. Findings of Fact and Conclusions of Law.

5. Final Decree entered April 23, 1935.

6. Petition for Appeal.

7. Assignment of Errors.

8. Order Allowing Appeal.

9. Citation on Appeal.

10. Bond on Appeal.

11. Narrative statement of evidence under Rule 75, and as part thereof, please also prepare six (6) copies of a Book of Exhibits and entitle the same Vol. II for use on said appeal, forwarding four (4) copies thereof to the Clerk of the United States Circuit Court of Appeals for the Ninth Circuit for filing, one (1) copy to the attorney for the plaintiff, and one (1) copy for the attorneys for

the defendants, reproducing therein the following documentary exhibits in the above case:

Plaintiff's Exhibit No. 1 – Bray patent in suit No.
1840432.

"   " 4 – Hofco Catalog.

The following portions of Defendants' Exhibit B (File Wrapper and contents of Letters Patent to Bray No. 1,840,432)

Pages 33740 – 2 to 337340 – 8 inclusive, and

Pages 337340 – 10 to 337340 – 21 inclusive.

The following exhibits received in evidence as a part of Defendants' Exhibit C.

1. Ex. No. 1.

2. Page 43, of Exhibit No. 2.

3. Exhibit No. 3.

4. Exhibit No. 4.

5. Exhibit No. 5.

6. Exhibit No. 6.

7. Exhibit No. 7.

8. Exhibit No. 8.

9. Exhibit No. 9.

10. Page 20 of Exhibit No. 10.

11. Page 7 and 8 of Exhibit No. 11.

The following exhibits received in evidence as part of Defendants' Exhibit D.

1. Exhibit 13A.

2. Exhibit 19.

The following exhibits received in evidence as part of Defendants' Exhibit D 1.

Exhibit 20.

Exhibit 21.

The following exhibits received in evidence as part of Defendants' Exhibit E.

Exhibit No. 15.

Exhibit No. 16.

Exhibit No. 17.

The following exhibits received in evidence as part of Defendants' Exhibit G.

Page 62 of Exhibit 18.

The following exhibit received in evidence as a part of Defendants' Exhibit H.

Exhibit 22.

The following Defendants' Exhibits

N', O', O², P', Q', R', S', T', U', V', W', X', and Y', inclusive.

Please mark the colored Exhibit N′, Ex. N and N′

" " " " " O′, Ex. O and O′

" " " " " P′, Ex. P and P′

" " " " " Q′, Ex. Q and Q′

" " " " " R′, Ex. R and R′

" " " " " S′, Ex. S and S′

" " " " " T′, Ex. T and T′

" " " " " U′, Ex. U and U′

" " " " " V′, Ex. V and V′

" " " " " W′, Ex. W and W′

" " " " " X′, Ex. X and X′

" " " " " Y′, Ex. Y and Y′

in accordance with the note appearing at page 56A of the Narrative Statement of the Evidence under Rule 75 on file herein.

Exhibits Z, AA, BB, CC, DD, EE, FF, GG, HH, II, MM, NN and QQ.

12. This Praecipe.

13. Stipulation and order omitting titles, etc., from all papers.

14. Order for Transmission of Exhibits.

Dated: Los Angeles, California, March 6th, 1936.

Respectfully,

Alan Franklin.

ALAN FRANKLIN

Attorney and solicitor
for Plaintiff

[Endorsed]: Filed Mar 7—1936.

[TITLE OF COURT AND CAUSE.]

CLERK'S CERTIFICATE.

I, R. S. Zimmerman, clerk of the United States District Court for the Southern District of California, do hereby certify the foregoing volume containing 415 pages, numbered from 1 to 415 inclusive, to be the Transcript of Record on Appeal in the above entitled cause, as printed by the appellant, and presented to me for comparison and certification, and that the same has been compared and corrected by me and contains a full, true and correct copy of the citation; bill of complaint; amended answer; statement of evidence; memorandum of conclusions and order for judgment; findings of fact and conclusions of law; final decree; petition for appeal; assignment of errors; order allowing appeal; bond on appeal; stipulation and order omitting titles from papers; order for transmission of original exhibits, and praecipe.

I DO FURTHER CERTIFY that the amount paid for printing the foregoing record on appeal is $ and that said amount has been paid the printer by the appellant herein and a receipted bill is herewith enclosed, also that the fees of the Clerk for comparing, correcting and certifying the foregoing Record on Appeal amount to.................. and that said amount has been paid me by the appellant herein.

IN TESTIMONY WHEREOF, I have hereunto set my hand and affixed the Seal of the District Court of the United States of America, in and for the Southern District of California, Central Division, this................ day of March, in the year of Our Lord One Thousand Nine Hundred and Thirty-six and of our Independence the One Hundred and Sixtieth.

R. S. ZIMMERMAN,

Clerk of the District Court of the United States of America, in and for the Southern District of California.

By

Deputy.

TOPICAL INDEX.

ii.

TABLE OF AUTHORITIES CITED.

iv.

No. 8171

In the United States
Circuit Court of Appeals
For the Ninth Circuit.

Patrick H. Bray, and Catherine Patricia Marquette,

Appellants,

vs.

Hofco Pump, Ltd., and D. W. Hoferer,

Appellees.

OPENING BRIEF FOR APPELLANTS.

PRELIMINARY STATEMENT.

This is an appeal from a final decree [Tr. p. 386] of the United States District Court, in and for the Southern District of California, Central Division, in favor of the defents (appellees), in a suit for infringement of Letters Patent for an invention, to-wit: a pump, the Court having found claim 1 of said Letters Patent null and void, and claims 2 and 3 good and valid in law, but limited in scope, and not infringed by the defendants.

STATEMENT OF CASE.

Patrick H. Bray, one of the plaintiffs-appellants herein, filed his bill of complaint [Tr. p. 3] against the defendants-appellees, Hofco Pump, Ltd., D. W. Hoferer, John Doe and Richard Roe, for infringement of United States Letters Patent No. 1,840,432 for Pump, granted January 12, 1932, to said Patrick H. Bray. The case was designated as Equity No. X-50-J and was filed in the United States District Court for the Southern District of California, Central Division, on July 13, 1932. After filing his bill, and before the trial of this case, said Patrick H. Bray assigned an interest in and to said Letters Patent to Catherine Patricia Marquette, who was thereupon joined as one of the plaintiffs in this case [Tr. p. 24]. After issue was joined the case proceeded to trial before said Court on July 6, 1934, and after hearing all the evidence of both parties, the case was submitted on oral argument and briefs of counsel. The Court thereafter, on April 16, 1935, signed and filed its Memorandum of Conclusions and Order for Judgment [Tr. p. 374] and, on April 23, 1935, the Court signed its Findings of Fact and Conclusions of Law [Tr. p. 379] and its Final Decree [Tr. p. 386], which findings and conclusions were filed, and which decree was filed, and entered and recorded, on April 23, 1935, by the clerk of said court.

STATEMENT OF FACT.

[The Complaint, Tr. p. 3.]

The bill of complaint, alleges, among other essential facts, the granting of United States Letters Patent, No. 1,840,432, for Pump to plaintiff (appellant), Patrick H. Bray, on January 12, 1932; ownership, by said plaintiff, of the entire right, title and interest in and to said Letters Patent, since the issuance thereof; wilful infringement of said Letters Patent by the defendants (appellees); notice to the public that the invention covered by said Letters Patent was patented by fixing thereon the word "Patented," together with the day and year that the patent was granted; and notice in writing to the defendants of said Letters Patent and of their infringement thereof; and concludes with the usual prayer for an injunction, an accounting, damages, costs, etc., against the defendants.

AMENDED ANSWER.

[Tr. p. 9.]

The amended answer admits that the defendant, Hofco Pump, Ltd., is a California corporation, having its office and principal place of business in Long Beach, County of Los Angeles, State of California, and a citizen of said State; that the defendant D. W. Hoferer, is a citizen of California and a resident of Long Beach, County of Los Angeles, State of California; that this is a suit arising under the patent laws of the United States; and that the defendants and each of them have been duly notified in writing of the Letters Patent in suit, No. 1,840,432, and the infringement thereof by said defendants, but denies infringement of the patent in suit by said defendants. Further answering, the amended answer sets up the usual defense of prior patents; prior printed publications; prior knowledge and use by others; prior public use; invalidity of the patent in suit, in view of the prior state of the art; and limitation of the claims of the patent in suit, in view of the file wrapper and contents thereof.

THE PATENT IN SUIT.

(Plaintiffs' Exhibit 1, Bray Patent No. 1,840,432, Book of Exhibits, page 1.)

The patent in suit covers a pump for wells, such as oil wells, or water wells, which will not choke up with sand, or "sand up" with the loose fine sand, which is present in the oil fields of California and will not form a "gas lock." The nature and object of the invention is stated in lines 1 to 49 inclusive of the patent as follows:

> "This invention relates to pumps and more especially to deep well pumps of the class including a *standing valve mounted on a standing fixed column* (or stem) provided with a packing means on which is reciprocated a travelling pump barrel carrying a head or *traveling valve.*
>
> A feature of this class of pumps is that the traveling barrel can be utilized as a means of pulling the standing column from a retaining lock when it is desired to remove the pump as a whole from the well without pulling the tubing.
>
> A common defect of this class of pump is that sand settles on the locking means and freezes the parts to the tubing so that the pump cannot be pulled without great risk of serious injury of parts and possible loss of the well by reason of blocking of the hole when it is impossible to fish out the obstructing parts below the break. For this reason one object of the invention is to maintain the sand in such constant state of agitation that it cannot pack down on the lock shoe box and lock shoe pin.

To that end means are provided to cause a *positive jet action* of liquid over the lock means and keep the sand in a state of suspension so that it will be carried off in the discharge, and therefore prevent 'sanding up.'

Another object is to provide means for positively interlocking the traveling barrel to the standing lock means to enable the screwing and unscrewing of parts of the rod string in event of need.

An additional object is to provide a pump of such structure and design that the traveling barrel can be reversed end for end after a period of use so that its useful life may be greatly extended.

Other objects, advantages and features of construction, and details of means and of operation will be made manifest in the ensuing description of the herewith illustrative embodiment; it being understood that modifications, variations and adaptations may be resorted to within the spirit, scope and principle of the invention as it is hereinafter claimed."

Referring more particularly to the specification and drawing of the patent, the pump includes generally a *stationary* piston or *plunger,* indicated by the numerals, 10, 13, 16, 17 and 19; a standing valve S on the *upper end* of said plunger; a traveling barrel B which fits over said plunger and slides up and down thereon; a head valve or *traveling valve* V on the upper end of said traveling barrel; a pin or *anchor* P seated in a lock box L below the lower end of said traveling barrel B; a *standing column,* comprising the column 10 and its lower section or pipe joint 11, which connects the lower end of said column 10 to the upper end of said anchor P; the part 3 (jet device),

ositive
ep the
arried
'sand-

sitively
ig lock
f parts

of such
can be
that its

of con-
tion will
1 of the
derstood
; may be
nciple of

tion and
nerally a
numerals,
upper end
over said
d valve or
:eling bar-
below the

on or pipe
column 10
jet device);

in the form of a bushing, with vertical *jet apertures* or ducts 5, screw-seated in the lower end of the traveling barrel B and freely surrounding the lower section 11 of said standing column, so as to move up and down over the standing column with said barrel; and a clutch comprising an upper clutch member in the form of jaws 4 on the lower end of said bushing or part 3, and a lower clutch member in the form a flat-sided head 2 on the upper end of said anchor P, which jaws 4 engage said head 2, when the barrel B is lowered sufficiently, to clutch and lock the barrel B to the anchor P against rotation.

The pump, as above described, is assembled in the usual pump tubing T, which extends from the surface of the earth to the lock shoe box L down in the well. The cage of the traveling valve V, at the upper end of the traveling barrel B, is connected to the lower end of a string of sucker rods (not shown) which extend upwardly through the pump tubing T and connect to the crank of the engine on the floor of the derrick, whereby the pump barrel B is reciprocated up and down by said engine for operating the pump.

An important feature of the patent is the particular type of pump which the invention embodies, including the *close relationship of the standing valve S and traveling valve V*, when the traveling barrel B is in its lowermost pumping position, as shown in Fig. 1 of the patent. [Tr. pp. 37 and 115.] Such close relationship of said valves S and V prevents the accumulation of any appreciable amount of gas in the pump chamber within the barrel B between said valves, whereby a *"gas lock"* between said valve is avoided. A "gas lock" in the traveling barrel B prevents functioning of the pump, in that the pressure of such gas holds the valve S on its seat, while the pressure of the oil upon the top of the valve V holds said valve on its seat against said gas pressure, which allows the pump barrel B to reciprocate up and down, expanding and compressing the gas therein, but prevents any oil passing through the pump chamber in the barrel, and thus prevents the pumping of any oil by and through the pump. Gas is a serious problem in pumping oil from the oil wells of California, and a pump with the valves close together as disclosed in the patent is the only type pump that will avoid a gas lock in these oil fields. [Tr. p. 245.] (See, also, Defendants' Exhibit No. 4, p. 8, under the title "Working Actions.")

The part 3, which is described in the patent as being in the form of a bushing, is known by several different names in oil-well pump practice. It is known as a "bushing," "puller nut," "guide," "lock nut," "sealer nut," etc. [Tr. p. 358.] It is immaterial what term is used in the patent to describe the element 3 or any other element, provided the *function* of the element is made clear by the context of the specification. The function of element 3 is made perfectly clear in the specification, page 2, line 31, of the patent, where it is further described as a *"jet device 3."*

> *Walker on Patents,* 6th Ed., Sec. 160;
>
> *Wheeler Salvage Co. v. Rinelli, et al.,* 295 F. R. 717, 727;
>
> 48 *Corpus Juris,* Sec. 340;
>
> *Carlson Motor Truck Co. v. Maxwell Brisco,* 197 Fed. 309, 315.

In pumps of the type disclosed in the patent in suit, with a standing plunger and traveling barrel, the function of the part 3, or "puller nut," is ordinarily to engage the lower end of the lowermost plunger guide 13, for pulling the anchor P out of its seat in the lock box L, and pulling the pump out of the pump tubing T for repairing the pump or cleaning sand out of the tubing, when the pump is not too badly "sanded up" in said tubing.

The annular space, designated EE on Defendants' Exhibits MM and NN [Book of Exhibits pp. 178 and 179; Tr. p. 172], in the lower end of the traveling barrel B, surrounding the lower section 11 of the standing column, between the part or jet device 3 and the lower end of the lowermost plunger guide 13, constitutes a *jet pressure-chamber,* whereby pressure is applied by said part 3 to the liquid in said chamber on the upward stroke of the traveling barrel B for forcing liquid out of said chamber through the ducts or jet apertures 5, in *strong downwardly projected jets,* directly against the top of the anchor P and the lower clutch member 2, which washes sand from said anchor and clutch member 2, so that the sand will not pack over said clutch member 2, and over the top of the anchor and seal it in its seat in the lock box L; and the washed sand is thus kept moving upwardly through and out of the upper end of the pump tubing T, so that the sand-free anchor can be readily pulled out of its seat in said lock box; and if any of the sucker rods should come apart at their connecting joints, the barrel B may drop down upon the sand-free top of the anchor P, while the upper clutch member 4 on the lower end of part 3 may engage the lower sand-free clutch member 2 on the top of said anchor and lock the barrel B against turning so that the disconnected and loose joints of the sucker rods may be screwed up tight, and securely connected together.

Another important feature of the pump disclosed by the patent is the plunger packing of such construction as to prevent leakage of oil past the plunger in either direction, and particularly in an upward direction from the jet-pressure-chamber EE (Book of Exhibits p. 178; Defendants' Exhibit MM) into the pump chamber in the traveling barrel B between the valves S and V on the upward stroke of the traveling barrel. The construction of the packing disclosed in the patent is described on page 1, lines 90 to 100, inclusive, and on page 2, lines 1 to 26, inclusive, and lines 35 to 44, inclusive, said last lines being as follows:

> "The guides 13 are of an alloy which allows the guide to be fitted and operated closer to the barrel than with any other known metal plunger and without heating and consequent freezing of the pump due to the expansion of the inner parts. In other words, this bronze plunger or guide is highly efficient as a packing, runs close and reduces slippage of oil past the guide and has long life and is non-heating."

The plunger-packing described in the patent is merely illustrative for the purpose of disclosing a complete and operative pump of the type which the invention embodies, the patent not being limited in any manner to any particular plunger-packing, since any suitable form of plunger-packing is comprehended by the claims of the patent.

A further feature of considerable merit, incidental to the construction of the pump covered by the patent, is the reversibility of the traveling barrel B, whereby the unworn section of the barrel may be positioned to take wear of the plunger, when the worn section is worn out by the plunger, thus doubling the length of usefulness of the traveling barrel. The reversibility of the traveling barrel B is made possible by the jet apertures or ducts 5 in the part 3 instead of in the sides of the lower portion of the barrel, as for example, in the O'Bannon patent, Fig. 1, indicated at 16', and the Northrup patent, Figs. 1 and 2, indicated at 32, said patents being Defendants' Exhibits R and R' and Defendants' Exhibits V and V', respectively. (Book of Exhibits pp. 111 and 129.) Elimination of openings in the side at one end of the barrel makes both ends of the barrel the same, so that either end may be used in place of the other. The reversibility of the traveling barrel B is described in the patent on page 2, lines 25 to 34, inclusive, as follows:

> "The barrel B can be reversed end for end to obtain full benefit by use of its unworn part and thus prolong its useful life. Ordinarily barrels of this class of pumps are perforated through the sides to allow fluid flow during operation on the standing packer, but by means of the jet *device 3* providing for the circulation in the present pump the barrel is imperforate and can, therefore, be reversed as above stated."

The three claims of the patent, all of which are in issue before this Honorable Court, are as follows:

> 1. The combination, with a pump of the class having a standing guide column with a working valve and a valved traveling barrel, running on said column; of a bottom guide, on said barrel, running

on said column and having a bottom face forming a coupler, and an anchor for the column presenting directly to said coupler face a top face forming a complementary coupling to be non-rotatively interlocked with the lower guide coupler.

2. The combination, with a pump of the class having a standing guide column with a working valve and a valved traveling barrel, running on said column; of a bottom guide, on said barrel, running on said column and having a bottom face forming a coupler, and an anchor for the column presenting a top face forming a complementary coupling to be non-rotatively interlocked with the lower guide coupler; said guide having a series of venting ducts which open directly toward the anchor coupler face so that during the strokes of the barrel, ejected liquid is impelled toward the anchor top face to aid in prevention of sand accumulation thereon.

3. The combination, with a pump of the class having a standing guide column with a working valve and a valved traveling barrel, running on said column; of a bottom guide, on said barrel, operating on said column and having a coupler part on its bottom end, and an anchor for said column having a top coupler part complementary to the coupler part of the said guide, whereby the engaged parts may be meshed in non-rotative interlock, and said guide being *apertured* for the ejection of liquid from the barrel directly toward the effective face of the lower, standing coupler part to aid in preventing sand settlement thereon.

Claim 1 covers a pump of the type disclosed, with a standing valve S on the *upper end* of the standing plunger and a traveling valve V on the traveling barrel, *in combination* with a clutch, comprising the upper clutch member 4, located on the part 3 on the lower end of the traveling barrel B, and the lower clutch member 2 located on the top face of the anchor P. In the patent drawing the valves S and V are shown close together for the purpose of preventing the accumulation of any appreciable amount of gas and the formation of a *"gas lock"* between said valves. *This close relationship of the valves,* as shown in the drawing, illustrates the *class or type* of pump described in the first paragraph of the patent specification, as follows:

> "This invention relates to pumps and more especially to deep well pumps of the *class* including a *standing valve* (S) mounted on a standing, fixed column (or stem) provided with a packing means on which is reciprocated a travelling pump barrel carrying a head or *travelling valve* (V)."

The claims of a patent are interpreted in the light of the specification, and the drawing, which forms a part of the specification. The specification is the dictionary of the claims.

> *Walker on Patents,* 6th Ed., Sec. 227, at p. 309.

> "The entire instrument, including the drawings and specifications, is to be considered in arriving at its intent and meaning."

> 48 *C. J.,* Sec. 335, p. 213.

In view of the specification and drawing of the patent in suit, the close relationship of the valves S and V must necessarily be read into claim 1 of the patent, and for the further reason that in the oil fields of California where loose sand is present, gas, under great pressure, is also present, and a pump designed to avoid "sanding up" in the oil wells of this state must necessarily be designed to avoid a "gas lock"; otherwise the pump would not be practically operative and would lack utility when used in the oil fields of this state, or similar oil fields. The patent is *prima facie* evidence of its operativeness and utility and presupposes a pump constructed and arranged to perform its intended function.

> *Remington Cash Register Co. v. National Cash Register Co.,* 6 Fed. (2d) 585.

The clutch between the lower end of the traveling barrel B and the anchor P, comprising the interlocking members 4 and 2, is a practical construction, for locking the barrel against rotation, so that the joints connecting the sucker rods may be screwed up tight.

Claim 2 covers a pump of the type disclosed in the specification and drawing, with the standing Valve S and the traveling valve V in *close relationship* to prevent a "gas lock," *in combination* with the bottom guide or part 3, on the lower end of the traveling barrel B, provided with a series of venting ducts 5, and the clutch comprising the coupler member 4, on the lower end of the part 3, and the coupler member 2, on the top face of the anchor P, to be non-rotatively interlocked, said ducts 5 opening directly toward the anchor coupler 2, so that during the upward strokes of the barrel B, ejected liquid is impelled or jetted through the ducts 5 toward the anchor top face and coupler 2 to aid in prevention of sand accumulation thereon.

Claim 3 recites the same combination of elements as covered in claim 2, but in broader terms, so as to cover varying forms of venting ducts or jet apertures, other than "a *series* of venting ducts" (5), as specified in claim 2. The last five lines of claim 3 are as follows:

> "said guide being *apertured* for the ejection of liquid from the barrel directly toward the effective face of the lower, standing coupler part to aid in preventing sand settlement thereon."

The term *"apertured"* in claim 3 is a broad term, which does not limit the claim to any particular arrangement or form of aperture, but comprehends any *suitable arrangement or form* of aperture, "for the ejection of liquid from the barrel directly toward the effective face of the lower, standing coupler part to aid in preventing sand settlement thereon."

This broad construction of the term "apertured" is fully supported by the patent specification, page 1, lines 41 to 49, inclusive, as follows:

> "Other objects, advantages and features of construction, and details of means and of operation will be made manifest in the ensuing description of the herewith illustrative embodiment; it being understood that *modifications, variations and adaptations* may be resorted to within the spirit, scope and principle of the invention as it is hereinafter claimed."

Appellants made two types of guides or jet device 3 for their pump [Tr. p. 38], one exactly like that disclosed in the patent in suit and specified in claim 2 of the pat-ent, with a *series* of venting ducts or jet apertures 5, and another with an *enlarged central opening* to provide an *annular space or jet* aperture between the standing column section 11 and the wall of said central opening [Plaintiffs' Exhibit 2; Tr. p. 39] in accordance with the terms of claim 3, because said *annular space* is a jet *aperture* within the broad language of claim 3, to-wit: "said guide being *apertured* for the ejection of fluid from the barrel," etc. The latter type of guide, or jet device (part 3) (Plaintiffs' Exhibit 2) was manufactured and sold by plaintiffs'appellants some time before defendants-appellees' infringing pump, with the same type of jet device (Defendants' Exhibits II and NN) was placed on the market. [Tr. pp. 41, 42 and 43.]

The principal advantages of the pump covered by the patent in suit will be summarized briefly as follows:

1. The *standing plunger* and *traveling barrel* enables the pump to lift the oil up through the pump tubing T on the *down* stroke of the barrel B under the influence of the *weight* of the string of sucker rods connected to the upper end of the barrel, instead of on the *up* stroke, by the *power* of the engine, as in the conventional pump with a standing or stationary barrel and traveling plunger. The *work* of appellants' pump in lifting the oil is done by the force of *gravity,* instead of by the power of the engine. The force of gravity under the weight of the sucker rods is amply sufficient to overcome the weight of the oil in the tubing T, to lift the oil therein, because the weight of the rods in a 5,000 foot well is approximately 15,000 pounds. [Tr. p. 47.] The greatest work of appellants' pump—that of lifting the oil through the tubing T to the surface of the earth—being done by *gravity,* great *economy in power* is effected. The only work done by *power* in appellants' pump is on the *up* stroke of the

traveling barrel B in lifting the sucker rods and traveling barrel, to lift enough oil from the bottom of the well through the standing column 11 and 10 to fill the pump chamber in the barrel between the standing valve S and the traveling valve V.

2.   The pump with the standing column and traveling barrel may be pulled entirely out of the pump tubing T for repairs, etc.   This is called a "dry job."   A conventional pump with a stationary barrel and traveling plunger cannot be pulled entirely, that is to say, the plunger may be pulled out of the barrel, but the barrel would be left in the tubing, and the whole tubing would ordinarily have to be pulled out of the well to remove the pump barrel therefrom.   In pulling the pump, the upward movement of the traveling barrel B causes the puller nut, or part 3, to engage the lower end of the collar 12 below the lowermost guide 13 of the plunger, which causes the anchor P to be pulled out of the lock box L and causes the plunger to be pulled with the barrel out of the tubing T.   The pulling of the pump, as above described, is a "dry job" and it is not a difficult job, and can be done in a short time, thus avoiding a "shut down" of the well for any serious length of time.   This a very great advantage over pumps which "sand up," under ordinary working conditions, and require a so-called "wet job" in pulling the tubing T.

3.   The washing of the sand from the top of the anchor P and lower clutch member 2 by the downwardly projected jets of oil through jet apertures 5 from the bottom of the barrel, prevents the sand from packing up in the bottom of the pump tubing T over the anchor P and clutch member 2, and prevents the pump from "sanding up,"

whereby the *pump,* including the traveling barrel B, plunger indicated 10, 13, 16, 17 and 19, and anchor P, *may be pulled* up out of the tubing T for repairing the pump, etc., leaving the tubing T in the well (the anchor P pulling out of the lock box L), and whereby the barrel B may be lowered to enable the clutch members 4 and 2 to interengage and lock the barrel against turning, for screwing up the disconnected and loose joints of the sucker rods.

The only thing that can be done with a pump when it "sands up" badly is to pull the whole tubing T, with the pump therein, out the well, and this is a very lengthy, dirty, wasteful and expensive job, a so-called "wet job" because the tubing has to be pulled up the length of one section at a time above the derrick floor, and each section unscrewed from the next lower section, and as each uppermost section, which is full of oil, is thus unscrewed, the oil therein pours out of the lower end of said section all over the derrick floor and around the derrick over a considerable area. When the tubing and pump are thus pulled together and the sand removed from the tubing around the pump, it requires considerable time and work to put the tubing and pump back down in the well, because the tubing has to be again screwed together and put down in the well one section at a time.

4. The close relationship of the standing valve S and traveling valve V prevents the formation of a "gas lock" in the pump chamber between said valves and thereby enables the pump to function properly against the heavy gas pressure in the oil fields of California.

5. The construction of the standing plunger with its metal oil-seal guides 13 and its spring-thrust packings 16,

provides a most effective seal against leakage past the plunger, and particularly against leakage of oil upwardly past the plunger from the jet pressure chamber EE (Defendants' Exhibit MM, Book of Exhibits p. 178), under pressure of the jet-pressure nut or part 3, on the upward stroke of the traveling barrel B, whereby the full upward pressure of said nut or part 3 against the oil in said chamber EE is utilized to force the oil out of said chamber through the ducts or apertures 5, in strong, downwardly projected jets, directly toward the top of the anchor P and the lower clutch member 2, thus effectively washing said anchor and clutch member and preventing the settlement of sand on said parts and in the bottom of the pump tubing T, and most effectively preventing the pump from "sanding up."

6. The location of the ducts or apertures 5 in the part 3 on the lower end of the traveling barrel B, instead of the sides of the barrel, as in the O'Bannon pump (16') (Defendants' Exhibits R and R') and the Northrup pump (32) (Defendants' Exhibits V and V'), makes it possible to reverse the barrel to utilize the unworn portion of the barrel in place of the worn out portion, thus doubling the length of use of the barrel and lengthening the life of the pump.

Appellants' novel pump is the *first in the art to combine* all of the above enumerated advantages in a *practical, efficient* and *commercially successful pump,* and consequently the *first in the art* to solve the serious sand-and gas problem in pumping oil wells in California. [Tr. pp. 279 and 286.] It is submitted that such a pump is a meritorious invention of high order, entitled to liberal treatment by the courts.

Walker on Patents, 6th Ed., Sec. 417.

While the pump art is considered by the lower court to be a crowded one, the claims of the patent in suit, nevertheless, cover *new combinations* of elements, which *accomplish new and useful results,* to-wit: prevent the pump from "sanding up" and prevent a "gas lock" in a pump of the type disclosed in the patent, including a *standing* plunger and a *traveling* barrel. A *new combination* which accomplishes a new and useful result is a patentable invention.

> *Loom Company v. Higgins,* 105 U. S. 580 (591).

> "A combination is a union of elements which may be partly old and partly new, or wholly old or wholly new. But whether new or old, the *combination* is a means—*an invention*—distinct from them * * *. In making the combination an inventor has the whole field of mechanics to draw from."
>
> *Leads & Catlin v. Victor Talking Machine Co.,* 213 U. S. 318, quoted in *Diamond Rubber Co. v. Consol. Tire Co.,* 220 U. S. 428.

A most vital and distinguishing feature of the invention, covered by the patent in suit, which stands out in bold relief from the prior art, hereinafter considered, is the arrangement of the clutch, comprising the members 4 and 2 on the lower end of the traveling barrel B and the upper end of the anchor P, respectively, *in combination* with the *vertical ducts* or jet apertures 5 extending through the jet device or part 3, in the lower end of the barrel B, through which apertures *strong vertical jets* of oil are projected directly downward under great pressure from the pressure chamber EE (Defendants' Exhibit MM, Book of Exhibits p. 178) toward the clutch member

2, on the upward stroke of the traveling barrel B, which jets most effectively wash all sand from said clutch member 2 and the anchor P, and keep said sand moving upwardly in the tubing T around the traveling barrel B, whereby the pump is prevented from "sanding up." Before this novel and highly meritorious feature was embodied in a practical and commercially successful pump, by its inventor, the appellant, Bray, there was no pump known, considering all the evidence before this Court, that would not "sand up" in the oil fields of California. This washing of the sand from the anchor and the clutch below the lower end of the traveling barrel, by the *strong downwardly-projected jets* of oil out of the lower end of the traveling barrel B from the pressure chamber EE (Defendants' Exhibit MM, Book of Exhibits p. 178) is an entirely *new function,* which places the invention in suit in the class of a *primary* or *pioneer invention* in an art which the lower court considered to be a crowded one.

> "A primary invention is one which performs a function never performed by an earlier invention."
> *Walker on Patents,* 6th Ed., Sec. 416, at p. 508;
> *Diamond Rubber Co. v. Consolidated Rubber Tire Co., supra.*

> "A patent may be a pioneer in a wide field or in *a narrow one.* * * *
> Columbus discovered half a world. Another man may be the discoverer of a minute coral atoll. Each is a pioneer—one in millions of square miles, the other in a couple of acres. The right of discovery belongs to each in all that he has discovered.

Coffield did not invent water motors. He did not invent water motors of the general type of his device. In those fields of invention he is not a pioneer. He is merely an improver. On the other hand *he did originate something.* He found that such a type of water motor could be made to work successfully if the valve stems and springs were so arranged that one would do the heavy work of unseating; the other complete the work of reseating. *In that field he is a pioneer, none the less because the field itself may be small.* He is entitled to protection for that which he has discovered."

A. D. Howe Mach. Co. v. Coffield Motor Washer Co., 197 F. R. 541 and 548 (C. C. A., 4th Cir.);

American Pneumatic Service Co. v. Snyder, 180 F. R. 725.

DEFENDANTS-APPELLEES' PUMP.

(Plaintiffs' Exhibits 3 and 4, and Defendants' Exhibits II and NN. See Book of Exhibits pp. 5-8, 177 and 179.)

Defendants-appellees' pump, known as the "Hofco Pump," contains the same elements, which function the same and accomplish the same results, as the elements of plaintiffs-appellants' pump, known as the "Bray Pump," claimed in the patent in suit, notwithstanding the fact that the jet aperture of the Hofco Pump is in the form of an annular space surrounding the standing column of the pump, because the shape of the jet aperture is merely a matter of form, and not of substance, and the patent in suit is not restricted to any particular form of jet aperture.

DEFENDANTS' DEFENSES.

[Amended Answer, Tr. p. 9.]

The defense of invalidity of the patent in suit was sustained by the lower court only as to claim 1, and the defense of no infringement was sustained as to claims 2 and 3. In sustaining the defense of no infringement the lower court gave claims 2 and 3 a very narrow construction, despite the broad language of the patent, and particularly of claim 3, which reads, *letter perfect,* on the Hofco Pump, and despite the fact that no pump of the prior art shows the combination of the clutch 2 and 4 and the vertical jet apertures 5 in the jet device 3, through which apertures *strong jets* of oil are projected from the pressure chamber EE directly *downward* toward the lower clutch member 2, on the up stroke of the traveling barrel B, whereby sand is washed from said lower clutch member and the top of the anchor P and kept moving upwardly in the pump tubing T, to prevent the pump from "sanding up," so that the pump barrel may be lowered to engage the clutch members 2 and 4, to lock the barrel against turning and so that the pump may be pulled out of the tubing T for repairing the pump or removing sand therefrom. The projection of the *strong jets* of oil *directly downward* from the *lower end* of the traveling barrel for *washing* sand from the *lower clutch member and the anchor,* is a *new function,* never performed before the advent of the Bray Pump covered by the patent in suit, which new function stamps the Bray Pump as a *primary* invention, entitled to a *liberal* construction rather than a narrow construction which the lower court gave the claims of the patent in suit.

Walker on Patents, 6th Ed., Sec. 416, at p. 508.

Although the pump art was considered by the lower court to be crowded, the Bray pump is nonetheless a *primary invention,* in view of its *new function* in projecting downward jets of oil to wash out the sand from the clutch and anchor, and its *new and useful result* of preventing the pump from "sanding up."

> *A. D. Howe Mach. Co. v. Coffield Motor Washer
> Co., supra;*
>
> *Diamond Rubber Co. v. Consolidated Rubber Tire
> Co., supra.*

There is nothing in the prior art which was before the Patent Office when the patent in suit was pending, nor is there anything in the file wrapper of the patent in suit, which defendants set up in their amended answer (Paragraph XV) in limitation of the patent claims [Tr. pp. 22-23], which, after all due consideration, can be fairly construed to narrow the scope of the patent claims to such extent as to enable the defendants-appellees to escape infringement of said claims, by merely using a jet aperture of slightly different *form* from that *shown, for illustrative purposes only,* in the patent drawing, but which form is clearly comprehended by the broad language of the patent specification, and by the particular terms of claim 3 of the patent in suit. No limitations can be read into the claims of the patent in suit, because claims 1 and 2 were allowed with only a few technical and not substantial, amendments, which do not narrow the scope of said claims, while claim 3 was allowed by the Patent Office as originally presented, and without amendment. (See File Wrapper, Defendants' Exhibit B, Book of Exhibits pp. 23 and 26.)

The affirmative defense of *inoperativeness* or *lack* of *utility* of the invention covered by the patent in suit is not set up in the amended answer, and this defense is therefore foreclosed to the defendants-appellees, according to Section 4920 of the Revised Statutes, Title LX (U. S. C. Title 35, Sec. 69):

> *Walker on Patents,* 6th Ed., Secs. 498, 554 and 636;
>
> *Ames & Frost Co. v. Woven Wire Mach. Co.,* 59 F. R. 705;
>
> *The Providence Rubber Co. v. Goodyear,* 9 Wallace 793 (1869).

Although not pleaded by defendants, a considerable amount of testimony was given by the defendants' expert, Edward T. Adams [Tr. pp. 166-211 and 220-232], in support of the defense of *inoperativeness* or *lack of utility* of the Bray pump, particularly the jet action covered by claims 2 and 3 of the patent in suit. The lower court in its memorandum of conclusions and order for judgment [Tr. p. 378], held as follows:

> "There was *strong* evidence offered as tending to show that the effect claimed for the jet action would not result in operation—this to the point of lack of utility. I am not disposed to conclude that the patent is void for that reason."

While the lower court did not sustain the defense of inoperativeness or lack of utility, and thereupon hold claims 2 and 3 of the patent in suit void, considerable weight must have been given by the court to the testimony of defendants' expert witness, Adams, concerning such defense, because the Court considered said testimony "strong evidence"—evidence ingeniously calculated to belittle and minimize the meritorious invention in suit—and it is submitted that the Court was thereby unduly influenced and prejudiced against the invention and patent in suit, giving claims 2 and 3 of the patent a narrow and unwarranted construction, which unjustly enabled the defendants to escape infringement of said claims; the holding of the Court as to claims 2 and 3 being as follows:

> "I hold that claims 2 and 3 are valid, limited closely to the feature of having a bottom guide running on the standing column, with a running fit, and the guide being provided with vents or apertures for the ejection of liquid directly toward the effective face of the lower coupler member; that the invention is to be strictly limited to the form described; that no infringement is proved."

Defendants' evidence supporting the defense of inoperativeness or lack of utility, which was not pleaded, was a surprise to plaintiffs, and prevented plaintiffs' expert, Fred H. Hayne, from giving sufficient study and time to said defense, in order to refute the fantastic theories and pseudo mathematics, set forth in the testimony of the defendants' expert, Adams.

From the final decree of the lower court [Tr. p. 386], holding claim 1 of the patent in suit null and void; strictly limiting claims 2 and 3; holding that defendants have not infringed any of the claims of the patent in suit; dismissing the bill of complaint; and awarding defendants their costs and disbursements, plaintiffs-appellants take their appeal to this Honorable Court on their assignment of errors. [Tr. pp. 389-400.]

CONSIDERATION OF ASSIGNMENT OF ERRORS.

The various assignments of error of the lower court, urged for consideration before this Honorable Court, may be grouped as follows:

1. Assignments of error IX, X, XI, XII, XIII, XIV, XIX, XXII, XXVII, XXVIII, XXIX, XXX, XXXI, XXXII, XXXIII, XXXIV, XXXV, XXXVI, XXXVII, XLI, LI, LII, LV and LX to the effect that the lower court erred in decreeing claim 1 of the patent in suit null and void and not infringed, and in not decreeing said claim valid and infringed by the defendants.

2. Assignments of error I, II, III, IV, V, VI, VII, VIII, XI, XIII, XIV, XV, XVI, XVII, XVIII, XIX, XX, XXI, XXIII, XXIV, XXIX, XXX, XXXI, XXXII, XXXVI, XXXVIII, XXXIX, XL, XLI, XLII, XLIII, XLIV, XLV, XLVI, XLVII, XLVIII, L, LIII, and LVI to the effect that the lower court erred in strictly construing or misconstruing the claims, and particularly claims 2 and 3 of the patent in suit, whereby the scope of the claims was unduly limited to such extent as to read out of the claims the defendants' infringing pump, and thereby enable the defendants unjustly to escape infringement of the patent in suit.

3. Assignments of error XVII, XVIII, XXI, XXII, XXIII, XXV, XXVI, XXXVII, XLVI, XLVIII, XLIX, L, LI, LIV, LVI and LVII to the effect that the lower court erred in decreeing that the defendants and each of them have not infringed any of the claims of the Letters Patent in suit and have not infringed said Letters Patent.

4. Assignments of error XX, XXV, LVIII, LIX, LX and LXI to the effect that the lower court erred in entering a decree in favor of the defendants (appellees) instead of in favor of the plaintiffs, as prayed.

ISSUES.

1. Is claim 1 of the patent in suit null and void?

2. Should the claims of the patent in suit, and particularly claims 2 and 3, be narrowly construed or limited to such extent as to read out of the claims the pump manufactured and sold by the defendants?

3. Do the defendants or either of them infringe the claims of the patent in suit or any of said claims?

4. Is the final decree of the lower court according to law?

ARGUMENT.

Issue 1.

Assignments of errors IX, X, XI, XII, XIII, XIV, XIX, XXII, XXVII, XXVIII, XXIX, XXX, XXXI, XXXII, XXXIII, XXXIV, XXXV, XXXVI, XXXVII, XLI, LI, LII, LV and LX.

The lower court erred in decreeing claim 1 of the patent in suit null and void and not infringed, and in not decreeing said claim valid and infringed by the defendants, because,

The granting of Letters Patent in suit is *prima facie* evidence of its validity, and raises a strong presumption of the validity of said Letters Patent, which can only be overthrown by proof to the contrary beyond a reasonable doubt.

> "Either Letters Patent, or such a copy thereof, is *prima facie* evidence of the validity of the Letters Patent." (Citing nine cases.)
>
> *Walker on Patents,* 6th Ed., Sec. 535;
>
> *Cantrell v. Wallick,* 117 U. S. 690, 6 Sup. Ct. 970, 29, L. Ed. 1017.

> "The patent in suit being *prima facie* evidence that the applicant for that patent was the first inventor, the *burden of proof* under this defense lies with a *degree of weight* upon the defendant."
>
> *Hopkins on Patents,* p. 420, Sec. 340.

"The burden of proof of a want of novelty rests upon him who avers it, and every reasonable doubt should be resolved against him. Novelty can only be negatived by proof which puts the fact beyond a reasonable doubt".

Walker on Patents, 6th Ed., Sec. 116;

Wilson & Willard Mfg. Co. v. Bole, 227 F. R. 607 (9th Circuit);

Bell Telephone Case v. American Telephone Co., et al., 22 Fed. Rep. 309;

Searchlight Horn Co. v. Victor Talking Machine Co., 261 F. R. 395 (9th Circuit).

The evidence upon which the lower court found claim 1 of the patent in suit null and void, not only falls far short of proof beyond a reasonable doubt, but the prior art fails absolutely to show the novel *combination* of elements of said claim. The nearest approach to claim 1 of the patent in suit is the Admore pump, Defendants' Deposition, Exhibits 6 and 7, "C" Book of Exhibits, pages 52 and 53, but a vital difference between this pump and the pump covered by claim 1 of the patent in suit lies in the relative position of the standing valve and traveling valve. In Fig. 1 of the patent in suit the standing valve S is shown at the upper end of the standing plunger and is in close proximity to the head valve or traveling valve V. On page 1, lines 86 to 88, of the patent specification, the position of the standing valve is described as follows:

"Within the barrel is a standing valve S which is screwed on the *upper end* of a 'stem' or column 10" (*standing plunger*).

Claim 1, therefore, interpreted in the light of the specification and drawing, covers a pump with a *standing valve S located on the upper end of the standing plunger,* or standing guide column as it is termed in the claim. Such arrangement brings the standing valve S close to the traveling valve V, and this close relationship of said valves reduces the space between said valves to a minimum, which prevents the accumulation of any appreciable amount of gas, under high pressure, between said valves, and thereby prevents a "gas lock" between said valves and enables the pump to function under the high gas pressure found in the oil fields of California.

In the Admore pump the standing valve 16 is not on the *upper end* of the standing plunger, but is located at the lower end of the standing plunger below the lower end of the traveling barrel 1, when said barrel is in its lowermost position. This arrangement of the standing valve (16) places said valve so far below the traveling valve (5) as to provide considerable space between said valves, in which space a large quantity of gas will accumulate and form a 'gas lock," which would prevent functioning of the pump, by preventing any liquid passing upwardly through said "gas lock" and out of the pump. While the pump covered by the patent in suit is not the first pump in which the standing valve and the traveling valve are positioned in close relationship to prevent a "gas lock," it is, however, the first in the art to combine, with this useful valve relationship, the part 3 (referring to the patent), the coupler or clutch member 4, at the lower end of the traveling barrel B, and the pin or anchor P, with a coupler or clutch member 2 on its upper face, which clutch members 4 and 2 may be non-rotatively interlocked, to enable the loose joints of the sucker rods to be screwed up tight. This novel, simple and useful combination of non-gas-locking valves and non-rotatively, interlocking clutch members, between the lower end of the traveling barrel and the top of the anchor, as claimed in claim 1 of the patent in suit, irrespective of the jet apertures 5, was properly considered, by the Patent Office, to be a patentable invention, and the lower court, in the absence of anticipating prior art, clearly erred in declaring claim 1 null and void.

The pump disclosed in Defendants' Deposition Exhibit 17, Book of Exhibits, pages 87 and 88, is more remote from claim 1 of the patent in suit than Defendants' Deposition Exhibits 6, "C", because the pump, Defendants' Exhibit 17, was not a practical pump, or did not function to the satisfaction of its manufacturer, and was superseded by the Admore pump, Defendants' Exhibit 6, "C". [Tr. p. 73.]

The Bramo pump, Defendants' Deposition Exhibit 1, Book of Exhibits, pages 29 and 30, offers no suggestion, whatever, of the pump covered by claim 1 of the patent in suit, because it is a pump of a different type, that is to say it has a reciprocating or *traveling plunger* A and a *stationary barrel* B, with a clutch C-D at the *upper* end of the stationary barrel, instead of a *standing* or *stationary plunger* (10-13-16) and a *traveling barrel* (B) with a clutch (4-2) at the *lower* end of the traveling barrel, as specified in claim 1 of the patent in suit. Pumps of the Bramo type have been superseded by the pump of the Bray type by about eighty per cent of the manufacturers of oil pumps [Tr. p. 38] in view of the advantages of the pump of the Bray type hereinbefore enumerated.

It may be here stated that none of the pumps set up in anticipation of the patent in suit of the Bramo type, which is the conventional type of pump with a *stationary* or *standing* barrel and a *reciprocating* or *traveling* plunger, have any bearing or anticipating effect on the pump of the patent in suit, which has a *stationary* or *standing plunger* and a *reciprocating* or *traveling barrel*, because the former conventional type of pump could not possibly function like the latter pump or accomplish the same results. The conventional traveling plunger and stationary barrel pump cannot lift the oil in the tubing T on the down-stroke of the plunger under the influence of gravity or the weight of the sucker rods, nor can such pumps be "pulled" out of the tubing T, because if the plunger were pulled up it would be pulled out of the upper end of the stationary barrel and leave said barrel down in the well. All of the pumps of the conventional type, with *stationary barrel* and *reciprocating plunger* are, therefore, entirely out of the legal picture, as prior art, in the case at bar.

Issue 2.

Assignment of Errors I, II, III, IV, V, VI, VII, VIII, XI, XIII, XIV, XV, XVI, XVII, XVIII, XIX, XX, XXI, XXII, XXIV, XXX, XXXI, XXXII, XXXVI, XXXVIII, XXXIX, XL, XLI, XLII, XLIII, XLIV, XLV, XLVI, XLVII, XLVIII, L, LIII, and LVI.

The lower court erred in not giving claims 2 and 3, of the patent in suit, a reasonably liberal construction, so as to comprehend the defendants' pump, and support the charge of infringement of said claims by the defendants, because

When Bray, the inventor of the pump in suit, came into the legal picture, there was no pump known in the art that would not "sand up," when used in the oil fields of California, and consequently all pumps had to be pulled frequently to clean out the sand. [Tr. pp. 244, 265, 270, 271 and 279.] Pulling the pump tubing is a dirty, wasteful and expensive job and requires a "shut down" of the well for several days, as hereinbefore stated. To provide a pump which would not "sand up" and require pulling of the pump tubing in the oil fields of California was the problem that confronted the inventor Bray, and he solved that problem by his conception and reduction to practice of his invention, as disclosed in his patent in suit. Bray embodied his invention in a pump of the so-called inverted barrel type, comprising a *standing plunger* and a *traveling barrel,* by placing his apertured part 3 (which is also designated in his patent specification as a "jet device," "bushing" and "bottom guide"), on the lower end of the pump barrel B, and locating his clutch, comprising jaws 4 and coupler head 2, between the bottom of the barrel B and top of the anchor P, so that, on the down-stroke of the barrel, oil in the well tubing T, passes from said tubing up through the apertures 5, in said part 3, into the jet pressure chamber EE (Defendants' Exhibit MM, Book of Exhibits, page 178), and, on the up-stroke of the barrel, oil in said chamber is positively projected therefrom directly downwardly, through said apertures 5, in strong jets toward the anchor P and coupler head 2, for washing away any sand, which would otherwise settle down through the oil, between the barrel B and well tubing T, upon said anchor and coupler head, whereby "sanding up" of the pump is prevented, so that the pump may be readily pulled out of the well tubing T, or the pump barrel B lowered to interlock the clutch members 4 and 2, and prevent rotation of the barrel, in order that the loose or disconnected joints of the pump sucker rods may be screwed up tight by a large wrench applied to a sucker rod at the top of the well.

Defendants' counsel have introduced in evidence a great array of prior art, comprising certain prior patents, prior publications and prior pumps which are not in point, in a vainglorious attempt to tear down the constructive work of an inventor of no mean magnitude, who has success-fully builded upon the shortcomings and failures of his predecessors and given to our important oil industry an invention of great merit. The Court will search the prior art, without success, for even a suggestion of the invention disclosed in the patent in suit. The prior art will be considered in its order.

Defendants' Exhibits 1 and 2, "Bramo" (Book of Exhibits, page 29). This pump is of the conventional type with standing barrel and traveling plunger, instead of the type with the standing plunger and traveling barrel, as specified in claims 2 and 3 of the patent in suit. It has no jetting chamber and no clutch on the lower end of a traveling barrel, and upper end of an anchor, and consequently no jetting action towards an anchor and a clutch can be produced by this pump as specified in said claims. Moreover the valve D at the upper end of the plunger and the valve at the lower end of said plunger are so far apart as to enable a considerable amount of a gas to accumulate between said valves and form a gas lock. This pump was built by the Bradford Motor Works of Bradford, Pennsylvania, for use in such oil fields as found in the State of Pennsylvania where there is no gas and no loose sand, such as found in the oil fields of California. [Tr. p. 245.] It might work very well in the oil fields of Pennsylvania, but it would not function in the oil fields of this state without a "gas lock" and without "sanding up," which is avoided by the Bray pump, as set forth in claims 2 and 3 of the patent in suit. There is no evidence that the Bramo pump was ever used in the oil fields of this state.

Defendents' Exhibits 3, 4 and 5, being photostatic copies of records of the Bramo pump, are, like the pump to which they relate, entirely incompetent, irrelevant and immaterial to the issue in question and of no legal effect.

Defendants' Exhibits 6 and 7, "Admore," Book of Exhibits, pages 52 and 53. This pump is of the inverted barrel type with standing plunger and traveling barrel, but its standing valve 16 is not on the *upper end* of the plunger in *close relationship* to the traveling valve 5, to avoid a "gas lock," and it has no vertical apertures in its puller nut 6 through which liquid might be projected out of the lower end of the barrel 1 directly downwardly toward the anchor 14 and lower clutch member to wash sand from said anchor and clutch member, as set forth in claims 2 and 3 of the patent in suit. The standing valve 16 is located below the lower end of the traveling barrel 1 at a considerable distance below the traveling valve 5, which arrangement provides considerable space between said valves which would fill up with gas and form a "gas lock," if this pump were used in the oil fields of California. This arrangement does not meet the terms of claims 2 and 3 which cover the close relationship of the standing valve S and the traveling valve V as shown in Fig. 1 of the patent in suit. The Admore pump has holes (not numbered) in the side of the traveling barrel 1 above the puller nut 6 to permit the escape of trapped oil in the lower end of said barrel, between the lower end of the plunger body 10 and the puller nut 6, when the traveling barrel moves upwardly, but the oil escaping through said side holes is projected laterally from the barrel directly against the side of the well tubing (not shown) in which the pump is located, and it is not possible for the oil to be projected directly downwardly through said holes out of the bottom of the barrel in strong jets against the lower clutch member and the anchor, as covered by claims 2 and 3 of the patent in suit. If any jet at all is produced the oil passing out of said holes on the side of the barrel and striking against the side of the well tubing will be immediately dissipated, and any such jet will be completely destroyed. Moreover, oil passing out of the holes in the side of the barrel will be acted upon by two forces which will *draw said oil upwardly* away from the anchor and clutch, first by the upward movement of the barrel, and second, by the differential in pressure of the oil in the well tubing, because the greater the depth of the oil in the well tubing the greater the pressure, and a jet of oil projected laterally into the body of oil in the well tubing would naturally go upwardly against the lesser pressure rather than downwardly against the greater pressure of said body of oil. It should therefore be obvious that it would be impossible for this Admore pump to project a jet of oil directly downwardly from the bottom of the barrel 1 toward the clutch and anchor on the up-stroke of the barrel to wash sand away from the clutch and anchor and consequently the Admore pump could not possibly function like the Bray pump, as covered by claims 2 and 3 of the patent in suit. Even if the Admore pump were constructed to project a downward jet, the jet would have to travel down over the standing valve 16 before it reached the anchor 14, and this greater travel of the jet through the body of oil in the well tubing, before reaching the anchor, due to the interposition of the standing valve 16 between the lower end of the barrel 6 and the anchor 14, would dissipate the jet and materially reduce its washing effect upon the anchor. This reduced effectiveness of the jet would allow the pump to "sand up" in a short time. This Admore pump was manufactured by the Bradford Motor Works, of Bradford, Pennsylvania, for use in the oil fields, such as found in the State of Pennsylvania, where there is no gas or sand, and could not function in the oil fields of California without "gas locking" and "sanding up."

Defendants' Exhibits 8, 9, 18, 19, 20, 21 and 22 are photostatic copies of records of job orders and sales sheets, etc., of the Admore pump, Defendants' Exhibits 6 and 7, are of no legal consequence whatever, since the Admore pump does not contain the elements of claims 2 and 3 of the patent in suit and it is quite immaterial whether the Admore pump was ever made or sold.

Defendants' Exhibits 10 and 11, Book of Exhibits, pages 78 and 79, are conventional pumps with a fixed or standing barrel, and reciprocating or traveling plunger. The standing valve and traveling valve are spaced far apart and a gas lock would form there between. The clutch is at the top of the *fixed barrel* and not at the bottom of a *traveling* barrel like the Bray pump as covered by the patent in suit. Said pumps have no jetting device for washing an anchor and a lower clutch member. There is no provision in this pump to prevent sanding up.

Defendants' Exhibits 13 and 13A, Book of Exhibits, page 81, represent a pump like the pumps of Defendants' Exhibits 10 and 11 with *fixed* or standing barrel and *traveling* plunger. The clutch is at the top of the barrel, not at the bottom, and the pump has no jetting device for washing an anchor and a lower clutch member below the barrel. There is nothing to prevent this pump from "sanding up."

Defendants' Exhibits 15 and 16, Book of Exhibits, pages 85 and 86, represent the O. F. S. Working Barrel and Bradford Plunger Liner respectively. These pumps have no bearing whatever on the issues of the case. They are the old style conventional pump with the fixed or standing barrel and reciprocating plunger. The clutch is at the top of the barrel. The valves are necessarily too far apart to prevent a gas lock. There is no jetting action in these pumps to wash an anchor or clutch member to prevent "sanding up" of the pump.

Defendants' Exhibit 17 is a photograph of defendants' physical Exhibits 17A and 17B [Book of Exhibits, pages 87, 88 and 89]. These exhibits represent a pump with a stuffing box, indicated 2 [Book of Exhibits, p. 89] which prevented any oil from being ejected from the lower end of the barrel and therefore had no jetting action to wash the anchor or the clutch to prevent the pump from "sanding up". There were no apertures at all in the lower part of the traveling barrel to permit oil to escape therefrom, and consequently, when oil finally leaked into the lower end of the barrel, the barrel could not be pulled up for its upward stroke, unless the oil could be compressed in the barrel between the puller nut, at the lower end of the barrel, and the lower end of the standing plunger, but oil, like water, is incompressible, according to Pascal's law, which is one of the basic principles of hydraulics and may be found in any text book on elementary physics. Pascal's Law is stated as follows:

> "Pressure exerted upon any part of an enclosed liquid is transmitted undiminished in all directions. This pressure acts with equal force upon all equal surfaces, and at right angles to them."

This Bramo pump [Defendants' Exhibits 17, 17A and 17B] was a perfect failure. Only 25 of them were made and, according to the admission of defendants' witness Morris, "They did not function to our satisfaction. In the design of this pump no satisfactory provision was made for the escape of oil trapped between the main tube and the plunger tube, and consequently the stuffing box principle was abandoned, and a pump built of the type later given the name 'Admore'. Exhibit 6 is of the type of device to which I just referred." [Tr. p. 73.]

This old Bramo pump [Defendants' Exhibits 17, 17A and 17B] was nothing but "an unsuccessful abandoned experiment," and an abandoned experiment is no anticipation of letters patent.

> "The rule of Section 101 will probably govern every case which justly comes within the doctrine that novelty is not negatived by any unsuccessful abandoned experiment * * *.

> If an experimental machine or manufacture was unsuccessful in the hands of its contriver, that fact must have been due either to one or more *faults of principle,* or to one or more faults of construction, or to one or more faults of each of these kinds. If partly or wholly due to any fault of principle, that very fact shows that the unsuccessful device was substantially different from subsequent successful patented things. For that reason alone it would have failed to negative the novelty of those things, even if it had not been unsuccessful."

Walker on Patents, 6th Ed., Section 102.

The Deis patent, No. 840,919, Defendants' Exhibits N and N' and Defendants' Exhibit EE, Book of Exhibits, pages 95-99 and page 173, is described as a balanced pump, but it is in the last analysis a *reciprocating plunger* and fixed barrel pump, because the reciprocating plunger, shown in red in Fig. 5, Defendants' Exhibit EE, is lifted by power to discharge oil from the pump through the spout d at the top of the *fixed barrel* designated 6. The statement of defendants' expert, Adams [Tr. p. 88], that the Deis patent is for a *traveling barrel* pump of the same type in that respect as the patent in suit, is simply not a correct description of the Deis pump. There is no pressure chamber in this pump equivalent to the pressure chamber EE of the Bray pump in suit [Defendants' Exhibit MM, p. 178, Book of Exhibits], for jetting oil out of the lower end of a traveling barrel on the up stroke of the barrel for washing an anchor and a clutch to prevent the pump from "sanding up", as covered by claims 2 and 3 of the Bray patent in suit. There are large valve-controlled openings t in the lower end of the reciprocating or *traveling plunger,* Fig. 5, Defendants' Exhibit EE, which openings are closed by the valve u on the up stroke of the plunger to prevent oil from passing downwardly through said openings and to direct the oil upwardly through the open valve n. Defendants' expert, Adams, states that the bottom guide S of the Deis pump is packed against the downward flow of oil [Tr. pp. 94 and 109] and consequently the Deis pump could not possibly have a downward jetting action. The openings t are too large, anyway, to produce a jet, even if the valves u and n were eliminated [Tr. pp. 42 and 43]. There is no suggestion of the Bray invention in the Deis patent.

The Cummins patent No. 1,323,352, Defendants' Exhibits O and O', O², FF and GG, Book of Exhibits, pages 100, 101, 174 and 175, is of the *stationary barrel* and *reciprocating plunger* type [Tr. p. 95]. The valve (23), at the upper end of the reciprocating plunger (15-16), and the valve 17 at the lower end of said plunger, are spaced apart the length of the plunger, which would provide considerable space between said valves to admit a large amount of gas to form a gas lock and prevent functioning of the pump [Tr. pp. 243 and 254]. The clutch, indicated 21 and 22, is at the top of the standing or fixed barrel and reciprocating plunger and not at the bottom of the reciprocating barrel, as in the Bray pump. There is no means in this pump to produce a jetting action to wash sand from the clutch and the anchor to prevent the pump from "sanding up".

Defendants' expert Adams has produced a drawing of a hypothetical pump, Defendants' Exhibit GG, in an idle attempt to duplicate the Bray pump in suit, but a mere reversal of the parts of the Cummins pump would not produce the Bray pump, because there is no jetting device in the Cummins pump to be reversed to produce the Bray jetting device, or the jetting action of Bray's device to wash the clutch and anchor to prevent "sanding up" of the Cummins pump. That there would be no jetting action in the hypothetical reversed Cummins pump, projected from the pressure chamber in the lower end of a traveling barrel, to wash sand from the clutch and the anchor, is admitted by defendants' expert, Adams [Tr. pp. 99, 101 and 104].

There is no evidence that Cummins' original pump as disclosed in Cummins' patent ever operated successfully in the oil fields of California or anywhere else and evidence of any operation at all of the hypothetical reversed Cummins' pump, Defendants' Exhibit GG, is zero minus. Anticipation of letters patent by prior art must be proved beyond a reasonable doubt.

> *Walker on Patents*, 6th Ed., Sec. 116;
>
> *Wilson Willard Mfg. Co. v. Bole, supra;*
>
> *Bell Telephone Case v. American Telephone Co., supra;*
>
> *Searchlight Horn Co. v. Victor Talking Machine Co., supra.*

A mere paper patent like Cummins' and the Cummins hypothetical pump drawing, Defendants' Exhibit GG, are certainly not proof which rises to the dignity of demonstration or proof beyond a reasonable doubt of anticipation of the letters patent in suit. A hypothetical reversed structure may appear to be operative on paper, but if such a structure were actually built and tried out it might not operate at all. Until the Cummins hypothetical reversed pump is actually built and put in operation it is of no value whatever as anticipating evidence of the Bray pump and patent in suit.

There are very grave doubts about the operation of Cummins' hypothetical reversed pump, Defendants' Exhibit GG. The lower section of the pump barrel 2 is obviously a bearing on the standing column 15, for centering the lower end of said barrel, because the plunger 16 is short, having little bearing surface, and the lower end of the barrel would wobble around said column if there were a loose fit between said lower section of the barrel 2 and the column 15 to provide sufficient clearance to enable the plunger 16 to force a jet downwardly all the length of said lower barrel section and out of the lower end thereof to wash sand off the anchor and lower clutch member.

There is a dead space between the lower end of the plunger 16 and the upper end of the lower section of the barrel 2, which if filled with air, the air would be compressed on the up-stroke of the barrel and would offer considerable resistance to the up-stroke of the pump, which would seriously impair the efficiency of the pump. Eventually said dead space would fill up with oil, due to leakage downwardly past the short plunger 16, and then the barrel 2 could not move upwardly at all, because the trapped oil in said dead space would form an hydraulic lock, which would not yield by compression between the lower end of the plunger 16 and the upper end of the lower section of the barrel 2, according to Pascal's law. If water is incompressible according to Pascal's law, oil is less compressible, because oil has a greater specific gravity than water and a much higher viscosity. The Cummins' hypothetical reversed pump structure, therefore, could not function at all, because of trapped oil forming an hydraulic lock in said dead space.

Another drawback to the Cummins' hypothetical pump structure is that it would not drop freely on the down-stroke, because the close bearing fit, between the lower section of the barrel 2 and the standing column 15, would provide no clearance between said barrel and column, or any jet aperture, through which oil could enter the lower end of the barrel, from the space between the column and the well tubing, on the down-stroke of the barrel.

Defendants' mechanical expert, Adams, has stated in answer to defendants' hypothetical question [Tr. p. 105] that in his opinion it would not have represented any act of invention, as distinguished from the exercise of mechanical skill, to have turned the Cummins' pump over in the year 1926. This answer is an attempt to define what constitutes invention, and is a rather ambitious statement for a layman to give, in view of the fact that the Supreme Court of the United States has declined to attempt to give any such definition.

> "The truth is, the word (invention) cannot be defined in such manner as to afford any substantial aid in determining whether a particular device involves an exercise of the inventive faculty or not.

In a given case we may be able to say that there is present invention of a very high order. In another, we can see that there is lacking that impalpable something which distinguishes invention from simple mechanical skill. Courts, *adopting fixed principles* as a guide, have by a process of exclusion determined that certain variations in old devices do or do not involve invention; but whether the variation relied upon in a particular case is anything more than ordinary mechanical skill is a question which cannot be answered by applying the test of any general definition."

McClain v. Ortmayer, 141 U. S. 427, 1891.

See *Walker on Patents,* 6th Ed., Sections 62 to 806, for process of exclusion of that which does not amount to invention.

While the question of invention may involve certain facts, for example, the state of the art, such facts are necessarily governed by legal principles and rules of law, to which no layman, like defendants' expert, Adams, is competent to testify, because such testimony would obviously require the application of the law to the facts.

What constitutes invention is a *mixed* question of fact and law, and only the Court, or one learned in the law, is qualified and competent to give an opinion concerning the same.

The Whitling Patent No. 1,391,873, Defendants' Exhibit P and P' discloses a conventional pump having a standing barrel and traveling plunger with a clutch at the upper end of the plunger, and with no means for producing a jet of oil for washing the clutch. Moreover, the valves 23 and 27 are spaced so far apart that a "gas lock" would form therebetween and prevent the pump from pumping oil from an oil well. A *chamber* is formed in the barrel 5, between the upper end of the plunger 16 and the lower end of the packing 30, which chamber, at first, would contain air, which would be compressed by the plunger and resist the up-stroke of the plunger. Such resistance to the up-stroke of the plunger would require more power than otherwise to operate the pump, and this would impair the efficiency of the pump. Eventually the chamber between the plunger 16 and packing 30 would fill up with oil, due to leakage past the packing and plunger, and this would form an hydraulic lock which would prevent any upward movement of the plunger, because oil in said chamber could not be compressed to allow such movement, according to Pascal's law. This would prevent functioning of the pump. There is no evidence that the Whitling structure ever operated anywhere. It certainly could not operate in the oil fields of California.

Defendants' Exhibit HH, Fig. 2, Book of Exhibits, page 176, is a drawing of a hypothetical pump structure in which the parts of the Whitling pump, Defendants' Exhibits P and P', are reversed, that is to say, the Whitling traveling plunger is shown as a standing plunger and the standing or stationary barrel is shown as a traveling barrel, with the clutch between the lower end of the barrel and the transverse partition 3. In this hypothetical reversed structure the valves 12 and 23 are spaced far apart and a "gas lock" would form therebetween, if this structure were used as a pump in the oil fields of California. It is immaterial whether the balls 12 and 23 move with relation to each other, like the standing valve 23 and the traveling 27, so far as a "gas lock" is concerned. A gas lock will form between two valves which are spaced equidistant apart at all times, like the valves 12 and 23 [Tr. pp. 254 and 255], and a gas lock between said valves would prevent the Whitling hypothetical pump [Fig. 2, Defendants' Exhibit HH] from pumping oil. In said reversed pump there would be *no jet* downwardly from the lower end of the barrel 5 to wash sand from the clutch 25 and 25ᵃ, because the lower end of the said barrel is sealed by the packing 31. This is admitted by defendants' expert, Adams [Tr. pp. 113, 114 and 116]. The jet action of the Bray pump which washes sand from the clutch and anchor, not being present in the Whitling reversed hypothetical pump structure, said structure fails to teach the Bray invention, because the jet washing action is a vital feature of the Bray invention. The Whitling hypothetical reversed pump is a perfect example of a pump which would "sand up" in a most effective manner and fail to function in view of the sand chambers 1 and 2 shown in Fig 2 of the Whitling patent, Defendants' Exhibits P and P', Book of Exhibits, page 103, which chambers are described in the patent, page 2, lines 5 and 11, as follows:

> "The section 2 of the well casing constitutes a sand chamber, as does also the section 1, and the sand may settle in the chambers 1 and 2 about the barrel 5 and about the stand pipe 8 between said valve (12) and stand pipe and the chamber wall."

If sand may settle in the chamber 1 about the traveling barrel 5, in the Whitling reversed hypothetical pump, said sand would pack tightly around said traveling barrel and the pump would "sand up" and prevent movement of the traveling barrel which would prevent functioning of the pump. The numerals 1 and 2 indicating the sand chambers are omitted from Defendants' Exhibit HH, and there is nothing in the reversed hypothetical Whitling pump (Fig. 2 of said exhibit) to indicate those sand chambers, which are fatal to the operation of said reversed pump. Why were the sand chambers omitted from the Defendants' Exhibit HH by defendants' expert Adams, who made the drawings forming said exhibit? The Court is entitled to all knowledge bearing on the construction and operation of an alleged anticipating reference. The Whitling patent is just another paper patent for an alleged pump which could not possibly function, either as it was originally conceived by the patentee or as reversed and revamped by the defendants' ingenious expert Adams, after he had acquired knowledge of the Bray invention. The hypothetical reversed Whitling pump, Fig. 2, Defendants' Exhibit HH, died before it was born, as an anticipation of the Bray invention and as an oil well pump it is a mechanical monstrosity.

The Downing patent, No. 211,230, Defendants' Exhibits Q and Q', has a standing plunger and a traveling barrel, but no clutch or jetting action, as in the Bray pump in suit. The standing valve is not mounted on the upper end of the plunger or standing column C, but is mounted in a cage B below the lower end of the plunger on the traveling barrel G. This arrangement places the standing valve and the traveling valve O so far apart that a gas lock would form between said valves and prevent the pump from pumping oil. Defendants' counsel admits that this Downing patent shows no clutch for holding the traveling barrel against rotation for tightening up the joints of the sucker rods. [Tr. p. 118.] There is a chamber in the lower part of the traveling barrel G between the plunger at the upper end of the standing column or tube C, and the nut H, at the lower end of the traveling barrel G, which chamber would at first be filled with air and then later with liquid, due to the leakage thereinto. The air would be compressed on the up-stroke of the traveling barrel and this would impair the efficiency of the pump. The liquid filling the chamber and being incompressible would prevent altogether any up-stroke of the barrel and this would prevent functioning of the pump. As illustrated in the patent, the Downing pump is inoperative. There are no apertures in the nut H at the lower end of the barrel, and said nut is shown with a close fit around the standing column or tube C, so that no liquid could go out of the lower end of the barrel on its up-stroke in the form of a jet, and consequently this Downing pump could not have a jetting action. This pump offers no suggestion of the Bray invention.

The O'Bannon patent, No. 1,454,400, Defendants' Exhibits R and R', is of the inverted type, but it has no clutch, nor any vertical jet apertures through its head member or puller nut 12 [Tr. p. 121] and consequently it could have no downward jet action for washing sand from a clutch. Oil escapes horizontally from the lower portion of the barrel on its up-stroke through openings 16' in the side of the barrel. Defendants' expert, Adams, speaks of a vacuum under the lower end of the barrel as it moves up, but any downward force of the oil flowing into said vacuum from the openings 16' would be neutralized by two upward forces, to-wit: the upward movement of the barrel, and pressure differential, the pressure of the oil outside the barrel being greater below than above said openings 16', and causing the oil flowing out of the barrel through said openings to be directed upward against the pressure of least resistance, and any possible jet action from said openings 16', would be negatived by said opposing forces. Furthermore, any jet of oil, flowing horizontally out of said openings 16', would strike against the well tubing (not shown) which closely surrounds the barrel B, and be dissipated and destroyed, thus eliminating any possibility of a downwardly projected jet action in this O'Bannon pump. The O'Bannon patent covers a packing for pumps, and there is nothing in said patent that teaches the Bray invention.

The Thurston patent, No. 1, 625, 230, Defendants' Exhibits S and S', Book of Exhibits, page 114, is of the inverted type. The standing valve 7 is not located at the upper end of the stationary plunger 9, but is located below the lower end of said plunger and the traveling barrel (jacket 15), at a considerable distance below the traveling valve (26 and 27), which arrangement provides considerable space between said valves in which a "gas lock" would form and prevent pumping of oil by the pump. This pump has no clutch at all, nor any jet aperture in the pull nut or guide 20 at the lower end of the traveling barrel, through which a jet could be projected to wash away the sand and prevent "sanding up" of the pump. The drawing of the patent shows the pull nut and guide 20 of the traveling barrel closely fitting the plunger 9 and the specification, page 1, lines 95 and 96, describes the "pull nut" and guide 20 which slidably *embraces* the extension 11 of the plunger. It is obvious that the patentee Thurston never intended a loose fit between the part 20 and the plunger extension 11 to provide an annular space through which a jet of oil might be projected for washing away sand. The Bray invention is certainly not taught by the Thurston patent.

The Thompson patent, No. 586,524, Defendants' Exhibits T and T', is of the inverted type, but has two valves e^4 and i spaced apart a considerable distance, in the standing plunger, which would admit a considerable amount of gas into the plunger and a gas lock would form between said valves and prevent the pump from pumping oil. [Tr. pp. 254 and 255.] Defendants' expert, Adams, testified on cross-examination concerning the operation of this Thompson pump [Tr. p. 127], as follows:

"Those orifices (t') allow the escape of the trapped oil in a direction *horizontal* from the standing column. They are in there for entirely *another purpose.* The pump is designed to allow escape in there with considerable force, carrying out a washing effect which the patentee had in mind. *That was not the washing of the clutch. There is no clutch, as a matter of fact, in this patent, and there is no device for washing the clutch.* These orifices to which I refer permit the escape of the trapped liquid in a horizontal direction from the traveling column; they show the ejection of liquid for the purpose of washing in a *horizontal direction* (the walls of the well, etc.) in this case, and *not downward* along the standing column."

If any sand should get into the lower end of the tubing i it would be trapped therein, and the pump would "sand up", because there is no downward jetting action in this pump to wash the sand upwardly, and even if there were such a jet action the sand would have no exit from said tubing except through the small orifices t' which would soon choke up with sand.

The Thompson patent specifies, page 2, line 52, that the sleeve k' does not fit closely around the tube f, but said loose fit is specified as allowing fluid to be admitted to the space between the working barrel k and the tube f. There is no place in the patent where it specifies that liquid is jetted from the barrel k through the space produced by the loose fit between the barrel and tube because said loose fit is nothing more than a sliding bearing fit, through which liquid may leak into said space between the barrel k' and the tube f, but such fit being insufficient to permit a jet to be projected therethrough. In lines 102 and 103, page 2 of said patent, it is specified that "upon the up-stroke of said piston u this fluid will escape through the orifices k²" *in the side of the barrel,* which would not produce a downward jet of fluid from the bottom of the barrel, but would direct the liquid directly against the side of the tubing r where it would be deflected and dissipated. The sleeve k' in the bottom of the barrel is at such a considerable distance above the anchor d, that even if a downward jet could possibly be projected from the bottom of the barrel, it would have to travel so far downward through the oil that it would be completely dissipated before it reached the anchor and could have no possible washing effect upon the anchor.

From the foregoing and particularly the admission of defendants' expert, Adams, it is clear that the pump of the Thompson patent is of different construction and operates on a different principle from that of the Bray pump, and that the Thompson pump teaches nothing concerning the Bray invention.

The Wright patent, No. 575,498, Defendants' Exhibits U and U', has a standing cylinder B and a slide barrel

D, but said barrel, in operation, is necessarily a traveling plunger, which is slidably and snugly fitted in the stationary tubing A, so that no oil can be forced up between said barrel and tubing on the down-stroke of said barrel. Oil is lifted in the tubing A by the barrel-plunger D on the up-stroke of said plunger by the power of the engine at the top of the well, the *up-stroke* being the pumping stroke. In the Bray pump the *down-stroke,* under weight of the sucker rods, is the pumping stroke, the oil being discharged from the pump on said down-stroke. In this Wright pump, the valves E and E' are at the extreme ends of the standing cylinder B and are so far apart as to receive a considerable amount of gas there between and form a "gas lock" which would prevent the pump from pumping oil, if the pump were used in the oil fields of California.

On page 1, lines 79 to 82, of the Wright patent, it is stated that the "aperture e is, however, of a *somewhat larger diameter* than the said member d' to admit of *free* entrance or exit of the fluid held in the space F', the lower end of e' of the working barrel operating as a piston for agitating and forcing the liquid in the space F through the passages B³ *out against the rock* to wash same at all times during the pumping operation."

If the aperture e is somewhat larger in diameter than the member d' to admit *free entrance or exit* of the fluid, there could hardly be a jet action through said aperture, because a jet action necessarily requires a *restricted* aperture. The aperture e is nothing more than a vent to permit displacement of liquid from one side to the other of the lower end e' of the plunger D, so that said plunger may reciprocate. The patent does not specify any jetting action through the aperture e on the *up-stroke* of the plunger D. The lower end e' of said plunger acts as a piston on its *down-stroke* for forcing the liquid in the space F through the passages b³ out against the rock to wash the same. The operation of the Wright pump if used in the oil fields of California would draw sand from the bottom of the well through the passages b³ into the space F, on the up-stroke of the plunger D and then force the sand back through said passages into the bottom of the well on the down-stroke of said plunger, the lower end e' of said plunger acting as a piston for the latter operation on its downstroke. There would be no point in pumping sand out of the bottom of a well and then pumping the sand back into the bottom of the well. Such an operation would be nothing more than "horseplay". There is *no clutch* in this Wright pump and consequently *no jetting action for washing sand from a clutch,* so that a traveling barrel could be lowered to interlock the clutch to prevent rotation of such barrel for screwing up the loose joints of the sucker rods, as covered by claims 2 and 3 of the patent in suit. The Wright pump might operate in oil fields where there is no gas or sand, as in Pennsylvania, and where the most primitive forms of pumps may be used [Tr. p. 245], but the Wright pump, operating on an entirely different principle from that of the Bray pump, would not function in the oil fields of California where we have a serious gas-and-sand problem with which to deal. The Wright patent is entirely foreign to the Bray invention.

. The Northrup patent, No. 1,378,268, Defendants' Exhibits V and V', Book of Exhibits, page 128, is of the inverted type, with standing plunger and traveling barrel, and with the close arrangement of the standing valve and traveling valve, "so positioned as to eliminate the accumulation of gas between them", and while this pump fails in other respects to meet the terms of the claims of the Bray patent in suit, or to anticipate or even suggest the Bray invention, it illustrates in a general way the general type of pump in which the Bray pump may be classified, and should aid the Court in interpreting the claims of the Bray patent in suit, and particularly the statement in the claims which reads as follows:

> "The combination, with a pump of the class having a standing guide column with a working valve and a valved traveling barrel."

This Northrup pump has apertures 32 in the side of the barrel 22 above the bushing or guide 24, through which apertures oil would be ejected horizontally against the inside of the well casing or tubing C and be dissipated, on the up-stroke of the barrel 22. The Northrup patent, page 2, lines 62 to 64, states that the ejection of oil through said apertures 32 will *agitate* the oil in the tubing to prevent the accumulation of sand upon the extension 8, but to have any appreciable washing effect on the extension 8, said agitation of the oil in the tubing C would have to force its way down past the bushing or guide 24 against the upward movement of the barrel 22 and against the greater pressure of the oil in the tubing C below the apertures 32 than above said apertures. The alleged washing effect of said agitation of the oil after meeting such effective resistance would be nil and would fail to prevent "sanding up" of the pump. There is quite a difference between the washing effect of mere *agitation* of the oil in the tubing, by oil ejected horizontally from the openings 32 *above the guide* 24 in the *upwardly moving* traveling barrel 22, against the inside of the well casing or tubing C, in the Northrup patent, and the washing effect of the *positive strong downwardly projected jets of oil from the bottom* of Bray's traveling barrel and jet device 3, *directly* against the anchor P and lower clutch member 2 of the Bray pump in suit. Oil, which is a dense liquid with a high viscosity, is not easily agitated, and particularly when the alleged agitation is produced by liquid ejected from the side of the traveling barrel, above the guide which is considerably above the extension 8 in the Northrup pump. Such remote agitation of the oil in the Northrup pump would have no washing effect on the extension 8 in the Northrup pump.

Counsel for defendants contend that a white space between the guide 24 and the pipe 2 in Fig. 1 of the Northrup patent was intended to be a jet aperture for projecting a downward jet, but there is no such space in Fig. 2 of said patent, and Fig. 2 represents the same pump as Fig. 1, the only difference between the two figures being that Fig. 2 shows certain parts in elevation. *The Northrup patent specification is silent regarding said white space,* and plaintiffs' expert, Hayne, who had formerly been an examiner in the Patent Office, and had examined perhaps 50,000 patent applications, could not state what said white space indicated. [Tr. p. 296.] He stated that, "It might be a draftsman's error." [Tr. p. 297.] Said white space not being shown in Fig. 2 of the drawing and not being described in the specification of the Northrup patent, is too uncertain to be of any value as an anticipating element of the Bray pump.

> "Novelty is not negatived by any prior patent or printed publication, unless the information contained therein is full enough to enable any person skilled in the art to which it relates, to perform the process or make the thing covered by the patent sought to be anticipated."

Walker on Patents, 6th Ed., Sec. 96, p. 120.

Regardless, however, of the foregoing analysis of the Northrup patent, said patent fails to meet the terms of the claims of the patent in suit, because said claims specify a coupling or *clutch between* the barrel and the anchor, to prevent rotation of the barrel, for screwing up the sucker rods, and there *is no clutch in the Northrup patent.* [Tr. p. 141.]

The Ellis patent, No. 1,513,699, Defendants' Exhibits W and W', Book of Exhibits, page 133, shows a pump of the inverted type. The cap nut 21 on the lower end of the barrel 19 is formed with vertical grooves or oil passages 22 intermediate the bearing surface 23 which slidably fit on the tubular standard 12. The plaintiff Bray used the same type of cap nut 21 on his pumps in 1926, but found it unsatisfactory, because the bearing surfaces 23 between the passages 22 wore out and enlarged the space between the standard 12 and the nut 21 which caused the oil to jet a little too much through said space. [Tr. pp. 40 and 42.]

The passages 22, in the nut 21, are vents and not jet apertures and are for the purpose *only* of permitting the free *flow* of liquid therethrough, to prevent the formation of a dead chamber 24 containing trapped liquid which will resist the free operation of the pump. (Page 2, lines 37 to 42 inclusive, Ellis patent specification.) Nowhere in the Ellis patent specification is found any suggestion of a jet action for washing sand from below the lower end of the traveling barrel to prevent "sanding up" of the pump. In Fig. 1 of the drawing of the Ellis patent broken lines extend across the well tubing 10 and the standard 12, between the lower end of the traveling barrel 19 and the top of the anchor 13, which indicates that a length of the pump between said broken lines is removed in order to place Fig. 1 on the patent drawing, which is of prescribed dimensions, without reducing the scale of said figure. The removal of said length of the pump brings the lower end of the barrel down near the top of the anchor 13, but such nearness of the bottom of the pump barrel to the top of the anchor as shown in Fig. 1 of Ellis' patent drawing is not the true relation of the bottom of the barrel to the top of the anchor. When the removed section of the pump is placed between the broken lines above described the length of the pump will be increased to its true length, and in the true length of the pump the lower end of the barrel 19 will be a considerable distance, while in its lowermost position, above the top of the anchor 13, and at too great a distance for the downward *flow* of liquid from the dead chamber 24, through the passage 22 in nut 21, to have any washing action on the anchor 13. The Ellis patent is concerned about preventing the entrance of sand in the dead chamber 24 in the lower end of the traveling barrel, and provides a valve 25 for this purpose, which closes the passages 22 in nut 21 on the down-stroke of the barrel, and opens said passages on the up-stroke of the barrel to permit the *free expulsion* of the oil in the dead chamber through said passages, but the patent is silent as to any jetting action through said passages for the purpose of washing sand away from the top of the anchor in the manner contemplated and provided in the patent, in suit. (See lines 17 to 45, page 1 of the Ellis patent specification.) The Ellis patent specification, page 2, line 39, speaks of a *free flow* of liquid through the passages 22 and not a jet of liquid. The idea of Bray's *jet for washing away sand from the top of the anchor* is not mentioned or suggested anywhere in the Ellis patent, and this fact taken in connection with the feature of eliminating the dead chamber, shows clearly that the Bray invention is not within the contemplation of the Ellis patent.

Furthermore, the Ellis patent fails to meet the term of the claim of the patent in suit, in view of the fact that the Ellis patent *shows no clutch,* and consequently no *jetting action for washing sand from a clutch*, as covered by the claim of the Brady patent. As an anticipation of the Bray invention the Ellis patent is of little, or no, value, whatever.

The Dickens patent, No. 1,486,180, Defendants' Exhibits X and X', Book of Exhibits, pages 136-140, discloses a pump of conventional type with a *stationary barrel* and *reciprocating plunger*. This pump has a clutch between the upper end of the standing valve and the lower end of the plunger. This pump could not be pulled, because if the plunger were pulled the standing valve and lower clutch member would be left in the barrel. The valves are spaced too far apart to prevent a "gas lock". There is no anchor in this pump and no jetting action for washing an anchor or the clutch. This Dickens patent is entirely foreign to the Bray invention.

The Neilsen *et al.* patent, No. 1,717,619, Defendants' Exhibits Y and Y', Book of Exhibits, pages 141 to 145 inclusive, has a standing column and a traveling barrel, but said barrel has a sliding fit against a lining in the well tubing, which prevents the barrel from forcing up the oil on its down stroke and converts the barrel into a reciprocating plunger which lifts the column of oil on its up-stroke. In this respect the Neilsen pump is of a different type from that of the Bray pump in suit. The valves 7 and 12 are spaced far apart and a "gas lock" would form therebetween. This is also true of the valves 31 and 3^b.

The Neilsen patent specification, pages 1 and 2, lines 106 to 108 and lines 2 to 5, respectively, states the operation of Fig. 1 of said patent as follows:

"* * * on the *up-stroke* of the plunger the spaces 15, 17 are filled with liquid simultaneous with the filling of the pump.

* * * * * * * * *

During the *down-stroke* of the plunger the liquid in the chambers 15, 17 is forced through the parts 18 into the cage 6, thus washing the ball 7."

On page 2, lines 47 to 49 and lines 52 to 55, the operation of Fig. 2 of the Neilsen patent is stated as follows:

"During the *upward* movement of the plunger the passage 40 and chamber 38 are filled with a liquid, * * * during the *down-stroke* of the plunger, the liquid in the passage 40 is forced through the parts 41, washing the ball 31."

From the above description, in the patent specification of the operation of the Neilsen pump, it clearly appears that said pump operates on an entirely different principle from that of the Bray pump in suit. There are no ball valves 7 or 31 in the Bray pump to be washed because such valves are between the standing valves at a considerable distance and would cause a "gas lock", if the Neilsen pump were used in the oil fields of California. One of the important features of the Bray pump is the close relationship of the valves to prevent "gas lock", which is necessary in the use of Bray's pump in the oil fields of California. The washing of the ball valves 7 and 31 in the Neilsen pump would, therefore, be a useless function, if said pump were used in the oil fields of this State.

The white space between the tube 8 and sleeve 16 at the lower end of the traveling plunger 14 in Fig. 1, and the white space between the stationary tube 32 and foot nipple 39 in Fig. 2 of the Neilsen patent, are nothing but *vent* openings of substantial size to permit the *free flow* of liquid into and out of the lower end of said plunger to prevent a hydraulic lock in the spaces 15 and 38 of Figs. 1 and 2 of the Neilsen patent, which hydraulic lock would of course prevent reciprocation of the plunger and functioning of the pump, in view of the incompressibility of liquid. Said white spaces, in view of their substantial cross-sectional area, are, therefore, not "jet apertures", and no jet action could be produced through said white spaces, to wash sand from the top of the valve cages 6 and 30 to prevent the pump from "sanding up". No such jet action is described or suggested in the Neilsen patent.

The sleeve 16 and foot nipple 39 of the Neilsen patent are not "jet devices" like the jet devices 3″ of the Bray patent in suit, because liquid travels *downwardly* through said white spaces above described on the *down-stroke* of the plunger of the pump, and *upwardly* through said white spaces on the *up-stroke* of the plunger, as will appear from the above quoted language of the Neilsen patent specification. The sleeve 16 and foot nipple 39 of the Neilsen patent act as plungers for forcing liquid in the well tubing downwardly on the *down-stroke* of said plungers, but in the Bray pump liquid is jetted out of the lower end of the traveling barrel on its *up-stroke,* through "jet apertures" 5, due to the positive *pressure* upon the liquid in the lower end of the barrel on its *up-stroke.* There is no pressure upon the liquid in the lower end of the pump barrels in the Neilsen pump, because the liquid *flows freely* into and out of the spaces 15 and 38 in said barrels, and consequently no liquid could possibly be *jetted* out of the lower end of the Neilsen pump barrels *under pressure.* The pressure of the liquid in the lower end of pump barrel, for positively *jetting* the liquid out of the bottom of the barrel through the jet apertures 5, for washing sand from the anchor and the clutch and preventing "sanding up" of the pump, is a vital feature of the Bray invention. This vital feature is missing in the Neilsen pump.

Another vital element of the Bray pump, which is missing in the Neilsen pump, is the clutch. *There is no clutch in the Neilsen pump, and consequently no jetting action for washing the clutch.*

As an anticipation of the Bray invention the Neilsen patent is of no legal force or effect.

The remaining patents of the prior art, set up by the defendants, are Savidge, Tierce, Franchi, Stephens and Barthel, Defendants' Exhibits Z, AA, BB, CC and DD, respectively, Book of Exhibits, pages 146 to 172, inclusive. These patents have no bearing on the Bray invention.

The Savidge patent covers a valve coupling for tubular wells. It has none of the elements of the patent in suit. The coupling of the Savidge patent is not the equivalent of the Bray clutch because it is used for an entirely different purpose.

The Tierce patent covers a standing valve puller, which is used in a conventional pump with a standing barrel and traveling plunger, but not in a pump of the type claimed in the patent in suit. The standing valve and plunger are pulled by the puller nut 3 on the lower end of the traveling barrel B of the Bray patent, and requires no tool like the Tierce valve puller.

The Franchi patent covers nothing but a piston for use in an internal combustion engine and contains no element of the Bray invention, since the specific structure of the Bray plunger is not claimed in the Bray patent in suit.

The Stephens patent covers a pump and method of packing the same. The pump is of a different type from that of the Bray pump. The liquid is lifted on the up-stroke of the plunger, by power, and not on the down-stroke by gravity, as in the Bray pump. The valves 5 and 16 are spaced far apart and a "gas lock" would form therebetween. There is no jet pressure chamber in the lower end of a traveling barrel, nor any jetting action for washing sand from an anchor or a clutch.

The Barthel patent covers a piston for an internal combustion engine and its specific construction is not claimed in the patent in suit. Not a single element of the Bray invention is shown in this Barthel patent.

From the foregoing analysis of the prior art, it will be apparent that the prior art fails miserably to anticipate the jet device 3 of the Bray patent in suit, for producing a *positive* and strong jet of oil *directly downwardly* from the pressure chamber EE (Defendants' Exhibit MM) through the vertical aperture 5 and out of the bottom of the traveling barrel B, directly towards the pin or anchor 3 and *coupler head or clutch member 2,* for washing sand from said anchor and *coupler head,* on the up-stroke of the traveling barrel, to prevent "sanding up" of the pump. The *function* of the jetting action and its *result,* the washing of the anchor and the *clutch* to prevent "sanding up" of the pump, are *new* and highly useful, and entitle the patent in suit to liberal treatment by the Courts.

Walker on Patents, 6th Ed., Sec. 417.

The *new function* of the downward jetting action from the bottom of the traveling barrel establishes the Bray patent in suit, in this respect, as a *primary* patent.

> *Walker on Patents,* 6th Ed., Sec. 416, at p. 508;
>
> *National Dump Car Co. v. Ralston Steel Car Co.,* 172 Fed. 393;
>
> *A. D. Howe Mach. Co. v. Coffield Motor Washer Co., supra;*
>
> *Diamond Rubber Co. v. Consolidated Rubber Tire Co., supra;*
>
> *American Pneumatic Service Co. v. Snyder, supra.*

"Pioneer (primary) patents are entitled to receive a broad and liberal construction and entitled to a liberal range of equivalents. This *rule is applicable also to a pioneer patent in a limited field.*"

Lyons v. Lewald, 254 Fed. 708;

Berry v. Fuel Economy Engineering Co., 248 Fed. 736.

"That a patent is not of pioneer character does not prevent its owner from invoking the doctrine of equivalents."

Seneca Camera Co. v. Gundlack etc. Optical Co., 236 Fed. 141;

Superior Skylight Co. v. August Kuhnla, Inc., 265 Fed. 282 (Aff. 273 Fed. 482);

Edwards Mfg. Co. v. National Fireworks Distributing Co., 272 Fed. 23.

"This examination of the prior art in connection with the Bruckman patent results in the conclusion that Bruckman, through using several known elements has added other new essential elements, has in a novel manner assembled and combined all in one machine operated through a single transmission of power, and has thereby *accomplished a new and desirable result. * * * This entitles his machine to be considered a pioneer* in the field of automatic production of sanitary batter cones. This conclusion has been reached in the district by a very able and careful judge in Bruckman *et al.* v. Stephens *et al.,* 268 Fed. 374, 376. The file wrapper and its limitations do not affect this conclusion, and is answered by the broad claims allowed by the Patent Office."

Roberts Cone Mfg. Co. et al. v. Bruckman et al., 226 Fed. 986, 990 (C. C. A. 8th Cir.).

"Where the invention, although for an improvement only, is of meritorious character, the range of equivalents is reasonably broad and fair."

Aeolin v. Schubert Piano Co., 261 Fed. 178;

Columbia Mach. etc. Corp. v. Adriance Mach. etc. Co., 226 Fed. 455;

David v. Harris, 206 Fed. 902, 124 C. C. A. 477;

Elbs v. Rochester Egg Carrier Co., 203 Fed. 705, 121 C. C. A. 661.

The question of whether the plaintiffs' patent is a primary patent or a secondary patent is of little consequence. Assuming for the purpose of argument only that plaintiffs' patent is only a secondary or improvement patent, it is nevertheless entitled to a liberal construction sufficient at least to read on the defendants' close imitation of plaintiffs' pump. We refer to *Walker on Patents,* 6th Ed., Sec. 417:

"The rule has been laid down that even though an invention is not generic, if it is *one of merit,* it is entitled to a *liberal construction and a fair range of equivalents.*"

"In construing improvement claims of a patent, consideration should be given to the *character of the improvements* introduced by the patentee and the change *in the art attributable to them. When they result in converting imperfection into completeness, and in producing the first practically and commercially successful machine, however simple the change appears,* the invention is entitled to liberal treatment *by the courts.*"

See:

> *Cent. Dig.,* Vol. 38, Patents, Sec. 249;
>
> *Wagner Typewriter Co. v. Wycoff, Seamans & Benedict,* 151 Fed. 585, 591 (2nd Cir. 1907).

"A patent which is not a pioneer patent but which is found at the head of a class, though in a well developed art, is entitled to a liberal range of equivalency."

> *Lamson Consol. Store Service Co. v. Hillman,* 123 F. 416, 59 C. C. A. 510 (7th Cir.).

"Courts look with favor upon patents for *primary improvements* which are novel and a manifest *departure from the principles of prior structure,* and which constitute the *final step necessary to convert* failure into success."

> *Wagner Typewriter Co. v. Wycoff, Seamans & Benedict, supra.*

Bray's pump certainly converted failure into success in preventing "sanding up." On page 279 of the transcript Bray testified as follows:

> "I found other pumps under the same conditions did sand up and mine did not."

Bray further testified concerning particular pumps which sanded up and were replaced by his pump [Tr. pp. 279 and 282], none of which have been pulled in four years for "sanding up" or for any other trouble. [Tr. p. 43.]

Defendants have failed to show a single pump which operated successfully without "sanding up" in the oil fields of California before the advent of the Bray pump in suit.

Bray's testimony as to the failure of other pumps and the success of his pump in preventing "sanding up" stands uncontradicted. Bray's testimony throughout is frank and honest and shows the practical knowledge of a practical oil man, with practical experience. No one can read his testimony without being impressed with his sincerity, and his testimony is of infinitely more value than the purely theoretical testimony of defendants' professional expert, Adams.

Nearly all the pumps of the prior art, set up by the defendants, are pumps produced by Eastern inventors, mostly of Pennsylvania, for Eastern oil fields, where oil field conditions and practice are vastly different from those of California. A comparison of oil field practice of California with that of Pennsylvania is like a comparison of modern times with the dark ages. There is a witticism among California oil men that "the farmers of Pennsylvania milk their oil wells like they milk their cows."

The prior art, set up to limit the fair scope and defeat the plain intent of the patent in suit, is nothing but "prior rot."

> "In short, in construing the patent statute, which was enacted to promote the useful arts, it is more important to study those developments of the art which are bright with use in the channels of trade than to delve into abandoned scrap heaps and dust-covered books which tell of hopes unrealized and flashes of genius quite forgotten. The somewhat meaningless terms 'inventive genius' and 'mechanical skill' must be clarified by an examination of the

article itself, and if the improvement be unusual, or if there be doubt, and the public has given its tribute, the judge should accord to the creator of the article the title of inventor."

Wahl Clipper Corporation v. Andis Clipper Co., et al., 66 Fed. (2d) 165.

The plaintiff Bray has testified concerning leading pumps, such as the O'Bannon and Neilsen, which "sanded up," and which he replaced with his own pump, which did not "sand up" [Tr. pp. 279-286], and Bray has further testified that he has put in about 500 of his pumps, and has not had a single pump "sand up," where he had to pull tubing to clear it, in approximately four years. [Tr. pp. 270, 45 and 43.] This is a remarkable record, which has no comparison.

The testimony of expert Adams is a labyrinth of inconsistencies and confusion. On page 177 of the transcript he testifies concerning the jet action of Bray's pump as follows:

"The liquid, as a whole, is stationary, but because of the restricted opening there is a velocity through there relative to the tube and that will persist, depending on the size and speed. *I do not say there is not any jet. I admit that there is a jet.*"

On page 179 of the transcript Adams testifies:

"Now then, is there a velocity there? Yes. What is the velocity? The velocity is the velocity of the flow as you pull the puller nut through the tube of liquid, of the flow past the restricted area of the puller nut. In that sense *there is a jet. It is a jet that is formed by an orifice that is moved through a fluid.*"

On page 201 of the transcript Adams testifies:

"In other words, *there is not in the Bray pump, in any of the other pumps that we have discussed, a jet action,* at any of the speeds which we have considered. Those average speeds and *the flow would not reach the dignity of a jet."*

In his vacillation from "jet" to "no jet," expert Adams flatly contradicts himself, and the only conclusion that can be drawn from such testimony is that the witness does not know what he is talking about.

On page 190 of the transcript Adams testifies in effect that the jet or "flow" of oil would be upwardly through the jet aperture into the pressure chamber in the lower end of the traveling barrel, on the up-stroke of the traveling barrel, and cites as an example, a man walking backwards on a moving train, but actually traveling forward with relation to the earth. Such sophistry is easily exposed. If a man ran backwards 15 miles per hour on a train going forward 5 miles per hour, the man would be traveling backwards 10 miles per hour with relation to the earth. It should be obvious to the Court that the liquid in the lower end of the traveling barrel will have to go downward and out through the restricted jet aperture 5, of the Bray patent, faster than the traveling barrel moves upward. A good example is the squirt gun. A slow movement of the plunger will cause liquid in the gun to squirt out of the restricted spout of the gun at a faster speed than the movement of the plunger. This is the jet action of the jet device of the Bray patent, and there is no question about it.

After citing the man walking backwards on the train, Adams concludes:

"So we have something of the same effect in this jet which is formed by an orifice moving backwards and a flow going past that. I do not want to carry the impression that there is absolutely no velocity there. *There is a velocity relative* to the puller nut when it is in motion, and if this velocity is of any great magnitude *it will persist somewhat and be carried downward.*"

On pages 183 and 184 of the transcript expert Adams grossly exaggerates the leakage upwardly from the bottom of the traveling barrel past the standing plunger into the pump chamber, and carries his fantastic theory to the point of assuming that when the traveling barrel moves up liquid will flow upwardly through the jet aperture into the bottom of the traveling barrel and up past the plunger into the pump chamber, thereby preventing any downward *jet action* through the aperture in the bottom of the traveling barrel. This is clearly a *reductio ad absurdum.* There is no such leakage in either the Bray or the Hofco pump. The Bray patent describes an efficient packing on page 2, lines 90 to 101, inclusive, and page 2, lines 1 to 24, inclusive, and the Hofco catalogue, Plaintiffs' Exhibit No. 4, page 6, Book of Exhibits, shows at the right a plunger with many packing rings, to prevent leakage past the plunger. See also Defendants' Exhibit II, Book of Exhibits, page 177. Only a negligible amount of liquid could get past the plunger in either the Bray or Hofco pump. If, however, the packing should become worn and leakage past the plunger should develop, it would be a simple matter to pull either pump and put in new packings in the plunger to pievent leakage. Expert Adams would have the Court believe that an efficient packing cannot be made, but such is not the case. The Court may take judicial notice of a hydraulic elevator plunger and an ammonia compressor in an ice plant. A few drops of liquid may be observed at times on the elevator plunger, but this is a very negligible quantity of leakage, and ammonia compressors are constructed with practically no leakage at all, because operatives in an ammonia plant would suffocate from the escape of ammonia gas.

Expert Adams testifies [Tr. pp. 198 and 199] that there are three forces in the bottom of the well tubing of the Bray pump which tend to keep the liquid and sand stirred up; the first force being the downward jets through the apertures 5 in the puller nut 3 at the bottom of the traveling barrel; the second being the plain churning action of the traveling barrel; and the third is the flow of liquid down between the traveling barrel and well tubing. Of said three forces Adams concludes that the force of the jet is the least important. If this were true *why did other pumps of the Bray type, without the Bray downward jets, "sand up"*, and *why have the defendants copied the downward jet feature of the Bray pump?* [Tr. pp. 279, 357 and 358.] See Defendants' Exhibit II, page 177, Book of Exhibits, white space between blue standing column A and red puller nut or jet device D. See also Plaintiffs' Exhibit 4, page 8, Book of Exhibits, under "Working Actions", which reads as follows:

> "Likewise the surge of the fluids in and out of the lower end of the reciprocating barrel keeps all sand and foreign sediment stirred and permits pulling of pump at any time without undue trouble."

The surge of the fluids in and out of the lower end of the Hofco traveling barrel is the jet from the lower end of the Bray barrel, and the Hofco catalogue, Plaintiffs' Exhibit 4, is the only exhibit or literature in the art, other than the Bray patent, that mentions a jet or *"surge* of the fluids in and out of the lower end of the reciprocating barrel (which) keeps all sand and foreign sediment stirred and permits pullling of pump at any time without undue trouble."

Adams admits, however, that, with the jet, sand is washed from the clutch [Tr. p. 198]:

> "I do not dispute that the coupler may be kept free of sand."

This is an admission of utility of the downward jet action of the Bray patent, because if other pumps "sanded up" without said jet action, and Bray's pump, with said jet action, does not "sand up", then the only conclusion that can be drawn from these facts is that the jet action is the new and contributing factor that prevents "sanding up" of the pump.

> "A patent is *prima facie* evidence of utility, and doubts relevant to the question should be resolved against infringers, because it is improbable that men will render themselves liable to actions for infringement, unless infringement is useful."
>
> *Walker on Patents,* 6th Ed., Sec. 125.

Adams further stated [Tr. p. 199] that he felt that it did not make much, if any, difference whether the flow was radial through the side of the barrel above the puller nut, or whether it passed through the puller nut, but this assumption is refuted by plaintiffs' witness, M. Rennebaum, a practical oil production man [Tr. pp. 357 and 358], as follows:

"The Neilsen pump is one of the old established pumps with *holes in the side. I found it wasn't very good in some wells;* in some wells it operated all right. *In other wells it didn't.*"

"When you put *holes on the side here you are much farther away from the sand. It hits up against your tubing here. You haven't the same force that you have if the oil comes straight down here,* because if it was washing over that standing valve every time it made a stroke it forced the oil down here and lets all of this up and oil would squirt out here on the side. Of course, it would drop down here. *It would not have the same force if it went straight down.*"

Adams considers loss by eddy currents and concludes that the friction of the jet against the fluid with which it comes in contact, is of such order that the entire energy of the jet would disperse within a very few inches of motion. [Tr. p. 200.] This fantastic theory, however, ignores the inevitable condition that when the traveling barrel moves up, liquid *must* go out of the bottom of the barrel through the restricted jet apertures 5, with *considerable force;* otherwise the liquid in the bottom of the traveling barrel, being incompressible, would form an *hydraulic lock* and prevent upward movement of the barrel; and when liquid flows into the vacuum under the lower end of the upwardly-moving traveling barrel and comes into contact with strong downwardly projected jets, we have *a meeting of two strong opposing hydraulic forces,* figuratively like the irresistible force meeting the immovable body, which would produce a *turbulence of the liquid directly between the lower end of the traveling barrel and the top of the anchor,* and this *turbulence* would necessarily stir up the sand and wash the sand from the anchor and clutch, and prevent the pump from "sanding up". *This turbulence of the liquid directly between the lower end of the traveling barrel and the top of the anchor necessarily resulting from the jet action is a new phenomena* in the art which made its first appearance in the Bray patent in suit, and there is nothing in the prior art which suggests this novel feature of the patent in suit.

The above considered testimony of defendants' expert Adams concerns the defenses of inoperativeness or lack of utility of the patent in suit, but since said defenses were not pleaded by defendants, all evidence concerning the same should be disregarded by the Court.

However, the preponderance of evidence that the patent in suit covers an operative and highly useful invention is decidedly in plaintiffs' favor. Against the testimony of one witness for the defendants stands the grant of the patent, after careful examination by experts in the Patent Office; the testimony of Patrick H. Bray [Tr. pp. 24-53 and 241-291]; the testimony of M. Rennebaum [Tr. pp. 356-363]; the testimony of Fred H. Hayne [Tr. pp. 293-355]; and those 500 Bray pumps, operating successfully over a period of four years without "sanding up" or requiring pulling—mute evidence of the operativeness and practical utility of the Bray invention more eloquent than all of the purely theoretical speculation of defendants' expert, Adams, to the contrary. [Tr. pp. 270, 279, 45 and 43.]

The lower court misunderstood the invention in issue and it is not surprising that the Court misapplied the law governing the case.

In the Memorandum of Conclusions and Order for Judgment [Tr. p. 374], the Court states the following:

> "A pump base is screwed into the pumping string at the bottom of the well."

The pump base (anchor P) is not screwed into the pumping string, but is slidably fitted in the lock box L and held in said lock box by a snap ring.

On page 375 of the transcript the Court states the following:

> "Oil or other fluid is sucked into the barrel on the *downward stroke* of the pump barrel, and, through an arrangement of valves, on the *upward stroke is expelled* into the pump casing."

Fluid is not sucked into the barrel on the *downward* stroke. It is sucked into the barrel, or pump chamber, on the *upward stroke*. And through an arrangement of valves, on the *upward* stroke oil is *not* expelled into the pump casing. It is expelled from the pump chamber, between the valves, on the *downward* stroke.

Further, on page 375, the Court states the following:

> "It was desirable to have a locking means between the pump barrel and the standing column base, whereby the barrel could be lowered until its base was in contact with the column fixture, and *when locked* the *base fixture could be unscrewed through the turning of the pumping rods,* allowing the entire pumping apparatus to be lifted from the hole."

The barrel is *not* lowered for the purpose of *unscrewing* the base fixture through a turning of the sucker rods to allow the entire pumping apparatus to be lifted from the hole. The base fixture (anchor P) is *not screwed* into the lock box L and is not unscrewed to lift the pump from the hole. *The pump barrel is lowered to clutch the barrel to the anchor to prevent turning of the barrel, so that the loose joints of the sucker rods may be screwed up tight.* It is not necessary to lower the barrel to lift the pump out of the well. All that is necessary to lift the pump out of the well, is to *pull straight upwardly* on the pump, so that the puller nut 3 will engage the lower end of the standing plunger and pull said plunger, together with the anchor, up and out of the well.

The lower court [Tr. p. 375] holds that plaintiffs' patent includes only two features: "An arrangement for *expelling sand* and fluid from the barrel, and an arrangement for locking the column to the barrel". This construction of the patent in suit construes the patent piecemeal and leaves out other vital elements and features of the patent claims, to-wit: the *close relationship of the standing valve and traveling valve* to prevent "gas lock" and *the downwardly projected jet* directly out of the traveling barrel *to wash sand from the clutch and the anchor* to prevent "sanding up" of the pump. The court instead of considering the *entire novel combination of elements* of the claims, has considered only *two features.* Even if the two features were old, that is no justification for giving the *entire novel combination of elements* of the claim a narrow construction, and especially when the complete new combination of elements has produced a novel pump which *produces a new function* and *accom-*

plishes a new result, to-wit: *washing the clutch and anchor with a direct downwardly projected jet, and preventing the pump from "sanding up".*

"In a combination patent there are no unpatented features in the sense that they are separable from patented ones, and no one of the elements is patented. They may all be old and not patentable at all unless there is some *new combination* of them. The point to be emphasized is that the *law looks not at the elements as factors of an invented combination as a subject for a patent, but only to the combination itself as a unit distinct from its parts,* and in such case there could be no comparison of patented and unpatented parts.

> *Yesbera v. Hardesty Mfg. Co.,* 166 Fed. 120 at p. 125 (C. C. A. 6th Cir.);
>
> *Leads & Catlin v. Victor Talking Machine Co., supra,* quoted in *Diamond Rubber Co. v. Consol. Tire Co., supra.*

The holding of the lower court concerning the jet device 3 [Tr. pp. 377-378] is obviously contrary to the teaching of the patent in suit. The language of the Court is as follows:

"The patentee contemplated that the *guide* used should have a *close* or *ordinary mechanical fit;* * * *. His invention is brought within a *narrow field and is not entitled to a wide range of equivalents. The teachings of the patent are specific in terms* as the patentee states: * * *. If perchance an operator does not employ a *guide with a running fit,* and so allow some discharge of fluid to get by between the guide and the circumference of the standing column, he is not, perforce, following the specifications and claims, hence is not infringing the plaintiffs' patent right. That is the most that can be asserted against the defendant in this case. * * * I hold claims 2 and 3 are valid, limited closely to the feature of having a bottom guide running on the standing column, with *a running fit*. * * * that the invention is to be strictly limited to the form described; that no infringement is proved."

The specification of the patent in suit does not employ the term "guide" although said term is used in the claims. The claims are to be interpreted in the light of the specification.

Walker on Patents, 6th Ed., Sec. 227, at p. 309;
48 *C. J.*, Sec. 335, p. 213.

The specification employs the terms "complementary part 3", "bushing" and "jet device 3", and there is nothing in these terms to suggest a *close or ordinary mechanical or running fit*. Claims 1 and 2 describe a bottom guide "running on said column" which is not necessarily a "close running fit."

The term "guide" is not the most appropriate term, but it is used loosely in the art like the term "puller nut", "bushing", etc. There is no statement in the specification that the part 3, bushing, or jet device 3, is used for guiding the lower end of the traveling barrel, because this is no feature of the invention in suit, and no such guide is necessary in the operation of the traveling barrel, because said barrel is adequately guided in its reciprocating movement by the long standing plunger, comprising the column 10, guides 13 and packing 16. The term "bushing" used in the specification and corresponding to the term "guide" used in the claims, means "something that reduces the diameter in the tube and changes the size of the hole" according to the testimony of Defendants' expert, Adams [Tr. pp. 143, 142 and 207]. We accept this definition of the term "bushing". In the plaintiffs' pump the bushing 3, which is called a guide in the claims, reduces the diameter of the lower end of the traveling barrel to the cross-sectional area of the jet apertures 5, so that the liquid will be jetted out of the lower end of the traveling barrel through said apertures on the up-stroke of the barrel, in the manner of a jet out of the nozzle of a squirt gun, upon movement of the plunger of the gun.

Claim 3 describes a bottom guide *"operating* on the standing column" and the term "operating" is a broad term which is certainly comprehensive enough to include a very loose fit of the jet device 3 around the standing column member 11, through which loose fit liquid could be jetted downwardly out of the lower end of the traveling barrel towards the anchor and lower clutch member. Claim 3 does not specify a series of jet apertures, but specifies the guide, meaning jet device 3, as being *"apertured"* for the ejection of liquid, and an annular space, formed by a loose fit between the guide and standing column, would certainly be an *aperture* in the guide, which would make the guide *apertured* in accordance with the terms of the claim.

There is nothing in the prior art, as clearly shown, or in the patent in suit, which limits the patent to the form particularly described therein. Page 1, lines 41 to 49 inclusive, of the patent are as follows:

"Other objects, advantages and features of construction, and details of means and of operation will be made manifest in the ensuing description of the herewith *illustrative* embodiment; it being understood that *modifications, variations* and *adaptations* may be resorted to within the spirit, scope and principle of the invention as it is hereinafter claimed."

"It does not necessarily follow, from the fact that a claim of a patent describes a specified form of construction of a machine or part, that the inventor is limited to that form; but it depends on his expressed intention and the scope of his actual invention."

Kings County Raisin & Fruit Co. v. U. S. Consol. S. R. Co., 182 Fed. 59 (C. C. A. 9th Cir.).

"The patentee having described his invention and shown its principles, and claimed it in that form which most perfectly embodies it, is in contemplation of law, deemed to claim every form in which his invention may be copied, unless he manifests an intention to disclaim some of those forms."

Western Electric Co. v. La Rue, 139 U. S. 601, 606; 11 S. Ct. 670; 35 L. Ed. 294.

"An inventor is entitled to all that his patent fairly covers, even though its complete capacity is not recited in the specifications and was unknown to the inventor prior to the patent issuing."

Diamond Rubber Co. v. Consol. Tire Co., 220 U. S. 428.

From the foregoing analysis of the prior art and the patent in suit, and the application of the law thereto, it is submitted that the lower court was clearly in error in construing claims 2 and 3 of the patent, narrowly, and that said claims should be *liberally construed,* particularly in view of the new function and new result of the Bray pump, *to read on the equivalent construction of the defendants pump.*

Issue 3.

Assignments of Error XVII, XVIII, XXI, XXII, XXIII, XXV, XXVI, XXXVII, XLVI, XLVIII, XLIX, L, LI, LIV, LVI and LVII.

The Lower Court erred in decreeing that the defendants, and each of them, have not infringed any of the claims of Letters Patent in suit and have not infringed said Letters Patent, and in not decreeing that the defendants and each of them have infringed each and all of the Claims of said Letters Patent, and have infringed said Letters Patent, because

The defendants' pump, Hofco Catalogue, Plaintiffs' Exhibit 4, Book of Exhibits, pages 5 to 8 inclusive, and Defendants' Exhibit II, Book of Exhibits, page 177, contains all of the elements of each of the claims of the patent in suit, which elements of defendants' pump cooperate in substantially the same manner, perform the same function and accomplish the same result as the corresponding elements of the respective claims of the patent in suit.

Applying claim 1 of the patent in suit to defendants' pump, the elements of said claim appear in defendants' pump as follows:

1. *Standing guide column marked A,* Defendants' Exhibit II [Tr. pp. 29-33], and marked 1, Plaintiffs' Exhibit 4, p. 7.

2. *Working valve* on the upper end of the standing column. marked 7, Hofco Catalogue, Plaintiffs' Exhibit 4, p. 7.

3. *Valved traveling barrel*, marked C and X, Defendants' Ex. II, p. 177, and marked 2 and 3, Plaintiffs' Exhibit 4, p. 7.

4. *Bottom guide on traveling barrel*, marked 4, Plaintiffs' Exhibit 4, p. 7, and Defendants' Exhibit II, marked D and P.

5. *Coupler formed on bottom face of guide*, marked E and P, Defendants' Exhibit II.

6. *Anchor*, marked 5 and 6, Plaintiffs' Exhibit 4, p. 7, and Defendants' Exhibit II, marked G and F, Figs 1 and 2, and H and F, Figs. 3 and 4.

7. *Complementary Coupling on top of anchor*, marked I and Y, Defendants' Exhibit II.

The coupling Y engages the bottom face of guide P in Fig. 4, Defendants' Exhibit II, to lock the traveling barrel A against rotation.

The close relationship of the standing valve 7 and traveling valve 3 are shown in Defendants' Hofco Catalogue, page 7, Plaintiffs' Exhibit 4. Only the traveling valve, marked X, is shown in Defendants' Exhibit II, page 177, the standing valve being omitted for reasons not explained by defendants' expert, Adams. The close relationship of said valves is for the purpose, already explained, of preventing gas lock in the pump. While the close relationship of the valves is not new *per se,* it is one of the vital elements of the *combination* of elements of claim 1, and of claims 2 and 3 of the patent in suit.

It is submitted that the defendants' pump infringes claim 1 of the patent in suit.

Applying claims 2 and 3 of the patent in suit, to defendants' pump the elements of said claims appear in defendants' pump, as follows:

1. *Standing column,* marked A, Defendants' Exhibit II, page 177, and Plaintiffs' Exhibit 4, page 7.

2. *Working Valve,* marked 7, Plaintiffs' Exhibit 4, page 7. [Tr. p. 31.]

3. *Valved Traveling Barrel,* marked C and X, Defendants Exhibit II, page 177, and Plaintiffs' Exhibit 4, page 7, marked 2 and 3.

4. *Bottom Guide on Traveling Barrel,* marked D, Defendants' Exhibit II, p. 177, and marked 4, Plaintiffs' Exhibit 4, page 7.

5. *Coupler* formed on bottom face of guide, marked E and P, Defendants' Exhibit II, page 177.

6. *Anchor,* marked 5 and 6, Plaintiffs' Exhibit 4, page 7, and Defendants' Exhibit II, marked G and F, Figs. 1 and 2, and H, and Figs. 2 and 3.

7. *Complementary Coupling on top of anchor,* marked I, Figs. 1 and 2, and marked Y, Fig. 4, Defendants' Exhibit II.

8. *Guide having venting duct or being apertured* for ejection of liquid directly toward the anchor top face and lower coupler, to prevent settlement of sand thereon. See white spaces between blue standing column A and red guide D, Defendants' Exhibit II, page 177. See also Plaintiffs' Exhibit 3, Hofco guide, and Defendants' Exhibit NN, white space (.638) between red guide marked FF and blue standing column marked 3.

No dimensions of the jet apertures 5 of the Bray pump are given in the Bray patent in suit, and the *patent is not restricted to* any particular size of jet apertures. The size of the apertures for producing an effective jet is determined according to the practical experience of the inventor and manufacturer Bray, who has installed 500 of his pumps, all of which have operated successfully over a period of four years. [Tr. pp. 277-279, 270, 45 and 43.]

The defendants have constructed the jet device D of their pump, Defendants' Exhibit II, page 177, with a jet aperture of substantially the same cross-sectional area as that of the plaintiffs' pump, because substantially the same amount of liquid is jetted out of the lower end of the traveling barrel of the defendants' pump as is jetted out of the lower end of the traveling barrel of the plaintiffs' pump. This is admitted by defendants' expert, Adams [Tr. p. 194] as follows:

> "An ordinary table drinking glass would hold something on the order of 12 to 15 cubic inches; so that in the case of Admore pump the quantity flowing downward which is relied on to form a jet, if there is a jet, would be somewhat over two glasses of fluid. *In the Hofco pump, the 49.5 cubic inches would be* about three full glasses; and *in the Bray pump, some little in excess of that."* [52.7 cubic inches, see Tr. p. 191.]

While not admitting the accuracy of expert Adams' figures as to the maximum amounts of liquid jetted by the two pumps, the proportionate amounts, as stated by Adams appear to be substantially correct. In Fig. 4 of the patent in suit a series of jet apertures 5 are shown extending around the central large opening in the jet device 3, through which opening extends a portion 11 of the standing column. The jet apertures accordingly extend *around* the standing column close to said column. In the defendants' Hofco pump the jet aperture is formed by enlarging the central opening in the jet device and thereby providing one continuous annular aperture extending *around* the standing column in the jet device, marked D, Defendants' Exhibit II. See also Plaintiffs' Exhibit 3. The difference between a plurality of closely arranged jet apertures and a single jet aperture extending *around* the standing column is a difference only of *form*, because the single aperture of defendants' pump performs the same function in substantially the manner and accomplishes substantially the same result as the plurality of apertures of plaintiffs' pump. Such a difference is not sufficient to avoid infringement of plaintiffs' patent.

"It involves no invention to cast in *one* piece an article which has formerly been cast in *two* pieces."
Carey v. Houston & T. C. R. Co., 37th Ed. 1039-1041.

"Infringement is not avoided by *dividing an integral element* of the patented machine into *two or more distinct* parts, so long as the function and operation remain substantially the same."
Kings County Raisin & Fruit Co. v. U. S. Consol. S. R. Co., 182 Fed. 59 (C. C. A. 9th Cir.).

"Change of *form* without change of function does not avoid infringement."
Triangle Kapok Mach. Corp. v. Solinger Bedding Co., 13 Fed. (2d) 494;
Daniel O'Connel, Inc. v. Riscal Mfg. Co., 228 Fed. 127;
Machine Co. v. Murphy, 97 U. S. 120.

The single annular jet aperture in the lower end of the traveling barrel around the standing column in defendants' Hofco pump is the *equivalent* of the plurality of jet apertures 5 in the lower end of the traveling barrel around the standing column of the Bray patent in suit, because the jet aperture in the defendants' pump performs the same function in substantially the same manner and accomplishes the same result as the jet apertures in the plaintiff's pump.

> "It is therefore safe to define an equivalent as a thing which performs the same function and performs that function in substantially the same manner of the thing of which it is alleged to be equivalent."
>
> *Walker on Patents,* 6th Ed., Sec. 415.

> "No substitution of an equivalent for any ingredient of a combination covered by any claim of a patent can avert a charge of infringement of that claim, *whether or not the equivalent is mentioned in the patent.*"
>
> *Walker on Patents,* 6th Ed., Sec. 412, at p. 501.

> "The *substantial equivalent of a thing* is, in the sense of the patent law, *the same as the thing itself.* Two devices which perform the same function in substantially the same way, and accomplish the same result, are, therefore, the same, though they *may differ in name and form.*"
>
> *Machine Co. v. Murphy,* 97 U. S. 120.

It is submitted that claim 2 of the patent in suit, which specifies a series of jet apertures, is clearly infringed by the defendants' pump, having a single jet aperture in the lower end of its traveling barrel.

If there were any doubt as to infringement of claim 2 of the patent in suit by the defendants' pump, there can be no doubt whatever about claim 3 being infringed by defendants' pump, because claim 3 specifies the "guide being *apertured* for the ejection of liquid from the barrel" and the part D, Defendants' Exhibit II, Fig. 1, which is a jet device, is certainly *apertured,* because the annular white space or strip between the red part D and the blue standing column A is certainly an *aperture.* Moreover, defendants' expert, Adams, admits that liquid goes out of the orifice in the lower end of the defendants' Hofco traveling barrel on the upward stroke of said barrel, and said aperture in part D of defendants' pump is the only opening in the lower part of the barrel through which liquid can escape. [Tr. pp. 194, 195 and 197.] There is nothing in the prior art or in the specification of the patent in suit to restrict the meaning of the term "apertured" in claim 3 of the patent.

In all of the voluminous prior art offered in evidence there is not a *single* pump structure which shows the *general* combination claimed in any of the claims of the patent in suit, and consequently said combinations, covering an invention that *is a practical success* and a *distinct advance in the art* are entitled to a liberal construction.

Walker on Patents, 6th Ed., Sec. 233.

> "When legitimate combination claims, such as those appealed, are rejected on *two or more* references, the trend of the best authorities indicates that at least *one* of the references ordinarily should show the *general* combination claimed."
>
> *Ex Parte McCullum,* 204 O. G. 1346;
> *Ex Parte Gee,* 261 O. G. 800.

"Instances of prior art showing individual elements of claimed combination do not show anticipation.

Anticipation structure cannot be built up from various instances of prior art, where individual elements of claimed combination are old.

Even if each element of patent was old, there was invention displayed in combining elements to produce an efficient boat, which no one else had done."

Welin Davit & Boat Corp. v. C. M. Lane Life Boat Co., Inc., 38 F. (2d) 685.

"It is not sufficient, in order to constitute an anticipation of a patented invention, that the device relied upon, might, by modification, be made to accomplish the function performed by that invention, if it were not designed by its maker, nor adapted, nor actually used for the performance of such function."

Topliff v. Topliff and Another, 145 U. S. 156, 36 L. Ed. 658.

That the patent in suit is *merely illustrative* of the invention claimed and is not *intended to be limited* to the exact details of construction shown in the drawing and particularly described in the specification, is clearly shown on page 1, lines 41 to 49, which states:

"Other objects, advantages and features of construction, and details of means and of operation will be made manifest in the ensuing description of the herewith *illustrative* embodiment; it being understood that *modifications, variations* and adaptations *may* be resorted to within the spirit, scope and principles of the invention."

In the recent case of *Oates v. Camp* (C. C. A. Va. 1936), 83 F. (2d) 111, a patent *claim was not required to be limited to the exact device* disclosed by the specification and drawings, since the *claims* of a patent and not its specification *measure the invention*.

Claim 3 of the patent in suit reads, letter perfect, on the structure of defendants' pump, and it is submitted that defendants' pump unquestionably infringes said claim.

Issue 4.

Assignments of Error, XX, XXV, LVIII, LIX, LX and LXI.

The above assignments of error, forming Issue 4 are a general objection to the final decree of the Lower Court in favor of the defendants and in view of the argument concerning Issues 1, 2 and 3, it is submitted that the final decree of the Lower Court is contrary to law.

CONCLUSION.

It is submitted that defendants-appellees have not made out their defenses; that each and all of the claims of the patent in suit are valid and infringed by the defendants-appellees; and that the decree of the Lower Court should be reversed in favor of the plaintiffs-appellants, as prayed in the Bill of Complaint herein.

Respectfully submitted,

ALAN FRANKLIN,
Attorney for Plaintiffs-Appellants.

No. 8171.

𝔍𝔫 𝔱𝔥𝔢 𝔘𝔫𝔦𝔱𝔢𝔡 𝔖𝔱𝔞𝔱𝔢𝔰
Circuit Court of Appeals
For the Ninth Circuit.

Patrick H. Bray and Catherine Patricia Marquette,

Appellants,

vs.

Hofco Pump Ltd., and D. W. Hoferer,

Appellees.

REPLY BRIEF OF APPELLEES.

BURKE & HERRON,
By MARK L. HERRON,
JOE C. BURKE,
Chapman Bldg., 8th and Broadway, Los Angeles,
Attorneys for Appellees.

FILED

OCT 15 193

Parker, Stone & Baird Co., Law Printers, Los Angeles.

PAUL P. O'BRIEN,
CLERK

SUBJECT INDEX.

Part A.

iv.

PART B.

THE BRADFORD DEPOSITIONS.

TABLE OF AUTHORITIES CITED.

No. 8171.

In the United States
Circuit Court of Appeals
For the Ninth Circuit.

Patrick H. Bray and Catherine Patricia Marquette,

Appellants,

vs.

Hofco Pump Ltd., and D. W. Hoferer,

Appellees.

REPLY BRIEF OF APPELLEES.

Foreword.

Appellants bring this appeal from a final decree of the United States District Court in and for the Southern District of California, in favor of defendants (appellees) adjudging claim 1 of United States Letters Patent No. 1,840,432, issued to one P. H. Bray, null and void for want of invention over the prior art in evidence, and adjudging claims 2 and 3 thereof good and valid in law, but limited in scope, and not infringed by the defendants.

THE ASSIGNMENTS OF ERROR.

1. Assignments I, II, III, IV, V, VI, VII, VIII, IX, X, XI, XII, XIII, XIV, XV, XVI, XVII, XVIII, XIX, XX, XXI, XXII, XXIII, XXIV, and XXV [Tr. pp. 389-396] assign error to the trial court in "finding" or in "failing to find" certain facts in its "Memorandum of Conclusions and Order for Judgment." [Tr. pp. 374-378.]

This document was *not* (see note below) adopted by the district judge as his findings of fact and conclusions of law. On the contrary, "Findings of Fact and Conclusions of Law" were duly proposed by counsel for defendants (appellees) and adopted and signed by the trial judge. [Tr. pp. 379-385.] Under these circumstances the "Memorandum of Conclusions and Order for Judgment" *is not a part of the record.*

> *Parker et al. v. St. Sure,* 53 Fed. (2d) 706, at p. 709;
>
> *National Reserve Insurance Co. v. Scudder,* 71 Fed. (2d) 884, 888.

And error is not assignable to such opinion of the court.

> *Stoody Co. v. Mills Alloys,* 67 Fed. (2d) 807, at 809.

All italics throughout this brief, except those appearing within quoted matters are ours unless otherwise noted. All italics appearing within quoted matters, which are followed by a star, are also ours.

2. Assignments XXXV, XL, LI, LX and LXI assign error to the trial court in failing to find certain facts. Plaintiffs (appellants) neither proposed findings of fact, nor requested any special findings of fact whatever, and necessarily failed to make or preserve any record of exception. In consequence, a review of this court may not extend to an examination of the evidence to determine whether the court so erred.

> *Chas. H. Lilly Co. v. I. F. Laucks, Inc.,* 68 Fed. (2d) 175, at 178, 179;

> *St. Paul Fire and Marine Insurance Company v. Tire Clearing House,* 58 Fed. (2d) 610, 613.

3. Assignments XXXVI, XLI and XLIV assign error to the trial court in "failing to heed or apply," or in "incorrectly applying" certain asserted principles of law. Plaintiffs (appellants) did not ask the trial court for any special declarations of law; wholly failed to propose such findings as they deemed proper in the light of these asserted principles; and have failed to make or preserve any record of exception.

The review of this court may not, therefore, extend to a review of the errors purportedly raised by these assignments.

> *Chas. H. Lilly Co. v. I. F. Laucks, Inc.,* 68 Fed. (2d) 175, at 178-179.

THE SPECIFICATIONS OF ERROR.

Specification 1.

As their first specification, appellants complain that the trial court erred in:

> "Decreeing claim 1 of the patent in suit null and void and not infringed and in not decreeing said claim valid and infringed by defendants." (Opening Brief p. 31.)

Appellants list by number, assignments IX, X, XI, XII, XIII, XIV, XIX, XXII, XXVII, XXVIII, XXIX, XXX, XXXI, XXXII, XXXIII, XXXIV, XXXV, XXXVI, XXXVII, XLI, LI, LII, LV and LX, and thus seemingly imply that they support the specification. We urge that this wholesale listing is not a compliance with Rule 24 of this court, requiring that *"the specification shall state as particularly as may be in what the decree is alleged to be erroneous,* and requiring appellants *to specify from the errors assigned in the court below,*—which in this case are particularly numerous,— *those upon which they will rely for reversal.*

> *Harrow Taylor Butter Co. v. Crooks, etc.* (C. C. A. 8), 41 Fed. (2d) 627, at p. 627.

This court "has definitely announced that Rule 24 must be strictly complied with, particularly stressing the necessity of *setting out in the brief* the specifications of error *relied upon." O'Brien's Manual of Federal Appellate Procedure,* 1936 Cum. Supp., p. 109, citing:

> *Gelberg, etc. v. Richardson, etc.,* 82 Fed. (2d) 314;
>
> *Gripton v. Richardson,* 82 Fed. (2d) 313;
>
> *Berry v. Earling, etc.,* 82 Fed. (2d) 317.

In *Barnett v. United States,* 82 Fed. (2d) 765, at 767, the court said:

> "Each specification of error in the brief should state the particular ruling claimed to be erroneous and the assignment or assignments upon which the specification is based."

An examination of the assignments so referred to by number discloses:

a. Assignments IX to XIV, inclusive, XIX and XXII, refer to alleged errors of commission or omission in the "Memorandum of Conclusions and Order for Judgment" and are nugatory for the reasons set out in subdivision "1", page 4 hereof.

b. Assignment XXXV assigns error to the trial court in "failing to find that said claim 1 of the patent in suit is not met in terms by any prior art structure, and that it cannot be properly anticipated by finding one similarity in one structure and another in another and still another in a third to build up the combination of the claims or any of them." We urge for the reasons set out in paragraph 2, page 5 hereof, that this assignment is nugatory.

c. Assignment XXXVI assigns error to the trial court "in failing in Findings of Fact, paragraph IX, to heed or apply the law to the effect that it is conclusively presumed that each element of the claimed combination is old whether in fact or not and that the only novelty in invention, under the law, resides in the combination as claimed." We urge that the assignment is nugatory for the reasons set out in paragraph 3, page 5 hereof.

d. Assignment of error XLI assigns error to the trial court "In assuming in such findings of fact as those of paragraphs IX and X that a proper or possible interpretation of a patent requires a combing through the art to discover what if any feature or element is novel, or asserted to be novel, and then substituting such alleged features of novelty in place of the claim as written for the purpose of passing upon questions of validity and infringement; and in failing to treat the claim as an indivisible unit in which each element is equally important for the purpose of interpretation." We urge this assignment is nugatory for the same reason as is assignment XXXVI immediately preceding.

e. We believe that assignments XXVII, XXVIII, XXIX, XXX, XXXI, XXXII, XXXIII, XXXIV, XXXVII, LI, LII, LV and LX tend to support the specification.

Specification 2.

As their second specification of error, appellants assert :

"That the lower court erred in strictly construing or misconstruing the claims, and particularly claims 2 and 3 of the patent in suit, whereby the scope of the claims was unduly limited to such an extent as to read out of the claims the defendants' infringing pump, and thereby enable the defendants unjustly to escape infringement of the patent in suit." (Appellants, Brief p. 29.)

Appellants, failing to *set out* the assignments upon which the specification is based, content themselves with referring to them by number, as follows: I, II, III, IV,

V, VI, VII, VIII, XI, XIII, XIV, XV, XVI, XVII,
XVIII, XIX, XX, XXI, XXIII, XXIV, XXIX, XXX,
XXXI, XXXII, XXXVI, XXXVIII, XXXIX, XL,
XLI, XLII, XLIII, XLIV, XLV, XLVI, XLVII,
XLVIII, L, LIII and LVI.

An examination of them discloses that:

a. Assignments I to XXIV, inclusive, refer to alleged
errors of commission or omission in the "Memorandum of Conclusions and Order for Judgment," and are nugatory for the reasons set out
in subdivision "1", page 4 hereinabove.

b. Assignment XXIX assigns as error the finding that
"the locking means between the pump barrel and
standing column base * * * does not in either
plaintiffs' or defendants' pumps, effect any functional relationship or mode of operation of the
parts different from the pumps found in the patented and unpatented prior art in evidence." This
assignment is directed to alleged error in the court's
finding as to the function of locking means both of
the patent and of the art; the specification to alleged
error in construction of claims. This assignment
clearly fails to support the specification.

c. Assignment of error XXX assigns as error the
finding that "the practice of using notched contact means between a pump barrel and the anchor
for the column or the column fixture was well
known in the patented and unpatented prior art
in evidence." This assignment is directed to alleged error in finding the state of the art; the
specification to alleged error in construing claims.
It fails to support the specification.

d. Assignment XXXI assigns as error the finding that "it would require merely the mechanical knowledge of any driller or pump operator skilled in the art to adopt such notched contact means to the designated use." This finding is not directed to alleged error in the construction of claims and therefore fails to support the specification.

e. Assignment XXXII assigns as error the finding that "the notched contact means * * * had been adapted by others to the designated use before patentee's invention of the subject matter in suit * * *" and thus fails to support the specification for the same reason.

f. Assignment XXXVI assigns error to the trial court in "failing * * * to heed or apply the law to the effect that it is conclusively presumed that each element of the claimed combination is old whether in fact or not and that the only novelty in invention, under the law, resides in the combination as claimed." As noted in paragraph 3, page 5, *supra*), the court was not asked to make any special declaration of law whatsoever. It cannot be ascertained moreover whether the court did or did not give heed to such asserted principle, and in any event the assignment does not support the specification.

g. Assignment XXXVIII assigns error to the trial court in finding "an alleged essence of invention, and in substance in *substituting such alleged essence of invention* for the claim as written." It is doubtful if this assignment supports the specification which is "that the court erred in *construing* or *misconstruing the claims* * * * whereby the scope was unduly limited," etc. (Italics ours.)

h. Assignment XXXIX assigns as error the finding that "the essence of invention claimed in claims 2 and 3 of the patent in suit resides in the arrangement for expelling fluid from the barrel, and in effect in *substituting this alleged essence of invention for the combination of the claims as written."** Again this asserted error of *substitution* fails to support the specification that the court erred in *construing* the claims.

i. Assignment XL assigns error to the court in failing to * * * recognize and apply the law that each element of a claim is conclusively presumed to be old and there is no such thing as a purported element, and that the issue is not whether any one or more or any principal element or feature is new or old but whether the combination of the claim as written is novel." We again note that the court was not asked for any such specific declaration of law (see paragraph 3, page 5 hereof); note that it cannot be ascertained whether the court did or did not recognize or apply this law; and likewise note that this assignment fails to support the specification that the court erred in construing and unjustifiably limiting the claims.

j. Assignment XLI assigns error to the trial court in "assuming in such Findings of Fact as those of paragraphs IX and X that a proper or permissible interpretation of a patent requires a combing through the art to discover what if any feature or element is novel, or asserted to be novel, and then substituting such alleged feature of novelty in place of the claim as written for the purpose of passing upon questions of validity and infringement; and in failing to treat the claim as an indivisible unit in which each element is equally important for the purpose of interpretation." Again the error assigned is that the court *substituted "alleged features of novelty for the claim as written"* while the specification is that the court erred in *strictly construing or misconstruing the claim* and thereby unduly limiting it. Again we urge the assignment fails to support the specification.

k. Specification of error XLV assigns error to the trial court "in such findings as that of paragraph XI of the Findings of Fact in effect substituting parts of the specification of the patent in suit for the claims as granted." This specification of error is specifically directed only to paragraph XI of the findings of fact in which the court finds in effect that the patentee contemplated that the guide should have a close or ordinary mechanical fit with the standing column, and that he intended to obtain the maximum of jet action to wash the coupler face of the column fixture, and that the specification so states. This finding does of itself purport to construe or limit the claims, and the assignment therefore fails to support the specification.

l. We believe that assignments XLII, XLIII, XLIV, XLV, XLVI, XLVII, XLVIII, L, LIII and LVI support the specification.

Specification 3.

As their third specification of error, appellants assert:

"* * * The lower court erred in decreeing that the defendants and each of them have not infringed any of the claims of the Letters Patent in suit and have not infringed said Letters Patent." (Opening Brief p. 30.)

Again appellants fail to set out the specific assignments of error urged to support the specification, and again merely make wholesale reference to them by number.

   a. An examination of those listed discloses that assignments XVII, XXII, XXIII, XXV, refer to alleged errors in the "Memorandum of Conclusions and Order for Judgment" and are nugatory.

   b. It is believed that assignments XVIII, XXI, XXVI, XXXVII, XXXVIII, XLVIII, XLIX, L, LI, LIV, LVI and LVII tend to support the specification.

Specification 4.

As their fourth specification of error appellants assert:
"* * * the lower court erred in entering a decree in favor of the defendants (appellees) instead of in favor of the plaintiffs." (Opening Brief p. 30.)

Again appellants merely list the specifications of error relied upon, and an examination of them discloses:

   a. Assignment of error XX assigns error to the trial court "in failing, contrary to the law, to construe the claims in suit and each of them in accordance with their respective terms and in implying into said claims details and incidents of the specification which specification described and illustrated but one particular embodiment." It is submitted that this assignment is so general as to be utterly insufficient.

   b. Assignment XXV assigns error to the trial court in its "Memorandum of Conclusions and Order of Judgment," which is not a part of the record.

   c. Assignments LVIII, LIX, LX and LXI are believed to support the specification.

MATTERS CONCEDED.

In the interest of brevity, appellees concede that the matters appearing under the headings "Statement of Case", "Statement of Fact", and "Amended Answer", appearing on pages 4 to 6 inclusive of appellant's brief are substantially correct. We take issue, however, with many of the asserted statements of fact beginning on page 7 of the brief, under the heading "The Patent in Suit."

The Patent in Suit and the Error of Appellants in Their Analysis Thereof.

1. The patent in suit is generally described in the specification of the patent as follows:

> "This invention relates to pumps and more especially to pumps of the class including a standing valve mounted on a standing, fixed column (or stem) provided with a packing means on which is reciprocated a travelling pump barrel carrying a head or travelling valve." (Spec. LL 1-7, Bk. Ex. p. 3.)

2. The patent specification makes it plain that *this class* of pumps was old in the art, and that it was recognized at the time of the filing of the patent in suit, that:

> "A feature of *this class of pumps** is that the travelling barrel can be utilized as a means of pulling the standing column from a retaining lock when it is desired to remove the pump as a whole from the well without pulling the tubing." (Spec. LL 8-13, Bk. Ex. p. 3.)

*Italics ours.

3. The patent specification further recites:

> "*A common defect of this class of pump is that sand settles on the locking means and freezes the parts to the tubing** so that the pump cannot be pulled without great risk of serious injury of parts and possible loss of the well by reason of blocking of the hole when it is impossible to fish out the obstructing parts below the break. *For this reason one object of the invention is to maintain the sand in such a constant state of agitation that it cannot pack down on the lock shoe box* and lock shoe pin.**" (Spec. LL 14-25, Bk. Ex. p. 3.)

4. The patent specification then states generally the means devised by the inventor to overcome this asserted defect of the class of pumps mentioned, in the following language:

> "To that end means are provided to cause a *positive jet action of liquid over the lock means** and keep the sand in a state of suspension so that it will be carried off in the discharge, and therefore prevent 'sanding up'." (Spec. LL 26-30, Bk. Ex. p. 3.)

5. Other objects of the invention are stated to be:

> "Another object is to provide means for positively interlocking the travelling barrel to the standing lock means to enable the screwing or unscrewing of parts of the rod string in event of need." (Spec. LL 31-35, Bk. Ex. p. 3.)

6. and:

> "An additional object is to provide a pump of such structure and design that the travelling barrel can

---

*Italics ours.

be reversed end for end after a period of use so that its useful life may be greatly extended." (Spec. LL 36-40, Bk. Ex. p. 3.)

It should be noted that such a reversible barrel was old in the art, as was conceded by appellants in the course of prosecution, (File wrapper and contents, Bk. Ex. p. 24) and that the element of a reversible barrel is *not* carried into any of the claims of the patent as an element of the claimed combination. The reference made by appellants to such reversibility, on page 13 of their brief, and elsewhere, merely tends to confuse the issue.

7. The specification refers to and the drawing illustrates certain packing means (LL 90-100, Bk. Ex. p. 3) and to their method of mounting (LL 1-24, Bk. Ex. p. 4), but the original claims, 7 to 14 inclusive and revised claims 1, 2 and 3 covering such features (Bk. Ex. pp. 14 to 15 and 20 to 21) were cancelled, (File wrapper, Bk. Ex. p. 18 and pp. 18 and 21) and none of these features are carried into the claims as elements of the combination of the patent.

Counsel for appellants moreover concede that the "plunger packing" described in the specification is:

"* * * merely illustrative for the purpose of disclosing a complete and operative pump of the type which the invention embodies, *the patent not being limited in any manner to any particular plunger-packing,* since any suitable form of plunger-packing is comprehended by the claims of the patent." (Opening Brief p. 12.)

---

*Italics ours.

8. It will thus be seen that although the only objects of the invention mentioned in the specification are:

a. To maintain the sand in a state of agitation, through means causing a positive jet action of liquid over the lock means. (Par. 3 & 4 above) and,

b. To provide means for positively interlocking the travelling barrel to the standing lock means to enable the screwing or unscrewing of parts of the rod string in case of need. (Par. 5 above.)

c. To provide a reversible barrel. (Not carried into any of the claims allowed.)

d. To provide certain packing means. (Not carried into any of the claims allowed),

Counsel for appellants vitiates much of his brief by predicating it upon the erroneous premise that:

"The patent in suit covers a pump for wells
* * * which * * * will not form a gas lock."
(Opening Brief LL 1-4, p. 7.)

We have carefully reread the patent specification. There is not one word therein referring to "gas lock", nor does any reference thereto, or to a positioning of the valves to minimize it, appear in the File Wrapper and Contents (Bk. Ex. pp. 9-28). It is not claimed that the pump of the patent is usable only in wells containing gas, and Mr. Bray would be the first to deny such a claim. The specification makes no reference to the distance at which the standing and working valves shall be mounted with reference to each other. The only mention of valves found in the specification is:

> "The invention *relates to the class of pumps** * * * including: A standing valve *mounted on a standing, fixed column (or stem),** provided with a packing means on which is reciprocated a travelling pump barrel *carrying a head or travelling valve**" (Spec. LL 3-7, Bk. Ex. p. 3), and

> "The upper end of the barrel is provided with the usual head valve V to the cage of which is connected the usual sucker rod (not shown)." (Spec. LL 82-85, Bk. Ex. p. 3), and

> "Within the barrel is a standing valve S which is screwed *on the upper end of a 'stem' or column 10** whose lower end screws into a pipe joint 11 screwed into the Pin P." (Spec. LL 86-90, Bk. Ex. p. 3.)

The claims of the patent do not claim any particular location of the valve on the standing column, or of the valve on the travelling barrel, as each claim broadly reads:

> "The combination, with a pump of the class having a standing guide column *with a working valve** and *a valved travelling barrel,** running on said column etc." (Spec. LL 46-49; LL 56-59 and LL 71-74, Bk. Ex. p. 4.)

It is at once apparent that this portion of each claim is in effect a preamble, and constitutes merely a reference ·to *the general type of pump to which the alleged inventive improvement contained in the body of each claim is to be*

---

*The italics are ours.

applied. It is further apparent that the general type of pump so referred to is that type of pump *which employs a standing column and a moving barrel,* both so valved as to be a pump, as distinguished from the type of pump *employing a standard barrel and moving plunger,* likewise valved.

This language cannot be construed to protect any particular location of the valve on the standing column, and Mr. Bray, the patentee, inferentially admitted this fact when he testified upon direct examination in his own behalf :

> "Q. Now, calling your attention to Exhibit 7 (Bk. Ex. p. 53) does this have what you call a valved travelling barrel?
>
> "A. Yes. That valved travelling barrel is marked 5 here. There is a working valve on the standing column. That is No. 16. It is not inside of the barrel but it is on the standing colunm * * *"
> [Tr. p. 260.]

for in that structure the valve "16" was pointed out by Mr. Bray as being near the bottom of the column.

The language of both the specification and the claims is on the contrary such as to indicate that both the applicant and the Examiner recognized that it was old in the art to locate travelling and working valves either close together or far apart, and recognized that every pump in the prior art employing a standing column was of necessity, provided with one or more valves placed in such position on the column as seemed best in the particular design. An examination of the prior art in the record shows:

a. Valves in the bottom portion of the standing column as in:

Downing Ex. "Q" (Bk Ex. p. 108)

Thurston Ex. "S" (Bk. Ex. p. 115)

and in the Bradford structures, Defendants Physical Ex. "12, 13, 14",

b. at the top, as in:

Northrup Ex. "V" (Bk Ex. p. 129)

O'Bannon Ex. "R" (Bk. Ex. p. 111)

Ellis Ex. "W" (Bk. Ex. p. 133)

c. and at both top and bottom, as in:

Thompson Ex. "T" (Bk. Ex. p. 119)

Wright Ex. "V" (Bk. Ex. p. 125)

and in the "inverted Bramo" of Bradford (Bk. Ex. 87, 88, 89, Physical Ex. "17A" and "17B").

The only matter appearing in the Bray patent from which, in our opinion, the erroneous contention of appellants could spring is the fact that in the drawing the standing valve "S" and the working valve "V" (Bk. Ex. p. 2) are illustrated in somewhat close proximity. We are, however, at a loss to understand how counsel for appellant, with any approach to integrity of logic, can contend, as he does on page 10 of his brief, that this incident of the drawing can be said to amount to a teaching that such location of the valves will prevent gas lock, much less carry such location of the values into the claims of the patent as a limitation, when it is not mentioned in either the specification or claims, and in the same breath assert (and we think correctly) that the particular packing means both illustrated in the drawings and described in the specification (See par. *7 supra*) is:

> "Merely illustrative for the purpose of disclosing a complete and operative pump of the type which the invention embodies, the patent not being limited in any manner to any particular plunger-packing, since any suitable form of plunger-packing is comprehended by the claims of the patent." (Opening Brief p. 12.)

We may concede that the "specification is the dictionary of the claims" and that "the entire instrument, including the drawings and specifications, is to be considered in arriving at its intent and meaning". (Op. Br. p. 15.) It does not follow, however, that features of design *disclosed in the specification or drawing* but not covered by the claims are "necessarily" to be read into them. (Opening Brief p. 15.) The law is to the contrary, and is that elements so disclosed and not claimed are dedicated to the public.

> *McClain v. Ortmayer,* 141 U. S. 419; 36 L. Ed. 801 at 802;
>
> *Gladding-McBean v. N. Clark & Sons,* (C. C. A. 9), 16 F. (2d) 50 at 51;

Walker on Patents, 6th Ed., par. 219, page 293, at
294, citing the following cases:

> *O. H. Jewell Filter Co. v. Jackson,* 140 F. R.
> 340, 1905;
>
> *Ball & Roller Bearing Co. v. Sanford Mfg.
> Co.,* 297 F. R. 163, 168, C. C. A., 2d Cir.;
>
> *Rip Van Winkle Wall Bed Co. v. Murphy
> Wall Bed Co.,* 1 F. (2d) 673, C. C. A.,
> 9th Cir.

Finally, each of the claims of the patent is a combination claim. There can be no possible cooperation of action between:

a. The elements of the claimed combination, designed
   to lock the barrel to the anchor, and to provide
   the more certain operation of such locking means
   by the provision of jet action to wash the sand
   from the anchor coupler face, and

b. The positioning of the working and standing valves
   of the pump in close proximity, in order to minimize gas "lock".

To have claimed such location of valves, for the function now claimed, in combination with the elements claimed
in the patent, would have rendered the claims invalid as
claims on an *aggregation* rather than a *combination.*

> *Hailes & Treadwell v. Van Wormer et al.,* 20 Wall.
> 353; 22 L. Ed. 241 at 248;
>
> *Walker on Patents,* 6th Ed., par. 70, page 84, and
> cases there cited.

This for the reason that the "clutch" and "jet" would
operate to perform their function, quite irrespective of

whether the valves were so placed as to diminish or increase the tendency toward gas lock, in the event that there happened to be gas in the well.

We have carefully searched the opening brief to see if appellants have anywhere hinted at any such cooperation. They have not done so. We find that on the contrary, they seek to justify their claim that the Court should read these elements into claims, upon the somewhat naive ground that:

> "the close relationship of the valves 'S' and 'V' *must necessarily be read into** claim 1 of the patent * * * for the further reason that in the oil fields in California where loose sand is present, gas under great pressure, is also present, and a pump designed to avoid 'sanding up' in the oil wells of this state must necessarily be designed to avoid 'a gas lock'; otherwise the pump would not be practically operative and would lack utility when used in the oil fields of this state or similar oil fields." (Opening Brief pp. 15-16.)

thereby confessing that they realize that a design to avoid "sanding up" and a design to avoid "gas lock" are separate and distinct features of design, wholly independent of each other; and further confessing that if it is desirable to incorporate both such independent features of design in one structure, such incorporation is desirable not because any cooperation of design is claimed as between them, but merely because a pump to possess utility in California oil fields, should in their opinion have both such features.

It seems to us an original, novel and erroneous theory to assert that because a pump has been "used" largely in California where it is assumed gas is present, elements not covered in the claims should be read into them, in order to make them read upon a structure adapted to the needs of the oil wells of this state. By the same token, had the pump been used largely in Pennsylvania, where appellants assume gas is not to be found, such elements should not be read into the claims. Such a theory of construction is palpably absurd.

While the statements of Mr. Bray relating to gas lock and its action [Tr. pp. 34, 241, 242, 243, 244 and 245] are far from clear and the term is in fact a misnomer, we believe it is plain in the light of the statement of appellants' counsel appearing on page 33 of their brief, that "gas lock" occurs when a substantial volume of gas or compressible fluid accumulates in the compression chamber of a pump, thus creating a condition under which the pressure in the pump chamber does not exceed the pressure of the liquid column against which the pump is operating, thereby causing the pump to, in effect, bounce on the resilient body of compressible fluid, and thus FAIL to completely drop.

Gas lock does not occur unless or until a substantial volume of gas accumulates in the compression chamber. The amount of gas which will accumulate in a compression chamber depends upon the shape or formation of the pump chamber and upon the stroke of the pump. It is recognized as desirable to form the pump chamber so that when the pump is moved to the extreme contracted position, the volume of the pump chamber is reduced to a minimum. This is obviously best accomplished if the valves

of the pump are so located as to come relatively close together, but it is equally important that the pump chamber
shall be free of extended cavities or pockets in which gas
can lodge. On the other hand, no matter how the valves are
located or how the pump chamber is constructed, to eliminate gas lock it is necessary to operate the pump to the
limit of its movement in the collapsed or contracted position. It therefore follows that the elimination of gas lock
in a particular pump or pump installation depends not only
upon the inherent structure of the pump, the arrangement
of valves and the general formation of the chamber, but
upon the manner in which the pump is operated.

It is further to be noted that Mr. Bray admitted that
it was common practice in a pump, employing valves
somewhat widely spaced in a gaseous well, to avoid any
possible gas lock through the use of an anchor below the
pump to make a separation of the gas at that point,
and thus prevent it coming through the pump. His statement to this effect is:

> "* * * with the ball on the bottom which they
> generally use in one of these, in this part here, the
> ball in there—that is, the bottom and this is the top—
> and another ball up here, makes a strong space for a
> gas lock; and in all California wells more or less gas;
> *you can't pump one of them that I know of success-*
> *fully without what they call an anchor below to make*
> *a separation of the gas, not have it come through the*
> *pump.*"* [Tr. p. 241.]

It is therefore apparent that such pumps can be successfully operated in the oil fields of California, where gas in

large quantities is present, merely through the use of such a common and well known gas anchor,—the statement of the appellants' counsel to the contrary notwithstanding.

In his attempt to avoid the force of certain of the prior patents and structures in the art, disclosed by the record, counsel for appellants attempts, by argument, to establish that these structures are inoperative because of what he terms gas lock and hydraulic lock. An example of such an attempt appears on pages 51 and 52 where, in discussing the Downing patent (Exhibits Q and Q', Bk. Ex. page 108), counsel asserts that the pump would be inoperative because gas lock might occur between the valves and because liquid might get into the space between the head on the standing column and the puller nut H. An examination of that structure will disclose that if appellants' theory of gas lock were sound, a gas pocket would quite as readily occur above the puller nut as between the valves of the pump. This is true because the space above the puller nut is a chamber which alternately contracts and expands. Gas so accumulating would render the "hydraulic lock" of counsel's imagining impossible. On the other hand, if a body of liquid totally free from gas accumulated in the chamber above the puller nut H (corresponding to the jet pressure chamber EE of the Bray patent) the liquid in the chamber between the valves will likewise be free from gas and hence free of "gas lock." We submit that in practice, however, neither extreme will be encounter in the average case. Some gas may accumulate in both chambers, and the efficiency of the pump be thereby somewhat lessened, but that the pump would be operative under average conditions which is all that is required to make it operative in point of law.

We have considered this error at some length, as the assumption that each claim is to be read as containing, as one of its elements, working and standing valves positioned in close proximity to each other for the purpose of minimizing or preventing gas lock, runs throughout appellants' brief, and *is made the sole ground of distinction between various prior art structures and the structure of the Bray patent.*

9. Another incorrect premise, upon which much of appellants' argument is based, results from the fact that appellants have incorrectly analyzed the action of the Bray pump. They state that the principal advantages of the pump are:

> "The *standing plunger* and *traveling barrel* enables the pump to lift the oil up through the pump tubing T on the *down* stroke of the barrel B under the influence of the *weight* of the string of sucker rods connected to the upper end of the barrel, instead of on the *up* stroke, by the power of the engine, as in the conventional pump with a standing or stationary barrel and traveling plunger. The *work* of appellants' pump in lifting the oil is done by the force of *gravity,* instead of by the power of the engine. The force of gravity under the weight of the sucker rods is amply sufficient to overcome the weight of the oil in the tubing T, to lift the oil therein, because the weight of the rods in a 5,000 foot well is approximately 15,000 pounds. [Tr. p. 47.] The greatest work of appellants' pump—that of lifting the oil through the tubing T to the surface of the earth— being done by gravity, great *economy in power* is effected. The only work done by *power* in appel-

lants' pump is on the *up* stroke of the traveling barrel B in lifting the sucker rods and traveling barrel, to lift enough oil from the bottom of the well through the standing column 11 and 10 to fill the pump chamber in the barrel between the standing valve S and the traveling valve V." (Opening Brief pp. 18-19.)

Analyzing the statement we find that appellants believe that:

a. The pump lifts the oil up through the pump tubing "T" on the *down* stroke of the barrel.

b. This lifting is done by the weight of the string of sucker rods connected to the upper end of the barrel, (*i. e.,* on the down stroke the weight of the sucker rods rests on the upper end of the pump barrel).

c. The force of gravity rather than the power of the engine does the work, and economy of power results, *i. e.,* the force of gravity acting upon the sucker rods overcomes the weight of the oil in the tubing and lifts the oil therein.

d. The only work done by power is on the *up* stroke of the traveling barrel, in lifting oil from the bottom of the well through the standing column 11 and 10 to fill the pumping chamber.

10. In order that the Court may correctly understand the action of the pump in suit, we respectfully ask that the Court turn to Ex. MM, page 178 of the Book of Exhibits. This is an enlargement of the drawing of the patent, with the standing parts of the pump colored blue, the moving parts red and the valves green. [Tr. p. 87.]

It should be noted *that the action* of this pump *is the same* as the action of all pumps of the class:

> "including a standing valve mounted on a standing fixed column (or stem) provided with a packing means on which is reciprocated a traveling pump barrel carrying a head or traveling pump valve." (Spec., lines 1-7, Bk. Ex. p. 3.)

and that Mr. Bray was not the original discoverer or inventor of this *class* of pumps.

All such pumps are mounted in the tubing which extends from the surface to the lock shoe Box "L" in the well. (Part "L" in the drawing of the patent.) The anchor "P" is seated in Box "L", below the lower end of the traveling barrel "B"; the standing column "10", "11" is connected with the upper end of the anchor or pin "P", and fitted with appropriate packing means "AA", "BB", "CC" and carries a standing valve "S" at some appropriate place thereon. The traveling barrel "B" carrying a traveling valve (marked "V" in the drawing of the patent) is fitted over the standing column and slides up and down on the packing.

The action of the pump is as follows:

a. When the space between the two pump valves, *i. e.,* inlet valve "S" and discharge valve "V", is increased by the upward movement of the traveling barrel, fluid to fill this increased volume will flow into the pump from the oil bearing formation through the bottom or inlet valve "S". As the barrel is lifted the volume of the pumping chamber within the pump is increased, and the pump as a whole elongates, and thus occupies a greater volume within the tubing.

b. The pump is mounted at the lower end of the tubing, and the tubing thereby closed by the lock Box "L" and the anchor "P" to the egress of any fluid, and to the ingress of any except through the inlet valve "S". Fluid, therefore, fills every part of the tubing not occupied by the pump and sucker rods. When the volume occupied by the pump is increased by the upward movement of the barrel, the capacity of the tubing to hold fluid is decreased in the same amount, and the fluid of necessity overflows at the top of the well. In exactly the same way, a level full glass of water will overflow if an object submerged in it is increased in size or volume thereby displacing more water.

The force causing the expansion of the pump, and the consequent lift or overflow of the fluid at the surface, is the *tension* on the sucker rods which, on the *up* stroke, *lifts* the barrel against the pressure of the fluid above it.

c. When the barrel has reached the upper limit of its stroke, it comes to rest, and as the volume of the pumping chamber is no longer being increased, gravity closes the inlet valve "S" and the fluid drawn from the well below the bottom of the tubing is thus trapped *within* the pump. The volume of fluid so trapped is equal to the increase in volume occupied by the pump as its barrel is lifted and is therefore equal to the amount of fluid which has been discharged by overflow from the top of the tubing.

d. The barrel is now allowed to descend. The tubing is full of liquid. The interior of the pump is also filled. The inlet valve is closed. The traveling or outlet valve on the barrel opens and there is free communication between the fluid in the pumping chamber and in the well tubing. The pump barrel and sucker rods sink down through the fluid in the tubing. The result is that the fluid trapped within the pumping chamber, passes from one side of the outlet valve to the other, *i. e.,* from the pumping chamber within the pump to the tubing; but, as the pump and all of its parts are filled with, and submerged in, fluid, the barrel and sucker rods sink down *without change in the total volume of the fluid in the tubing,* just as a pebble submerged in a full glass of water may sink deeper, or be otherwise changed in position therein, without causing an overflow.

It is thus apparent that appellants are mistaken in their understanding of the *operation of the pump of the patent in suit.* The fluid is *not* pumped by gravity, *i. e.,* by the downward pressure exerted by the weight of the sucker rods, but rather by the tension exerted on them by the engine, and the asserted "economy in power" is not effected.

11. Appellants, moreover, fail to comprehend that the uncontradicted evidence in the record is that the weight of the sucker rods does *not rest* on the traveling barrel of the pump. Mr. Adams testified:

"* * * Let us consider the pump as standing in equilibrium so that we cut out the question of the inertia which comes in with the motion, and we will say that the plunger barrel has been lifted appreciably, say a foot or a foot and a half, so that the entire weight of the barrel, column of the fluid and the sucker rods clear up to the polish rod all hang on the walking beam and *they hang in equilirium at that time.** Now, then, the sucker rods are stretched; they are a spring, just as much as if they were a coiled spring; they stretch and that stretch is governed by the same laws that govern the stretch of a spring, one being tension and other being torsion. *It stands in equilibrium and is theoretically affected by any resistance that you put at the bottom, despite all that tremendous weight up there.** If you should increase that by an amount that you could lift with your finger, theoretically, not, practically, you have by that much lessened the tension of that spring. Now, if you will just come to what we know practically, we know that any resistance down here shortens the stroke of the pump. I referred to fluid packed pumps which had an annular space which was not vented and said those pumps did not make their full stroke, but only made a portion of a stroke. Suppose this pump was sanded up, and suppose you got enough resistance in here so that the plunger is sanded up, it is well known that the sucker rods simply buckle in the tubing and that the plunger would remain practically stationary. Therefore, between the touching that would theoretically resist the free dropping in a certain amount, and the actual stopping of the pump by sand, you have all the degrees of resistance that lie between the two, and one of them is that degree which will hinder the free drop of the pump and will shorten the stroke. Now, does that answer?

Q Then, as I understand it, your testimony to this point has been that, despite the weight of the rods, and the counterbalancing of the engine, that any friction down here would have some effect?

A I testified that any friction down here would shorten the stroke of the pump despite the 5,000 or 10,000 feet of rods going down. Any friction down there would shorten the stroke by the amount of that friction, that is, be proportional to it.

Q If the strength of this friction, whatever the proper term for it is, was less than the force of the 5,000 feet of rods, which would win?

A *The 5,000 feet of rods are not on the pump* * * * *They* * * * *are hung up at the top and you are just touching the pump.** If you have a weight down at the pump hung by a spring, the total weight carried by the spring is up here on the sucker rod a mile in the air; that weight is there. *That weight is up where you take hold of the walking beam,** up 4500 feet or a mile away from where you have the pump. We are now talking about friction down at the bottom of the well. Any friction there would reduce the tension of the spring. The tension of the sucker rod is the spring which it has and any friction there would reduce the tension of the spring. If you reduce the tension of a spring, as a rubber band, why, it slaps back and by that much you shorten the stroke.

Q By how much?

A I wouldn't give a figure for it, wouldn't begin to." [Tr. pp. 228 to 230.]

and it therefore follows, that the fluid could not be lifted:

"under the influence of the *weight* of the string of sucker rods connected to the upper end of the barrel" (Opening Brief p. 18.)

even if the pumping action, as appellants have erroneously asserted occurred on the *down stroke* of the barrel.

12. On page 11 of their brief, appellants point out:

"In pumps of the type disclosed in the patent in suit, with a standing plunger and traveling barrel, the function of the part 3, or 'puller nut,' is ordinarily to engage the lower end of the lowermost plunger guide 13, for pulling the anchor P out of its seat in the lock box L, and pulling the pump out of the pump tubing T for repairing the pump or cleaning sand out of the tubing, when the pump is not too badly 'sanded up' in said tubing."

and on page 19 of their brief, state:

"the pump with the standing column and traveling barrel may be pulled entirely out of the pump tubing T for repairs, etc."

They further state that this so-called "dry job" is of great advantage over a wet job, *i. e.,* the repairing of a pump which cannot be pulled, save by pulling the tubing from the well.

This is true. The advantage, however, is an *advantage of the class of pumps* employing a standing column and moving barrel, as distinguished from *the class* employing a fixed barrel and moving plunger. It is not an advantage peculiar to the pump of the patent but is a well known advantage of that class of pumps which Bray assertedly improved.

13. On page 19 of their brief appellants assert:

"The washing of the sand from the top of the anchor P and lower clutch member 2 *by the downwardly projected jets of oil through jet apertures 5* *from the bottom of the barrel,** prevents the sand from packing up in the bottom of the pump tubing T over the anchor P and clutch member 2, and prevents the pump from 'sanding up'."

Concerning this jetting action and the structure designed to produce it, the trial court in effect found, upon evidence which amply supported the finding (See pp. 90 to 98 *infra*), that the essence of the invention claimed in claims 2 and 3 of the patent (claim 1 does not purport to cover the jet device) resides in the arrangement for producing the jet; that the patentee in order to obtain the maximum jet action taught the use of a guide making a close or ordinary mechanical fit with the standing column in order to generate the necessary pressure; that the patent was specific in teaching that the guide was to be provided with a series of ducts or apertures designed for the ejection of liquid directly toward the anchor and the clutch face thereon; that the alleged infringing pumps did not employ this tight fitting guide, nor was the guide of such structures provided with such ducts or apertures; and that the pumps of defendants (appellees) did not, therefore, infringe upon either claims 2 or 3 which were closely limited to the features named. [Findings IX to XV, Tr. pp. 382 to 385.]

14. That the Court may understand the action of the Bray pump in producing the "jets" of the patent, we ask that the Court turn to Defendants' Exhibit MM, Book of Exhibits, p. 178.

As the barrel "B" descends, the annular space "EE" surrounding the stem "11", between the lower most packing "AA" and the guide or bushing "DD" expands in volume. Fluid therefore flows through the ducts 5 into the space "EE" and fills it. As the barrel ascends, the fluid trapped in this confined space "EE" is ejected through the "ducts" 5, or in the language of the specification:

"* * * the bushing is provided with a series of ducts 5 from top to bottom so that as the barrel slides up from the bottom position, liquid *will be forced down in strong jets** and wash sand from the top of the pin P so that this can be readily pulled from its seat in the lock box L." [Spec., lines 75-81, Book of Exhibits, p. 3.]

15. That a structure designed to provide this strong jet action is the very essence of appellants' asserted invention is made plain by the additional statement in the specification:

"A common defect of this class of pump is that sand settles on the locking means and freezes the parts to the tubing so that the pump cannot be pulled without great risk of serious injury of parts and possible loss of the well by reason of blocking of the hole when it is impossible to fish out the obstructing parts below the break. *For this reason one object of the invention is to maintain the sand in such a constant state of agitation that it cannot pack down on the lock shoe box and lock shoe pin."** (Spec. lines 14-26, Book of Exhibits p. 3); and,

This is in effect confessed by appellants who state in their brief:

"The annular space, designated EE on Defendants' Exhibits MM and NN [Book of Exhibits pp. 178 and 179; Tr. p. 172], in the lower end of the traveling barrel B, surrounding the lower section 11 of the standing column, between the part or jet device 3 and the lower end of the lowermost plunger guide 13, constitutes a *jet pressure-chamber,* whereby pressure is applied by said part 3 to the liquid in said chamber on the upward stroke of the traveling barrel B for forcing liquid out of said chamber through the ducts or jet apertures 5, *in strong downwardly projected jets,* directly against the top of the anchor P and the lower clutch member 2, which washes sand from said anchor and clutch member 2, so that the sand will not pack over said clutch member 2, and over the top of the anchor and seal it in its seat in the lock box L; and the washed sand is thus kept moving upwardly through and out of the upper end of the pump tubing T, so that the sand-free anchor can be readily pulled out of its seat in said lock box." (Opening Brief p. 11.)

16. Despite this admission that the space "EE" is a "jet pressure-chamber," appellants in describing the structure of the Bray patent erroneously state the guide or "jet device" is:

"* * * in the form of a bushing, * * * screw-seated in the lower end of the traveling barrel B *and freely surrounding the lower section 11 of said standing column** so as to move up and down over the standing column with said barrel; * * *" (Opening Brief pp. 8-9.)

This statement is directly contrary to finding XI of the trial Court that:

"The Court finds that the patentee contemplated that the guide used should have a close or ordinary mechanical fit with the standing column and that he intended to obtain the maximum of jet action to accomplish the desired effect of washing off the coupler face of the column fixture, and further finds that the teachings of the patent are specific in terms, as the patentee states 'the bushing is provided with a series of ducts 5 from top to bottom so that as the barrel slides up from bottom position liquid will be forced down in strong jets and wash sand from the top of pin P so that this can be readily pulled from its seat in lock bock L.'" [Tr. p. 383.]

and is obviously incorrect, for it is plain that if the annular space "EE" is to function as a *"jet pressure-chamber"* and result in the production of

"strong downwardly projected jets of oil out of the lower end of the traveling barrel B from the pressure chamber "EE." (Opening Brief p. 23.)

the guide must make a "close or ordinary mechanical fit with the standing column" and thus by efficiently sealing the annular space "EE" to the passage of fluid except through the vertical ducts 5, build up such pressure therein as will do the work which the patentee states is accomplished by his invention.

The specification, the drawings, and claims 2 and 3 of the patent (which alone include the jet device), are *consistent* with the finding of the trial court and totally at variance with the claim of appellants that the guide "freely surrounds" the standing column. (See p. 99 to p. 171 *infra*.) Yet this *basic misconception of fact* is made the predicate for much of the argument that appears in appellants' brief.

17. Appellants further state:

"Appellants made two types of guides or jet device 3 for their pump [Tr. p. 38], one exactly like that disclosed in the patent in suit and specified in claim 2 of the patent, with a *series* of venting ducts or jet apertures 5, and another with an *enlarged central opening* to provide an *annular space or jet* aperture between the standing column section 11 and the wall of said central opening [Plaintiffs' Exhibit 2; Tr. p. 39] in accordance with the terms of claim 3, because said *annular space* is a jet *aperture* within the broad language of claim 3, to-wit: 'said guide being apertured for the ejection of fluid from the barrel,' etc. The latter type of guide, or jet device (part 3) (Plaintiffs' Exhibit 2) was manufactured and sold by plaintiffs-appellants some time before defendants-appellees' infringing pump, with the same type of jet device (Defendants' Exhibits II and NN) was placed on the market. [Tr. pp. 41, 42 and 43.]" (Opening Brief pp. 17-18.)

In this counsel for appellants (who did not try the case below) is mistaken. Mr. Bray's testimony is in many matters vague, discursive, and exceedingly hard to follow, but the record discloses that he testified:

(a) That when he first started making pumps in 1926, he made a guide with a smooth face with "half holes" bored in it forming a sort of rosette-like structure. He produced one of these guides and at defendants' request it was marked Defendants' Exhibit A. [Tr. pp. 40-41.]

(b) That he changed from this guide "because the tube (standing column) running through this guide wore these (the little ribs which bore on the standing column) out" [Tr. p. 42], because there was not "solid foundation enough for" the column to run upon; that he "changed to a solid foundation," that is to the guide of the patent with its series of vertical venting ducts [Tr. p. 42]; that when the ribs of the rosette guide wore out it left a little too much space between the standing column and the inside of the guide, because *to get a jet you have to have a smaller* opening through which the dropped (trapped) oil is passed." [Tr. p. 42.]

(c) That he also made a pump (without ducts or channels), with a clearance of three thirty-seconds (3/32) of an inch between the standing column and the guide. [Tr. p. 37.] Concerning this structure, the court asked:

"You don't plan to have oil go through in that clearance?"

And Mr. Bray answered:

"Yes, a little bit, not much though. You have to run it pretty loose, anyway. You cannot afford to have it too tight, if you do it is not going to work very good."

The Court:

"It is more, therefore, a guide?"

A. "Yes, the clearance is all right for making the hole." [Tr. p. 37.]

It is thus plain that Mr. Bray, first made a guide of the rosette type, found that the ribs wore out, then made a guide with ducts rather than channels, and also made a guide without channels or ducts, but with a clearance of three thirty-seconds (3/32) of an inch between it and the standing column, through which he planned to have "a little bit" of the oil pass.

18. Mr. Bray recognized that the *greater the clearance* between the standing guide column and the guide the *"smaller"* the jet will be. He testified:

"you can put more oil out (of a larger clearance) *but you wouldn't have the pressure to throw it down.** If you have got a jet like that, or a smaller place, you have got to figure on how much oil goes in. *In other words, to get a jet you have to have a smaller opening through which the dropped (trapped) oil is passed."** [Tr. p. 42.]

But admitted *that he did not know, even at the time of trial, the amount of clearance which would result in the failure of the structure to produce a jet, stating:*

"I don't know how great the aperture can be between the standing column of the pump and the guide and yet produce the jet action which I mention in my patent. The only thing I can figure on is my practical experience and my pump, the way it drops, the way the plunger drops free, and I found out that I have no trouble, sand trouble; that my standing valve pulls free without any interference from sand, and I figure I have got about what I wanted. *I think any greater clearance than that indicated on my blueprint would, in the light of my experience, teach me that I wouldn't get a jet action with any greater clearance. I intend to imply by my answer, that I wouldn't get any jet action if the clearance was larger. The larger it is the less pressure you would get."** [Tr. p. 43.]

The evidence discloses that the clearance between the standing column and the puller nut of appellees' structure was 8/32 of an inch, and hence 5/32 of an inch greater than the 3/32 of an inch to which Bray referred. (See pars. 2 and 3, pp. 48-49 *infra*.)

It further discloses that this greater clearance results in the absence of such jet action as Bray taught. (See pars. 11 to 14, pp. 117 to 125 *infra.*)

Counsel for appellants ignores the testimony of the appellant-patentee and this undisputed difference in the amount of such clearance, and blandly speaks of the device of defendants-appellees as being equipped:

"with the same type of jet device." (Opening Brief p. 18.)

This we deny, and on the contrary, assert that the function of the clearance in appellants' pump is wholly different from that of the clearance in the Bray pump mentioned, as is fully developed on pages 120 to 131 *infra.*

19. The statement of appellants that the type of Bray pump employing a guide without ducts or apertures, but provided with clearance between the guide and the standing column :

"* * * *was manufactured and sold** by plaintiffs-appellants *some time before defendants-appellees' infringing pump,* with the same type of jet device (Defendants' Exhibits II and NN) *was placed on the market.** [Tr. pp. 41, 42 and 43.]" (Opening Brief p. 18.)

finds no support in the record. We have carefully examined the pages cited and find that the only testimony remotely bearing on the matter is Mr. Bray's statement:

"I said that prior to my application for patent *I had built** a pump with clearance like Mr. Hoferer's." [Tr. p. 42.]

This is a far cry indeed from the statement of appellants' brief, that this type of pump was manufactured and sold some time before defendants-appellees' infringing pump with the same type of jet, was placed on the market, and from the implied innuendo. We deprecate it.

20. On page 16 of their brief, appellants state:

"The clutch between the lower end of the traveling barrel B and the anchor P, comprising the interlocking members 4 and 2, is a practical construction, for locking the barrel against rotation, so that the joints connecting the sucker rods may be screwed up tight."

This we do not deny. That the court may fully understand its operation we ask the court turn to Ex. MM, Book Ex., p. 178. The specification of the patent states the operation of the clutch to be as follows:

"The pump is shown as assembled in the usual pump tubing T having a suitable shoe lock means L to detachably lock a shoe pin P. A feature of this pin is that it has a clutch or coupler head 2 to be non-rotatively interlocked with a complementary part 3 on the bottom end of a traveling Barrel B. The part 3 is in the form of a bushing having side jaws 4 to mesh with the flat-side head 2 of the pin P thus coupling these parts against rotation when the bushing is lowered to close the coupler parts."
(Spec. lines 62-73, Book Ex. p. 3.)

and its intended function is made clear by the specification:

"Another object is to provide means for positively interlocking the traveling barrel to the standing lock means to enable the screwing or unscrewing of parts of the rod string in case of need." (lines 36-40, Bk. Ex. p. 3.)

We did not in the court below, and do not now, assert that this is not a "practical construction", but urge that the burden is on appellants to establish that the finding IX of the trial court:

> "* * * that said interlocking feature claimed in said claim 1 does not present any novelty or amount to any invention over the prior art in evidence both patented and unpatented, * * *" [Tr. p. 382.]

is unsupported by the evidence.

21. On page 21 of their brief, appellants state:

> "* * * Appellants' novel pump is * * * the first in the art to solve the serious sand-and-gas problem in pumping oil wells in California." [Tr. pp. 279 and 286.]

An examination of the pages cited discloses that all that Mr. Bray asserted was:

> "* * * I found other pumps under the same conditions did sand up and mine did not. * * * We will say Neilsen is one (which did sand up) and O'Bannon pump was another one." [Tr. p. 279.]

Mr. Bray then testified to but *one* instance observed by him where a Neilsen pump stuck [Tr. pp. 279, 281] which occurred some six or seven months before Mr. Bray testified [Tr. p. 288], *i. e.,* long after the filing of the action; and further testified that he didn't know whether the well in which the O'Bannon pump stuck, had been shut down before the sticking occurred. [Tr. p.

281.] The significance of this answer is that if the well had been shut down, the sand in suspension in the oil in the tubing would gradually settle, accumulate about the anchor and stick the idle pump, quite irrespective of its type or character. [Tr. pp. 280-281.]

We have searched the record and find that Mr. Bray did *not* claim that he was first in the art to solve the "serious * * * gas problem of California." Quite to the contrary he testified upon *direct examination* by Mr. Sobieski:

> "Q. Can you state, from your experience in operating a pump in this district, whether in deep well pumps in which gas is encountered, the practice is to use pumps of a design having two valves at the top close together?
>
> A. Well, I *probably** was *one** of the first who started this in this county here. There were two other pumps of that kind. One of them was about a year after I started *and one was* here *about the time I started. That was the Ellis pump.** They worked the same way, that is, they worked on a traveling barrel different designed and different made. [Tr. pp. 243-244.] The first time I saw the Ellis pump may have been in 1926 or 1927. [Tr. p. 246.] I think it was in the Spring of 1927." [Tr. p. 258.]

In the light of the prior art mentioned on page 20 above, and in the light of the fact that there is no evidence in the record that Mr. Bray's pump ever met with wide acceptance in the art, to expand his testimony into the claim:

> "that appellants' novel pump * * * *is the first in the art** to solve the serious sand and gas problem in pumping oil wells in California,"

is in our opinion wholly unjustifiable. While the claim predicated upon this expansion that:

> "the invention in suit (is) in the class of a primary or pioneer invention in an art which the lower court considered to be a crowded one." (Opening Brief p. 23.)

is even more extravagant.

We desire to note in passing that the statement appearing on page 23 of appellants' brief:

> "Before this novel and highly meritorious feature was embodied in a practical and commercially successful pump, by its inventor, the appellant, Bray, there was no pump known, considering all the evidence before this Court, that would not "sand up," in the oil fields of California."

is simply unsupported by their record. It is true that Bray testified to certain instances where he asserted he had replaced pumps which did "sand up" with his pump, and that his pump operated successfully, but there is no testimony to the effect that prior to his advent into the field there was no pump known * * * that would not "sand up" in the oil fields of California.

SUMMARY.

It follows from the foregoing that the statement of appellants appearing on page 21 of their brief:

> "Appellants' novel pump is the first in the art to combine all of the above enumerated advantages in a practical, efficient and commercially successful pump, and consequently the first in the art to solve the serious sand-and-gas problem in pumping oil wells in California." (Opening Brief p. 21.)

is contrary to the findings of the trial court; is predicated upon a misunderstanding of the action of the pump of the patent, upon the erroneous supposition that the weight of the string of sucker rods rests on the travelling barrel, and upon erroneously reading into the claims, the elements of a reversible barrel and valves so located as to minimize "gas lock"; and is further predicated upon a complete disregard of the effect of the greater clearance between the standing column and the puller nut of appellees' structures, and upon a misreading of the record in the other respects pointed out above.

The Pump of Defendants-Appellees.

1. The pumps of defendants-appellees alleged to infringe the Bray Patent are illustrated in Defendants' Ex. II (Bk. Ex. p. 177). It was stipulated at the trial that appellees make the one general type of pump illustrated in Fig. 1; that it is made in various dimensions, some 70 to

80 per cent being of the 2½ inch size, and that this general type of pump is varied only [Tr. p. 166]·

a. In the types of packing employed, which are illustrated by Fig. 2, Fig. 3, Fig. 4 and Fig. 5. (These differences in the types of packing were, as we believe, properly considered by both parties not to be an issue in the Court below. [Tr. p. 166.])

b. In the type of device used for the purpose of screwing up rods. One type is the jaw type of clutch, as illustrated by the parts E and the co-operating part I of Fig. 1, and Fig. 2. The other type of clutch is illustrated in Fig. 4 at "Y" and consists of a flat surface on part "P" (Defts' Physical Ex. JJ) a flat surface on top of Part "H" (Defts' Physical Ex. KK) and a notched ring "Y" resting over the standing column. (Defts' Physical Ex. LL.) [Tr. p. 167.] This ring is made of harder metal than the parts "P" and "H" and its teeth are so designed that it *frictionally locks* the barrel against rotation in *one direction only*. [Tr. p. 36.] Figure 3 is merely intended to indicate the character of the surfaces at "P" and "H" before the ring "Y" is interposed. No finished pumps of this type are made without the ring. [Tr. p. 167.]

2. The dimensions of the so-called 2½ inch pump were stipulated to be typical. [Tr. p. 168.] It was stipulated that the larger pumps differ from this pump only in that their parts are proportionately larger [Tr. p. 51],

and that the dimensions of the significant parts of the 2½ inch pump are:

a. The interior diameter of the nut, part D, Fig. 1 and Fig. 2 was originally one and twelve sixteenths; later following a conference of the code authorities, a tentative diameter of one and eleven sixteenths was agreed upon. A few of that diameter were made, but appellants were making, at the time of the trial, and it was stipulated that they intend to make the interior diameter one and twelve sixteenths, *i. e.*, one and three-quarters inches. [Tr. p. 51.]

b. The outside diameter of the puller nut part D, Fig. 1 and 2, is two and one-quarter inches.

c. The outside diameter of the standing column or puller tube "A" is one and one-half inches.

d. The A. P. I. inside diameter of so-called 2 inch tubing (with which the so-called 2½ inch pump is used) is 2.441 inches. [Tr. p. 168.]

3. It is thus plain that the clearance between the inside of the puller nut in the Hofco structure and the outside of the standing column over which it rides is four-sixteenths, *i. e.*, eight thirty-seconds, of an inch.

THE ARGUMENT OF APPELLANTS AND FALLACIES THEREIN.

Foreword.

On page 31 plaintiffs begin that portion of their brief entitled "Argument." As we understand it, neither the assignments of error nor the specifications of error of the Opening Brief raise the issue that the Findings of Fact do not support the Conclusions of Law and the Decree, but are rather directed to certain findings of fact asserted to be erroneous, which findings it is asserted result in an erroneous decree. (See pages 4 to 14 *supra.*)

Because of the facts that appellants urge only four exceedingly general specifications of error, have listed in support of some one or more of these general specifications *every* assignment of error appearing in their "Assignments of Error" [Tr. pp. 389 to 405], and have not in any manner other than by this wholesale listing, directed their argument to the specific assignments of error upon which they rely, counsel for appellees are in serious doubt as to how fully this court may deem it necessary or proper for us to analyze the evidence, in order to fully meet this inchoate and rambling attack.

We have decided, however, to set out each finding of fact which seems even inferentially to be disputed by appellants, and then to marshall the evidence supporting the finding, noting and answering, such of the arguments of appellants as may be cognate to the evidence marshalled.

The Law.

In following this analysis we respectfully urge that the Court have in mind the principles of law, that:

1. An appeal from a decree in equity is not a trial *de novo*.

    Youngs Rubber Corp., Inc. v. C. I. Lee & Co., Inc., et al. (C. C. A. 2), 45 Fed. (2d) 103, at 106.

2. The findings of the trial court are presumptively correct and will not be disturbed unless a serious mistake of fact appears; and where there is substantial evidence to support the finding of the trial court, it is immaterial that the appellate court might differ with the process of reasoning employed to reach the finding.

    O'Brien Manual of Federal Appellate Procedure, 1936, Cumulative Supplement, p. 59, citing:

    Conqueror Trust Co. v. Fidelity & Deposit Co. of Maryland (C. C. A. 8), 63 F. (2d) 833, 837;

    Woods-Faulkner & Co. v. Michelson (C. C. A. 8), 63 F. (2d) 569, 570;

    Stromberg Motor Devices Co. v. Zenith-Detroit Corp. (C. C. A. 2), 73 Fed. (2d) 62, at 64;

    Sherman et al. v. Bramham et ux. (C. C. A. 4), 78 Fed. (2d) 443, at 444.

3. Findings of the trial court in a suit in equity based on conflicting testimony, taken in open court, will not be disturbed on appeal.

    O'Brien Manual of Federal Appellate Procedure 1936 Cumulative Supplement page 59, citing:

    Clements v. Coppin (C. C. A. 9), 61 Fed. (2d) 552, 557;

McCullough v. Penn. Mutual Life Ins. Co. of Phila.
(C. C. A. 9), 62 Fed. (2d) 831, at 831;

U. S. etc. v. McGowan (C. C. A. 9), 62 Fed. (2d)'
955, at 957;

Collins et al. v. Finley (C. C. A. 9), 65 Fed. (2d)
625, 626.

The Asserted Issues.

I.

On page 30 of their brief, appellants state issue 1 to be:

"Is claim 1 of the patent in suit null and void?"

We believe the issue should properly be restated as:

Did the trial court err in its findings of fact, supporting the Conclusion of Law that claim 1 of the patent in suit is null and void, and in entering its decree to that effect.

1. Examining the issue as so restated it is to be noted that claim 1 of the patent reads:

"What is claimed is:

The combination, with a pump of the class having a standing guide column with a working valve and a valved travelling barrel, running on said column; of a bottom guide, on said barrel, running on said column and having a bottom face forming a coupler, and an anchor for the column presenting directly to said coupler face a top face forming a complementary coupling to be non-rotatively interlocked with the lower guide coupler." (LL 45 to 55, Bk. Ex. p. 4.)

2. Reducing the findings relating to this claim to their elements, it is clear that the trial court found:

Finding VIII [Tr. p. 382]:

A. That the "locking means between the pump barrel and the standing column base whereby the barrel could be lowered until its lower end was in contact with the column fixture and be thereby locked, and which is used in defendants' pumps, does not in either plaintiffs' or defendants' pumps effect any functional relationship or mode of operation of the parts different from pumps found in the patented and unpatented prior art in evidence."

B. That the "practice of using notched contact means between a pump barrel and the anchor for the column or the column fixture was well known in the patented and unpatented prior art in evidence."

C. That "it would require merely the mechanical knowledge of any driller or pump operator skilled in the art to adapt such notched contact means to the designated use."

D. "That it had been so adapted by others before the plaintiff sought to use it." [Tr. p. 382.]

Finding IX [Tr. p. 382]:

E. "That the essence of the invention claimed by Claim I * * * lies in the interlocking feature described in finding VIII hereof."

F. That Claim I "presents no other feature of asserted novelty."

G. "That said interlocking feature claimed in said claim I does not present any novelty or amount to any invention over the prior art in evidence both patented and unpatented."

H. That "the Court finds all of the elements of claim I, * * * in the same or equivalent combinations in the prior art in evidence both patented and unpatented."

3. We will now marshall the evidence supporting each element of these findings in the order in which we have listed them.

A. That the "Locking Means Between the Pump Barrel and the Standing Column Base Whereby the Barrel Could Be Lowered Until Its Lower End Was in Contact With the Column Fixture and Be Thereby Locked, and Which Is Used in Defendants' Pumps Does Not in Either Plaintiffs' or Defendants' Pumps Effect Any Functional Relationship or Mode of Operation of the Parts Different From the Pumps Found in the Patented and Unpatented Art in Evidence."

In order that the court may have a comprehensive understanding of the prior art against the background of which the trial court made this finding, we believe we should note:

1. That as Mr. Hayn, the expert witness for appellants, admitted in one of his few frank and non-evasive statements, as distinguished from the reluctant admissions to which the record shows he was driven, every element of the Bray patent is of itself old in the art. He stated:

"Every element that is found in the Bray patent here has been used time and time again." [Tr. p. 294.]

2. The most cursory examination of the prior art at once establishes:

    A. That pumps of the class having a standing barrel and a moving plunger are legion in the art.

        Examples are: Cummins Ex. O. (Bk. Ex. p. 100); Whitling Ex. P (Bk. Ex. pp. 103-104); Dickens Ex. X (Bk. Ex. p. 137); etc.

    B. That pumps employing a moving barrel and a standing fixed plunger are likewise numerous. It was not, of course, our purpose to burden the record with countless cumulative examples of each such type.

        Examples are: Deis Ex. N (Bk. Ex. p. 96); Downing Ex. A (Bk. Ex. p. 108); O'Bannon Ex. R (Bk. Ex. p. 111); Thurston Ex. S (Bk. Ex. p. 115); etc.

    C. It is obvious from a glance at the exemplars of each class introduced, that since pumping action depends upon alternate contraction and expansion of a pumping chamber, suitably provided with valves, that this pumping action may be produced by either moving the plunger with relation to the barrel, the barrel with relation to the plunger, or both with relation to each other. [Tr. pp. 96-97.]

3. We would, therefore, expect to find patents of which Thompson, Ex. "T" (Bk. Ex. p. 119) is an example, showing that the inventor realized that his device could equally well employ either of these modes of operation, and that he disclosed and

taught that the alternative mode of operation could be secured by turning the pump upside down. [Tr. p. 126.]

4. It accordingly cannot be claimed that there was anything new, at the time of Mr. Bray's asserted invention, in the idea of a traveling barrel pump, nor in the idea of producing one by turning a pump of the fixed barrel type upside down, or end for end.

5. It is also true that in every prior art structure of the traveling barrel type the structure employed a standing column, with relation to which the barrel moved.

6. Such standing column, was of necessity in all such prior structures, provided with one or more valves placed in such position on the column *as seemed* desirable in the particular design. As we noted in paragraph 8, page 17 *supra,* the contention of appellants that the standing and working valves of the Bray patent are located in close proximity to prevent "gas lock" is a figment of the imagination of Counsel.

7. It is also true that in any traveling barrel type pump, the barrel must of necessity be valved, and every prior art patent of this type in evidence was so valved, as are the Bradford structures.

8. It is likewise true that in every traveling barrel pump, the valved traveling barrel must "run on" the column, because there must be. a tight fit between the barrel and the piston or packing portion of the column to produce the pumping action.

9. In the prior art we find working barrel pumps with various types of puller nuts located on the traveling barrel, and in this connection desire to observe:

    A. That there was nothing new in puller nut so located. That it was accepted in the art that a traveling barrel, provided with a puller nut, could be used to pull the pump was recognized by Bray himself, his Specification, lines 9 to 11 (Bk. Ex. p. 3) reading:

    "A feature of this class of pumps is that the traveling barrel can be utilized as a means of pulling the standing column from a retaining lock when it is desired to remove the pump as a whole from the well without pulling the tubing."

    B. The claims of the Bray patent are entirely without reference to any claim that the guide is to be used to pull the pump, nor do they purport to cover a structure designed for that purpose.

    C. The puller nuts of the prior art, mounted on the bottom of the traveling barrel, may be broadly classed as:

    (1) Puller nuts running loosely over, but not on the column, serving *not* as guides, though perhaps under some circumstances functioning as stops. [Tr. pp. 137 to 140.] This class is illustrated by:

    (a) Thompson, Ex. "T" where the clearance is specifically referred to in lines 50 to 53; (Bk. Ex. p. 121)

    "Fluid may also be admitted to this above mentioned space through the sleeve K, which does not fit closely around the tube f."

(b) Wright, Ex. "U", lines 80 to 83; (Bk. Ex. p. 126):

". . . which aperture e is, however, of a somewhat larger diameter, than the said member b to admit of a free entrance or exit of the fluid held within the space F, . . .", and in

(c) Neilsen, Ex. "Y", while not mentioned in the specification, such clearance is obviously shown in the drawings, both in Figs. 1 and 2, being the space between parts 32 and 39, which fact Mr. Hayn at last admitted. [Tr. pp. 335-343.]

(d) The Admore pump of Bradford (Bk. Ex. p. 52; Defts' Physical Ex. 12, 13 and 14 [Tr. pp. 209-210].)

(2) Puller nuts running on the column, and thus serving as a guide are illustrated by:

(a) Ellis, Ex. "W", the specification of which clearly teaches this guiding fit, lines 90 to 93 (Bk. Ex. p. 134):

"At the lower end of the working barrel 19 a cap nut 21 is threaded therein and has a *bearing upon* the standard 12."

(b) Thurston, Ex. "S", lines 93 to 96 (Bk. Ex. p. 116):

". . . 19 has threadedly connected thereto a combined pull nut and guide 20 which *slidably embraces* the extension 11 of the plunger, . . ."

(c) This close fit is clearly shown in the drawings of O'Bannon, Ex. "R", (Bk. Ex. p. 111) in Downing, Ex. "Q" (Bk. Ex. p. 108) and in Deis, Ex. "N" (Bk. Ex. p. 96), and is emphasized by the presence of a packing gland "3" shown between the parts "V" and "I".

(d) The "Inverted Bramo" pump of Bradford. [Bk. Ex. pp. 87, 88 and 89; Defts' Physical Exhibit 17A, Tr p. 213.]

10. It is likewise true that clutches were old in the pertinent art.

A. The jaw type clutch shown in the Bray patent is disclosed in:

Deis, Ex. "N", parts 4 and Z (Bk. Ex. p. 96);

Cummins, Ex. "O", parts 21 and 22 (Bk. Ex. p. 100);

Whitling, Ex. "P", parts 25 and 25a (Bk. Ex. p. 103).

B. Nor was there anything new in locating such a clutch at the bottom of the barrel, with its corresponding part located on the top of the anchor. Deis "N" (Bk. Ex. p. 96) disclosed this in 1907, and over 300 pumps with the clutch so located had been manufactured and sold by Bradford before the clutch was so located by Bray. [Tr. p. 60.]

C. Clearly this form of clutch is shown located at the upper end in Cummins, Ex. "O" (Bk. Ex. p. 100), and in Whitling, Ex. "P" (Bk. Ex. p. 103).

D. That it would not involve invention to turn a pump of the traveling plunger type over, thus making it a pump of the traveling barrel type, must follow from the fact that this reversal is specifically taught by Thompson, when he says, Ex. "T" (lines 27 to 31, Bk. Ex. p. 122):

"In Fig. 4 of the drawings I have illustrated another form of my invention in which the *smaller working barrel v is inverted** and secured to the anchor d, whereby it remains stationary."

and follows this by teaching in the succeeding paragraphs of the specification, the necessary reversal of the valves and similar details.·

E. That a simple and literal turning over of the pump of Cummins or Whitling would result in a structure carrying a clutch at the bottom of the barrel is apparent. The extreme obviousness and simplicity of the reversal is shown by defendants' large drawings, Exs. F.F. (Bk. Ex. p. 174), and G.G. (Bk. Ex. p. 175). That such literal reversal would result in a structure substantially the same as Deis, Ex. "N" (Bk. Ex. 96), and the Inverted Bramo and "Admore" pumps, is plain, and makes the fact that the Patent Office failed to cite Deis, Cummins or Whitling in its consideration of the Bray application of controlling significance. (See File Wrapper and contents—Bk. Ex. pp. 9-28.)

11. Every pump of the traveling barrel type employs of necessity an anchor, to hold the standing column in position in the tubing. In Deis, Exs. "N" (Bk. Ex. p. 95), and "EE" (Bk. Ex. p. 173), this anchor literally:

"presents to said coupler face a top face forming a complementary coupling * * *" (Parts W, Y with Z).

This is also the case in the reversed Cummins, top of 19 with 21, Ex. "GG" (Bk. Ex. p. 175), and in the case of the reversed Whitling, top of 25 with 25a, Fig. 2, Ex. "HH" (Bk. Ex. p. 176).

12. Any of the clutches mentioned in paragraph 10, *supra,* when located on the bottom of the barrel, with the complementary part mounted on the anchor, are of necessity when engaged together, literally "non-rotatively interlocked" with the lower guide or puller nut coupler.

13. Now it cannot be contended that there was anything new in using such a clutch means to provide for the tightening of the joints in the string of sucker rods from which the barrel was hung, because:

A. Any device resulting in holding the bottom of a string of rods against rotation, will result in tightening the joints of the rods if the polish rod at the top is rotated, in the direction to result in turning up the threads.

B. The use of a jaw type clutch for this specific purpose was taught in 1904 by Dickens, Ex. "X", lines 92 to 95 (Bk. Ex. p. 139).

"In pumping the well sometimes the sucker rod 21 becomes unscrewed at one of its joints or from the valve 20. In such case the traveling valve 20 will ordinarily descend by its own weight and rest on the head 11, the tongue 23 seating in the groove 12, and the traveling valve will thereby be held against rotation when it is attempted to reconnect the disconnected joint of the sucker rod 21, said traveling valve thus serving as an anchor to permit the tightening up of the joint or joints of the sucker rod."

14. From the analysis thus far made, it is apparent that the trial court's finding that the "locking means * * * does not effect any functional relationship or mode of operation of the parts different from the patented or unpatented art in evidence" is clearly supported by the record, for the clutch, in each of the cases considered above and in the structure of the patent, serves only the function of forming a coupler non-rotatively locking the traveling barrel to the lower guide coupler and thus permitting screwing up of rods in case of need. Moreover this finding is *not attacked in the opening brief of appellants,* and assignment of error XXIX assigning error to it is therefore deemed to be abandoned.

B. THAT THE PRACTICE OF USING NOTCHED CONTACT MEANS BETWEEN THE PUMP BARREL AND THE ANCHOR FOR THE COLUMN OR THE COLUMN FIXTURE WAS WELL KNOWN IN THE PATENTED AND UNPATENTED PRIOR ART IN EVIDENCE.

The discussion of the evidence presented under point "A" immediately preceding makes the correctness of this finding plain. Appellants do not attack this finding in their brief, nor make any attempt to show wherein they contend it to be erroneous, and assignment XXX assigning error to it must likewise be deemed abandoned.

C. THAT IT WOULD REQUIRE MERELY THE MECHANICAL KNOWLEDGE OF ANY DRILLER OR PUMP OPERATOR SKILLED IN THE ART TO ADOPT SUCH NOTCHED CONTACT MEANS TO THE DESIGNATED USE.

There was a direct conflict of testimony upon this issue between the witnesses for plaintiffs and defendants. This conflict the trial court resolved in favor of defendants-appellees.

1. Mr. Hayn testified that to add to the top face of part 6 of Northrup (Ex. "V", Bk. Ex. p. 129) the male part 22 of Cummins (Ex. "O", Bk. Ex. p. 100) and to the bottom of part 24 of Northrup, the corresponding female part 21 of the Cummins clutch, would amount to "invention of a high order as distinguished from mere mechanical skill." [Tr. p. 299.]

2. Mr. Adams testified this change would not involve invention [Tr. p. 132], and further testified it would not involve invention to reverse in the year 1926 Whitling as in Fig. 2, Ex. "HH" (Bk. Ex. p. 176), thus bringing the clutch parts at the bottom as in Bray. [Tr. p. 105.] Neither Mr. Hayn nor any witness for plaintiffs-appellants controverted this testimony as to the Whitling reversal.

3. The finding is thus based on conflicting testimony as it relates to the Northrup-Cummins combination, and upon the uncontradicted testimony of Mr. Adams as to Whitling, and under the authorities cited in paragraph 3, pages 51-52, *supra*, will not be disturbed.

4. The depositions taken in Bradford demonstrate that Mr. Adams' testimony was correct, as they show that the moving plunger type "Bramo" (Bk. Ex. p. 29) manufactured by that concern (obviously a pump essentially like Cummins), was in point of fact simply inverted into the inverted Bramo, Ex. 17 (Bk. Ex. pp. 87-88-89; Defts'. Physical Exs. 17-A and 17-B #) and into the Admore pump (Bk. Ex. p. 52), producing structures carrying the clutch at the bottom rather than at the top of

---

#The photographs are bound in the appellants' Book of Exhibits with the anchor of the pump part 3, appearing at the top rather than the bottom of the page.

the barrel, by the firm's employees in the ordinary course of their work. The testimony that such an inversion or reversal would amount to nothing but the application of ordinary mechanical skill, is thus borne out by what was done by ordinary mechanics at Bradford. [Tr. pp. 69 to 74.]

5. On page 47 of their brief, appellants, in an attempt to meet this issue, make the following rather curious argument:

> "Defendants' mechanical expert, Adams, has stated in answer to defendants' hypothetical question [Tr. p. 105] that in his opinion it would not have represented any act of invention, as distinguished from the exercise of mechanical skill, to have turned the Cummins pump over in the year 1926. This answer is an attempt to define what constitutes invention, and is a rather ambitious statement for a layman to give, in view of the fact that the Supreme Court of the United States has declined to attempt to give any such definition.

> " 'The truth is, the word (invention) cannot be defined in such manner as to afford any substantial aid in determining whether a particular device involves an exercise of the inventive faculty or not. In a given case we may be able to say that there is present invention of a very high order. In another, we can see that there is lacking that impalpable something which distinguishes invention from simple mechanical skill. Courts, *adopting fixed principles* as a guide, have by a process of exclusion determined that certain variations in old devices do or do not involve invention; but whether the variation relied upon in a particular case is anything more

than ordinary mechanical skill is a question which cannot be answered by applying the test of any general definition.'

> *McClain v. Ortmayer,* 141 U. S. 427, 1891.

"See *Walker on Patents,* 6th Ed., Sections 62 to 806, for process of exclusion of that which does not amount to invention.

"While the question of invention may involve certain facts, for example, the state of the art, such facts are necessarily governed by legal principles and rules of law, to which no layman, like defendants' expert, Adams, is competent to testify, because such testimony would obviously require the application of the law to the facts.

"What constitutes invention is a mixed question of fact and law, and only the Court, or one learned in the law, is qualified and competent to give an opinion concerning the same."

Appellants have obviously misunderstood the statement of the Supreme Court that invention "cannot be defined" and are contorting it to mean that the presence or absence of invention cannot be the subject of expert testimony. The rule on the subject is summarized in 48 *Corpus Juris* on page 96, Section 98, as follows:

"Expert Opinion. The opinion of persons who by reason of professional, scientific, or technical training, or practical experience, may be regarded as experts in the particular art to which the method or device relates, may be received on the issue whether the exercise of invention was required for its production. It has been asserted that a defense of. lack of.

invention should be supported by the testimony of expert witnesses where the patent relates to a complex subject."

And in the case of *Dunbar v. Meyers,* 94 U. S., pp. 187-202; 24 L. Ed., pp. 34, at 38, the Supreme Court considered and gave effect to such testimony.

In the court below, appellants were of a different opinion as to the propriety of such testimony, as their expert, Mr. Hayn, was asked by appellants' attorney to give, and did give, testimony of precisely this sort. [Tr. pp. 295, 301, 334 and 345.]

In any event, appellants have not presented a proper record of objections and exceptions with relation to such testimony, and have made no pretense of complying with Rules 11 and 24 of this court, requiring the full substance of testimony which it is urged was wrongfully admitted to be quoted in their assignment of errors and in the specification of errors of their brief.

6. Appellants, moreover, do not specifically attack this finding, and assignment of error XXXI must therefore be deemed abandoned.

D. THAT IT (THE NOTCHED CONTACT MEANS) HAD BEEN SO ADAPTED BY OTHERS BEFORE THE PLAINTIFFS SOUGHT TO USE IT.

The record discloses that the patent to Deis, Ex. "N" (Bk. Ex. p. 95) is an example of such adaptation, at parts "4" and "Z", while the *uncontradicted* testimony of the witnesses whose depositions were taken in Bradford fully support the finding. Moreover it is not attacked in plaintiffs' brief and assignment of error XXXII must therefore be deemed abandoned.

E. THAT THE ESSENCE OF THE INVENTION CLAIMED BY CLAIM 1 * * * LIES IN THE INTERLOCKING FEATURE DESCRIBED IN FINDING VIII HEREOF. And

F. THAT CLAIM 1 "PRESENTS NO OTHER FEATURE OF ASSERTED NOVELTY."

1. Counsel for appellants seem inferentially to attack these findings by asserting:

"Claim 1, therefore, interpreted in the light of the specification and drawing, covers a pump with a *standing valve S located on the upper end of the standing plunger,* or standing guide column as it is termed in the claim. Such arrangement brings the standing valve S close to the traveling valve V, and this close relationship of said valves reduces the space between said valves to a minimum, which prevents the accumulation of any appreciable amount of gas, under high pressure, between said valves, and thereby prevents a "gas lock" between said valves and enables the pump to function under the high gas pressure found in the oil fields of California." (Opening Brief p. 33.)

2. As we have pointed out heretofore (see pages 18 to 21, *supra*), the distance separating the valves is not mentioned in the patent, and the relative positioning of the valves in the patent drawing is purely illustrative of the general type of structure to which the alleged invention contained in the body of each claim is applied. The references in the specification to "a standing valve S located on the upper end of the standing plunger" is purely fortuitous. *It is an element not carried into any of the claims.* Yet it is the sole ground relied upon to distinguish the Admore pump from the structure claimed in

claim 1 of the patent. This attempted distinction is without merit for the reasons advanced at pages 17 to 27, *supra*.

On page 41 of their brief, appellants state:

"Defendants' Exhibits 13 and 13A, Book of Exhibits, page 81, represent a pump like the pumps of Defendants' Exhibits 10 and 11 with fixed or standing barrel and traveling plunger. The clutch is at the top of the barrel, not at the bottom, and the pump has no jetting device for washing an anchor and a lower clutch member below the barrel. There is nothing to prevent this pump from 'sanding up.'"

These exhibits are sketches of parts of the Admore pump, [Physical Exhibit 13, Tr. p. 62], also illustrated on pages 52 and 53 of the Book of Exhibits, and are of the movable barrel and fixed plunger type. [Tr. pp. 62 to 64.] This statement of appellants is therefore a sheer misstatement of fact. We do not question that it is made inadvertently.

3. Appellants evidence a complete misunderstanding of the significance in the record of such structures as the Bramo pump, and other pumps such as those of Cummings and Whitling, all of which employ a stationary or standing barrel and a reciprocating or traveling plunger. They state:

"The Bramo pump, Defendants' Deposition Exhibit 1, Book of Exhibits, pages 29 and 30, offers no suggestions, whatever, of the pump covered by claim 1 of the patent in suit, because it is a pump of a different type, that is to say, it has a reciprocating or *traveling plunger* A and a *stationary barrel* B, with a clutch C-D at the *upper* end of the stationary barrel, instead of a *standing* or *stationary plunger* (10-13-16) and a *traveling barrel* (B) with a clutch (4-2) at the *lower* end of the traveling barrel, as specified in claim 1 of the patent in suit. * * * It may be here stated that none of the pumps set up in anticipation of the patent in suit of the Bramo type, which is the conventional type of pump with a *stationary* or *standing* barrel and a *reciprocating* or *traveling* plunger, have any bearing or anticipating effect on the pump of the patent in suit, which has a *stationary* or *standing plunger* and a *reciprocating* or *traveling* barrel, because the former conventional type of pump could not possibly function like the latter pump or accomplish the same results." (Opening Brief pp. 34-35.)

These prior art structures were not introduced for the purpose of attempting to establish that they could "function like the latter (traveling barrel) pump, or that they accomplished the same results," but for the purpose of establishing that in the light of the art, it was *not* invention to invert such pumps and thereby convert them into traveling barrel pumps with a clutch at the bottom. The trial court, upon conflicting testimony as to whether such a change amounted to invention, held that it did not do so. [Finding VIII, Tr. p. 382.] As the question of invention or non-invention is a question of fact;

> *Proudfit Loose Leaf Co. v. Kalamazoo Loose Leaf & B. Co.* (C. C. A. 6), 230 Fed. 120 at 127, and cases there cited.

this finding will not be disturbed. (See cases cited in paragraphs 2 and 3, pages 51 and 52, *supra*.)

It therefore follows that the attempts of appellants on pages 37 and 38 of their brief to deny the significance in the record of the Bramo pumps; on page 41 the O.F.S. working barrel and the Bradford plunger liner; on pages 44 and 45 the Cummins pump; on pages 48 and 49 the Whitling pump; on page 62 the Dickens Pump, and on page 66 the Tierce pump, on the grounds that they are of the standing barrel and moving plunger type, a fact that *appellants have never disputed,* amounts merely to an attempt to demolish a straw man.

Their further attempt to distinguish each of these pumps, with the exception of the Tierce, on the ground that the valves of these pumps are so relatively located as not to prevent gas lock, is without merit for the reasons set forth at length herein.

G. That the Said Interlocking Feature Claimed in Claim 1 Does Not Present Any Novelty or Amount to Any Invention Over the Prior Art in Evidence Both Patented and Unpatented.

1. Appellants do not attack this finding in their brief and we cannot see how they could do so in the light of the record as analyzed in paragraph 10 pages 59-60, *supra.*

2. It is to be noted that appellants assert, as stated under Points E and F, *supra,* that claim 1 presents a feature of patentable novelty in addition to "said interlocking feature," *i. e.,* the close proximity of the traveling and standing valves to minimize gas lock. That is, however, far different from the assertion that the "interlocking feature claimed," *itself presents novelty over the prior art.* Assignments XXX and XXXII,

in so far as they assign error to the portion of finding IX under discussion, must therefore be deemed to be abandoned.

H. THE COURT FINDS ALL OF THE ELEMENTS OF CLAIM 1, * * * IN THE SAME OR EQUIVALENT COMBINATIONS IN THE PRIOR ART BOTH PATENTED AND UNPATENTED.

1. From the foregoing analysis of the evidence in the record it is apparent that the trial court was correct in finding the following elements, being all of the elements of claim 1:

(1) Standing guide column (Par. 5, p. 56, *supra*)

(2)   with working valve (Par. 6, p. 56, *supra*).

(3) Valved traveling barrel (Par. 7, p. 56, *supra*)

(4)   running on the column (Par. 8, p. 56, *supra*).

(5) Bottom guide on the barrel (Par. 9, p. 57, *supra*)

(6)   running either on or over the column (Par. 9, p. 57, *supra*)

(7)   having a bottom face forming a coupler (Par. 10, p. 59, *supra*).

(8) An anchor for the column (Par. 11, p. 60, *supra*)

(9)   presenting directly to said coupler face a top face (Par. 11, p. 60, *supra*)

(10)   forming a complementary coupling (Par. 11, p. 60, *supra*)

(11)   to be non-rotatively interlocked with the lower guide coupler (Par. 12, p. 61, *supra*).

in the "same or equivalent combination in the prior art both patented and unpatented."

2. At pages 16 and 76 of their brief, appellants state that a "patent is *prima facie* evidence of utility, and doubts relevant to the question should be resolved against infringers because it is improbable that men would render themselves liable to actions for infringement unless infringement is useful." This may be conceded. We fail however to see the pertinency of this observation for the reason that the court did not find any of the claims void for lack of utility, although counsel for appellants admits in his brief that the court felt there was strong evidence to that effect.

On pages 31 and 32 of their brief, appellants assert that the action of the trial court, in finding claim 1 of the patent null and void, is erroneous because the grant of the Letters patent in suit is *prima facie* evidence of the validity of said Letters patent, which can only be overthrown by proof to the contrary beyond reasonable doubt.

It is true that general language to this effect is to be found in the cases, and we do not dispute that the general rule is as appears in *Stoody Company v. Mills Alloys*, 67 Fed. (2d) 807, at page 809, and *Anraku v. General Electric Co.*, 80 Fed. (2d) 958 at page 960:

> "The general rule is that a patent is presumptively valid. This presumption arises from the grant of the patent by the Patent Office after the application has been examined *thoroughly** by the Examiner."

The general rule is, however, as stated in the *Stoody Co.* case at page 810, subject to the limitations noted by the late Judge Rudkin of this Court, in the case of *Wilson*

& Willard Mfg. Co. v. Bole (227 Fed. 607 at 609), as follows:

> "But this rule is founded in reason. It presupposes an adjudication by the Patent Office of every fact essential to the validity of the patent, and one who attacks that adjudication in a collateral proceeding must establish his claim by clear and satisfactory proof, or, as is often said, by proof beyond a reasonable doubt. But where it appears that *there has been no such adjudication by the Patent Office, as where the patent has been issued through inadvertence or mistake, the reason upon which the rule is founded ceases,** and the rule ceases with it."

· Another leading case on the subject is *American Soda Fountain Co. v. Sample* (C. C. A. 3), 130 F. 145, 149, 150, in which Judge Gray used the following language:

> "We do not agree with the contention, that the fact that the file wrapper discloses the patent to have been granted as first applied for, without any reference, adds any force to the presumption of novelty arising from the grant. *On the contrary, we think the force of that presumption is much diminished, if not destroyed, by the lack of any reference by the Examiner to, or consideration of, the 'Clark' patents. It does not seem likely that an expert examiner would pass them by, without notice or consideration, if they had been called to his attention.* We feel compelled, therefore, to the conclusion, that the first and fifth claims of the patent in suit are invalid for want of patentable novelty."

It is to be noted that in the case at bar neither the Deis, Cummins or Whitling patents were cited, nor was the existence of Bradford constructions, which are before this Court, known to the Examiner. We submit that, as was said of the "Clark" patent, they would not have been passed by had they been called to the Examiner's attention. On page 26 of their brief appellants assert:

> "No limitations can be read into the claims of the patent in suit, because claims 1 and 2 were allowed with only a few technical and not substantial amendments, which do not narrow the scope of said claims, while claim 3 was allowed by the Patent Office as originally presented, and without amendment."

This statement, while literally true, is, in effect, false, because it was not until after all of the claims of the original application (Bk. Ex. pp. 14, 15 and 16) had been argued and amended and not until "Amendments C and D" were filed (Bk. Ex. pp. 23 and 26) that the present claims of the patent *drawn in approximately their final form to meet the references and objections* of the Examiner, were presented.

3. It is to be noted that the claim is void of all reference to the jet action or to the means of producing that action. It claims a tight guide but claims no vents, ducts or apertures to produce the jet. It reads directly on:

Deis, Ex. "N", Bk. Ex. p. 96

Cummins reversed, Ex. "F. F." Bk. Ex. p. 174

Whitling reversed, Ex. "H. H." Bk. Ex. p. 176

and upon the combination of:

O'Bannon Ex. "R", Bk. Ex. 111

Thurston " "S", Bk. Ex. p. 115

Ellis " "W", Bk. Ex. p. 133

Downing " "Q" Bk. Ex. p. 108

Thompson " "T", Fig. 1, Bk. Ex. p. 119

Wright " "U", Bk. Ex. p. 125 or

Neilson " "Y", Bk. Ex. p. 142

with a clutch such as is taught in Deis "N" or Cummins "O" (Bk. Ex. p. 100) or Whitling "P" (Bk. Ex. p. 103) or Dickens "X" (Bk. Ex. p. 137).

Mr. Hayn, out of his Patent Office experience of over 25 years, his service as Acting Examiner in several Divisions, and his astounding performance in having read and interpreted patents, sometimes at the rate of 50,000 a year [Tr. p. 293], testified as heretofore noted that it would be invention of a high order—which he seems so readily to find—to add the clutch of Cummins to Ellis [Tr. pp. 301 and 327]; to one who finds it impossible to recognize a clearance shown in a drawing, without written text in the patent to tell him it is a clearance [Tr. p. 311] to one who could not visualize the cutting off of the valve cage of Ellis, without entirely eliminating the part shown in Fig. 5 [Tr. p. 302] to one who finds it would involve invention, also of a high order, to plug the holes of the Admore pump [Tr. p. 348] thereby automatically increasing the discharge downward through the puller nut, such an opinion may be natural, but surely it is not surprising that the trial court by its findings impliedly found it to be of no value.

It is interesting to note in the light of such testimony that the file wrapper and contents show that claims 4, 5 and 6 of the original application were rejected on Ellis taken with Dickens, the Examiner making the common sense and laconic observation:

> "To provide Ellis with an anti-rotating device would not be invention in view of Dickens." (Paper No. 6, Bk. Ex. p. 22.)

How much less is it invention in the light of Deis, which patent the Office search did not disclose, or in the light of the obvious reversibility of Cummins, which patent was also overlooked? The expedients to which Mr. Hayn resorted in seeking to avoid a frank and definite answer to questions on cross-examination, his haste to say that plugging the holes of Admore would amount to invention (though he admitted it could not if its effect on the jet action was "frivolous", [Tr. p. 355] and likewise admitted he did not know what effect such plugging would have in terms of increased jet action) [Tr. p. 355] all so thoroughly discredit his opinions, that it is profitless to speculate why he found "a high degree of invention" with such frequency and alacrity.

4. Examining claim 1 in the light of Bradford Motor Works constructions and without again listing the elements (see p. 71, *supra*) it is plain:

(a) that every element of claim 1 read in any possible way, is found in one or the other of these structures.

(1) If the patent is read as the court found it must be, as covering a tight fitting guide, it is found element for element, in exactly the same combination in the Inverted Bramo. (Physical Ex. p. 17-A and 17-B, Bk. Ex. pp. 87, 88 and 89.)

On pages 34, 41, 42 and 43 of their brief, appellants attempt to avoid the force of this reasoning by arguing that the inverted Bramo pump was an "unsuccessful, abandoned experiment." The depositions taken in Bradford disclose that not more than twenty-five of these pumps were built, that they did not function to the satisfaction of the manufacturer, due to the fact that, because of the tight stuffing box located on the bottom of the barrel (Part 2, Bk. Ex. p. 87) and bearing on the standing column (part shown between spaced jaws of clutch, Bk. Ex. p. 88) escape of the oil trapped in the space corresponding to the annular space EE of Ex. MM (Bk. Ex. p. 178) was prevented; that this stuffing box was accordingly eliminated with the result that the Admore was produced. [Tr. p. 73.]

It is to be noted that this pump contained precisely the same clutch as that of the Bray patent, located between the bottom of the barrel and the top of the anchor; and that since claim 1 of the patent makes no reference whatsoever to any ejection of fluid for the washing of the anchor, this structure completely anticipates the combination purportedly covered by such claim irrespective of whether or not there was escape of oil between the standing column and the guide.

On page 213 of the transcript, Mr. Adams testified that the Bramo structure would be operative at speeds dependent upon the rate of leakage between the column and the stuffing box. Mr. Bray testified, on page 241, that while in his opinion the pump would not be operative in a deep well, such a pump would be operative to the extent that the plunger could move up and down as against the tendency of the air compressed in the space EE to resist compression, [Tr. p. 252] and that he did not want to be understood as testifying further than that the pump would not work in deep gaseous wells. [Tr. p. 253.]

Since neither invention nor utility is dependent upon the ability of a pump to operate in deep wells where gas is present, but is measured by its ability to operate under any conditions where pumps are used, *i. e.,* also in shallow wells where gas is not present, it follows that the pump is not as counsel contends, in the eyes of the law, inoperative.

(2) If the patent is read as covering a loose fitting guide, as appellants contend it should be, every element of it is found in the Admore pump, in precisely the same combination. Any possible contention that the presence of holes in the barrel of Admore differentiates that structure from the Bray device, can have no bearing on this claim, silent as it is as to jetting action. This claim covers no apertures in the guide, claims no fluid either admitted thereby or ejected therefrom—and it is of course only such functions that the holes in the barrel of Admore could possibly affect.

At page 33 and at pages 38, 39 and 40 of their brief, appellants attempted to distinguish the Admore pump from the structure of the Bray patent. These attempted distinctions are:

I. The standing and working valves are separated by a considerable space and in consequence would tend to be subject to gas lock. (Opening Brief pp. 33, 38 and 39.)

II. The Admore pump has no vertical aperture in its puller nut 6 through which liquid might be projected out of the lower barrel to wash the anchor and lower clutch member.

III. The Admore has holes in the side of the traveling barrel above the puller nut to permit the escape of oil trapped in the space corresponding to the annular space EE of Bray.

We have repeatedly pointed out that this attempt to read into any of the claims of the patent, such spacing of the valves as would tend to prevent gas lock is wholly erroneous, and have pointed out hereinabove that since Claim 1 does not cover the jet device, the presence or absence of the vertical apertures in the puller nut is of no consequence in considering the validity of this claim, nor is the presence or absence of holes in the barrel to allow the escape of fluid from the space corresponding to the annular space EE of Bray of any consequence. It accordingly follows that since the only distinction attempted, which has any relevancy to the elements claimed, (*i. e.,* the positioning of the valves to prevent gas lock), is not even purportedly protected by any of the claims of the patent, appellants have wholly failed to distinguish the Admore pump from the structure of the Bray patent. The finding of the trial court that Claim 1 is anticipated by the structures found "in the unpatented prior art" is therefore fully supported by the evidence.

(b) The "inverted Bramo", and the "Admore" pumps of the Bradford Motor Works were made and used in 1926. The depositions taken in Bradford establish that the Inverted Bramo, Ex. "17a", and the Admore pump, Ex's. "12, 13, 14", were made and used in 1926.

In that year, some 25 of the Inverted Bramos were made [Tr. p. 73] and 308 Admores made and sold. About 14,000 of the Admores have been made and sold since that time. [Tr. pp. 60 and 77.]

(c) The file wrapper discloses that application was not made for patent until Feb. 4, 1929.

5. Let us consider the defenses pleaded in the amended answer. Under the terms of the patent statute, if the invention of the Bray patent:

(a) Had been in public use or on sale in this country for more than two years prior to the said Feb. 4, 1929, that is to say prior to Feb. 4, 1927—the affirmative defense pleaded in Par. XIV [Tr. p. 22] of our answer is made out. In this connection we call the court's attention to the following principles of law and the authorities supporting them:

    (1) "A patent is void if the invention covered thereby was in public use or on sale earlier than two years before the application for that patent."

        Walker on Patents, 6 Ed. Vol. 1, p. 163, citing:

        Eastman v. Mayor, etc., of City of New York, 134 F. R. 844; and

        Twyman v. Radiant Glass Co., 56 Fed. (2d) 119.

    (2) "The two years contemplated by this law are ascertained by measuring backward from the date of the filing in the Patent Office."

        Walker on Patents, 6 Ed. Vol. 1, p. 165, citing:

        Campbell v. N. Y., 35 F. R. 504.

(3) "To constitute public use, it is not necessary that more than one specimen of the thing invented should have been publicly used, nor that more than one person should have known of that use."

> *Walker on Patents,* 6 Ed. Vol. 1, p. 165, citing:
>
> *Consolidated Fruit Jar Co. v. Wright,* 94 U. S. 94; 22 L. Ed. 68;
>
> *National Cash Register Co. v. American Cash Register Co.,* 178, F. R. 79;
>
> *A. Schrader's Sons v. Wein Sales Corp.,* 9 Fed. (2d) 306 (C. C. A., 2d Cir.).

(b) Likewise, if the invention claimed in the Bray patent had been described or shown in a printed publication, or patent prior to Feb. 4, 1927 (as we assert it was shown by the Bradford catalogue, Ex. 7, Bk. Ex. p. 53), the affirmative defense pleaded in Par. XII [Tr. p. 14] of the amended answer has been established. In this connection we call the court's attention to the following principles of law and the authorities supporting them:

(1) "Constructive abandonment of any invention claimed in a patent applied for after 1897 will be found to have occurred if its inventor did not make application for that patent . . . until more than two years after that subject was described in some printed publication somewhere."

> *Walker on Patents,* 6 Ed. Vol. 1, p. 174, citing:
>
> *Statutes at Large,* p. 692, Chapter 391, Secs. 1, 3 and 8 construed together.

(2) "A printed publication is anything which is printed, and, without any injunction of secrecy, is distributed to any part of the public in any country, and such a publication may negative novelty."

> *Walker on Patents,* 6 Ed. Vol. 1, p. 118, citing:

> *Rosenwasser, et al., v. Berry,* 129 U. S. 47; 32 L. Ed. 628, at p. 630.

(c) Likewise if the invention of the Bray patent had been known or used, even in a single instance prior to the dates of his invention—the defense of Par. XIII [Tr. p. 16] has been established. This date is, of course, the date of application for patent, that is to say, Feb. 4, 1929, for no attempt has been made in this case to carry the date of invention back of the filing date. In this connection we call the court's attention to the following principles of law and the authorities supporting them:

(1) "Novelty is negatived by prior knowledge and use in this country by even a single person of a thing patented."
> *Walker on Patents,* 6 Ed. Vol. 1, p. 137, citing among other cases:
> *Sewall v. Jones,* 91 U. S. 171, at pp. 179 and 180; 23 L. Ed. 275, at p. 276.

(2) "Novelty is also negatived by evidence that even one specimen of the thing patented ex-

isted and was known in this country prior
to its invention by the patentee even though
it was not used prior to that time."

> *Walker on Patents,* 6 Ed. Vol. 1, p. 139,
> citing:
> *Stitt v. Railroad Co.,* 22 F. R. 650.

(3) "In the absence of other evidence, the date
of the application for a particular patent,
is evidence of invention at the date of the
application of all matters disclosed therein."

> *Walker on Patents,* 6 Ed. Vol. 1, p. 210,
> citing, among other cases:
> *Melbourne Co. v. Davis-Bouronville Co.,*
> 270 U. S. 390, 70 L. Ed. 651.

(4) "Burden is on plaintiff to carry date of invention back of the date of application if
he seeks to establish such earlier date."

> *Automatic Weighing Machine Co. v.*
> *Pneumatic Scales Corporation,* 166 F.
> R. 288.

(d) All of this would be true, even though we had
not established, as the Court found that we had,
the defense pleaded in paragraph XV [Tr. p. 22]
of the amended answer, *i. e.,* that the alleged invention did not constitute patentable knowledge or
invention within the meaning of the patent laws—
in view of the prior state of the art, and in view
of what was common knowledge on the part of
those skilled in the art.

It was not questioned in the Court below nor is it
questioned in the opening brief of appellants that the

date of manufacture and use of the Bradford structures has been amply established. Upon this point the appellants simply state:

> "The evidence upon which the lower court found claim 1 of the patent in suit null and void, not only falls far short of proof beyond a reasonable doubt, but the prior art fails absolutely to show the novel combination of elements of said claim." (Opening Brief p. 32.)

without anywhere in their brief pointing out where they contend the proof so falls short.

We shall, therefore, not burden this argument with a detailed marshalling of the evidence bearing on this seemingly conceded fact. For the convenience of the Court, and to the end that this brief may be complete, we have prepared such an analysis and it appears as "Part B" of this brief.

6. That these findings, if supported by the evidence, are sufficient to support the "Conclusions of Law" and the "Decree" is not questioned in the opening brief, and we therefore assert that assignments of errors LI, LII, LV, LVII, LVIII, LIX, LXI [Tr. pp. 402 to 405] are without merit, and urge that the record discloses that the trial court could have found, likewise supported by the evidence, that claim 1 of the patent is void upon each of the following grounds:

a. Public use and sale for more than two years preceding application.

b. Description in a printed publication, more than two years preceding application.

c. Public use prior to date of invention.

II.

On page 30 of their brief appellants state that Issue 2 is:

"Should the claims of the patent in suit, and particularly claims 2 and 3, be narrowly construed or limited to such extent as to read out of the claim the pump manufactured and sold by the defendants?"

1. This so-called issue requires restatement and division if it is to be a fair statement of the issue suggested, and is not to be hopelessly confusing because compound. So restated it presents two questions:

I. Did the trial court err in construing claim 2 as being limited as found in the findings, and hence err in finding that the alleged infringing structures of of appellees did not infringe?

II. Did the trial court err in construing claim 3 as being limited as found in the findings, and hence err in finding that the alleged infringing structures of appellees did not infringe?

2. We will consider under the heading "Questions I and II" all matters common to both questions, and discuss under the heading "Question I" or "Question II" only those matters which are solely applicable to such questions.

3. Claim 2 of the patent reads:

"The combination, with a pump of the class having a standing guide column with a working valve and a valved traveling barrel, running on said column; of a bottom guide, on said barrel, running on said column and having a bottom face forming a coupler, and an anchor for the column presenting a top face forming a complementary coupling to be non-rotatively interlocked with the lower guide coupler; said guide having a series of venting ducts which open directly toward the anchor coupler face so that during the strokes of the barrel, ejected liquid is impelled toward the anchor top face to aid in prevention of sand accumulation thereon." (Bk. Ex. p. 4, lines 56 to 70.)

4. Claim 3 of the patent reads:

"The combination, with a pump of the class having a standing guide column with a working valve and a valved traveling barrel, running on said column; of a bottom guide, on said barrel, operating on said column and having a coupler part on its bottom end, and an anchor for said column having a top coupler part complementary to the coupler part of the said guide, whereby the engaged parts may be meshed in non-rotative interlock, and said guide being appertured for the ejection of liquid from the barrel directly toward the effective face of the lower, standing coupler part to aid in preventing sand settlement thereon." (Bk. Ex. p. 4, lines 70 to 85.)

5. Reducing the findings relating to claims 2 and 3 to their elements, it is clear that the trial court found:

FINDING X [Tr. p. 382].

A. "that the essence of the invention claimed in claim 2 and in claim 3 * * * resides in the arrangement for expelling fluid from the barrel, for which the patentee claimed originality in this, that he asserted his method secured the result of washing clean the interlocking part at the column fixture, which part would otherwise be covered with sand and prevent proper interlocking contact when the barrel was lowered;" and

B. "that said claims present no other feature of asserted novelty except the interlocking feature which is the subject of findings VIII and IX hereof."

FINDING XI [Tr. p. 383].

C. "that the patentee contemplated that the guide used should have a close or ordinary mechanical fit with the standing column and that he intended to obtain maximum of jet action to accomplish the desired effect of washing off the coupler face of the column fixture.

D. "that the teachings of the patent are specific in terms, as the patentee states the 'bushing is provided with a series of ducts 5 from top to bottom so that as the barrel slides up from bottom position liquid will be forced down in the strong jets and wash sand from the top of pin P so that this can be readily pulled from its seat in lock box L.'"

FINDING XIII [Tr. p. 384].

E. "that claims 2 and 3 are limited closely to the feature of having a bottom guide running on the standing column, with a running fit, the said guide being provided with vents or apertures for the ejection of liquid directly toward the effective face of the lower coupling member or column fixture and that the invention is to be strictly limited to the form as so described."

FINDING XII [Tr. p. 383].

F. "that neither of the said defendants as shown by the evidence employs in his, or its, pump a guide with a close, ordinary mechanical or running fit."

G. that "they and each of them, thus allow some discharge of fluid to get by between the guide and the circumference of the standing column."

H. "that neither of said defendants is following the specification of claims 2 and 3 of patent No. 1,840,432."

FINDING XIV [Tr. p. 384].

I. "that none of the pumps in evidence, manufactured, used or sold by either of said defendants employs a guide which is provided with vents or apertures for the ejection of liquid directly toward the effective face of the lower coupler member or column fixture."

FINDING XV [Tr. p. 384].

J. "that none of the pumps of defendants made like those in evidence infringe upon either claim 2 or claim 3 as so limited."

We shall now marshall the evidence supporting each element of these findings in the order in which we have listed them, and at the same time answer such observations as appellants have made concerning each element.

A. THAT THE ESSENCE OF THE INVENTION CLAIMED IN CLAIM 2 AND IN CLAIM 3 * * * RESIDES IN THE ARRANGEMENT FOR EXPELLING FLUID FROM THE BARREL, FOR WHICH THE PATENTEE CLAIMED ORIGINALITY IN THIS, THAT HE ASSERTED HIS METHOD SECURED THE RESULT OF WASHING CLEAN THE INTERLOCKING PART AT THE COLUMN FIXTURE, WHICH PART WOULD OTHERWISE BE COVERED WITH SAND AND PREVENT PROPER INTERLOCKING CONTACT WHEN THE BARREL WAS LOWERED.

Questions I and II.

1. We assert that the evidence amply supports this finding, as the only semblance of proof that the Bray pump accomplished any new and useful result is found in the contention of plaintiffs that the structure of the Bray patent is so designed as to result in the production of jets of such magnitude and endowed with such velocity as to do useful work in washing the face of the anchor and the interlocking parts, because:

a. A reading of the specification discloses that what Mr. Bray thought he had invented, and what he therefore sought to protect by claims 2 and 3, was a mechanism designed to produce such jet action. Examining the specification we find, lines 14 to 18 (Bk. Ex. p. 3), reads:

"A common defect of this class of pump is that sand settles on the locking means and freezes the parts to the tubing so that the pump cannot be pulled without great risk of serious injury * * *"

Lines 21 to 31:

> "* * * For this reason one object of the invention is to maintain the sand in such a constant state of agitation that it cannot pack down on the lock shoe box and lock shoe pin."

> "To that end means are provided to cause a positive jet action of liquid over the lock means and keep the sand in a state of suspension so that it will be carried off in the discharge and therefore prevent sanding up."

Lines 75 to 81:

> "Further, the bushing is provided with a series of ducts 5 from top to bottom so that as the barrel slides up from bottom position, liquid will be forced down in strong jets and wash sand from the top of the pin P so that this can be readily pulled from its seat in the lock box L."

b. We thus find the problem, which Bray said was presented by the preceding pumps of this class, was that sand settled on the means locking the anchor of the standing column to the tubing, thereby preventing the pulling of the pump from the tubing—and the solution provided was the use of jets to wash the anchor and its cooperating part free from sand.

c. The only other objects mentioned in the Bray specification are clearly old, and are, moreover, not covered by the claims. They are:

> (1) Means for screwing and unscrewing the rods, the specification reading:

> > "Another object is to provide means for positively interlocking the traveling

barrel to the standing lock means to enable the screwing or unscrewing of parts of the rod string in event of need." (Bk. Ex. p. 3, lines 30 to 35.)

As we have pointed out, this thought is not new, but was specifically mentioned in the specification of Dickens, Ex. "X", in 1923, as follows:

"In pumping the well sometimes the sucker rod 21 becomes unscrewed at one of its joints or from the valve 20. In such case the traveling valve 20 will ordinarily descend by its own weight and rest on the head 11, the tongue 23 seating in the groove 12, and the traveling valve will thereby be held against rotation when it is attempted to reconnect the disconnected joint of the sucker rod 21, said traveling valve thus serving as an anchor to permit the tightening up of the joint or joints of the sucker rod." (Spec., Bk. Ex. p. 139, lines 93 to 105.)

It is inherent in the Deis structure of 1906, Ex. E. E. "N", Bk. Ex. p. 173, in which case the clutch is on the bottom of the barrel, 7 at "4" and "Z", and is to be seen in both the Inverted Bramo Bk. Ex. pp. 87, and "4" and the Admore (Bk. Ex. p. 52, parts 6 and 15).

(2) A design such that the barrel may be reversed, the specification reading:

> "An additional object is to provide a pump of such structure and design that the traveling barrel can be reversed end for end after a period of use so that its useful life may be greatly extended." (Bk. Ex. p. 3, lines 36 to 41.)

Such a reversal is not new. In the file wrapper Ex. "B", paper No. 7, counsel for Mr. Bray stated specifically of the barrel of Ellis:

> "We concede that his barrel is reversible end for end, though he does not discuss it." (Par. 2 of Remarks.) (Bk. Ex. p. 24.)

This reversibility of barrel is not covered by the claims and with it we are not concerned.

2. The problem as stated by Bray is that of keeping the top of the anchor free of sand, so that the standing column can be pulled from it * * * the asserted solution * * * by causing

> "* * * a positive jet action of liquid over the lock means, and keeping the sand in a state of suspension so that it will be carried off in the discharge and therefore prevent 'sanding up'." (Bk. Ex. p. 3, lines 26 to 30.)

that is to say, by such means:

> "to maintain the sand in such a constant state of agition that it cannot pack down on the lock shoe box and the lock shoe pin." (Bk. Ex. p. 3, lines 21 to 26.)

This is to be accomplished through providing the bushing with a

> "Series of ducts from top to bottom so that as the barrel slides up from the bottom position liquid will be forced down in strong jets and wash sand from the top of the pin 'P' so that this can be readily pulled from its seat in the lock box L." (Bk. Ex. p. 3, lines 75 to 80.)

and the claims cover, and cover only, a structure embodying a guide, described in claim 2 as:

> "said guide having a series of venting ducts which open directly toward the anchor coupler face so that during the strokes of the barrel, ejected liquid is impelled toward the anchor top face to aid in prevention of sand accumulation thereon." (Bk. Ex. p. 4, lines 65 to 70.)

It is to be noted that in Bray, as in Northrup, the specific teaching is *not that the jets are of value in keeping the clutch face clear of sand,* thereby facilitating engagement of the clutch, although the trial court found this would occur, *but rather that* the jets of Bray like those of Northrup, *wash sand from the top of the anchor* so that it can be drawn from its seat in the tubing. The Bray specification reads (Bk. Ex. p. 3, lines 79 to 81):

> "wash sand from the top of the pin P. so that this can be readily pulled from its seat in the lock box L."

3. We ask the court to note that claims 2 and 3 stress the force of the jet:

> Claim 2. "* * * said guide having a series of venting ducts which *open directly* toward the anchor coupler face so that during the strokes of the barrel, *ejected liquid is impelled toward* the anchor top face to aid in prevention of sand accumulation thereon." (Bk. Ex. p. 4, lines 65 to 70.)

> Claim 3. "* * * said guide being apertured for the *ejection* of liquid from the barrel *directly toward* the effective face of the lower, standing coupler part to aid in preventing sand settlement thereon."

4. This idea of washing the top of the anchor clear of sand so that it can be pulled from its seat, is not new in the art. It is specifically taught in Northrup, Ex. "V" (Bk. Ex. p. 131, lines 59 to 70):

> "Upon ascent of the working barrel, this oil is forcibly ejected through the openings 32 so as to effectively wash all sand and the like therethrough, as well as agitating the oil in the tubing to prevent the accumulation of sand upon the extension 8 which supports the pipe 1 and seals the tubing C against the escape of oil except through said pipe. The sand is thus held in suspension in the oil and discharges upwardly with such oil, through the well tubing, instead of again settling into the well. * * *"

and it follows that since this idea of washing the anchor was itself old, as were structures designed to accomplish it, the trial court was of necessity compelled to seek invention in the particular means shown by the patent to produce it.

5. Appellants in their Opening Brief state Bray's asserted problem and his asserted solution in practically the same way, i. e.:

"To provide a pump which would not 'sand up' and require pulling of the pump tubing in the oil fields of California was the problem that confronted the inventor Bray, and he solved that problem by his conception and reduction to practice of his invention, as disclosed in his patent in suit. Bray embodied his invention in a pump of the so-called inverted barrel type, comprising a standing plunger and a traveling barrel, by placing his apertured part 3 (which is also designated in his patent specification as a 'jet device,' 'bushing' and 'bottom guide'), on the lower end of the pump barrel B, and locating his clutch, comprising jaws 4 and coupler head 2, between the bottom of the barrel B and top of the anchor P, so that, on the down-stroke of the barrel, oil in the well tubing T passes from said tubing up through the apertures 5, in said part 3, into the jet pressure chamber EE (Defendants' Exhibit MM, Book of Exhibits page 178), and, on the up-stroke of the barrel, oil in said chamber is positively projected therefrom directly downwardly, through said apertures 5, in strong jets toward the anchor P and coupler head 2, for washing away any sand, which would otherwise settle down through the oil, between the barrel B and well tubing T, upon said anchor and coupler head, whereby 'sanding up' of the pump is prevented, so that the pump may be readily pulled out of the well tubing T, or the pump barrel B lowered to interlock the clutch members 4 and 2, and prevent rotation of the barrel, in order that the loose or disconnected joints of the pump sucker rods may be screwed up tight by a large wrench applied to a sucker rod at the top of the well." (Opening Brief pp. 36-37.)

This statement is itself a confession of the soundness of the finding under discussion, and in the light of it assignments of error XXXVIII and XXXIX must be deemed abandoned.

B. THAT SAID CLAIMS PRESENT NO OTHER FEATURE OF ASSERTED NOVELTY EXCEPT THE INTERLOCKING FEATURE WHICH IS THE SUBJECT OF FINDINGS VIII AND IX HEREOF.

Questions I and II.

The mere reading of appellants' statement of the problem and its asserted solution set out in paragraph 5, immediately preceding, makes clear that this finding is correct. Yet, despite the fact that the statement amounts to an admission that the solution of "gas lock" was no part of the problem which confronted Bray, and the more important fact that, as we have pointed out at pages 22, 23, *supra,* there is no possible co-operation of the action between the "arangement for expelling liquid from the barrel" to wash the anchor and the positioning of the valves so as to best minimize "gas lock," and that such wholly

independent elements to perform this wholly independent
function could not therefore be combined to form a valid
combination claim, appellants state on page 16 of their
brief:

> "Claim 2 covers a pump of the type disclosed in
> the specification and drawing, with the standing valve
> S and the traveling valve V in close relationship to
> prevent a 'gas lock,' in combination with the bottom
> guide or part 3, on the lower end of the traveling
> barrel B, provided with a series of venting ducts 5,
> and the clutch comprising the coupler member 4, on
> the lower end of the part 3, and the coupler member
> 2, on the top face of the anchor P, to be non-rota-
> tively interlocked, said ducts 5 opening directly
> toward the anchor coupler 2, so that during the up-
> ward strokes of the barrel B, ejected liquid is impelled
> or jetted through the ducts 5 toward the anchor top
> face and coupler 2 to aid in prevention of sand accu-
> mulation thereon."

This obvious attempt to read into this claim an element
not mentioned therein, performing a function wholly
unrelated to the function of the claimed combination, in
order to meet an asserted problem not even mentioned in
the specification is plainly without justification. An examination of the assignments of error discloses, moreover,
that appellants have wholly failed to assign as error the
action of the trial court in making the finding considered
in this paragraph, and for that reason the alleged failure
of the evidence to support it cannot be questioned by them.

C. THAT THE PATENTEE CONTEMPLATED THAT THE GUIDE USED SHOULD HAVE A CLOSE OR ORDINARY MECHANICAL FIT WITH THE STANDING COLUMN AND THAT HE INTENDED TO OBTAIN THE DESIRED EFFECT OF WASHING OFF THE COUPLER FACE OF THE COLUMN FIXTURE.

D. THAT THE TEACHINGS OF THE PATENT ARE SPECIFIC IN TERMS, AS THE PATENTEE STATES THE "BUSHING IS PROVIDED WITH A SERIES OF DUCTS 5 FROM TOP TO BOTTOM SO THAT AS THE BARREL SLIDES UP FROM BOTTOM POSITION LIQUID WILL BE FORCED DOWN IN THE STRONG JETS AND WASH SAND FROM THE TOP OF PIN P SO THAT THIS CAN BE READILY PULLED FROM ITS SEAT IN LOCK BOX L."

E. THAT CLAIMS 2 AND 3 ARE LIMITED CLOSELY TO THE FEATURE OF HAVING A BOTTOM GUIDE RUNNING ON THE STANDING COLUMN, WITH A RUNNING FIT, THE SAID GUIDE BEING PROVIDED WITH VENTS OR APERTURES FOR THE EJECTION OF LIQUID DIRECTLY TOWARD THE EFFECTIVE FACE OF THE LOWER COUPLING MEMBER OR COLUMN FIXTURE AND THAT THE INVENTION IS TO BE STRICTLY LIMITED TO THE FORM AS SO DESCRIBED.

Questions I and II.

1. It is obvious that, at a given speed of operation, the strength of the jet produced on the up stroke of the barrel will vary with the capacity of what we have termed the annular space, relative to the capacity of the openings through which the liquid is forced. To state that it is desirable to prevent sanding up, and that it might be prevented by jets of liquid agitating the sand, without show-

ing the mechanism by which jets capable of so agitating the sand could be produced, would not be invention. It would merely state a problem, and indicate a possible solution. It would be like saying, "It is desirable to travel to the moon, a rocket may be provided to do it." Invention? Not at all.

2. As the solution or, failure of solution, of Bray's problem turned on the efficacy of the jets to remove sand from the top of the anchor, it was necessary—if any of the claims were to be held valid—for the trial court:

a. To find in the specification a disclosure and description of a structure capable of producing that sand removing type of jet.

b. To find this structure claimed in one or more of the claims.

This the court did, and found that one element of such a structure was the use of a guide having "a close or ordinary mechanical fit."

3. If the Bray specification and claims teach anything concerning a structure capable of producing such sand removing jets, it is that the structure must employ a *"guide" running or operating* on the column in such fashion as to make a running fit on the column—thereby providing a compression chamber for the generation of the force necessary to eject liquid in such jets,—and that this tight fitting guide shall be provided with a series of ducts or be otherwise so apertured to cause the "positive jet action" or the "strong downward jets" referred to in the patent.

Precisely this structure is shown in the patent—no other is shown nor suggested. It is significant that element 3 of Bray is actually a guide, having a series of apertures described as being designed for the ejection of liquid, and differs in design from the guides of the prior art (with the exception of Ellis).ᐧ The fact that the apertures in the guide were designed for the *direct downward* ejection of liquid was, in the prosecution of the patent, relied upon as the *raison d'etre* of the patent (File Wrapper, Bk. Ex. p. 23). It is only in this tight fitting apertured guide, adapted to cause forcible ejection downward, that any distinction over the ordinary plain puller nuts, sleeves, bushings, heads, slush nuts or other like structures can be contended to lie—for all manner of puller nuts, both loose and tight, have been shown to exist in the prior art. (See par. 9, p. 57, *supra.*)

In the specification (Bk. Ex. p. 4, lines 35 to 45), Mr. Bray, in speaking of the *fit* between the plunger portion of the standing column and the barrel, states as follows:

> "The guides 13 are of an alloy which allows the guide to be fitted and operated closer to the barrel than with any other known metal plunger and without heating and consequent freezing of the pump due to expansion of the inner parts. In other words this bronze plunger or guide is highly efficient as a packing, runs close and reduces slippage of oil past the guide and has long life and is non-heating." (Bk. Ex. p. 4, lines 35-40.)

The fact that he here uses the apt and accurate word *"guide"* to describe a part fitting so closely, that only a particular sort of metal is suitable to make it, is demonstrative of the fact that he meant an exceedingly close fit when he used the same word in describing another part, likewise calculated to produce compression. Obviously he cannot be heard to say that, having used the word *"guide"* to express the thought of a compression member having such a fit, he used the same word in designating another compression member—intending thereby to indicate as loose fitting a member as he might desire to claim in order to assert infringement. His drawing, moreover, buttresses and reinforces his written description, for he shows no clearance at all in either case.

It is also to be noted that the valve traveling barrel, with its special alloy fit, is spoken of in the claims as *"running on* the column," and the puller nut guide is likewise referred to as *"running on* the column" in claims I and II, and *"operating on* the column" in claim III.

Can there be a difference in the meaning conveyed by the words *"running on* said column" when used in succeeding phrases in the same claim? To say language could be so elastic, would be to say it was incapable of conveying thought.

4. If from what we have already said the necessity of this interpretation is not at once apparent, recourse may be had to the file wrapper and contents, Exhibit "B."

The original claim 6 of the application for patent, as amended by paper No. 5, reads:

> Claim 6. "A standing lock member supporting a standing stem and a jet bushing member on a travelling barrel operatively coupled with the stem to pull it and the locking member, said members having intermeshing coaction to prevent relative rotation." (Bk. Ex. p. 14.)

This claim clearly sought, in the use of the term "jet bushing member," etc., protection for any kind of bushing, whether loose or otherwise, so long as it produced a jet, and was used in connection with the clutch, the traveling barrel and the standing column of the patent.

This claim was rejected by the patent office in paper No. 6 (Bk. Ex. p. 22). This rejection was acquiesced in without argument. The claim was cancelled at request of applicant in paper No. 7, Bk. Ex. p. 23, and the present claims of the patent, narrowed and restricted to "a *guide running* on the column," then filed. Yet plaintiff is here contending for a breadth of construction which would give him a monopoly in the use of the very structure denied him by the rejection and cancellation of the broader claim.

5. If Bray does not teach the production of his jet through the compression caused by a tight fitting guide *running on* the column, he has not taught how to produce it. If the patent in suit is not read as requiring a guide "running or operating on the column" in the sense of making a running fit thereon, the patent fails to describe or claim a structure capable of producing the jet action necessary to eliminate the sanding up which Bray decries. So read, the patent would amount merely to the suggestion that by some possible combination of a loose fitting guide, jets might be produced. The question of the amount of clearance of the guide would be left open for the determination of the person attempting to follow up the suggestion. Clearly if this clearance is great enough no jet would be produced. Mr. Bray has testified that in the structure he makes, as distinguished from the structure shown in the drawing of the patent, he has reached the maximum of clearance—in order to obtain a free

dropping barrel—which will at the same time produce the desired jet, and that he determined this maximum clearance from his practical experience. (See p. 41, *supra.*)

To attempt to give to the Bray patent a breadth which would cover any structure that experiment might show would produce jets of the required magnitude, as distinguished from an interpretation covering the structure described and shown, would be equivalent to a confession that the specification and claims do not cover a described and determinable device, but rather indicate a suggested field for experimentation. The patent so read would not have solved the problem stated by Mr. Bray through describing a structure producing sufficient compression to result in the strong downward jets which he contends do such useful work, and by the same token the claims so read would not define invention.

Question I.

1. As we understand appellants' brief, they do not assert that the court erred in construing claim 2, as containing the limitation of a "bottom guide on the lower end of the traveling barrel "B" provided with a *series* of venting ducts 5," for despite the confusion necessarily arising from their failure to separate their argument in its application to each of these claims, they do specifically state on page 16 of their brief:

> "Claim 2 covers a pump of the type disclosed in the specification and drawing * * * *in combination* with the bottom guide or part 3, on the lower end of the traveling barrel B, *provided with a series of venting ducts 5,* and the clutch * * * said ducts 5 opening directly toward the anchor coupler 2,

so that during the upward strokes of the barrel B, ejected liquid is impelled or jetted through the ducts 5 toward the anchor top face of coupler 2 to aid in prevention of sand accumulation thereon." (Opening Brief p. 16.)

and give unmistakable point to this admission by adding:

"Claim 3 recites the same combination of elements as covered in claim 2, but in broader terms, *so as to cover varying forms of venting ducts or jet apertures, other than 'a series of venting ducts (5) as specified in claim 2.'"** (Opening Brief p. 17.)

2. Since liquid would not be jetted through the *series of ducts 5* if the compression necessary to so impel the liquid, was not generated by a tight fitting guide, and since there is no contention in the brief of appellants, that Bray contemplated or taught that his structure should have both the *series of ducts* and an *annular clearance* between the column and the guide, or that such a structure is covered by claim 2, it follows that the findings under consideration are not attacked insofar as they refer to claim 2, and that assignments of error XLII, XLIII, XLVI, XLVII and XLVIII in so far as they refer to claim 2, must be deemed abandoned.

Question II.

1. It seems to be the contention of appellants that the evidence fails to support this finding as it relates to claim 3. Their argument is as follows:

"Claim 3 recites the same combination of elements as covered in claim 2, but in broader terms, so as to cover varying forms of venting ducts or jet apertures, other than a *'series of venting ducts'* (5), as *specified in claim 2.* The last five lines of claim 3 are as follows:

> 'said guide being *apertured* for the ejection of liquid from the barrel directly toward the effective face of the lower, standing coupler part to aid in preventing sand settlement thereon.'

The term 'apertured' in claim 3 is a broad term, which does not limit the claim to any particular arrangement or form of aperture, but comprehends any *suitable arrangement or form* of aperture, 'for the ejection of liquid from the barrel directly toward the effective face of the lower, standing coupler part to aid in preventing sand settlement thereon.'

This broad construction of the term 'apertured' is fully supported by the patent specification, page 1, lines 41 to 49, inclusive as follows:

> 'Other objects, advantages and features of construction, and details of means and of operation will be made manifest in the ensuing description of the herewith illustrative embodiment; it being understood that *modifications, variations, and adaptations* may be resorted to within the spirit, scope and principle of the invention as it is hereinafter claimed.' " (Opening Brief, p. 17.)

2. It is first to be noted that no substantial difference in the device of the patent is suggested or intended by the slight change in wording of the claim. The file wrapper and contents makes this plain (Paper No. 9, "Remarks"; Book of Exhibits p. 26) reading:

> "Claim 17 (claim 3 of the patent as issued) is based on features as of allowed claim 16 (claim 2 of the patent as issued) but in different structural terms."

3. The only difference of any materiality at all is that claim 2 limits protection to a combination employing a guide "having a series of venting ducts," while claim 3 seeks to protect a combination likewise employing a "guide," which guide is to be "apertured for the ejection of liquid."

4. Appellants argue:

"Claim 3 describes a bottom guide *operating* on the standing column' and the term 'operating' is a broad term which is certainly comprehensive enough to include a very loose fit of the jet device 3 around the standing column member 11, through which loose fit liquid could be jetted downwardly out of the lower end of the traveling barrel towards the anchor and lower clutch member. Claim 3 does not specify a series of jet apertures, but specifies the guide, meaning jet device 3, as being 'apertured' for the ejection of liquid, and an annular space, formed by a loose fit between the guide and standing column, would certainly be an aperture in the guide, which would make the guide apertured in accordance with the terms of the claim." (Opening Brief p. 83.)

5. This, in our opinion, is exceedingly loose reasoning. In the first place, according to the appellants' own statement:

"claim 3 * * * specifies the guide * * * as being 'apertured' for the ejection of liquid."

The portions of the claim material to this point read:

"The combination . . . of a bottom guide; on said barrel, operating on said column . . . and said guide being apertured for the ejection of liquid from the barrel. . . ."

It will be noted that the claim calls for two distinct things. It calls:

(1) for a bottom guide on said barrel;

(2) for said guide being apertured.

As we pointed out at the trial and in our brief below, a block of metal might in all other respects be similar to a guide, but it could never *be a guide nor a puller nut* until a hole had been drilled in it through which the standing column can pass. Counsel seemingly assumes that the very hole essential to produce "a bottom guide on said barrel" may likewise satisfy the requirement of the claim reading, "and said guide being apertured." This, in our opinion is an error. The claim clearly requires that there be first a guide, second, that said guide be apertured. In defendants' structures there is but one hole. Through this hole the standing column passes. The guide so formed is not in addition apertured. No attempt is made by appellants to anticipate this pertinent observation."

At pages 17, 24 and 26 of their brief, appellants state in varying terms that:

"the term 'aperture' in claim 3 is a broad term which does not limit the claim to any *particular arrangement or form of aperture.**"

We did not assert below, and do not now assert, that this claim is limited to any "particular arrangement or form of aperture" but do assert that there must first be a guide with a running or mechanical fit on the column, and second, that that guide must be apertured with some appropriate arrangement or form of aperture, *i. e.,* either ducts or channels, or like passages, so that the strong downwardly ejected jets of the patent are projected through such apertures. The error of counsel lies in his assumption that because the claim is not limited to "any particular arrangement or form of aperture", that it follows that the claim is satisfied by a structure employing a guide which has no ducts or apertures at all, but is merely provided with the opening through which the standing column passes, which opening makes the block of metal out of which the guide is fashioned, a "guide." An opening of this kind with varying clearances, has been shown at pages 57 to 59 (*supra*) to be old in the art and to be characteristic of, and necessary to, every traveling barrel pump.

6. Appellants claim a construction that is clearly too broad:

    (a) Because it ignores the necessary distinction between a guide which is a device having a guiding action and a puller nut which is a device by which one element is pulled by another. (See Par. 9, pp. 57 to 59, *supra*.)

    (b) Because such a construction is shown in Thompson, Ex. "T", (Bk. Ex. p. 119)
Wright, Ex. "U", (Bk. Ex. p. 125) and
Neilsen, Ex. "Y", (See Par. 9, Subdivisions (a), (b) and (c), pp. 57 to 59, *supra*.)

    (c) Because it ignores the fact that the Admore pump has identically the same means for ejecting a portion of the trapped liquid downward through the top part of the locking device. (See Par. 7, pp. 110 to 113, *infra*.)

    (d) Because it cannot be invention to decrease the size, or close up the holes in the barrel of Admore, which is all that Bray has done if his patent is so read.

A very illuminating indicia of the lack of fairness of the approach of plaintiffs' expert Hayn, and of the value to be given to his testimony, is to be found in the fact that he readily asserted, on pages 347 and 348 of the transcript, that to simply close these holes would involve invention. He was later forced to admit that he did not know what, if any, increase of jet action would result from such closure [Tr. pp. 350-353]; and stated that he had no opinion at all as to whether any increase of flow, no matter how slight, would amount to an inventive act; and later qualifying even this statement by admitting that invention could not result if the result were "frivolous". [Tr. p. 355.]

(e) Because Bray, in the proceedings in the Patent Office, cancelled after rejection a claim of the same breadth as this construction, and filed the present claims, limited to a close fitting compression producing "guide" in place thereof. (See Par. 4, pp. 102-103, *supra*.)

7. At pages 39 and 40 of their brief, appellants seek to distinguish the Admore pump from the pump of the Bray patent by claiming that since the Admore pump has holes in the side of the travelling barrel above the puller nut to permit the escape of oil, trapped in what we have heretofore termed the annular space EE, that the oil escaping through said side holes:

"* * * is projected latterly from the barrel directly against the side of the well tubing in which the pump is located and it is not possible for the oil to be projected directly downwardly through said holes out of the bottom of the barrel in strong jets against the lower clutch member and the anchor as covered by claims 2 and 3 of the patent in suit."

Counsel then attempts an analysis of the effect produced by oil passing through said holes in the barrel and at page 40 of their brief, state:

> "It should, therefore, be obvious that it would be impossible for this Admore pump to project a jet of oil directly downwardly from the bottom of the barrel 1 toward the clutch and anchor on the up stroke of the barrel to wash sand away from the clutch and anchor and consequently the Admore pump could not possibly function like the Bray pump as covered by claims 2 and 3 of the patent in suit."

This observation is simply untrue, and springs, no doubt, from the fact that counsel has not carefully examined the Bradford Admore pump. We ask the Court to check the structure itself, and to note that the clearance between the inside of the puller nut and the standing column is substantial and of such magnitude, that the area of the opening created thereby is approximately the same as the area of the four holes in the barrel, *i. e.,* the area of the holes is approximately 4/7 and the area of the annular clearance approximately 3/7 of the total area available for the escape of fluid. [Tr. p. 211.] It follows that the most that counsel could say with accuracy as to the differences between the structures, is that, paraphrasing counsel's language:

> "The Admore pump has holes * * * in the side of the travelling barrel * * * to permit the escape of *a portion** of the trapped oil therethrough". (Op. Br. p. 39.)

It is clear that they should add,—and their failure to do so is most significant, the *remaining and equal portion of said trapped oil is projected directly downwardly be-*

*tween the standing column and the inside of the puller
nut so as to directly wash the upper face of the anchor
and the lower jaw of the clutch.*

On page 39 of his brief counsel for appellants in an
attempt to distinguish the Admore pump *completely
ignores the escape of fluid through this annular clearance,* and advances a rather unique theory as to the action
of the fluid discharged from these lateral holes. He
states that any jet will be "immediately dissipated" and
"completely destroyed" by striking against the tubing and
that such oil will be drawn upwardly by the "upward
movement of the barrel" and "the differential in pressure
of the oil in the well tubing".

Counsel's suggestion that the lateral jets will be dissipated by striking the wall of the tubing is based upon the
erroneous assumption which runs throughout his entire
brief, that the jets produced by the structures considered
are of a high velocity and hence comparable to such a jet
as issues from the nozzle of a garden hose. This is not
the case, as has been pointed out on pages 121 to 123
hereof. They are in fact of low velocity and are rather
comparable, as Mr. Adams stated to "a very slow flowing
river, say, two miles per hour, to be on the very conservative side." [Tr. p. 197.] If a high velocity jet issued
from an orifice and impinged upon an opposite wall, a
substantial dissipation of the jet and a consequent loss of
its energy might be expected, but a fluid moving at the
rate of a slow moving river will strike an obstruction such
as the wall of the tubing and be bodily deflected in direction with very little loss of velocity or energy just as a
river may make a sharp turn without impeding its flow.

It is plain that the "upward movement of the barrel" and the "differential in pressure of the oil in the well tubing" is present in both the Bray and the Admore pumps. Mr. Franklin does not attempt to explain why these asserted jet dissipating and destroying forces do not operate in precicesly the same way upon a downwardly projected jet, as he asserts that they do upon a laterally projected one, and his theory obviously lacks any basis for such a distinction. We would, therefore, be entirely justified in accepting the theory advanced by him despite its questionable physics, and, standing upon his theory, assert that it results in the dissipation and destruction of the jets of the Bray patent,—the effectiveness and usefulness of which he has elsewhere in his brief repeatedly asserted. We prefer, however, to call the attention of the Court to the real action of the fluid ejected from the holes of Admore, *i. e.*: Upon the upward movement of the barrel an area of reduced pressure below the guide is created, and the fluid issuing from the lateral ports is immediately deflected downwardly, relative to the lower end of the barrel, to fill this area, precisely as fluid flows downwardly in the Hofco pump with its annular clearance. We believe that this analysis likewise makes clear, that any differential or difference in head pressure of fluid standing in the well tubing is offset or negatived by the reduced pressure, created immediately below the guide or puller nut 6 of Admore, so that any such differential pressure has no influence whatever upon the building up or dissipation of any jet produced at the bottom end of the traveling barrel, irrespective of whether the jet openings are vertical or lateral.

8. From the foregoing it is clear that the use of the word "apertured" can at the most have merely the effect

of extending protection to a *true guide* "apertured" in the sense that instead of individual bores as shown in the drawing of Bray, distinct channels are cut out on the inner face of the guide, as in Fig. 2 of Ellis (Bk. Ex. p. 133) or in the rosette like guide of plaintiff, Defendants' Physical Exhibit "A".

That plaintiff did not contemplate that his device covered a loose puller nut, as distinguished from a true guide is shown by his statement, that he first employed this rosette like guide but found that the bearing ribs between the channels wore away and then adopted the guide with ducts in order to present a larger bearing surface. (See pages 39 and 40, *supra.*) It is true that Mr. Brady testified that before he made application for his patent he had "built a pump," employing a loose fitting puller nut instead of a guide [Tr. p. 42], but he *produced no such structure,* and although a blue print was mentioned [Tr. p. 42] plaintiffs-appellants did not introduce it in evidence. Instead Mr. Bray produced the rosette like guide and admitted that when, through wear, it approached in character such a loose fitting puller nut, he adopted a structure designed to eliminate this wear by resorting to the duct type guide. In other words, *he deliberately sought and found a structure designed to overcome the automatic production of the very structure he now contends his claim covers.*

9. Moreover, this contention of appellants, *necessary though it is to their claim that appellants' structures employing a loose puller nut or stop infringe the Bray patent,* leads them into the dilemma mentioned in par. 5, page 103, *supra,* i. e., such a suggested construction of the Bray claims is equivalent to a confession that the specification

and claims do not cover a described and determinable device but rather indicate a suggested field for experimentation.

10. A study of the actual operation of appellants' pump in producing the so-called "jets" demonstrates that a close fitting guide is an essential element if any semblance of a useful, work accomplishing jet is to be produced.

(a) On pages 27 to 28 of their brief, appellants complain that although the affirmative defense of inoperativeness or lack of utility of the Bray invention was not set up in the amended answer, a considerable amount of evidence was given by defendants' witness Adams in support of such defense; and that although the lower court did *not sustain* the defense, considerable weight must have been given by the court to the defense because the court considered it strong evidence:

"tending to show that the effect claimed for the jet action would not result in operation—this to the point of lack of utility." (Op. Br. p. 27.)

Appellants further stated that this evidence was:

"ingeniously calculated to belittle and minimize the meritorious invention in suit, and it is submitted that the court was thereby unduly influenced and prejudiced against the invention and patent in suit, giving claims 2 and 3 of the patent a narrow and unwarranted construction, which unjustly enabled the defendants to escape infringement of said claims." * * * (Op. Br. p. 28.)

and that:

> "Defendants' evidence supporting the defense of inoperativeness or lack of utility, which was not pleaded, was a surprise to plaintiffs, and prevented plaintiffs' expert, Fred H. Hayn, from giving sufficient study and time to said defense, in order to refute the fantastic theories and pseudo mathematics, set forth in the testimony of the defendants' expert, Adams." (Op. Br. p. 28.)

This complaint is wholly without merit because:

(1) While it is true that defendants-appellees did not plead "lack of utility" as a defense, the court did not find such lack of utility.

(2) The testimony was admitted without objection or exception.

(3) Plaintiffs did not advance at the trial any such apology for the total failure of their expert, Mr. Hayn, to meet the testimony of Mr. Adams, and we submit it is now made by them without any support in the record.

(4) If the testimony of Mr. Adams had in fact been a surprise to plaintiff, and had it prevented plaintiffs' expert from "giving sufficient study and time to refute the fantastic theories and pseudo mathematics" of Mr. Adams, the remedy was for plaintiffs to ask a continuance if necessary to meet it. This they did not do.

(5) The testimony of Mr. Adams, which appellants correctly state was considered by the court to be "strong evidence" that the effect claimed for jet action "would not result in utility in operation— this to the point of lack of utility", was highly

material for the purpose of showing that even in a device constructed in strict accordance with the teachings of Bray, the so-called "jet action" was inconsiderable, and that in a device constructed as were the alleged infringing devices of appellees, there was no such jet action as would accomplish even this inconsiderable effect; and that if the claims were to be so construed as to claim a device that was patentable as possessing "utility", they must be read as covering only a device constructed in strict accordance with the disclosure of Bray. *This was in fact precisely the holding of the trial court upon the record.*

11. We will now analyze the evidence advanced by appellants bearing upon this asserted jet action.

(1) We believe that Mr. Bray was honest in his expressed opinion, that the structure claimed in his patent, as well as the structure which he actually builds and sells, has a jet action accomplishing useful work. But his testimony amounted to nothing but an expression of opinion by an interested party—who attempted no analysis of the possibility of such action, but based his opinion wholly on the fact that in actual operation his pumps do not, he says "sand up". He admitted that in reaching this opinion he ignored the effect of all other factors operating to prevent the sand from settling [Tr. p. 278], such as the turbulence caused by the churning up and down of the travelling barrel; the flow of fluid down between the tubing and barrel and upward, due to the inevitable leak in the seal of the piston; the eddying of the fluid,

filling up the area vacated by the barrel as it travels up; and erroneously ascribes to his supposed jet the results obtained as a combination of all these factors. [Tr. pp. 198-199.]

The only thing concretely to be gathered from the observation of the actual working of the pump negatives his claim of any effective jet action. He has testified that when the last packing of his pump fails (and parenthetically, the fact that Mr. Bray lets his love of his brain child color his opinion of its functioning and of the results it accomplishes, is found in the fact that he claims his packing is 100% effective in preventing leakage up to the very moment that the last ring of non-metallic packing fails)—when this packing fails, Mr. Bray testified his pump immediately fell off some 20 to 50 per cent in effective operation [Tr. p. 289], and that his pump often operated in that condition for weeks at a time, without sanding up. [Tr. p. 277.]

Since in a pump running in this condition the jet action would be negligible, if not entirely absent, it follows that the freedom from sanding up cannot be due to any such jet action. Why do we say this? When the packing fails to such an extent, the always present leak of the fluid [Tr. p. 213], standing in the tubing, downwardly between the tubing and the barrel, and upward through the clearance between the puller nut and the column [Tr. pp. 182-183], is at once sharply increased. [Tr. p. 289.] This is true because, as Mr. Adams explained, the difference between the pressure in the pumping chamber which is only rock pressure, (since it communicates through the hollow standing column with the pressure in the formation) and the total pressure of the fluid standing would in California run to 4000 and upward feet of head in the tubing [Tr. p. 173]. As Mr. Bray admits, in the case of failure of the piston seal this leakage at once greatly increases in amount. [Tr. p. 289.] This large upward leakage neutralizes the jet action to the extent of the leakage, so that the flow in the case of a pump in such a defective condition must be up through the puller nut into the pumping chamber, rather than downwardly through the puller nut. [Tr. p. 183.] It is obvious that this upward flow will continue, until the pressures are equalized. The barrel will then move downward and upward through a fluid of equalized pressure, neither pumping nor producing a jet. It is likewise clear that, because of the limited volume of the annular space above the "guide," all jet action would have been lost long before this equalization occurs. Mr. Bray stated that as a matter of fact under such circumstances his pumps do not sand up, which is simply to say, that since they do not sand up when there can be no jet, it is not the supposed jet action which prevents "sanding up" at any time.

(2) Another observed fact, that shows that failure to sand up, is *not* due to jet action, in the case of the Hofco pumps at least, is the fact that such pumps do sometimes sand up. Mr. Hoferer testified the barrel may in such cases be reciprocated up and down, but that such movement will not result in removing the sand. [Tr. p. 365.] He testified that the barrel of the pump must, in such cases, be lowered until the puller nut is in contact with the sand, and the sand actually

drilled out by mechanical pounding [Tr. p. 365], which pounding and drilling may be felt by the vibrations transmitted to the polish rod above. [Tr. p. 370.]

(3) Lastly, if there were in fact any jet action, of great magnitude, driving the jets laden with sand in suspension downwardly against the top of the anchor part, thereby washing away such sand as would otherwise settle there, the continual impact or washing of such sand carrying liquid, forcibly brought into repeated contact with the same place on the anchor part, could not fail to abrade it and thus leave positive evidence at the point of such impact. Appellant Bray testified he had operated his pumps for years. [Tr. p. 40.] If such evidence existed he could not have overlooked it, and it would have been produced in the Court below.

12. Mr. Adams, a witness for defendants, made a careful examination of the forces involved in the operation of the various structures in suit. It will be remembered that the record discloses that plaintiffs did not, in the trial court, dispute either the correctness of the natural laws concerning which Mr Adams testified, nor the accuracy of his arithmetical calculations, although the measurements involved were stipulated into evidence long before the close of the case.

In their opening brief appellants content themselves with branding Mr. Adams' testimony with such epithets as:

"fantastic theories and pseudo mathematics"

without showing wherein the theories were "fantastic," or the mathematics pseudo. Such few specific criticisms of Mr. Adams' testimony as appellants indulge in (Op. Br. pp. 73 to 78), will be considered at the conclusion of our analysis of Mr. Adams' testimony. That testimony may be analyzed and summed up as follows:

(a) That the so-called jet action was often the minor sand stirring force, and a large part of the time non-existent; [Tr. pp. 171-172]

(b) That such downward flow as resulted was produced only on the upstroke of the pump, and in point of quantity of fluid emitted, amounted to but an ordinary water glass or a glass and a half of fluid every three seconds, when the pump was operated at the normal speed of twenty strokes per minute. [Tr. p. 195.]

(c) That due to the motion of the crank, downward flow was non-existent when the barrel was near the lower end of its stroke; started as the barrel gained appreciable velocity; slowly increased as the velocity increased, reaching maximum intensity when the guide (originally offset some 6 inches from the top of the anchor) had traveled a considerable distance upwardly away from the anchor; then decreased until at the upward end of the stroke there was again no downward flow. [Tr. pp. 185 to 188, 194-195.]

(d) That any such downward flow is counteracted by the upward movement of the guide which of necessity moves upward and away from the parts to be washed in order to produce downward flow. [Tr. p. 188.]

(e) That in every pump there is leakage upward between the plunger and the barrel. This leakage
is a constant factor. At slow speeds the only
flow from the annular space is upward, because
the volume of fluid escaping by leakage is greater
than the volumetric reduction of the annular space
caused by the upward movement of the barrel.
[Tr. pp. 182-183.]

(f) That there was, therefore, no downward flow
whatever when the guide was near the part intended to be washed. [Tr. p. 198.]

(g) That there was a turbulance or churning action
occasioned by the displacement of fluid as the bar-
rell moved downward. [Tr. pp. 180, 199.]

(h) That as the barrel moves upward, fluid both
from the opening in the guide and from the
column of oil between the pump and the tubing,
flows into the space vacated by the upward
travelling barrel. [Tr. pp. 179-180.]

(i) That any downward flow was into a body of oil
of considerable viscosity, thickened and made of
greater density by sand in suspension, and this
down flow must move through this sand laden and
viscous medium. [Tr. pp. 200-201.]

(j) That any downward flow is lessened by the eddy
action resulting from all those interacting forces.
[Tr. p. 200.]

(k) And finally, disregarding all of these elements,
tending to counteract downward flow, except the

factor of upward leakage or slippage past the plunger, Mr. Adams arrived at the following figures in feet per second [Tr. p. 197]:

> For the structure shown by the Bray patent   11.76
>
> For the structure actually made by Bray   4.46
>
> For the structure made by Hofco   1.36

He testified in substance that these figures give the average flow downward during one stroke of the barrel and are proportional to the effectiveness of the "jet", that there are instantaneous values greater and less than the average, the maximum at midstroke is about 1.57 times the average [Tr. p. 196], which is an instantaneous value soon lost and balanced by the periods at each end of the stroke when the flow is into the annular chamber instead of downward from it. The average is therefore a true measure of efficiency and this velocity acting during one stroke of the barrel represents the stirring action achieved. When reduced to miles per hour this velocity in the Bray structure is three miles per hour and in the Hofco structure less than one mile per hour. Three miles per hour is a velocity less than that of a slow flowing river, which velocity cannot possibly be called a jet. [Tr. pp. 197, 198.]

13. While the Bray patent mentions "positive" and "strong" jet action, and a study of the comparative jet velocities is therefore useful, the factor with which we are really concerned is the relative energy of the jets or their capacity for doing work.

The Court will take judicial notice that energy as defined by Webster, or as given in any Engineering text, is equal to M $V^2/2$, or, energy equals one-half the mass times the square of the velocity. Now mass is simply weight divided by gravity, and since the quantity of liquid emitted is practically the same in all three cases, it follows that the weight and mass is also the same, and the only variable is the velocity. Therefore, the capacity for doing work is exactly proportional to the square of the velocity of flow in each case. These relative values are tabulated below:

	Velocity	Capacity for doing work
Bray of patent	11.76	138.
Bray as now built	4.46	19.9
Hofco as built	1.36	1.85

We have not attempted to carry this calculation to the point of expressing energy in terms of foot pounds (the usual method of expressing energy). In this case we are only interested in the relative capacity of the three structures for doing work. The above table discloses that the pump of the Bray patent has 74.59 times the capacity for doing this work that is possessed by the Hofco structure. It immediately becomes apparent that the Hofco pump is *not* designed to do the work that Bray asserts his patent teaches.

14. The design of the Hofco pump is, on the contrary, such as to produce a free dropping barrel, as distinguished from a barrel the free dropping of which is retarded by the deliberate building up of compression between the bottom of the piston and the puller nut, in order to gen-

erate the jets. [Tr. pp. 220 to 226.] Mr. Adams so testified, and his testimony is corroborated by the Hofco circular, Exhibit 4 (Bk. Ex. p. 8), which reads:

"DESIGNS"

"The design of all three of these pumps are along the same lines, all being of the removable loose-fitting, free-dropping, maximum capacity per stroke make up.

It is worthy to note that the loose fit of these pumps offer positive action of the valves (balls and seats) at all times and thus keeps the rods taut, this eliminating the snapping of rods. *The fit is so loose that rod pick-up is often hardly discernible.*"*

and under the heading, "WORKING ACTIONS",

"as has been mentioned, the working actions of the Hofco Pumps are such that a *loose, free dropping fit is always maintained,*"*

It is significant that the Hofco circular nowhere speaks of any jet action, but simply states:

"likewise, the surge of the fluids in and out of the lower end of the reciprocating barrel keeps all sand and foreign sediment stirred and permits pulling of pump at any time without undue trouble."

We have this "surge". The above table furnishes proof that it adds little, if anything, to the churning action of the barrel and to the displacement flow. It was old in 1919 with Northrup. It is present in the Admore. It is present, in a greater or less degree, in each of the prior art traveling barrel pumps employing a loose puller nut.

15. One of the most amusing inconsistencies of the appellants' brief is that although they argue that since we have a clearance between the puller nut and the standing column, it follows that we necessarily have a jet producing mechanism and state,

> "It is immaterial what term is used in the patent to describe the element 3 or any other element, provided the *function* of the element is made clear by the context of the specification. The function of element 3 is made perfectly clear in the specification, page 2, line 31 of the patent, where it is further described as a *'jet device 3'*." (Opening Brief p. 10.)

they as readily assert, (when seeking to distinguish structures of the prior art having varying forms and degrees of clearance) that such clearance is merely "a vent" to admit of *free* entrance or exit of the fluid held in the space, corresponding to the annular space EE of the Bray pump. (Defendants' Exhibit MM, Bk. Ex. p. 178.) Examples of such argument is to be found relating to the Wright patent.

> "On page 1, lines 79 to 82, of the Wright patent, it is stated that the 'aperture e is, however, of a somewhat larger diameter than the said member b' to admit of free entrance or exit of the fluid held in the space F', the lower end of e' of the working barrel operating as a piston for agitating and forcing the liquid in such space F through the passages b³ out against the rock to wash same at all times during the pumping operation. If the aperture e is somewhat larger in diameter than the member b' to admit free entrance or exit of the fluid, there could hardly be a jet action through said aperture, because a jet action necessarily requires a restricted aperture. *The aper-*

ture e is nothing more than a vent to permit dis-placement of liquid from one side to the other of the lower end e' of the plunger D, so that said plunger may reciprocate.*" (Opening Brief p. 56.)

Relating to the Ellis patent:

"The Ellis patent, No. 1,513,699, Defendants' Exhibits W and W', Book of Exhibits, page 133, shows a pump of the inverted type. The cap nut 21 on the lower end of the barrel 19 is formed with vertical grooves or oil passages 22 intermediate the bearing surface 23 which slidably fit on the tubular standard 12. The plaintiff Bray used the same type of cap nut 21 on his pumps in 1926, but found it unsatisfactory, because the bearing surfaces 23 between the passages 22 wore out and enlarged the space between the standard 12 and the nut 21 which caused the oil to get a little too much through said space. [Tr. pp. 40 and 42.]

The passages 22, in the nut 21, are vents and not jet apertures and are for the purpose only of permitting the free flow of liquid therethrough, to prevent the formation of a dead chamber 24 containing trapped liquid which will resist the free operation of the pump.* (Page 2, lines 37 to 42 inclusive, Ellis patent specification.)" (Opening Brief pp. 60 and 61.)

and relating to the Neilsen patent:

"The white space between the tube 8 and sleeve 16 at the lower end of the traveling plunger 14 in Fig. 1, and the white space between the stationary tube 32 and foot nipple 39 in Fig. 2 of the Neilsen patent, *are nothing but vent openings of substantial size to permit the free flow of liquid into and out of the lower end of said plunger to prevent a hydraulic lock* *in the spaces 15 and 38 of Figs. 1 and 2 of the Neilsen patent*,* which hydraulic lock would of course prevent reciprocation of the plunger and functioning of the pump, in view of the incompressibility of liquid. Said white spaces, in view of their substantial cross-sectional area, are, therefore, not 'jet apertures', and no jet action could be produced through said white spaces, to wash sand from the top of the valve cages 6 and 30 to prevent the pump from 'sanding up'. No such jet action is described or suggested in the Neilsen patent. * * * There is no pressure upon the liquid in the lower end of the pump barrels in the Neilsen pump, because the *liquid flows freely into and out of the spaces 15 and 38 in said barrels,** and consequently no liquid could possibly be jetted out of the lower end of the Neilsen pump barrels under pressure. *The pressure of the liquid in the lower end of pump barrel, for positively jetting the liquid out of the bottom of the barrel through the jet apertures 5, for washing sand from the anchor and the clutch and preventing 'sanding up' of the pump, is a vital feature of the Bray invention. This vital feature is missing in the Neilsen pump."** (Opening Brief pp. 64 and 65.)

The impossibility of performing the task assumed by appellants, who seek to differentiate the structures of the prior art from that of the Bray patent, without by the same token conceding that the alleged infringing structures of appellees are not covered by the claims, springs from the fact that as Mr. Bray testified the pump of his invention was:

"* * * one which will produce the maximum jet action* consistent with an operative pump" [Tr. p. 45]

the Neil.
of course
nctioning
of liquid.
:ial cross-
ures', and

alve cages
up'. No

els in the

'rels,* and
:ted out of
rels under

t apertures

, is a vital
l feature is
ning Br et

assumed by
:ures of the
:out by the
iging struc-
ims, springs
pump of his

e maximu::
pump" [Tr.

and that in his pump, made without ducts or channels and with a clearance of *only three thirty seconds of an inch between the puller nut and the standing column* as indicated on his blueprint [Tr. p. 37], he had provided the *maximum* clearance which would result in jet action. He testified:

> "I think any greater clearance than that indicated on my blueprint would, in the light of my experience, teach me that *I wouldn't get a jet action with any greater clearance."**

> "I intend to imply by my answer, that *I wouldn't get any jet action if the clearance was larger.** The larger it is the less pressure you would get." [Tr. p. 37.]

Mr. Adams corroborated this testimony, stating:

> "There are two theories under which a pump can be designed, and *Mr. Bray in his design very skillfully worked on one theory and Mr. Hoferer in his design has worked on an entirely different theory.** * * * Between Hofco and Bray there is a broad distinction. [Tr. p. 182.] * * * *If you want to create a velocity through this puller nut relative to the puller nut, you would make this orifice 5* (Ex. MM, Bk. Ex. p. 178) *as small as Mr. Bray has made it in his design with the object of getting a jet action.** * * * Whatever jet action you get is paid for by building up pressure in the annular space ("E" of Ex. MM) and retarding the drop of the barrel, so that the engineer of the Bray design very skillfully made a *guide which was tight on the standing tube.** He had orifices, as shown by that puller nut shown in evidence with eight three-sixteenth inch holes. *That would build up quite a material pressure*

in the annular space, and restrict the downward stroke of the barrel.* As you build up pressure there to gain a velocity through these orifices, you by that much restrict the capacity of your pump, because the plunger drops under gravity and the weight of the rods above it. Anything that retards that motion shortens the stroke and that in turn lessens the amount of fluid displaced and the quantity pumped.

We have therefore the Bray type of design in which this clearance is restricted; and another type of design, that of Hofco, which is shown to scale here (Defts. Ex. NN, Bk. Ex. p. 179), *in which the opening through the puller nut is made the maximum amount consistent with having an upper surface left with which the parts may be pulled from the well.** The difference in area is (not) very great but the theory back of it is that if this is made a perfectly large opening, perfectly free, and if you disregard * * * any possible jet action which you get from the interior of the puller nut, you would make that orifice as large as you could and with a low rate of flow through it." [Tr. pp. 188-189.]

Mr. Hayn, the expert of plaintiffs-appellants, *did not deny* this fundamental distinction between the structures of appellants and appellees. This testimony of Mr. Bray and Mr. Adams stands uncontradicted in the record, and counsel for appellants in arguing that the clearance between the guide and the column, in Wright, Ellis and Nielsen, produces not a jet but a "free flow" of liquid, impliedly recognize it. This so-called "free flow" is nothing more or less than the "surge" spoken of in the Hofco circular and we may, with entire truth, paraphrase the

language of counsel for appellants in speaking of the
Nielsen patent and assert that in appellees' structure:

> "The liquid flows freely in and out" of the space EE
> (Ex. MM, Bk. Ex. p. 178) "and consequently no
> liquid could possibly be jetted out of the lower end
> of the" Hofco "pump barrel under pressure. The
> pressure of the liquid in the lower end of pump bar-
> rel, for positively jetting the liquid out of the bottom
> of the barrel through the jet apertures 5, for wash-
> ing sand from the anchor and the clutch and prevent-
> ing 'sanding up' of the pump is a vital feature of the
> Bray invention. This vital feature is missing in the"
> Hofco "pump".

Counsel cannot urge a reading of the Bray claims broad
enough to cover this surge without conceding anticipation
by the structures named.

It follows, therefore, that both the evidence of that
which occurs in the operation of the Bray pump, and the
uncontradicted calculations based on the structure of the
Bray patent itself, discloses that it is questionable if a
structure built strictly according to the Bray patent results in a structure producing jets capable of doing work,
and that if it is made according to the design of the alleged
infringing structures, the record discloses *that there is no
such jet action* and it is therefore plain, that the finding
under examination was the only one which the Court
could make upon the evidence, and save the claim.

16. Beginning on page 72 of their brief, appellants
state that the testimony of Mr. Adams is

> "a labryinth of inconsistencies and confusion."

and by quoting short and disconnected excerpts from his testimony reached the conclusion that:

> "In his vacillation from 'jet' to 'no jet' expert Adams flatly contradicted himself, and the only conclusion that can be drawn from such testimony is that the witness does not know what he is talking about." (Opening Brief p. 73.)

A reading of the witness' entire testimony will disclose that there are no such inconsistencies as appellants assert. The position of Mr. Adams was that while there was a jet at the point under discussion in the excerpt quoted by counsel from page 177 of the transcript, and while he stated in the excerpt quoted from page 179 of the transcript, that there was a jet in the sense that a jet was formed "by an orifice that moved *through a fluid,**" he nevertheless reached the conclusion, after a full analysis of all of the opposed forces which decreased the effectiveness of this jet:

> "there is not in the Bray pump (or), in any of the other pumps that we have discussed, a jet action, at any of the speeds which we have considered. Those are average speeds and the flow would not reach to the dignity of a jet." [Tr. p. 201.]

Counsel for appellants conspicuously omit to quote the next sentence of Mr. Adams, which makes the consistency of his testimony plain. It is:

> "I might note that the jet from a faucet is about 80 to 90 feet per second as compared to 1.36 feet per second from the Hofco pump." [Tr. p. 201.]

We submit that such a flow *does not* "rise to the dignity of a jet," as that term is used in the patent.

Since we assume that counsel for appellants is entirely honest in his expressed belief that there is inconsistency in the testimony of Mr. Adams, we feel it wise to summarize this testimony as it relates to jet velocity. The testimony of Mr. Adams was in effect, that because the orifice in the pump nut is moved through a quiescent fluid, instead of itself being stationary, and the fluid discharged through it [Tr. pp. 178-188] the matter of "relative velocity" becomes material. [Tr. p. 190.] He pointed out that velocity, as commonly considered, is velocity relative to the earth, but that the "relative velocity" of two bodies related to the velocity of such bodies with reference to each other, rather than with reference to the earth or any other body. Mr. Adams illustrated his meaning with the example of a moving train. [Tr. p. 190.] Let us consider a puller nut located in the tubing which is full of fluid, quiescent except for the motion of the barrel and its attached puller nut. As the puller nut is drawn through this fluid, the nut is moving relative to the earth and the anchor of the pump, but the fluid is at rest relative thereto. [Tr. p. 188.] If the nut is moved at the rate of one foot per second, the velocity of the nut relative to the fluid is one foot per second, and the "relative velocity" between the fluid and the nut is correctly said to be one foot per second, although the fluid is actually stationary relative to the earth. Mr. Adams is therefore entirely correct in his so-called "conflicting statements." When the nut is moved, the fluid, although stationary relative to the earth, has a velocity relative to the nut, but because the fluid has no velocity relative to the earth, it has no velocity toward the anchor other than the velocity set up by the friction of the nut on the fluid due to movement of the nut [Tr. p. 188], which may persist to a modest extent as Mr. Adams stated. [Tr. p. 190.] The error in the reasoning of appellants is that they consider the problem to be one in which the fluid moves in relation to a fixed nut, rather than one, as is the actual condition, in which the puller nut is moved through a quiescent fluid.

17. On page 73 of their brief, appellants state:

"On page 190 of the transcript Adams testifies in effect that the jet or 'flow' of oil would be upwardly through the jet aperture into the pressure chamber in the lower end of the traveling barrel, on the upstroke of the traveling barrel, and cites as an example, a man walking backwards on a moving train, but actually traveling forward with relation to the earth. Such sophistry is easily exposed. If a man ran backwards 15 miles per hour on a train going forward 5 miles per hour, the man would be traveling backwards 10 miles per hour with relation to the earth. It should be obvious to the Court that the liquid in the lower end of the traveling barrel will have to go downward and out through the restricted jet aperture 5, of the Bray patent, faster than the traveling barrel moves upward. A good example is the squirt gun. A slow movement of the plunger will cause liquid in the gun to squirt out of the restricted spout of the gun at a faster speed than the movement of the plunger. This is the jet action of the jet device of the Bray patent, and there is no question about it."

This testimony of Mr. Adams, at which this somewhat caustic criticism is directed, is summarized in paragraphs (d), (e) and (f), pages 121 to 122, *supra*. It is the only evidence in the record bearing on the matter. It was not challenged by appellants' expert witness, Mr.

Hayn. It may seem to appellants a natural comparison to state that the action of the Bray pump on the fluid trapped in the space "EE" within the barrel and above the nut is that of a squirt gun. That Mr. Bray secures any such action by reason of his assertedly unique packing and his multiple perforations of the nut is not satisfactorily disclosed by the record (see pages 117 to 123, *supra*), but, however this may be, the alleged infringing structures have no such action. It must be remembered that both the Bray and the Hofco pumps are wholly immersed in a fluid column some 4,000 feet high. [Tr. p. 182.] Fluid from a squirt gun ejected into the atmosphere has a place to go free from any substantial impediment. Fluid from the space EE of a pump in a well is a part of the column of crude viscous oil in the well, is inherently a fluid difficult to squirt, and the conditions under which it is discharged do not favor squirting. Under such conditions no atmosphere is present into which such fluid can be freely discharged. The only space into which it may flow freely is the space being vacated by the barrel and nut as these move upward, thereby leaving what would be a vacuum if liquid did not flow into the space so vacated. The opening through the puller nut in the Hofco structure is greater than the opening in the pump built by Bray (see pages 41 to 42, *supra*) and the flow therethrough is of a very low velocity. (See page 123, *supra*.) The action, therefore, more nearly resembles that of a pencil being withdrawn from a glass of water [Tr. p. 176] than it does the action of a squirt gun.

18. On page 74 of their brief, appellants assert:

"On pages 183 and 184 of the transcript expert Adams grossly exaggerates the leakage upwardly from the bottom of the traveling barrel past the standing plunger into the pump chamber, and carries his fantastic theory to the point of assuming that when the traveling barrel moves up liquid will flow upwardly through the jet aperture into the bottom of the traveling barrel and up past the plunger into the pump chamber, thereby preventing any downward *jet action* through the aperture in the bottom of the traveling barrel. This is clearly a *reductio ad absurdum.*"

We have examined the pages of the transcript cited, and find no justification whatever for this statement of appellants. There is no evidence anywhere in the record that Mr. Adams grossly or at all exaggerates the leakage. Mr. Hayn, plaintiffs' expert, did not so claim. Mr. Adams' testimony *is the only testimony in the record on the matter,* and it is inherently logical and reasonable. He stated on page 183:

"Then if there is any material leakage, if the pump is worn, we will say, to a third of its life, in the one stroke it not only 'can' but does happen that the leakage from the annular space, as the pump moves upward, is at a rate greater than the decrease in volume of the space, and so during that entire time the flow is upward through the orifices or openings in the puller nut into the annular space, and is not downward at any time, *provided that rate of leakage past the plunger is greater than the reduction of area in the annular space.**"

On page 184, Mr. Adams gave his assumed conditions and testified as to the leakage at different rates of speed, concluding:

> *"So we can say broadly that up to a speed of about 5 or 6 strokes per minute*, as pumps are operated on the present low production rate, you would have no jetting action from the orifices in the puller nut at all.*"*

If this conclusion—which is a matter of mathematical calculation, based on an assumed and stated stroke, head, gravity, viscosity, volumetric efficiency, and upon the dimensions of the pumps in evidence—was erroneous, it would have been a simple matter for appellants' counsel to *point out the error.* Much too much of appellants' brief consists of the hurling of such epithets as "fantastic theory", "sophistry", "gross exaggerations", "purely theoretical speculations", and "pseudo mathematics", which, while colorful to read, tender no substantial issue.

Appellants' argument continues:

> "There is no such leakage in either the Bray or the Hofco pump. The Bray patent describes an efficient packing on page 2, lines 90 to 101, inclusive, and page 2, lines 1 to 24, inclusive, and the Hofco catalogue, Plaintiffs' Exhibit No. 4, page 6, Book of Exhibits, shows at the right a plunger with many packing rings, to prevent leakage past the plunger. See also Defendants' Exhibit II, Book of Exhibits, page 177. Only a negligible amount of liquid could get past the plunger in either the Bray or Hofco pump." (Opening Brief, p. 74.)

This statement reflects merely the opinion of plaintiffs' counsel. It is not only without support in the record, but the uncontradicted testimony of Mr. Adams is to the contrary, and it therefore tenders no issue.

Mr. Franklin continues:

"Expert Adams would have the Court believe that an efficient packing cannot be made, but such is not the case. The Court may take judicial notice of a hydraulic elevator plunger and an ammonia compressor in an ice plant. A few drops of liquid may be observed at times on the elevator plunger, but this is a very negligible quantity of leakage, and ammonia compressors are constructed with practically no leakage at all, because operatives in an ammonia plant would suffocate from the escape of ammonia gas." (Opening Brief, p. 75.)

We do not question that a seal can be or has been developed which will be practically leak-proof, as we may assume is the case in a hydraulic elevator or in an ammonia compressor. The error of counsel lies in assuming, contrary to the uncontradicted evidence, that the Bray or Hofco structures employ such a device, or that they *could* do so.

He entirely overlooks the fact that a seal in a pump must be of such a character as *to result in an operative pump*. One of the essential requirements of a pump is that the *barrel shall drop*. The testimony of Mr. Bray on direct examination, by which appellants are bound, in reference to the clearance of his guide was:

"* * * you have to run it pretty loose, anyway. You cannot afford to have it too tight, if you do it is not going to work very good." [Tr. p. 37.]

and upon cross-examination:

"Q. You testified, did you not, that in the designing of the pump for which you applied for patent you were designing a pump which would furnish the maximum amount of jet action consistent with a reasonably free operation of the pump; is that correct? A. Well, I answered that question before. I will answer it again. I figured through my practical experience—that is how I manufactured them, through the practical experience with the pumps I worked with. I have probably worked 500.

Q. It was that type of pump which you sought to construct, was it not, one which would procure the maximum jet action consistent with an operative pump? A. Yes." [Tr. pp. 44 and 45.]

and again:

"You have got to have *your plunger* (barrel) *drop free,** if it don't it will crystallize and break your rod." [Tr. p. 45.]

and, speaking of the Ellis and O'Bannon pumps he stated at page 244:

"They worked the same way, that is, they worked on a traveling barrel, different designed and different made, and I find that most all the people that are pumping wells successfully have come to the upside down pump product. They find them easier on the rods. *It is very seldom you have rod trouble with a free traveling plunger. I claim I have got a jet and claim Mr. Hoeferer has not,** and the consequence is, if your plunger drops free it goes down to the bottom and it don't whip the rod; in other words, the plunger don't stick; the rods drop down with the weight pretty perfect. They don't buckle. They don't go into any corkscrew. *That is, when* *the barrel drops freely without any undue friction.** If the barrel sticks—if the plunger sticks in the barrel, of course, part way down, it causes the rods to bend all the way down, see, and when they come up they come up with a whip. But if you do that often enough and the plunger sticking there, you are going to break off that rod or break off the pin or break off the rod itself, which happens very often in a plunger that sticks." [Tr. p. 244.]

and finally on cross-examination:

"Q. I think you said in your direct testimony that you consider that you had a free dropping barrel to your pump, is that correct?

A. A free dropping barrel, yes.

Q. And that if you did not have a free dropping barrel the sucker rods would tend to 'buckle or corkscrew.' Is that what you mean?

A. Well, if the plunger didn't drop.

Q. Freely?

A. That is, going down somewhere, the rods are going to bend every place all the way down.

Q. You said it would set up a whipback and the rods would buckle or bend?

A. They ain't going to be like the spring because they wouldn't last long. They don't bend much.

Q. You mean they bend to some degree or buckle or corkscrew?

A. They will. Every rod all the way down will bend a little bit, so it makes it a whole lot when you get down." [Tr. pp. 262-263.]

and, after describing the method of setting the pumps, stated:

"After you get that set once, if your plunger drops free there is no reason why the rods would buckle or anything else; *but if it don't drop free, if the plunger sticks away up at the top after it comes up there, that will do it.**

Q. Exactly. In consequence, in the pump you built you tried to get away from that?

A. Yes. I know from my own experience I have got one that will." [Tr. p. 263.]

and after discussing a fluid pack pump replaced by one of his own stated:

"Q. And that proves to you—

A. (Interrupting.) That my plunger was dropping freely.

Q. In other words, that your barrel was dropping freely?

A. Yes.

Q. In consequence, you did not break them?

A. No." [Tr. pp. 265-266.]

It is thus obvious that a seal constructed so tight as to practically stop all leakage, would prevent the free drop of the barrel, and hence result in an inoperable pump. The error of counsel for appellants in this regard is no doubt due not only to his failure to appreciate the effect of this pertinent testimony but also to his erroneous impression that the full weight of the rods rests on the pump, and therefore in his assumption that such weight could force the barrel to move over a packing of the elevator plunger or the ammonia compressor type. The weight does not so rest. (See paragraph 11, pages 31 to 34, *supra*.)

Counsel for appellants then correctly states:

"Expert Adams testifies [Tr. pp. 198 and 199] that there are three forces in the bottom of the well tubing of the Bray pump which tend to keep the liquid and sand stirred up; the first force being the downward jets through the apertures 5 in the puller nut 3 at the bottom of the traveling barrel; the second being the plain churning action of the traveling barrel; and the third is the flow of liquid down between the traveling barrel and well tubing. Of said three forces Adams concludes that the force of the jet is the least important." (Opening Brief p. 75.)

and then asks:

"If this were true why did other pumps of the Bray type, without the Bray downward jets, 'sand up,' and why have the defendants copied the downward jet feature of the Bray pump?" [Tr. p. 75.]

In paragraph 21, (page 44, *supra*), we have pointed out the exceedingly meager evidence relative to the sanding up of other pumps. We pointed out in paragraph 11 (2), (pp. 119, 120, *supra*), that the Hofco pumps do sand up, and we assert that the finding of the trial court, supported by the evidence as heretofore analyzed, disproves the assumption contained in counsel's question that we "copied the downward jet feature of the Bray pump."

19. On pages 77 and 78 of their brief, appellants assert:

"Adams considers loss by eddy currents and concludes that the friction of the jet against the fluid with which it comes in contact, is of such order that the entire energy of the jet would disperse within a very few inches of motion. [Tr. p. 200.] This fantastic theory, however, ignores the inevitable condition that when the traveling barrel moves up, liquid must go out of the bottom of the barrel through the restricted jet apertures 5, with considerable force; otherwise the liquid in the bottom of the traveling barrel, being incompressible, would form an hydraulic lock and prevent upward movement of the barrel; and when liquid flows into the vacuum under the lower end of the upwardly-moving traveling barrel and comes into contact with strong downwardly projected jets, we have a meeting of two strong opposing hydraulic forces, figuratively like the irresistible force meeting the immovable body, which would produce a turbulence of the liquid directly between the lower end of the traveling barrel and the top of the anchor, and this turbulence would necessarily stir up the sand and wash the sand from the anchor and clutch, and prevent the pump from 'sanding up.' This turbulence of the liquid directly between the lower end of the traveling barrel and the top of the anchor necessarily resulting from the jet action is a new phenomena in the art which made its first appearance in the Bray patent in suit, and there is nothing in the prior art which suggests this novel feature of the patent in suit." (Opening Brief pp. 77 and 78.)

"Adams considers loss by eddy current and concludes that the friction of the jet against the fluid with which it comes in contact is of such order that the entire energy of the jet would disperse within a very few inches of motion. [Tr. p. 200.]"

Mr. Franklin first states that this testimony of Mr. Adams is a "fantastic theory" and then proceeds to attempt to meet this testimony by making statements which find no support in the record, thus in effect attempting to substitute his unsworn statements for the sworn testimony of Mr. Adams. [Tr. p. 77.] His argument is, however, pregnant with the admission of the correctness of Mr. Adams' testimony. An examination of it will disclose that Mr. Franklin *assumes* that in order to prevent a "hydraulic lock" the liquid "must go out of the bottom of the barrel through the jet aperture 5, *with considerable force*," whatever that may mean. The undisputed testimony of Mr. Adams, who measured the forces, is that it was not of such magnitude as to rise to the "dignity of a jet." [Tr. p. 201.]

Mr. Franklin admits, as Mr. Adams testified (see par. "h", page 122, *supra*), that on the upward movement of the barrel

> "liquid flows into the vacuum under the lower end of the upwardly-moving traveling barrel." (Opening Brief p. 77.)

and then assuming the *existence* of the "strong downwardly projected jets" states:

> "We have a *meeting of two strong opposing* hydraulic forces, figuratively like the irresistible force meeting the immovable body, which would produce a *turbulence of the liquid directly between the lower end of the traveling barrel and the top of the anchor,* and this turbulence would necessarily stir up the sand and wash the sand from the anchor and clutch, and prevent the pump from 'sanding up.' *This turbulence of the liquid directly between the lower end*

of the traveling barrel and the top of the anchor necessarily resulting from the jet action is a new phenomena in the art which made its first appearance in the Bray patent in suit, and there is nothing in the prior art which suggests this novel feature of the patent in suit." (Opening Brief pp. 77-78.)

We find it difficult to believe that counsel for appellants realizes the effect of this admission. It amounts to a confession that the jets meeting the inflow of the fluid into the vacuum created by the upwardly moving barrel *lose all identity as jets before they reach the anchor to be washed,* and *result merely in the creation of a turbulence below the anchor and the bottom of the traveling barrel.*

If this is true, it inevitably follows that the structure of the Bray patent does not have the "means * * * to cause a *positive action of jet liquid over the lock means*"* (Bk. Ex. p. 3, Spec. lines 26 to 28), nor will "liquid * * * be forced down in *strong jets** and wash sand from the top of pin 'P' so that it can be pulled," etc. (Bk. Ex. p. 3, Spec. lines 78 to 81), which, as we have pointed out in paragraph 3, pages 100 to 101, *supra,* was the *raison d'etre* of the patent.

Certainly if this is true of the structure discussed by Mr. Franklin, provided with the "restricted jet apertures 5," it is more true of the Hofco structures which do not have such restricted jet apertures, *and a confession of non-infringement by such structures is inescapable.* It was no doubt a recognition of this fact which caused Mr. Bray to testify, *"I claim I have got a jet and claim Mr. Hoeferer has not."* [Tr. p. 244.] This is quite another matter from the defense of "inoperativeness or lack of utility" to which defense counsel erroneously assumes the testimony and admission can alone refer. (Opening Brief p. 78.)

On pages 47 to 49 of this brief we have pointed out that the dimensions of the alleged infringing pumps were stipulated into the record. An examination of those dimensions discloses that the clearance between the inside of the puller nut in the Hofco structure, and the outside of the standing column over which it rides is 8/32 of an inch. As we noted at page 41 of this brief Mr. Bray admitted that a pump made with a larger clearance than 3/32 of an inch at this point would not produce any jet action.

The difference of 5/32 of an inch in diameter may without analysis appear small, but it results in the elimination of the asserted jet of Bray, and in the production of the maximum annular opening consistent with retaining an area of metal on the puller nut sufficient to lift the pump, and the puller nut of appellees therefore is purely a lifter of the type long known in the art.

The surprising "about face" between the position of Mr. Franklin set out on page 22 of his brief that:

"A most vital and distinguishing feature of the invention, covered by the patent in suit, which stands out in bold relief from the prior art, here-inafter considered, is the arrangement of the clutch, comprising the members 4 and 2 on the lower end of the traveling barrel B and the upper end of the anchor P, respectively, in combination with the vertical ducts or jet apertures 5 extending through the jet device or part 3, in the lower end of the barrel B, *through which apertures strong vertical jets of oil are projected directly downward under great pressure from the pressure chamber EE**

(Defendants' Exhibit MM, Book of Exhibits p. 178) toward the clutch member 2, on the upward stroke of the traveling barrel B, *which jets most effectively wash all sand from said clutch member 2 and the anchor P, and keep said sand moving upwardly in the tubing T around the traveling barrel B, whereby the pump is prevented from 'sanding up.*'" (Opening Br. pp. 22 and 23)

and his statement just analyzed that:

"This *turbulence of the liquid** directly between the lower end of the traveling barrel and the top of the anchor necessarily resulting from the jet action *is a new phenomena** in the art which made its first appearance in the Bray patent in suit, and there is nothing in the prior art which suggests this novel feature of the patent in suit."

is but one example of the inconsistencies and lack of logic which characterizes the brief.

20. Another example of the inconsistent position to which appellants are driven in their vain attempt to distinguish the structures of the prior art, is to be found on page 40 of their brief where they assert:

"Even if the Admore pump were constructed to project a downward jet, the jet would have to travel down over the standing valve 16 before it reached the anchor 14, and this greater travel of the jet through the body of oil in the well tubing, before reaching the anchor, due to the interposition of the standing valve 16 between the lower end of the barrel 6 and the anchor 14, would dissipate the jet and materially reduce its washing effect upon the anchor. This reduced effectiveness of the jet would allow the pump to 'sand up' in a short time."

If a jet produced by a structure such as the Admore pump, Exhibit 6, which structure is admittedly analogous to the Bray structure, is so weak that it will not be effective for a distance greater than the height of the cage 15 in the Admore structure, it is likewise so weak as to be of little or no real effect in any pump. It stands uncontradicted in the record that no appreciable jet or jet velocity is obtained until the traveling barrel has gained a substantial upward velocity and therefore no appreciable jet occurs until the jet creating device on the lower end of the barrel has moved a considerable distance from the anchor that it is intended to wash. (See par. (c), page 121, *supra*.) The stroke of a well pump is in the order of several feet. Mr. Bray has testified that the stroke of his pump is 56 inches. [Tr. p. 29.] It likewise stands uncontradicted in the record that the maximum jet action occurs at about the middle of the stroke [Tr. p. 196], i. e., after the jetting device has been moved about 2½ feet away from the bottom end of its stroke. If appellants are to contend that the Bray structure creates a jet which will penetrate the body of fluid between the bottom end of the barrel and the puller nut, which body of fluid is some 2½ feet deep,—as they do when such argument is deemed by them to be to their advantage— such a jet will not be retarded, or its effect materially reduced, by the presence of the cage 15 in the Admore pump which extends only a few inches above the anchor of the pump. It is to be regretted that the labor of appellees in writing this brief and the time of this Court in reading it, has not, to some extent at least, been minimized by some consistency of argument in the opening brief.

21. Scattered throughout pages 31 to 67 of their brief appellants attempt, largely by assertions unsupported by testimony, to avoid the effect of various structures, both patented and unpatented, which the record shows to be part of the prior art. This they do by arguing that the structures are inoperative.

With respect to the patented structures, appellants blandly ignore the principle which they assert at page 16 of their brief, that:

> "The patent is *prima facie* evidence of its operativeness and utility and presupposes a pump constructed and arranged to perform its intended function."

They base their attack principally upon the assertion, that the structures disclosed will not operate because of "gas lock" or "hydraulic lock." In view of the fact that we have previously considered the error of this theory (see pages 17 to 27, *supra*), we will not again answer it except to note those cases in which the structure, by the terms of appellants' own argument, can have no such lock.

They further argue, as in the case of Cummins' patent (Opening Br. p. 45), that because there is no proof of actual successful operation of the pump "in the oil fields of California or anywhere else" the patent is of no effect as an anticipation. This is obviously unsound in principle, as a patent is a printed publication and as such requires no proof that its teaching has ever been actually carried out.

> *Fulton Co. v. Bishop & Babcock* (C. C. A. 6), 284 Fed. 774, at 777;
>
> *Keene v. New Idea Spreader Co.* (C. C. A. 6), 231 Fed. 701, at 708.

22. We will now consider the alleged inoperativeness of the following specific structures:

(a) Deis, Ex. N, N′ and EE (Bk. Ex. pp. 95-99 and 173). Appellants boldly state that this patent

"is in the last analysis a reciprocating plunger and fixed barrel pump" (Opening Brief, p. 43)

but do not state why this is contended to be true. Mr. Adams testified that this patent discloses a pump of the traveling barrel type, and this testimony was not controverted by plaintiffs' expert Hayn.

Defendants' Exhibit "EE", Fig. 4, shows clearly the standing element of the Deis pump. Fig. 5 shows the traveling element, which is plainly a shell or barrel that fits over and runs on the standing column shown in Fig. 4 and therefore in spite of the "last analysis" by counsel for plaintiffs, the fact remains that the Deis patent shows a traveling barrel type of pump.

Appellees do not contend that Deis discloses a pressure chamber such as "EE" of the Bray patent in suit, causing a downward jet of oil during the upward movement of the traveling barrel. It is contended, however, that Deis discloses a traveling barrel type of pump in which there is a cooperating clutch construction, parts Z and 4, Defendants' Exhibit EE, which clutch is washed by the action of the pump to prevent the standing column 1 from becoming sanded up. The only fundamental difference, between the Deis patent and the structure of the Bray patent, is that in Deis the pumped fluid washes in through holes 5 during the up-stroke of the traveling barrel to wash around the clutch parts and the anchor, while in the Bray construction it is intended that jets be created to wash down onto the anchor from the interior of the barrel. In both structures there is a traveling barrel working over a standing column, the necessary valves, the cooperating clutch parts, and an arrangement and formation of parts whereby the clutch parts are washed during the upward travel or movement of the traveling barrel.

(b) CUMMINS, Ex. O and O' (Bk. Ex. p. 100). The suggestion that a "gas lock" will occur between the upper and lower valves of the reciprocating plunger in the Cummins patent (Op. Br. p. 44) is made with entire disregard of the fact that the space between these valves is not a pumping chamber in which "gas lock" can be detrimental, and in further disregard of the fact that, as the valves 17 and 23 are mounted at the opposite ends of a tube and operate in the same direction, the upper valve 23 could be eliminated if gas accumulated between these valves.

(c) CUMMINS REVERSED, Ex. O² and GG (Bk. Ex. p. 101 and 175). Appellants' principal argument against "Cummins Reversed" is based on the premise that because the Cummins patent was not proved to have been successfully operated, it is of no effect as an anticipation. (Opening Br. p. 45.) As this premise is unsound, (see page 149, *supra*), it follows that the argument against "Cummins Reversed", based upon it is likewise unsound.

The obvious reversibility of the Cummins structure was pointed out by Mr. Adams and its operation, when so reversed, clearly explained [Tr. pp. 96 to 99]. Plaintiffs—(appellees) did not contend that the structure of "Cummins Reversed" anticipated the apertured guide of Bray's claims 2 and 3, but rather asserted that in light of it, and of other prior art cited, claim 1 of Bray did not amount to invention.

The suggestion, by counsel for appellants, that "Cummins Reversed" might not work because "the plunger 16 is short" and the barrel might "wobble" if sufficient clearance were left to gain a jet (Op. Br. p. 46) is, in effect, an admission that sufficient clearance might be provided between the standing column 15 and the part 2 in "Cummins Reversed" to gain such jet, without the exercise of invention. This is made plain by the fact that counsel for appellants recognized that such jet may easily be provided despite the fact that Mr. Adams explained that he had literally reversed Cummins without other modification, and did not testify that the further slight modification suggested by counsel would produce a jet.

Counsel for appellants has stated that the "dead space" below the plunger head 16 of "Cummins Rerversed" would, if filled with air, impair the efficiency of the pump (Op. Br. pp. 46 and 47). How much he does not state. We are unable to find any evidence to this effect in the record, or to conjure up a theory that will make it so. He further states that if oil leaks past the plunger into this so-called "dead space" the pump will be locked and become inoperative because oil cannot be compressed. Counsel has apparently overlooked the fact that any opening or clearance present in "Cummins Reversed", that will permit oil to leak into and fill the dead space, will likewise allow such oil to leak out of the "dead space." As counsel for appellants has suggested, this leakage might occur between the parts 2 and 15. Mr.

ed that in
l of Bray

hat "Cum-
he plunger

ance might
15 and the
1 jet, with-
plain by the
rat such jet
Mr. Adams
iins without
the further
uld produce

"dead space"
Rerversed"
ency of the
uch he does
lence to this
ory that will
eaks past the
ie pump will
oil cannot be
oked the ias
Cummins R:
1 fill the dea:
of the dea:
iggested this
and 15. Mr.

Franklin attempts to meet this fact by arguing that in such a case the barrel of the pump might "wobble", although there is no evidence in the record, or basis in reason to suggest that wobbling of the barrel would render the pump inoperative. Moreover, adopting for the sake of argument, appellants' suggestion that the structure should be considered as being operated in California, where appellants seemingly believe "gas lock" occurs wherever there is the slightest possibility of it, we see no reason why gas would not accumulate in this "dead space." Gas in the pump chamber may impair pumping action but its cushioning action in such a "dead space" will do no harm.

It is further suggested that in "Cummins Reversed" the barrel will not drop freely because of the close bearing fit between the lower plunger section 2 and the standing column 15. (Op. Br. p. 47.) It is to be assumed that any one building a pump would build it so that it will operate, and that any mechanic would provide appropriate clearance between the working parts 2 and 15. In fact, as the bearing between the parts 2 and 15 in "Cummins Reversed" is long and might tend to retard the necessary free dropping of the barrel, it is obvious that a mechanic constructing it would allow sufficient clearance between these parts to allow free dropping, and thus automatically produce such jetting action from the bottom of the barrel as counsel for appellants has assumed in his argument relating to "wobble."

(d) WHITLING, Exhibits P and P' and HH. [Tr. pp. 103, 104 and 176.]

In answer to the contentions appearing on pages 48 and 49 of appellants' brief, we will first point out that the valves 23 and 27, which they state to be so far apart that a gas lock will form between them, are both discharge or working valves operating in the same direction, and that either one of them can therefore be eliminated at the will of the operator.

The chamber formed in the barrel 5 between the plunger 16 and packing 30, if filled with air or gas, will as in the case of "Cummins Reversed" have little or no influence upon the general action of the pump other than to cushion its motion. If liquid finds its way into the chamber to the total exclusion of gas, it will of necessity have to be forced from the chamber through the channels by which it entered. Even under such condition the pump would *not* be subject to "gas lock" *between the valves 23 and 27,* and when the liquid had been forced from the chamber between the plunger 16 and the packing 30 the pump would be highly efficient, in fact, ideal.

(e) WHITLING REVERSED, Exhibit HH, (Bk. Ex. 176.) The reversal of Whitling results in a structure in which the valves 23 and 27 are relatively close together, valve 23 being at the upper end of the standing column 22, and thus prevents the occurrence of "gas lock" in the pump chamber upon appellants' own hypothesis. The statement of appellants that "gas lock" will occur between valves 23 and 12 (Opening Brief, p. 49) or that the pump will "sand up" around member 8 and valve 12 (Opening Brief p. 50) is beside the point, as valve 12 is simply a supplemental or auxiliary valve having the same action as valve 23. This valve 12, together with the tube 8, the element 6 and the casing section 2 may, therefore, be detached and eliminated without in any way altering the action of the pump, as the structure remaining after such parts are removed is of itself complete and operative.

In the Whitling structure literally reversed as shown in Fig. 2 of Ex. HH, the packing at 30-31 would prevent downward jetting of fluid in proportion as such packing is efficient. As the packing between the working barrel and the column becomes worn and clearance develops, fluid will enter the chamber formed between the working barrel 5 and the column 22, and will be forced or jetted out through the clearance so developed to jet downwardly toward the anchor face or upwardly past the packing 17.

In the Whitling construction as shown in the patent a chamber termed a "sand chamber" will form around column 5. When the structure is reversed this chamber is no longer subject to being sanded up as the fluid accumulating therein is kept in a constant state of agitation due to the plunging and churning action of the lower end of the barrel which reciprocates in said chamber. Such churning action, accompanied by whatever leakage there may be out of the lower end of the working barrel, would keep sand from settling around column 22, contrary to appellants' contention appearing on pages 50-51 of their brief. The innuendo contained in counsel's query as to why the sand chambers were omitted from Defendants' Ex. HH (Opening Brief p. 51) which was introduced as a simplified exhibit, is, in view of the fact that the patent was itself introduced and the presence of the sand chambers thereby disclosed for such use as appellants' expert might care to make of it, as unworthy of answer as is their branding of the structure as a "mechanical monstrosity".

Appellees introduced the Whitling patent and pointed out the reversal of the Whitling structure, that it might be considered as a teaching of the general state of the art, and because "Whitling Reversed" is substantially identical with the Bray pump, including in detail, the clutch parts and excluding only the jet apertures.

(f) DOWNING, Ex. Q & Q' (Bk. Ex. p. 108). On pages 51 and 52, appellants urge that this pump is inoperative because gas lock might occur between the valves, and because liquid might get into the space between the head on the standing column and the puller nut H. The matters of gas lock and of liquid becoming trapped in the space above the puller nut have been heretofore discussed at page 26.

(g) O'BANNON, Ex. R & R' (Bk. Ex. p. 111). Counsel for appellants, on pages 52 and 53 of his brief, discusses the O'Bannon patent, and points out that this patent discloses laterally opening ducts at the lower end of the traveling barrel and is without cooperating clutch parts.

The theory advanced with reference to the laterally opening ducts 16' is the same as that advanced with reference to the laterally opening ducts in the Admore pump, and is in our opinion completely answered on pages 111 to 113, *supra*.

The absence of clutch parts on the O'Bannon pump is admitted; however, as we have repeatedly pointed out, the addition of clutch parts, old and well known in the art, to a pump such as is shown by O'Bannon, does not amount to invention.

(h) THURSTON, Exhibits S and S' (Bk. Ex. pp. 114 and 115). Appellants point out that the standing valve 7 is shown at the lower end of the standing column 9-11, and assert that the puller nut 20 on the lower end of the working barrel 19 could not possibly have jet action such

as Bray teaches, because Thurston refers to it as a
"guide", it being counsel's contention that:

> "Thurston never intended a loose fit between the
> part 20 and the plunger 11 to provide an annular
> space through which a jet of oil might be projected
> for washing away sand. The Bray invention is cer-
> tainly not taught by the Thurston patent." (Open-
> ing Brief pp. 53 and 54.)

The contention of appellants that Thurston does not
contemplate clearance between the puller nut 20 and the
standing column 11 is most significant. Counsel says
that this is true because Thurston, in his specification,
states that the member 20 is a "combined pull nut and
guide". (Lines 94 and 95, Bk. Ex. p. 116). This is entirely consistent with appellee's contention that Bray
when referring in his patent to "a guide" and when defin-
ing his invention in the claims as embodying a "guide,"
meant a part having a close running fit on the standing
column just as is shown in the patent to Thurston. How
can appellants admit that a member such as the part
20 of Thurston should be construed as confined to a
close running fit, because it is referred to as a "guide,"
and yet urge that the corresponding part 3 in the Bray
patent, which is also referred to as a "guide," should be
construed as covering a loose fit such as characterizes
appellees' devices? We submit that a comparison of the
Thurston and the Bray patents makes it plain that Bray
must, on the basis of appellants' own argument, be confined to a member 3 having a close running fit on the
standing column, and that his jet action must, of necessity, be gained through the apertures in the guide 3, which
apertures are separate and distinct from the opening
which forms and constitutes the guide.

The comment of appellants with reference to the location or spacing of the valves in the Thurston pump, is based on appellants' theory of gas lock and totally disregards the clear teaching throughout the prior art, that standing columns can be provided with standing valves at either end, or at both ends. See Thompson, Defendant's Ex. T and T' (Bk. Ex. 119); Wright, Defendant's Ex. U and U' (Bk. Ex. p. 125); Northrup, Defendant's Ex. V and V' (Bk. Ex. p. 129).

(i) THOMPSON, Ex. T and T' (Bk. Ex. p. 119). Counsel for appellants would have it appear that the patent to Thompson means practically nothing. We point out that this patent was issued in 1897, and represents a comprehensive embodiment of the prior art relating to well pumps as that art applies to the alleged invention of Bray.

Appellants first state that this patent discloses the inverted type of pump (Opening Brief p. 54), whereas it, in fact, discloses such a pump in Figs. 1 and 2, and in Fig. 4 illustrates a reversal of that structure, i. e., the standing barrel type of pump in which the plunger or column reciprocates. Thus in 1897 Thompson taught the reversibility of plunger and barrel, thus specifically disclosing to all who followed in the pump art that pumps can be reversed, as we have pointed out in the case of Cummins and Whitling, *supra*.

Counsel for appellants attempts to apply his familiar gas lock theory to the Thompson patent. He *cannot* say that gas lock can occur in the pump chamber of the Thompson patent, Fig. 1, because the valve i is at the upper end of the standing column and the valve o is on the traveling barrel adjacent the valve i. In nowise discouraged, he suggests, however, that a gas lock might occur between the two standing valves i and e^4 of the column, despite the fact that these valves operate in the same direction, and are not in the pumping chamber. Obviously, gas or any fluid that might accumulate in the standing column between the valve e^4 and the valve i will flow upwardly out of the upper end of the column into the pumping chamber, when the barrel is moved upwardly and a reduced pressure thereby brought about in the pumping chamber between the valve i' and the valve o. It is, therefore, clear that a "gas lock" cannot occur between the two valves on the standing column, as counsel for appellants would have it appear. Moreover, as the valves i and e^4 are both standing valves and both operate in the same direction it is perfectly obvious that one of these valves, for instance, the valve e^4, can be removed or eliminated in case it should be found, for any reason whatever, that gas accumulated in the column between these valves and interfered with the action of the pump. It is, of course, a well-established principle of law that an element such as the valve e^4 and its corresponding function can be eliminated from a structure without the exercise of invention.

> *Walker on Patents,* 6th Ed., par. 73, p. 90, and cases there cited.

Counsel for appellants attempts to lead away from the pertinent structural features of the Thompson patent by referring to lateral jets t' in the well tubing, as distinguished from in the pump parts proper. The presence or absence of the parts t' has no bearing whatever upon the operating parts of the pump with which we are concerned. It is obvious, referring to Fig. 1 of the Thompson patent, that we are in this case interested in the anchor at d, in the standing column f, in the reciprocating barrel k operating over the column, in the standing valve i at the *upper end* of the standing column, in the working valve o which is located at the *upper end* of the standing column, in the working barrel, and in the puller nut k' which fits loosely around the column f, thus admitting fluid into the space above the puller nut and correspondingly allowing fluid to be jetted from the space above the puller nut. The fact that Thompson discloses numerous other features in connection with his construction, such as the jets t' in the well tubing, does not militate against the pertinency of the elements just mentioned.

The sleeve k', which is a puller nut threaded to the bottom end of the traveling barrel, is described as fitting loosely around the standing column. Thompson refers to the sleeve k^2 in the following language:

> "Fluid may *also** be admitted to this above mentioned space (that is, the space between the standing column and the working barrel) through the sleeve k' which *does not fit closely around tube f.**" (Page 2, lines 50 to 53, Bk. Ex. p. 121.)

and further, on page 2, lines 97 to 107:

> *"A portion** also of the fluid admitted through the orifices t' and valve t^2 will, upon the down stroke of the piston 1 pass up through the orifices k^2 into the space between the working barrel k and the tube f and upon the upstroke of said piston 1 *this fluid** will escape through the orifices k^2. It is apparent that in case the sleeve k' is not used the orifices k^2 would be superfluous as the lower end of the working barrel would be open to admit the fluid."

The space shown in the patent drawing between the standing column and the sleeve k' is a substantial space. It is definitely shown in Fig. 1, in which figure the part k' is very small, and from the drawings it is plain that the space is substantially larger than the orifices k² referred to by Thompson. As the sleeve k' is described as loose and is so illustrated, fluid will enter the space between the barrel and the standing column *through the sleeve k'* as well as through the orifices k², and when the barrel moves up *fluid will be ejected downwardly through the sleeve k' as well as outwardly through the orifices k²*. Thompson has suggested that the orifices k² are superfluous if the sleeve k' is not used, thus indicating that in proportion as the bottom end of the barrel is open to admit fluid into, and out of the space between the barrel and standing column, the orifices k² are not necessary. It is also significant that the sleeve or fitting k² in the Thompson patent is applied to the lower end of the barrel and is related to the standing column, exactly as the puller nut is applied to the barrel and related to the standing column in appellee's device. The relative sizes and proportions of the parts and clearances are substantially the same. Counsel's statement that the jets t' will choke up with sand is not only without significance but is not supported by a single word of testimony in the record. It is perfectly plain that the jets t' will no more sand up than will the apertures in the jet devices of the Bray patent. Moreover, if and when the orifices k² become sanded up the entire flow from the space between the barrel and column would have to be out through the clearance provided between the sleeve k' and the standing column.

Counsel for appellants makes the statement, on page 55 of his brief, that the loose fit of the sleeve k' is nothing more than "a sliding bearing fit". We find no reference in the specification of the Thompson patent to the sleeve k′ as being a "sliding bearing fit" on the standing column and the drawing of the patent definitely shows a clearance between the sleeve k′ and the standing column, and, therefore, we submit that this is simply another sheer misstatement of fact.

The comment that the sleeve k′ at the bottom of the barrel is so far removed from the anchor d that the downward jets would be dissipated before reaching the anchor indicates that appellants are grasping at straws. As we have heretofore noted, in the Bray construction the traveling barrel must have moved a substantial distance, in fact, about half the length of its stroke before any appreciable jetting action is obtained; at which time the jet device is far removed from the parts intended to be washed. Thompson shows a break in the parts between the sleeve k′ and the anchor indicating clearly that it is a matter of choice as to how much distance is left between these parts. It could be much or little and obviously in proportion as an operator wished jetting action to approach the anchor he would make the distance small.

We submit that the patent to Thompson discloses every element of the Bray patent if it is read as employing a loose fitting guide, with the exception of the cooperating clutch parts between the anchor and the sleeve k. We submit, as the Court found, it did not amount to invention to add such clutch parts in the light of the prior art set out in detail in our analysis of claim 1. (See pages 62 to 64, *supra*.)

(j) WRIGHT, Ex. U and U′ (Bk. Ex. p. 124). The Wright pump is not in operation as appellants assert, a traveling plunger type of pump. (Opening Brief p. 56.) An examination of the patent discloses that the traveling barrel is shown slidably fitting the tube A. It is plain, from the colored parts (Bk. Ex. p. 125), that the pumping action is gained through cooperation of a standing column B and a traveling barrel D which works over the standing column; the standing or foot valves EE' being carried by the standing column and the working valve G being carried by the traveling barrel. The traveling barrel is shown slidably fitted in the tube A but this *does not change* the working action of the pump in so far as the essential parts are concerned. The fit of the working barrel in the tubing A simply causes a pumping or surging action to occur at the anchor B, the anchor having appertures b^3 to pass the fluid circulated by the cooperation of the working barrel and tubing A. The fluid thus circulated is outside of the pump proper, as the opening b^3 in the anchor opens to the space below the foot valve E. Traveling barrel D can be made to fit the tubing A closely, as shown in the Wright patent, or freely as shown in the Bray patent, without in any way influencing or modifying the operating principle of the pump; that is, without changing the fact that the barrel D is the traveling part operating over the standing column and without altering the fact that fluid, trapped in the space between the lower reduced part of the standing column and the working barrel, will be jetted downwardly through the clearance provided at aperture E during the upward movement of the traveling barrel, precisely as fluid is jetted downwardly in the Bray construction. This pump discloses every detail that Bray claims, except the apertures in the flange at the lower end of the traveling barrel to which cooperating clutch parts could obviously be applied if desired without the exercise of invention.

(k)   NORTHRUP, Ex. V and V¹ (Bk. Ex. p. 129). The patent to Northrup is admitted by counsel for appellants to be of the same general type as the Bray patent. (Opening Brief pp. 57-60.) Counsel's comments with reference to the Northrup device are characterized by statements as to what it will not do, but as such statements are not supported by the evidence there is no point in discussing them in detail.

Fig. 1 clearly shows space between the puller nut 24 and standing column. It may be noted that a similar space is disclosed by the patent to Thompson (Ex. T and T', Bk. Ex. p. 119). It is significant to note that appellants cannot raise the cry of "gas lock" when discussing the Northrup patent for the valves 21 and 28 are in the identical arrangement shown in the Bray patent. He cannot raise the cry concerning the great distance between the puller nut and the anchor as Northrup has shown the parts almost touching. In fact, his only defense to this patent is based upon the testimony of appellants' expert Hayn that he was unable to understand or, in fact see, the space clearly pictured between the puller nut 24 and the column 1. (Opening Brief pp. 59-60.) The fact that the draftsman went to the trouble to illustrate such space in Fig. 1 indicates clearly that if any mistake was made it occurred when he failed to illustrate the corresponding space in Fig. 2, and in any event whether the disclosure is the result of accident or otherwise, it is nevertheless a disclosure. In point of fact the showings of the puller nut in Figs. 1 and 2 are not necessarily inconsistent. If we view the puller nut as being of the rosette type which appellants urge is the equivalent of the uniform bore type, and which is old in the art (see Ellis Ex. W-W', Bk. Ex. p. 133). It is plain that Figs. 1 and 2 may simply illustrate the puller nut 24 in different rotative positions.

In commenting on the action necessarily resulting from the presence of this space between the puller nut and the standing column, counsel for appellants lays great weight upon the fact that Northrup gives no detailed description of, and does not explain in so many words, that the fluid would flow out through the space obviously provided. [Tr. p. 59.] In thus attempting to avoid the disclosure of the Northrup patent, appellants entirely overlook the fact that they have urged inoperativeness of numerous prior art patents by reason of "gas lock" and have urged that the elimination of gas lock is an underlying principle of the Bray patent although, as we have pointed out, there is not a single word in the Bray patent referring to the asserted fact or even remotely suggesting, that Bray was in any way concerned with "gas lock", thus affording another example of the inconsistency of argument which has made this brief so long.

It is perfectly obvious that in view of Fig. 1 of the Northrup patent, the Bray patent cannot amount to invention, as the puller nut 24 and anchor face may be provided with the well known cooperating clutch parts of the prior art. These are obviously applicable to the Northrup patent and so if applied would not, as the court found, amount to the exercise of invention. It is further to be noted that the Northrup structure of Fig. 1 would result in a washing of whatever is present at the anchor face, if there is, in fact, any washing obtainable in any pump having a jet device, and that the presence or absence of a clutch device to be so washed would not alter the jet action obtained. It follows that the combination of jet action and clutch device is, in the last analysis, a mere aggregation of elements that have no real cooperative relationship.

(1) ELLIS, Ex. P (Bk. Ex. 133). Counsel for appellants, referring to the Ellis patent, points out that Mr. Bray found that the rosette type of puller nut *after it is worn* may allow a little too much oil to jet out. (Opening Brief p. 60.) This seems an admission that before it is worn it must work satisfactorily. We can, therefore, assume that appellants admit that the Ellis type of puller nut, or cap nut, as Ellis terms it, will work satisfactorily until it is worn. We call attention to the fact that Fig. 3 discloses that the ducts 22 are spaced apart circumferentially in the wall of the bore of the puller nut, so that there are substantial bearing surfaces 23 between the ducts giving the puller nut the necessary wearing qualities.

Ducts such as 22, when used by Mr. Bray in his rosette type of puller nut, are, according to appellants, "jet apertures" (Opening Brief p. 17), whereas, when such ducts are found in the prior art patents counsel for appellant would have it that they are "vents". We might inquire as to where the line of distinction lies between ducts that are vents and ducts that are jet apertures. It is clear that in light of counsel's theory Bray's invention is confined to the field of "jet ducts". Otherwise it must, of necessity, be specifically anticipated by the ducts of Ellis. Bray admitted, quite frankly, that he himself did not know where the line of division is, if in fact there is one, when he testified [Tr. p. 43]:

> "I don't know how great the apertures can be between the standing column of the pump and the guide and yet produce the jet action which I mentioned in my patent."

When the inventor of the patent in suit is unable to give any concrete idea as to what dimensions or proportions are contemplated within his invention, how can counsel say that the ducts 22 in Ellis are vent ducts and not jet ducts, and are, therefore, not anticipations of Bray's invention?

Ellis states that flow through the ducts 22 is "free" but does not confine his description to such a free flow as will not amount to a "jet". To say that flow through a duct or orifice is "free," simply means that there is no obstructing means retarding passage through the duct or aperture, but does not mean that such flow may not be of such velocity as to constitute a jet. We submit that it is obvious, from the proportions of the parts illustrated in Ellis, that fluid in the chamber 24 would, during the normal operation of such a pump, jet down through the ducts toward the anchor in much more clearly the same manner and range of action taught in the Bray patent than found in the structure of appellees.

Ellis specifically teaches that the valve 25 may be eliminated:

> "Under some conditions the valve ring may be omitted as the passages in the cap nut permit the free flow of liquid and prevent the formation of a

dead chamber containing trapped liquid which will resist the free operation of the pump." (Lines 37 to 42, Bk. of Ex. p. 135.)

And it is obvious that as Mr. Adams testified, the valve holder 26-27 would be eliminated with the valve, so that the down flow or jets issuing from the ducts 22 would flow directly toward, and wash, the anchor face, and any parts there present.

Counsel for appellants would have it appear that, because a break is shown in the Ellis drawing below the cap nut 21, Ellis does not anticipate operation of the pump in such wise as that the cap nut closely approaches the anchor face. (Opening Brief p. 61.) He has characteristically neglected to point out that there is also a break shown in the tubing and traveling barrel between the valves 18 and 20, indicating that the parts as illustrated are not necessarily at either end of their stroke. Obviously, the two breaks mentioned, together with the break illustrated between the numerals 12 and 24, Fig. 1 of Ellis, make it clear that Ellis anticipated adding any desired length at these points, and that any ordinary mechanic would add such length at these points as to gain any desired stroke of the barrel relative to the standing column, and any desired spacing of the cap nut 21 from the anchor when the pump is in the fully collapsed position.

The Ellis construction operated with the valve 25 and its cage removed, and proportioned to bring the cap nut 21 reasonably near the anchor when the pump is in the collapsed position, would obviously result in a downward jetting of fluid through the ducts 22 toward the anchor to wash any part at the anchor, if a construction of this general type, such as is shown in the Bray patent, is capable of any washing action.

To add appropriate cooperating clutch parts to the Ellis construction is an obvious adaptation of an old and well known feature to the Ellis patent, which would not, as the Trial Court found, rise to the dignity of invention.

(m) DICKENS Ex. X and X' (Bk. Ex. pp. 136-150). The patent referred to at page 62 of the opening brief is of the stationary barrel and reciprocating plunger type and was introduced in evidence merely to illustrate the fact that such pumps have been provided with jaw clutch devices of the general character hereinabove discussed, and such as Bray has incorporated in his pump at the bottom of the traveling barrel.

(n) NEILSEN Ex. Y and Y' (Bk. Ex. pp. 141-145). Appellants' comments with reference to the valves in the Neilsen pump are somewhat misleading. (Opening Brief p. 63.) Appellants seemingly refer only to such valves of the structure as can be made, upon cursory examination, to seem to fit the theory which they are seeking to advance, and ignore the existence of the other valves which are shown in the Neilsen disclosure. For instance, they refer to valves 7 and 12 in Fig. 1 of Neilsen as being far apart, but neglect to point out that these valves simply supplement each other and are both valves in the standing column, whereas the *active valves* of the pump are valves 12 and 20 *which are shown in very close proximity to each other*. As we have heretofore pointed out with reference to other pump structures, the lower or supplemental standing valves, such as 7, may be eliminated entirely from the structure at the will of the operator. In the same fashion, referring to Fig. 2, counsel for Appellants refers to valves 31 and 3^b as being far apart, whereas the *actual operating valves* of the pump are the valves 3^b and 43, *which are in close proximity to each other.* Valve 31 is a bottom standing valve in the column supplementing valve 3^b and is capable of being eliminated entirely, if desired.

We believe that counsel for appellants has somewhat misconstrued the language of the Neilsen patent in claiming:

> "* * * on the up-stroke of the plunger the spaces 15, 17 are filled with liquid simultaneous with the filling of the pump."

We submit that in actual operation as the barrel of the Neilsen patent 14 moves upwardly the spaces 15 and 17, Fig. 1, will be filled with fluid, and will remain at all times filled with fluid, the fluid passing from space 15 into space 17 during the upward movement of the plunger. It is plain that fluid from space 15 must be expelled as the barrel moves up. It has only one place to go and that is downwardly between the standing column 8 and the puller nut into the space 17 toward the anchor face. As the barrel moves downwardly fluid from space 17 will pass upwardly into the space 15, the spaces 15 and 17 remaining filled at all times.

The Neilsen patent discloses a number of features in addition to those disclosed by the Bray patent; for instance, it provides for a pumping or flushing action in connection with valves 7 and 31, and it provides a liner in which the plunger fits to gain such pumping action.

It is plain that these features can be eliminated or disregarded, leaving only the essential working parts of the Neilsen structure, as colored on page 142 of the Book of Exhibits. The parts so colored constitute a structure which completely anticipates every detail of the Bray patent save the series of jet apertures in the puller nut and the clutch on the puller nut and anchor face. To provide cooperating clutch jaws in the Neilsen construction would obviously amount to no more than the exercise of mechanical skill, in light of the prior art patents.

In analyzing the character of flow obtained through the orifice of the puller nut of Neilsen, counsel for appellants again states that such flow is a "free flow," and that the opening is a "vent" as distinguished from a "jet aperture." Despite these statements we again inquire where the line of division or distinction lies between a "vent opening" and a "jet aperture." It is clear that when asserting infringement appellants recognize no such line of distinction.

During operation of the Neilsen pump fluid will obviously pass in and out of space 15 or 38, through the bore of the puller nut and such fluid will be jetted or made to flow downwardly toward the anchor face, depending upon the speed at which the pump is operated. To claim that this is "free flow" and to term a like flow in the structure of appellees a "jet," is to claim a distinction without a difference.

F. THAT NEITHER OF SAID DEFENDANTS EMPLOYS IN HIS, OR ITS PUMP A GUIDE USED WITH A CLOSE, ORDINARY MECHANICAL OR RUNNING FIT.

G. THAT THEY AND EACH OF THEM ALLOW SOME DISCHARGE OF FLUID TO GET BY BETWEEN THE GUIDE AND THE CIRCUMFERENCE OF THE STANDING COLUMN.

H. THAT NEITHER OF SAID DEFENDANTS IS FOLLOWING THE SPECIFICATION OF CLAIMS 2 AND 3 OF PATENT NO. 1,840,432.

Questions I and II.

1. On page 24 of their brief, appellants state:

"* * * the jet aperture of the Hofco pump is in the form of an annular space surrounding the standing column of the pump."

It is clear that appellants do not contend that appellees' structures employ a guide "with a close ordinary mechanical or running fit," nor do they dispute that appellees allow some discharge of fluid to "get by between the guide and the circumference of the standing column." Such a claim would be impossible in view of the fact that the *character and dimensions of defendants' structures were stipulated,* and that such stipulated facts disclose the clearance between column and guide. (See par. 2, pages 47-48, *supra.*)

It follows that the evidence amply supports the findings lettered F and G, and that a concession of the correctness of finding H must logically follow.

I. THAT NONE OF THE PUMPS IN EVIDENCE, MANU-
FACTURED, USED OR SOLD BY EITHER OF SAID DE-
FENDANTS, EMPLOYS A GUIDE WHICH IS APER-
TURED WITH VENTS OR APERTURES FOR THE EJEC-
TION OF LIQUID DIRECTLY TOWARD THE EFFECTIVE
FACE OF THE LOWER GUIDE COUPLER.

Question I.

From the foregoing discussion on pages 104 to 105, it
is clear that appellants do not question the soundness of
this finding in so far as it pertains to claim 2, and that
Assignment of Error XLIX in so far as it refers to
claim 2 must be deemed abandoned.

Question II.

The analysis of the evidence on page 99, *supra,* demonstrates that the finding of the trial court that claim 3 of
the Bray patent is limited to a "bottom guide running on
the column, with a running fit," and that this guide "must
be provided with * * * apertures for the ejection of
liquid directly toward the face of the lower coupling member, are fully supported by the evidence. It is likewise
perfectly clear that the Hofco structures employ not a
guide, but a puller nut running *over* the column, and so
mounted on the barrel as to leave an annular space between the puller nut and the standing column; that this
hole in the puller nut is necessary to make it a puller nut,
and that it is not in addition apertured. It follows that
the evidence supports this finding in so far as it relates to
claim 3, and that Assignment of Error XLIX as directed
to claim 3 is without merit.

J. THAT NONE OF THE PUMPS OF DEFENDANTS MADE
   LIKE THOSE IN EVIDENCE INFRINGE UPON CLAIM 2
   OR CLAIM 3 AS SO LIMITED.

Questions I and II.

As the foregoing analysis has established that the findings previously analyzed, are supported by the evidence,
it follows that the above finding is a necessary corollary
thereto and that Assignment of Error L is without merit.

That these findings, if supported by the evidence, are
sufficient to support the "Conclusions of Law" and the
"Decree" is not questioned in the opening brief, and we,
therefore, assert that as they are so supported, Assignments of Error LIII, LIV, LVI, LVIII, LIX [Tr. pp.
403 to 405] are without merit.

III.

On page 85 of their brief, appellants state that issue
3 is: :

> "The Lower Court erred in decreeing that the de-
> fendants, and each of them, have not infringed any
> of the claims of Letters Patent in suit and have not
> infringed said Letters Patent, and in not decreeing
> that the defendants and each of them have infringed
> each and all of the claims of said Letters Patent, and
> have infringed said Letters Patent."

They assert that this follows because

"The defendants' pump, Hofco Catalogue, Plaintiffs'
Exhibit 4, Book of Exhibits, pages 5 to 8 inclusive,·

and Defendants' Exhibit II, Book of Exhibits, page 177, contains all of the elements of each of the claims of the patent in suit, which elements of defendants' pump cooperate in substantially the same manner, perform the same function and accomplish the same result as the corresponding elements of the respective claims of the patent in suit."

and then read claim 1 of the patent in suit on the appellees' structure. This argument ignores the fact that the holding of the trial court was that claim 1 *was void for lack of utility over the prior art in evidence,* and that it follows that no structure could infringe this claim.

On page 87 of their brief, appellants attempt to read claims 2 and 3 on the structure of appellees. They do not apply the claims separately, but rather attempt to collectively apply them, despite the fact that the claims are admitted by them to be somewhat different.

In their attempt to read claim 2 on appellees' structures, they ignore the limitation of the claim that the guide is to have "a series of venting ducts," apparently convincing themselves, by a curious process of reasoning, that the annular space around the column can be said to be such a "series of venting ducts."

In attempting to read claim 3 on appellees' structure, appellants repeat their contention that there is no distinction between such an annular clearance as is necessary to make the block of metal of which the guide is composed into a "guide," and a "guide" answering the limitation of the claim which is "said guide being apertured for the ejection of liquid," etc. They advance the specious argument that,

> "The difference between a plurality of closely arranged jet apertures and a single jet aperture extending around the standing column is a difference only of form,"

thus, completely ignoring the finding of the trial court that the structure of the patent was limited to a tight fitting guide so apertured as to result in the production of the work accomplishing jets of the Bray patent. We do not question that where form is not of the essence, *i. e.*, the change in form does not result in a difference in the functioning of the parts so changed in form, that a difference in form may sometimes be disregarded through the application of the doctrine of equivalents. Appellants apparently recognized that it is only in such cases that a limitation of form may be disregarded, as they state:

> "The difference * * * is a difference only of form because the single aperture of defendants' pump performs the same function in substantially the manner and accomplishes substantially the same result as the plurality of apertures of plaintiffs' pump. Such a difference is not sufficient to avoid infringement of plaintiffs' patent." (Opening Brief p. 89.)

They thus beg the very question which the Trial Court upon ample evidence found against them, and thus assume as a premise for their chain of reasoning, the existence of a fact which the record disproves. The cases cited by them relating to the doctrine of equivalents, are therefore not pertinent to the issue.

We cannot do more to disprove the fundamental error which runs throughout their brief, than to requote the last case quoted by them. They state:

> "In the recent case of Oates v. Camp (C. C. A. Va. 1936), 83 F. (2d) 111, a patent claim was not required to be limited to the exact device disclosed by the specification and drawings, since the claims of a patent and not its specification measure the invention." (Opening Brief p. 93.)

As the *claims* of the patent, and *not the specification* measures the invention, it is plain that the ability of the Bray structure to "minimize gas lock" is not a measure of the invention, while the claims, limited as they are to a guide so apertured as to provide a useful jet, cannot be said to include within their measure a structure which is not so apertured, and which the record discloses does not do this assertedly useful work. We have hereinbefore analyzed the evidence supporting the finding of the Trial Court that the Bray patent is limited to a tight fitting guide with ducts or apertures, and have pointed out that such finding is sound for the obvious reason that this type of guide and these apertures are necessary limitations of the claims in order to distinguish the structure claimed by the patent from the pertinent structures of the prior art. If, as counsel urges, the construing of claims 2 and 3 to cover only such a structure was erroneous, the decree would nevertheless be right for the reason that claims 2 and 3 would, if not so limited, be void because of the anticipating prior art.

IV

On page 93 of their brief, appellants assert:

"The above assignments of error, forming Issue 4 are a general objection to the final decree of the Lower Court in favor of the defendants and in view of the argument concerning Issues 1, 2 and 3, it is submitted that the final decree of the Lower Court is contrary to law."

We contend they have shown no error in any of the findings criticized by them or in the Conclusions of Law, and therefore submit that for the reasons contained in the preceding portions of our brief, the Final Decree of the Lower Court should be sustained.

PART B.

THE BRADFORD DEPOSITIONS.

The depositions taken in Bradford, Pennsylvania, on May 8th to 10th, 1934, establish:

[Morris, Tr. p. 69]:

That George B. Morris was President of the Bradford Motor Works, hereinafter referred to as "Bradford," and had first become connected with the company in 1911.

The O. F. S. Working Barrel, Ex. 15. (Bk. Ex. p. 85.)

[Morris, Tr. pp. 69-70]:

1. That the O. F. S. Working Barrel had been manufactured by the predecessor of Bradford, and when Bradford started business in 1911, the manufacture of this pump was continued. Morris, by reference to Ex. 15, described this pump as:

> "The O. F. S. Working Barrel is of the inserted type, the lower end of which is seated in the top of a conventional or standard working barrel. The tube, part 1, is stationary. Reciprocating within the tube is a plunger fitting, attached to a valve stem, part 2, which in turn is connected to the sucker rods by a substitute, part 3. Fluid is discharged from the working barrel into the tubing through holes in the crown, part 4. The crown is fitted at its upper end with a notch, or groove, part 5. A corresponding tongue, No. 6, is formed on the substitute, No. 3. By the engagement of tongue, No. 6, in slot, No. 5, rotation of part 3 is prevented, thus making it possible to screw up loose or disconnected sucker rods. This was a well recognized and valued feature of the unit."

Moving Plunger Type—Clutch.

2. It will be noted that this pump is of the moving plunger type, and its combination includes, as "a well recognized and valued feature," a clutch (parts 3 and 5, Ex. 15), to screw up loose and disconnected rods. [Ingleright, Tr. pp. 61 and 62]:

Fay L. Ingleright testified that he was a machinist and Superintendent of Bradford; that when he entered the employ of Bradford on February 25th, 1926, they were making the O. F. S. Working Barrel, and that it contained the clutch at the point marked "B" on page 8, of Ex. 11. (Bk. Ex. p. 80.)

Catalogues Frick and Lindsay, 1915, Ex. 10.

[Morris, Tr. p. 70]:

Mr. Morris testified that this pump was listed in the catalogue of "Frick and Lindsay Company, marked 'Copyright 1915'," Ex. 10, at page 20 (Bk. Ex. p. 78); that he first saw this catalogue in 1915 or 1916 and saw copies of it in various supply stores throughout the country.

[Johnson, Tr. pp. 53-54, 59]:

Sirvertus H. Johnson, Secretary of Bradford, and in its employ since 1920, testified that this catalogue was in the office when he first entered the employ of the company and that he pasted in it the flyleaf which is written in his handwriting, on October 8, 1923.

Bradford Catalogue, Ex. 11.

[Morris, Tr. pp. 70, 71]:

3. Mr. Morris testified that he was instrumental in the preparation of catalogue Ex. 11, "Pumping Supplies for Oil Wells—Bradford Motor Works," and saw the first shipment of them when they came from the printer in 1921; that the O. F. S. pump is illustrated on page 8; that between five and ten thousand copies of this catalogue were printed and an effort made to put at least one copy of it in every oil well supply store in the United States, and in the hands of production men in all of the oil companies in the United States; that those catalogues were received in several shipments in 1921, and paid for September 19, 1921, as shown by "Cash Disbursements, page 30"; that this distribution was made by mail and through the salesmen of the company.

[Johnson, Tr. p. 59]:

Johnson testified he helped make up this catalogue (Ex. 11), and caused it to be printed about August, 1921; that it was circulated and distributed under his direction to all Bradford customers, either by mail or through its salesmen; that not less than 5000 were so distributed.

[Ingleright, Tr. p. 61]:

Floyd Ingleright testified he saw an exact copy of this catalogue in February, 1926, which copy was sent to him at Grand Rapids, Michigan, by Mr. Morris, immediately prior to the time he entered the employ of Bradford, and that he saw in said catalogue the illustrations of the O. F. S. Working Barrel shown on page 8.

Extent of Sales.

[Morris, Tr. p. 70]:

4. Morris testified that although they are not now manufacturing the O. F. S. pump in quantities and advertising it, they continue to furnish repair parts, and at intervals have been supplying complete units; that more than 1,500 O. F. S. pumps were made and sold.

The Bradford Plunger Liner, or I. X. L. Pump. Ex. 16. (Bk. Ex. p. 86.)

[Morris, Tr. pp. 70, 71]:

1. Mr. Morris testified that in 1912 they began the manufacture of the Bradford Plunger Liner, also called the I. X. L. Pump. He described it as:

> "This is a pump of the inserted type, and is operated without the customary cups. The suction is provided by reciprocation of a plunger tube, No. 1, through a stuffiing box, No. 2, fitted interiorly with packing. The groove, No. 3, and tongue, No. 4, operate in exactly the same manner as the groove, No. 5, and tongue, No. 6, on Exhibit No. 15, and for the same purpose."

Moving Plunger Type—Clutch.

2. It should be noted that this is also a moving plunger pump, and is provided with the clutch (Ex. 16, parts 3 and 4).

[Ingleright, Tr. p. 61]:

Ingleright testified that when he entered the employ of Bradford in 1926, they were making and selling this pump; and that the pumps then manufactured contained the clutch illustrated at the point marked "A", on page 7, Ex. 11. (Bk. Ex. p. 79.)

Bradford Catalogue, Ex. 11.

(The printing and distribution of this catalogue has been established by the testimony set out in Par. 3, *supra.*)

Extent of Sale.

[Morris, Tr. p. 71]:

3. Morris testified at least 500 of these Bradford Plunger Liners, or I. X. L. pumps, were made and sold.

The Bramo Pump, Ex. 1. (Bk. Ex. p. 29.)

[Morris, Tr. p. 71]:

1. Mr. Morris testified that Bradford began to construct this pump, and the first one was completed and installed on a well in March, 1917. He described it as:

> "The structure of this pump embodies a combination of the essential features of the pumps on Exhibit No. 16 and Exhibit No. 15. It has the cup structure and lower valve, like Exhibit No. 15, and the plunger tube and stuffing box, like Exhibit No. 16. It also has the lock feature common to both Exhibit No. 15 and Exhibit No. 16, and indicated by letters 'C' and 'D' on Exhibit No. 1, the tongue appearing on part 'D' and the groove on part 'C'."

[Johnson, Tr. p. 54]:

Mr. Sirvertus H. Johnson testified he entered the employ of the company May 1, 1920, and that at that time the company was manufacturing the Bramo pump, which was then being used and on sale; that Ex. No. 1 is a true and correct representation of the pump as it was then being constructed. He described this pump as:

> "The Bramo is an oil well pump, commonly known as a liner barrel, designed to seat in a special seat or

in a common working barrel. The pump has a reciprocating plunger tube, designated by 'A' on Exhibit No. 1, the outside tube, which is designated by 'B' on Exhibit No. 1, remaining stationary. Part 'C' shown on Exhibit No. 1 is a stuffing box gland in which a groove is milled to engage a milled section in the upper connection, which is designated as 'D' on Exhibit No. 1. Part 'C' is at the upper end of an assembly at the upper end of part 'B'."

Moving Plunger Type, With Clutch.

2. It is to be noted that this is also a moving plunger type pump, employing the lock feature, or clutch (Ex. 1, parts C and D).

Catalogues Frick and Lindsay, 1919, Ex. 2.

[Morris, Tr. p. 72]:

3. Mr. Morris testified this Bramo pump is illustrated in Frick and Lindsay catalogues dated May, 1919 (Ex. 2, p. 43; Bk. Ex. p. 30), which he first saw some time in 1919, and has seen in various supply stores. [Johnson, Tr. pp. 54-55]:

Mr. Johnson testified this catalogue was in the office when he started to work May 1, 1920, and had remained in the files ever since.

Extent of Sales.

[Morris, Tr. p. 72; Johnson, Tr. p. 55]:

4. Mr. Morris testified they are still making this pump, and had made and sold not less than 15,000 of them. Mr. Johnson testified they had made and sold not less than 10,000 pumps since he entered its employ.

Inverted Bramo, Ex. 17a.

[Morris, Tr. pp. 72, 73]:

1. Mr. Morris testified that Bradford manufactured a device called the "Inverted Bramo Working Barrel"; that it was illustrated by Ex. 17 (Bk. Ex. pp. 87-89), and its structure and operation was:

> "In the original Bramo Working Barrel the stuffing box and main tube were stationary while the pump was in operation. The plunger fitting was reciprocated within the tube and connected by a plunger tube to the sucker rods. In the Inverted Bramo Barrel the main tube, No. 1, and the stuffing box No. 2, reciprocate over a stationary plunger fitting within the tube. This plunger fitting is connected to the standing valve, No. 3, by means of a plunger tube, also stationary. The tongue and groove assembly, indicated at No. 4, is the same as that used on the original Bramo Working Barrel, but is located at the lower end of the pump, . . ."

2. He further testified that they shipped an "Inverted Bramo Working Barrel" to Mr. Wm. H. Maxwell in Los Angeles, and that Exhibit 17 consisted of photographs of it. (Mr. Maxwell testified he received the pump so shipped; that Ex. 17 is made up of three photographs of it. This pump was received in evidence and marked Ex. 17a.)

Pump of Traveling Barrel Type With Clutch.

[Morris, Tr. pp. 72 and 73]:

3. It will be noted that this pump was of the traveling barrel type, and Mr. Morris stated:

> "The tongue and groove assembly (the clutch) indicated at No. 4, is the same as that used on the origi-

nal Bramo Working Barrel, but is located on the lower end of the pump, due to the inserted construction of the pump. As a matter of fact, we have used this construction on every type of inverted pump we have ever manufactured and the only deviation from a general standard is due to the necessity for thread changes in order to adapt this feature to different types of pump."

Use.

[Morris, Tr. p. 73]:

4. He further testified that the company made not more than 25 of these pumps; that they did not function to its satisfaction, as in this design satisfactory provision was not made for the escape of oil trapped between the main tube and the plunger tube; and that consequently the stuffing box principle was abandoned, and a pump built of the type later given the name "Admore."

The Admore Pump, Ex. 6. (Bk. Ex. p. 52.)

[Morris, Tr. p. 73]:

1. As to this Admore pump, Mr. Morris testified, referring to Ex. 6:.

"The Admore Pump is of the inverted type, involving a stationary plunger tube, part No. 2, a' stationary plunger fitting located at the upper end of said plunger tube, and a standing valve screwed to the lower end of said plunger tube. It also involves the reciprocating parts No. 6-1-3-4 & 5. The reciprocation of tube, No. 1, creates a suction above the plunger fitting. The locking device is positioned at the same place as in Exhibit No. 17, and in construction and purpose identical with Exhibit No. 17."

[Johnson, Tr. p. 57]:

2. Johnson testified that Exhibit 6 illustrated the Admore pump made by the company. He testified this device was first made early in the year 1926, and described the device as follows:

> "The Admore is an oil well pump, commonly known as a liner barrel, which is designed to seat in a special seat or common working barrel. The standing valve, comprised of parts 14 to 19, inclusive, to which is attached the plunger tube, part 2, and plunger fitting, which is made up of parts 13-B-8-13-11-12-7-9 & 10, remains stationary during operation, and the outside tube, designated as part 1, together with parts 3-4-5 & 6, reciprocate with the pumping action. Part 15, which is known as a closed crown, is designed with milled tongues which engage in part 6 which is known as the tube guide, part 6 being milled with slots which engage with the tongues on part 15, part 6 being assembled at the lower end of the outside tube, shown as part 1 in illustration."

[Flickinger, Tr. p. 78]:

3. Raymond C. Flickinger testified that he went to work for Bradford in February, 1913, left December 12, 1925, and returned in the latter part of January, 1927; that Ex. No. 6 represents the type of pump which was being manufactured in the plant at the time of his return, and that the locking device (parts No. 6 and No. 15 on Ex. 6) was included in the pump being manufactured upon his return; that all of the Admore pumps manufactured since that time have carried that locking device.

[Ingleright, Tr. pp. 62, 63, 64]:

4. Fay L. Ingleright testified that at the time he entered the employ of Bradford (Feb. 25, 1926) this pump had been designed, and the first such pump was actually built under his direction. That drawing, Ex. 13a, (Bk. Ex. p. 81), represents the construction of the pump so built, with the exception that the first Admore Liner Barrel Pump built had:

a. One lock nut at the point marked A on Ex. 13a.

b. A lock washer at the point D, between the nut and the part E.

c. One piston ring is missing in illustration 13a, at the point B and C, which rings were present in the structure originally built.

5. Referring to Ex. 6, he stated that the first Admore pump built had:

a. A standing valve similar to that shown on Ex. 6, parts 14-15-16-17-18 and 19.

b. A similar plunger tube, part 2.

c. A similar upper valve assembly, parts 3-4 & 5.

d. An exactly similar outside tube, part 1.

That the plunger fitting of the early Admore differed from that shown in Ex. 6, and is illustrated by Ex. 13a. It consisted of:

a. A body F,

b. Four cups G,

c. Three cup rings H,

d. One lock nut on top at A,

e. Grooves cut in the body to receive four piston rings I.

The plunger fitting shown on Ex. 6 consists of a body, Part 10, upon which are mounted four cups and three rings. He also stated that the early pump carried a shake proof lock washer mounted on the upper end of the plunger fitting.

He further described the structure illustrated by Ex. 6, as follows:

". . . . Directly above that (the plunger fitting) another member is attached, carrying a female coupling which in turn has screwed into its upper end another tube and upon which is mounted a spring, part No. 12, a packing follower, part No. 11, a coil of packing, part No. 13, a top nut, part No. 8, and a lock nut, part No. 13-B. . . . In running a liner barrel of the type shown in Exhibit No. 6 and in Exhibit No. 13-A the lower valve, parts No. 14 and 19, inclusive, form a standing valve which is run down in the top of an old working barrel or special seat, seating at the lower edge of part No. 15. The upper end of part No. 15 is milled to form a tongue at its upper end which engages a female slot in part No. 6, same being screwed into the lower end of part No. 1. When the well is being pumped, a string of sucker rods attaches to the upper end of part No. 3 to give the reciprocating pumping action. In the event of rods becoming unscrewed, the upper, movable member, consisting of parts No. 1-3-4-5 & 6, drop down by their own weight. The operator at the surface of the ground may then lower the rods until they engage, and a slight twist of the rods from above causes the notch or female groove in part No. 6 to become engaged with the tongue on part No. 15. This locks the lower end of the string of sucker rods, and because of the lower valve, namely, parts No. 14 to 19, inclusive, being held in a fixed position in the working barrel or special seat, allows the operator to tighten up the loose joints of sucker rods. After hooking onto the power the movable members, parts No. 1-3-4-5 & 6, are spaced in their proper pumping relation. This disengages the lock between parts No. 6 and No. 15 of Exhibit No. 6. . . ."

Moving Barrel Type With Clutch.

6. It will be noted that this pump was also of the moving barrel type, and that it carries a clutch on the bottom of the barrel; concerning this Mr. Ingleright testified:

[Ingleright, Tr. p. 64]:

". . . The design of the lock between part No. 6 and No. 15 of Exhibit No. 6, and of part No. 6 and No. 15 of Exhibit No. 13-A, has been unchanged since the early manufacture, except for some minor details of dimensions. The position of parts No. 6 and No. 15, in respect to the other parts of the pump, has been unchanged."

[Morris, Tr. p. 74]:

Mr. Morris stated that no change had been made in the general construction of the locking device from the original Admore, testifying:

". . . its general construction is identical with that used on the first pumps made."

The Physical Exhibits.

Exhibit 12.

[Johnson, Tr. pp. 59-60]:

7. Johnson identified Ex. 12, and testified he first saw this "plunger assembly and tube guide of an Admore liner barrel" on the lease of T. P. Thompson, at Red Rock, Pennsylvania. That Mr. William Kolbe, Mr. W. H. Maxwell and he were looking for samples of Admore Barrels which had been manufactured among the first Bradford had put out, and that they discovered it among the junk which had accumulated in a barn on this lease; and that it was in the same condition at the time he gave his deposition as when he first discovered it.

Exhibits 13 and 14.

[Tr. p. 60]:

He also identified Ex. 13 as a plunger assembly and tube guide of an Admore liner barrel, and Ex. 14 as the outside tube, upper connection, crown, ball and seat; he stated that these exhibits were found by the same persons and at the same time and place as Ex. 12, and that at the time they were discovered Exs. 13 and 14 were assembled as one unit.

Exhibit 12.

[Kolbe, Tr. p. 75]:

The witness William P. Kolbe, corroborated Johnson's testimony, stating that he was an oil field worker, a pumper; that he had been for five years employed on the Thompson lease; that about the middle of August, 1933, Ex. 12 was located in the presence of Mr. Johnson and Mr. Maxwell in a barn on that lease; that it was in substantially the same condition then as it was at the time he gave his testimony, a "little dirtier if anything."

Exhibits 13 and 14.

[Tr. p. 75]:

He also identified Exs. 13 and 14 as having been located and examined at the same time and place, and testified that they also were a little bit cleaner than when discovered, but were otherwise the same.

Date of Manufacture of Exhibit 12.

[Ingleright, Tr. pp. 64, 65]:

8. Ingleright identified Ex. 12 as being a plunger fitting assembly of an Admore liner barrel, practically the same as the one illustrated in Exhibit No. 13-A, and comprising a standing valve, No. 15 on Exhibit No. 13-A; a tube guide, No. 6, a plunger fitting parts "A" to "I", inclusive, and a plunger tube, No. 2, and testified he was able from the design and various characteristics of that exhibit to fix the approximate date of its manufacture. He testified it was assembled *during the summer of 1926.* He stated it was of the second lot manufactured because:

"It has two lock nuts on top while the earliest plunger fitting made had one lock nut; the plunger fittings which had two lock nuts had no threads in the member directly below the two lock nuts, *i. e.,* the member carrying the two piston rings; in the first lot the upper member carrying the piston rings had a tapped hole. The upper member with piston rings, corresponding to E on Ex. 13-A, was made to be slidably mounted on the plunger fitting body."

He was able to further fix the date of manufacture by the following characteristics of Ex. No. 12, stating:

a. ". . . The Shakeproof lock washer between the two lock nuts on Exhibit No. 12 was found by experience to be faulty, as it had a tendency to break up, scoring the working barrel tube and the piston rings. The last Shakeproof lock washers were assembled in the latter part of 1926, and none were used thereafter. . . .

b. The piston rings were also discontinued during the latter part of 1926.

c. The plunger tube (3' 6" long), corresponding to part No. 2, Ex. 13-A, is made from black iron pipe, . . . The material was afterwards changed to seamless steel tubing.

d. The closed crown, corresponding to part No. 15-A, Exhibit 13-A, measures 1⅞" O.D. which is of a size that was discontinued after the second run of Admore Closed Crowns.

e. The tube guide, corresponding to part No. 6 of Ex. 13-A, is of a design used in making up the first two lots of Admore pumps, and is identified by the shortness of the length of the large outside diameter, this being afterwards lengthened to provide a better wrench hold."

Date of Manufacture of Exs. 13-14.

[Ingleright, Tr. p. 66]:

9. Ingleright testified the approximate date of manufacture of Ex. 13 was the same as Ex. No. 12. He stated the fact that Ex. No. 13 does not have a shakeproof lock washer simply means it was there and was lost off.

He also testified Ex. No. 14 was made during either early 1926, or along in the summer of that year, because the part corresponding to part No. 3 of Ex. 6 is known as the long pattern valve cage. The two earliest runs of Admore liners were equipped with this length valve cage.

The Date of Manufacture as Fixed by the Contemporaneous Records of Bradford.

[Ingleright, Tr. pp. 66, 67, 68]:

10. Ingleright testified that the contemporaneous records of the Bradford Motor Works fixed the date of manufacture of Exs. 12, 13 and 14. He pointed out by reference to Ex. 8 (Bk. Ex. pp. 54 to 67) that certain of the parts found in these structures were only incorporated in the pumps manufactured during a portion of 1926, and that the manufacture of these parts were discontinued during that year. He stated, referring to Ex. 8:

"*A.* Sheet 3 (Bk. Ex. p. 56), dated April 30, 1926, specified 250 Admore Closed Crowns (part 15-A, Ex. 13-A) to be made out of $1\frac{7}{8}''$ O. D. round cold rolled steel. It shows they were completed June 25, 1926. Sheet No. 6 specifies 400 of the same parts to be made from the same size steel, and shows completion Aug. 27, 1926. That is the last date of record of any Admore Closed Crowns being made having the same outside diameter as that member on Exhibit 12 and 13, for Sheet 14 (Bk. Ex. p. 67) specified the next lot of Admore Closed Crowns were to be made with an outside diameter of 1-27/32" cold rolled steel, and were completed Jan. 28, 1927.

B. Sheet 7 (Bk. Ex. p. 60) is a record of the last job order of Admore Plunger Fitting Bodies of the piston ring type, corresponding to part F on Ex. 13-A, and found on Ex. 12 and 13. It is shown to have been completed Aug. 13, 1926.

This design was changed at the completion of this run of parts, and a new run of a totally different design was started and completed Oct. 11, 1926, as shown by job order files Ex. 19. (Bk. Ex. p. 82.)

C. Sheet 10 (Bk. Ex. p. 63) is a record of the mating part for the Admore Plunger Fitting Body, corresponding to E on Ex. 13-A, and is a record of the last run of those parts. These Admore Cup Nuts were started August 9, 1926, and completed August 13, 1926.

On succeeding Admores this part was replaced by parts of standard design and carried regularly in stock, and a search of our records shows that no more of the Admore Cup Nuts referred to were made.

D. Sheet 2 (Bk. Ex. p. 55) is a record of the second run of ¾″ extra heavy pipe plunger tube, corresponding to the plunger tube of Ex. 12 and 13. The last run of this material was completed Aug. 27, 1926, as shown by job order No. 4679 of our job records. (Ex. 20.) (Bk. Ex. p. 83.)

When our supply of these ¾″ pipe plungers was exhausted our design was changed to specify a seamless steel tube 1-1/16″ O. D. x 13/16″ I. D. x 4′6″ long. The first run of these tubes was completed November 19, 1926, as shown by our job order record No. 5022." (Ex. 21.) (Bk. Ex. p. 84.)

[Iverson, Tr. pp. 81, 82, 83]:

George F. Iverson testified he was employed by Bradford March 2, 1925, that he served the first year as billing clerk, and from then on as planning clerk. He fixed the date of Ex. 12, 13 and 14 as follows:

a. "The lock washer on Ex. 12, shown at point 'A' on Exhibit 13-A was last used approximately November 11, 1926.

b. The last run of piston ring type plunger assemblies such as those found on Ex. 12 and 13 (Parts A-D-E-B-C-H-C-F and I) was found on August 13, 1926, as shown Ex. 8, sheet No. 7. The first run of cup and packing type was finished on Oct. 11, 1926, as shown by this book record Ex. 19. (Bk. Ex. p. 82.)

c. The plunger tubes of Ex. 12 and 13 are iron pipe, and the use of iron pipe was discontinued August 27, 1926, as shown by the book record marked Ex. 20. (Bk. Ex. p. 83.)

   In place thereof we used seamless steel tubing. The first run of plunger tubes from seamless tubing was finished Nov. 19, 1926, as shown by job order #5022, marked Ex. 21. (Bk. Ex. p. 84.) I remember this change on account of the tubing we received being soft and hard to thread, and we put the ends of it into the furnace to harden it some so that we could thread it easily.

d. The last run of closed crowns of the dimensions of Ex. 12 and 13, shown as part 15-A on Ex. 13-A, made from 1-7/8″ stock was finished Aug. 27, 1926, as shown on page 6, Ex. 8 (Bk. Ex. p.

59), and the first run made from 1-27/32″ stock was finished Jan. 28, 1927, as shown on page 14, Ex. 8." (Bk. Ex. p. 67.)

He summed up the matter by stating:

"I can't state the exact date, but I know they were made during the year 1926."

Catalogue of Bradford, Ex. 7.

[Morris, Tr. p. 73]:

11. Morris testified that the Admore pump is illustrated in the one-sheet catalogue Ex. 7 (Bk. Ex. p. 53); that the company ordered and received 5700 of these circulars, which were, under his direction and to his knowledge, distributed in the usual manner to the supply companies and oil companies.

[Johnson, Tr. p. 57]:

Ingleright testified that Ex. 7 illustrated the Admore pump; that this catalogue was published in March of 1926, and was distributed generally to the trade by mail and by salesmen of the Bradford Motor Works.

[McCutcheon, Tr. p. 79]:

James G. McCutcheon, Jr., testified he is a printer and was engaged in that business in 1926. He stated he printed Ex. 7; that he had a record of doing that work, and referring to that record at page B-35 of Ledger No. 1, stated the entry in the book reflecting this transaction showed:

"Our job number was No. 2567, the order was placed on March 16th, 1926, and the order was for 5700 folders—Admore Liner Barrel. The price paid was $55.00."

He testified, upon his attention being directed to Ex. No. 7, that he had kept that sheet as a retained or filed copy of the document he had printed, and that approximately two weeks elapsed between the time the order was placed and the time he printed it.

Early Uses.

[Morris, Tr. p. 74]:

12. Mr. Morris testified that one of the first Admore pumps installed was in March, 1926, on a property near the Bradford factory and owned by Emery & Mason; that he personally saw this working barrel put in the well and watched it operate at periodic intervals for a month or two; that the records show (Ex. 9, Sheet #4) (Bk. Ex. p. 71) this pump was taken by us from our shop to the property on March 30, 1926. It was sold on a trial basis, and the invoice was not mailed until the customer expressed approval of the pump and willingness to pay for same; this it did the first part of August, 1926, and this invoice is dated August 6, 1926. Following these early uses all through 1926, we made consistent sales efforts, both in the Eastern fields and in the Mid-continent fields, and by the end of the year had a rather wide distribution and a consistent demand.

[Eidson, Tr. p. 76]:

Gilbert T. Eidson testified that Ex. 13 and 14 when assembled made up the pump known to him as the Admore; that he installed or caused such a pump to be installed in wells Nos. 140 and 145, on the Melvin lease, in 1926; that he had a record book, Ex. 18 (Bk. Ex. p. 90), which shows on page 62 thereof a record of these installations; that the pumps which he installed worked.

[Perkins, Tr. pp. 78-79]:

Carl M. Perkins testified that in 1926 he was employed at the South Penn Oil Co. at Bradford, Pennsylvania. That in the forepart of the summer of 1926 he installed an Admore pump in well 149 of the South Penn Oil Co. at Bradford, Pennsylvania. That Lewis Rounds assisted him in that installation and that Rounds died January 18, 1927. He described the Admore pump which he installed in well 147 as:

> "It had four cups on the bottom, and a plunger which goes up into the barrel. The barrel works on the outside of the plunger, and it has a plunger which has four cups and piston rings. The standing valve has four cups and a ball and seat. There is a guide bushing on the bottom of the outside tube that works over the standing plunger tube, and when the outside tube is set down the guide bushing locks with the standing valve crown so the rods can be tightened. The plunger in Admore pumps now has a spring and no piston rings."

Records of Early Sales of Admore Pumps, Ex. 9.

[Johnson, Tr. pp. 58, 59]:

13. Johnson identified these sheets as photostatic copies of sales sheets upon which Bradford billed its customers for purchases of Admore liner barrels. These sheets disclosed:

> Sheet No. 1 (Bk. Ex. p. 68)—Record of a charge dated May 4, 1926, for 6—2" x 6' Admore liner barrels to Frick-Reid Supply Co., Tulsa, Oklahoma.

Sheet No. 2 (Bk. Ex. p. 69)—Record of a charge dated June 7, 1926, for 1 Admore liner working barrel to Harland Oil Co., Wolco, Oklahoma.

Sheet No. 3 (Bk. Ex. p. 70)—Record of a charge dated July 7, 1926, for 1—2″ x 6′ Admore liner working barrel to Forest Oil Corporation, Bradford, Penna.

Sheet No. 4 (Bk. Ex. p. 71)—Record of a charge dated August 6, 1926, for 1 Admore liner working barrel to Emery & Mason, Bradford, Penna.

Sheet No. 5 (Bk. Ex. p. 72)—Record of a charge dated Sept. 7, 1926, for 1—2″ x 6′ Admore liner working barrel to South Penn. Oil Co., Bradford, Penna.

Sheet No. 6 (Bk. Ex. p. 73)—Record of a charge dated October 6, 1926, for 1—2″ x 6′ Admore working barrel to South Penn. Oil Co., Bradford, Penna.

Sheet No. 7 (Bk. Ex. p. 74)—Record of a charge dated November 5, 1926, for 1—2″ x 6′ Admore liner working barrel to Frick-Reid Supply Co., Tulsa, Okla.

Sheet No. 8 (Bk. Ex. p. 75)—Record of a charge dated December 3, 1926, for 1—2″ x 6′ Admore liner barrel to South Penn. Oil Co., Bradford, Penna.

Sheet No. 9 (Bk. Ex. p. 76)—Record of a charge dated January 7, 1927, for 1—2″ x 6′ Admore liner barrel to Frick-Reid Supply Co., Tulsa, Oklahoma.

Sheet No. 10 (Bk. Ex. p. 77)—Record of a charge dated February 3, 1927, for 1—2″ x 6′ Admore liner working barrel to Oil Well Supply Co., Tulsa, Oklahoma.

Exhibit 22.

Johnson also identified certain original sales sheets taken from the records of Bradford and testified they show the following sales:

a.   Sheet 123 (Bk. Ex. p. 93), April 12, 1926, 2 Admore working barrels to the Carter Oil Co., Casper, Wyoming.

b.   Sheet 128 (Bk. Ex. p. 94), April 15, 1926, 1 Admore liner barrel to A. L. & L. M. Lilly of Rixford, Penna.

c.   Sheet 308 (Bk. Ex. p. 92), April 26, 1926, 1 Admore liner to Sloan & Zook Company, Bradford, Penna.

d.   Sheet 327 (Bk. Ex. p. 91), April 27, 1926, 1 Admore liner barrel to the South Penn. Oil Co., Bradford, Penna.

Manufacture and Sales of Admore.

[Morris, Tr. p. 74]:

14.   Morris testified that since 1926 Bradford has continued to make and sell Admore pumps and is now making and selling them.

[Johnson, Tr. p. 60]:

Johnson testified that 308 Admore pumps were made and sold in 1926, and that about 14,000 have since been sold.

Summary.

In the light of this testimony, all of which stands uncontradicted, we submit that the date of manufacture, use and sale of each of these structures has been clearly established, Appellants' assertion to the contrary notwithstanding.

CONCLUSION.

We believe that it is abundantly apparent from the foregoing discussion that the findings of the Trial Court are
supported by the evidence, and that they support the
"Conclusions of Law" and the "Decree."

There is so much repetition of broad assertions unsupported by the record, scattered throughout the brief of
appellants, that we cannot in a brief of any reasonable
length, hope to answer them seriatim. A great portion
of appellants' brief approaches the problems involved in
this appeal, as if it were a trial *de novo,* or a trial in the
District Court, and argues the weight of conflicting testimony. Much of their argument is difficult to answer specifically because it is not specifically put. Much of their
analysis of the prior art patents and structures cannot be
followed through to a practical application because appellants failed to point out the application of such analysis
to the asserted issues. There is an almost total lack—
although this is essentially an appeal based on the alleged
insufficiency of the evidence to support the findings—of
reference to the respects wherein it is asserted, the various
findings of the Trial Court lack support. We nevertheless have fully and fairly, though perhaps too exhaustively,
considered all of the matters even inferentially raised by
appellants' brief, as we have best been able to puzzle out
its meaning.

We submit sincerely and earnestly that the analysis contained in our brief, discloses that each finding of the Trial Court finds proper support in the evidence that the findings support the Conclusions of Law, and that the "Final Decree" of the District Court should therefore be affirmed.

Respectfully submitted,

BURKE & HERRON,

By MARK L. HERRON,

JOE C. BURKE,

Attorneys for Appellees.

No. 8171.

In the United States
Circuit Court of Appeals
For the Ninth Circuit.

10

atrick H. Bray, and Catherine Patricia Marquette,

  Appellants,

vs.

ofco Pump, Ltd., and D. W. Hoferer,

  Appellees.

REPLY BRIEF OF APPELLANTS.

ALAN FRANKLIN,
114 W. 3rd St., Los Angeles,
Attorney for Appellants:

Parker, Stone & Baird Co., Law Printers, Los Angeles.

TOPICAL INDEX.

TABLE OF AUTHORITIES CITED.

CASES.   PAGE

Text Books and Encyclopedias.

No. 8171.

In the United States
Circuit Court of Appeals
For the Ninth Circuit.

Patrick H. Bray, and Catherine Patricia Marquette,

Appellants,

vs.

Hofco Pump, Ltd., and D. W. Hoferer,

Appellees.

REPLY BRIEF OF APPELLANTS.

OPENING STATEMENT.

In the short time allotted to Appellants it would be impossible and a waste of time to attempt to meet categorically all of the various trivial and irrelevant issues raised in the voluminous reply brief for Appellees. The real issues of this cause are quite clear and simple and we will confine our reply brief strictly to them, believing that in so doing this Honorable Court will the more readily see the justice of Appellants' meritorious cause.

Only four issues are raised in Appellants' opening brief, which are set forth on page 309 of said brief.

ASSIGNMENTS OF ERROR.

While counsel for Appellees have attempted to show that said issues are not supported by some of the assignments of errors, he concedes that said issues are supported by certain of our vital assignments of error, which are abundantly sufficient to support the issues of this appeal. (See Reply Brief for Appellees, pp. 8, 12 and 13.) While we do not agree with counsel for Appellees concerning our assignments of error we do not consider it worth while to quibble over this matter.

It is true that there are many assignments of error for this appeal, but Appellants' present counsel is not responsible for them, because they were prepared by other counsel before he was retained by Appellants, shortly before the time limit of appeal expired. The grouping of assignments of error in equity suits before this Court is not original with counsel for Appellants. Any number of equity appeals to this Court could be cited where assignments of error were grouped, instead of being set out separately as required in actions at law. If the assignments of error in this case had to be set out and argued separately Appellants' opening brief would have run up into several thousand pages. We believe that our opening brief covers the issues raised by the assignments of error thoroughly and clearly and that we have greatly reduced the time and work of the Court in reading and understanding this simple case of patent infringement, but which has been made rather complicated by the elaborate defense of counsel for Appellees.

THE PATENT IN SUIT.

In considering the novel *combinations* of elements of the claims of the Bray patent in suit, we brought out in our opening brief the fact that each of the three claims brings into its combination "a pump of the class having a standing guide column with a working valve (standing valve S), and a valved traveling barrel, running on said column (the valve on the traveling barrel being the traveling valve V in the Bray patent). The standing valve S is described in the patent specification, page 1, lines 86-88, as being on the upper end of a "stem" or column 10 and is so illustrated in the patent drawing. The traveling valve V is described in the specification, page 1, lines 83-85, as provided at the upper end of the traveling barrel and is so illustrated in the patent drawing with the traveling barrel B at the lower end of its stroke and the traveling valve V in its lowermost position in close relationship to the standing valve S. This close relationship of the standing valve S and traveling valve V at the lower end of the stroke of the traveling barrel B is an important feature of the class of oil well pump described, shown and claimed in the patent in suit, and the specification, drawing and claims are necessarily required to be considered together in interpreting the patent and its claims.

> *Walker on Patents,* 6th Ed., Sec. 227 at p. 310;
> 48 *C. J.,* Sec. 335, p. 213.

It should be borne in mind that the patent in suit covers a *pump* of the *class* disclosed and not any particular part of the pump, and the claims cannot be construed piecemeal, like the lower Court has construed them, but must be construed as a whole, that is to say, that the complete combination of elements of each claim must be considered, as the measure of the invention, and not any particular element or elements.

> *Leeds & Catlin v. Victor Talking Machine Co.,*
> 213 U. S. 318;
>
> *Diamond Rubber Co. v. Consol. Tire Co.,* 220
> U. S. 428;
>
> *Yesbera v. Hardesty Mfg. Co.,* 166 Fed. 120, at
> p. 125 (C. C. A. 6th Cir.).

"The patentee makes all the parts of a combination material, when he claims them in combination and not separately."

Walker on Patents, 6th Ed., Sec. 411, p. 499.

It is true that the patent does not state that an object of the invention is to prevent "gas lock", but the pump of the class described, shown and claimed, is obviously a pump that must necessarily prevent gas lock, because the pump is the invention of a practical California oil pump manufacturer and the evidence shows that the pump has been used primarily in the California oil fields, and in these oil fields gas is a serious problem as well as sand, and any practical pump used in these oil fields, no matter what other improvements it may have, should be constructed to prevent gas lock. It was not necessary to state in the patent that the pump would prevent gas lock, because this feature of the pump alone is not new, but is only one of the old elements of the pump, and in stating an old element in a patent, it is not necessary to describe it in detail, but only necessary to refer to it in general terms, as it is referred to in the patent specification and claims, as "a pump of the class having a standing guide column with a working valve and a valved traveling barrel", and to illustrate the class of pump referred to, as it is illustrated in the drawing of the patent in suit. The effect of the pump structure disclosed in the patent necessarily prevents gas lock, and the patentee obviously intended to claim an operative pump to prevent sanding up, because gas is a serious condition to be met in oil fields where there is sand, as in California. There is nothing unusual in reading the patent to cover an operative pump, which in the light of the disclosure of the patent and of the evidence, necessarily includes a pump structure that will prevent gas lock.

> "Patents are to be liberally construed, so as to secure to an inventor the real invention which he intends to secure by his patent, and there are cases in which an element described in the specification has been read into the claim from the specification."
>
> *Walker on Patents,* 6th Ed., Sec. 226, p. 307 and cases cited.

The close relation of the valves has another important function in that it effects a quicker action in the displacement of the oil. [Tr. p. 259.]

There is no valid reason why the valve structure disclosed in the patent and referred to and included, as an element in the claims should be read out of the claims, since the reading of such element in the claims narrows the claims instead of broadening them, and this is not against the public interest.

The contention of counsel for Appellees that the inclusion of the valve structure in the patent claims makes the claims aggregations is without merit, because, obviously, each of the elements of each claim cooperates with the other elements of the claimed combination to the extent that it contributes to a common purpose, to-wit: to *pump oil from wells.* The sand difficulty met by Bray's pump is only one of the difficulties that must be met by a practically operative pump in the California oil fields, and unless the pump as a whole meets all of such difficulties, including "gas lock" the pump would not be practically operative and would be of little utility. The grant of letters patent carries with it the presumption of operativeness and utility, and Appellees have offered no reliable evidence to overthrow that presumption. Moreover, the conclusions of the lower Court is to the effect that claims 2 and 3 of the Bray patent cover an operative and useful pump. [Tr. p. 378.]

The issuance of a patent raises the presumption that the device covered thereby is a valid combination, as distinguished from an aggregation.

> *Galvin Electric Mfg. Co. v. Emerson Electric Mfg. Co.,* 19 Fed. Rep. (2d) 885 (C. C. A. 8, 1927).

This presumption cannot be overcome by showing that an expert with the patent before him might be able to build up the structure covered by the patent by selecting and adapting appliances old in the art.

> *McMichael & Wildman Mfg. Co. v. Ruth,* 128 F. R. 706 (C. C. A. 3, 1904);
>
> *Vortex Mfg. Co. v. Ply-Rite Contracting Co.,* 33 F. R. (2d) 302.

Counsel for Appellees has gone to considerable trouble to show that Bray, the patentee in suit, was not the first to put a clutch, in an oil well pump, to enable screwing up of the sucker rods of the pump, but no such claim is made by Bray. It is Bray's novel *combinations* of elements, forming his novel pump, that he claims as his invention, and the clutch is only one of the elements of the combinations, but it is combined in a different arrangement with relation to the other elements than is found in the prior art.

ISSUE I.

In considering Issue 1 concerning the error of the lower Court in decreeing claim 1 of the patent in suit invalid in view of the prior art we will consider only the Admore pump (Bk. Ex. pp. 52 and 53) which is the nearest approach in the art to said claim. If this pump does not anticipate claim 1 it is useless to consider any other pump of the prior art. This Admore pump is the only pump of the prior art which has a standing column or plunger, a traveling barrel, a standing valve on the plunger, a traveling valve on the traveling barrel, and a clutch at the lower end of the traveling barrel, the upper member of which clutch being on the bottom guide running on the standing column. This Admore pump, however, does not meet the terms of claim 1 in its entirety, and we have shown that a combination claim is an entirety, because this pump has its standing valve *below* and outside of the traveling barrel, and not on the *upper end* of the standing column within the traveling barrel and pump chamber, as described in the Bray patent (p. 1, lines 86-88), and as shown in the Bray patent drawing. This position of the Bray standing or working valve S is a feature of "a pump of the *class* having a standing guide column with a working valve", etc., which *class* of pump is included in the combination of elements covered by claim 1, and also claims 2 and 3 of the Bray patent. The position of the Bray standing valve brings said valve close to the traveling valve V, when the traveling barrel B is at the lower end of its stroke, as shown in Fig. 1 of the Bray patent, and this close relationship of the valves forces all the gas out of the pump chamber between the valves, upwardly past the traveling valve V into the column of oil in the well tubing T, and prevents a gas lock in the pump. The standing valve 16 of the Admore pump (Bk. Ex. pp. 52 and 53) being below the lower end of the traveling barrel is a considerable distance below the traveling valve 5, at the lower end of the stroke of the traveling valve, and in this elongated space between the valves 16 and 5 would accumulate a considerable amount of gas and form a "gas lock", which would prevent oil from passing through the pump chamber and prevent pumping of oil by the pump. This Admore pump is a very useful pump in the oil fields of Pennsylvania where there is no gas, but it would not meet the high pressure gas conditions of the oil fields of California and would not be a practical pump in these oil fields. There is no evidence that this pump ever operated successfully in the oil fields of this state or any other oil fields where there is a high pressure gas condition.

Claim 1 further specifies "an *anchor* for the column presenting *directly* to said coupler face (4) (on the lower end of the traveling barrel) a top face (2) forming a complimentary coupling to be non-rotatively interlocked with the lower guide coupler" (4) (Bray patent). This is a structural difference between the Bray pump and the Admore pump, and may be an advantage in the Bray pump, in that in the event of a break in the sucker rods, the barrel would drop under the influence of the weight of the sucker rods, which might amount to 15,000 pounds, in a deep well, until the lower end of the barrel rested upon the anchor P, the strongest element in the pump structure. A break in the sucker rods of the Admore pump would cause the barrel, under the influence of the weight of the sucker rods, to drop upon the valve cage and the valve cage might be broken. From the evidence we do not know whether there is any merit in this observation or not, but the location of the lower coupler *on the anchor* is a term of the claims of Bray's pump, and the Admore pump does not meet this term. The presumption, however, from the grant of the patent is that a term in a claim has merit or utility and all doubts are resolved in favor of such presumption.

Walker on Patents, 6th Ed., Sec. 125.

The Dies clutch 4-Z (Bk. Ex. p. 96) is for the purpose of screwing the stud Y into the plug W and unscrewing the stud from the plug. If the plug is turned in one direction the stud will be unscrewed from the plug and if there were sucker rods in this pump the whole string of said rods might be allowed to turn altogether and the joints of the sucker rods might not be unscrewed, as in the Bray pump. As there are no jointed sucker rods in this Deis pump the clutch has no function concerning the same. The function of the Dies clutch 4-Z is therefore not the same as that of the Bray clutch. This Dies pump is a traveling plunger pump and lifts fluid on the *up-stroke* and not on the *down-stroke* like the Bray pump. The Dies pump is therefore not a pump of the *class* embodying Bray's invention.

The Dickens pump (Bk. Ex. p. 137) is a *traveling plunger* pump, and while it has a clutch, it is not a pump of the *class* specified in the claims of the Bray patent having a *standing plunger* and *traveling barrel*. Bray's *entire pump* may be pulled by the puller nut or guide 3, which is not present in the Dickens pump. Only the *plunger* of the Dickens pump can be pulled, leaving the lower clutch coupler 11 and the standing valve 8 down in the well tubing. It requires power to lift the Dickens plunger to lift the fluid, while Bray's pump lifts the fluid and performs its major work on the down-stroke under the influence of the weight of the sucker rods.

The clutch of the Whitling pump (Bk. Ex. pp. 103-106) is not on a guide and an anchor as specified in the Bray claims, and the Whitling pump is of the traveling *plunger* class and not of the traveling *barrel* class, as specified in Bray's claims. The Whitling pump cannot be pulled in its entirety like the Bray pump. Only the plunger can be pulled out of the barrel, leaving the barrel and the standing valve 12 down in the well. The valves are spaced far apart at the respective ends of the pump chamber and would form a "gas lock".

What has been said of the Whitling pump may be said of the Cummins pump (Bk. Ex. pp. 100-101). It is a traveling *plunger* pump and not of the same class as specified in the claims of the Bray patent. It lifts oil on the *up-stroke,* not on the down-stroke.

The Stephens pump (Bk. Ex. pp. 160-167) has a clutch, but the upper clutch coupler is not on a guide or puller nut traveling over a standing column as specified in Bray's claims. This pump has no guide or puller nut like Bray's guide or puller nut 3 and the pump cannot be pulled in its entirety like the Bray pump. The Stephens pump does not meet the terms of Bray's claims.

We have analyzed all of the pumps having clutches and found them wanting. They do not meet the terms of Bray's claims and do not anticipate the Bray invention as embodied in claim 1.

We have pointed out above why none of the prior art pumps have a clutch construction which meets the specific terms of Bray's claim 1, and for this reason, irrespective of the close relationship of the valves, the validity of claim 1 is established. Claims 2 and 3 include the jetting device for washing sand from the clutch, and we will show that this feature is not found in the prior art and that this feature, irrespective of the close relationship of the valves, distinguishes claims 2 and 3 from the prior art. However, we do not have to distinguish claims 2 and 3 from the prior art, so far as the validity of these claims is concerned, since the lower court has found these claims valid, and that decisively distinguishes them from the standpoint of validity.

Counsel for Appellees on page 27 of his brief takes issue with our statement of operation of the Bray pump, and on page 29 states that Bray was not the original discoverer or inventor of this class of pumps. This issue as to operation of the pump is only one of the many irrelevant issues raised in Appellees' brief. We have not asserted that Bray is the original inventor of this general class of pumps, but we do maintain that Bray was the first to embody this class of pump in his novel *combinations* of elements and is the first to produce a pump that will not sand up. The basic operation of this general class of pump, which we concede to be old, whichever way it operates, is immaterial, so far as the validity of the patent is concerned, but the operation, as we have stated it, has advantages over the traveling plunger class of pump, and those advantages are utilized in the novel combinations of elements forming the Bray invention.

We again and emphatically maintain that the Bray pump operates as we have stated its operation on pages 18 and 19 of our opening brief, to-wit: that it *lifts* the oil through the well tubing on its *down-stroke* under the influence of the *weight* or part of the weight of the string of sucker rods. The traveling barrel B (Bray patent) must necessarily go down under the influence of the weight of the sucker rods, and not under the influence of power applied to the sucker rods by the engine on the derrick, because if power were so applied to push the rods and the traveling barrel down, on the down-stroke of the pump, the sucker rods would only bend, and buckle against the resistance of the oil under the lower end of the traveling barrel. As for the lifting of the oil on the down-stroke, this is necessarily true in view of the fact that additional oil is introduced into the tubing T (Bray patent) surrounding the pump barrel, only on the down-stroke of the traveling barrel, when the traveling valve V opens and discharges the oil from the pump chamber in the upper end of the traveling barrel. On the up-stroke of the pump the upper end of the traveling barrel, of course, displaces oil and tends to push oil above it upwardly in the tubing T, but, as fast as the upper end of the traveling barrel

displaces oil in the tubing T, an equal amount of oil flows
from the tubing T into the space vacated under the lower
end of the upwardly-moving traveling barrel, and this
operation amounts to nothing more than transferring the
displaced oil at the upper end of the traveling barrel to the
space vacated under the lower end of the barrel by the
upward movement of the barrel, and since, on the up-stroke
of the barrel, the traveling valve V remains closed, and
no oil is discharged from the pump chamber in the upper
end of the barrel, no additional oil is added to the oil in
the tubing, and consequently no oil is lifted in the tubing
or discharged therefrom on the up-stroke of the pump.
If it were not for the *opening* of the traveling valve V
on the *down-stroke* and the *closing* of said valve on the
up-stroke of the traveling barrel the traveling barrel could
be moved either up or down in the oil in the tubing T
without pumping any oil out of the upper end of said
tubing, because in such case oil would merely be displaced
from the upper end to the lower end of the barrel and
vice versa, just as a pebble in a glass full of water could
move up or down without pushing any of the water over
the upper edge of the glass.   But Bray's pump is different
from a pebble because it is not solid like a pebble, but
has a pump chamber in the upper end of the traveling
barrel, controlled by the standing valve S and the traveling
valve V, so that when the traveling barrel moves down
the traveling valve V opens and oil is discharged from
the pump chamber into the oil in the tubing T, and since
the discharged oil must go somewhere and cannot go down,
it must go up into the oil in the tubing and force an equal
amount of oil out of the upper end of the tubing, assuming
the tubing to be full of oil.   If a hollow sphere full of
water were substituted for the pebble in a glass full of

water, and the sphere could discharge its water when it sank, into the water in the glass, the discharged water from the sphere would push an equal amount of water from the full glass over the upper edge of the glass. From the above analysis of the operation of the Bray pump, which applies equally to the Hofco pump, we believe the Court will understand the operation of both pumps as we have stated it.

Bray testified [Tr. p. 47] concerning the operation of his pump as follows:

"When your tubing is always full of oil, or gas and oil, supposed to be full of oil, and it is filled down here to the shoe, that fills it until it cannot go any further. After the oil is by-passed now, this down here, you understand, that 7 feet is full of oil below the plunger. Now your displacement of this oil shows that this pump *pumps with the weight of the rod*. It *pushes the oil up* in place of *pulling* it up like a traveling plunger will. With a pump of this kind at 5,000 feet there are probably 15,000 pounds of rods, you understand, but as I say, the displacement is on the outside of this wall. We will say this is full; it is two and a half inches in diameter and there is an eighth clearance between the tubing and the pump and of course that is full of oil. *As this* (the barrel) *goes up, that oil* (above the barrel). *comes down here and fills this place* (space below barrel). *There is your displacement.* In other words, *your barrel, as that goes up, don't lift none of that oil, it runs down here and fills this here* (space below barrel) *as it comes down.* This barrel has got to have room and this (the barrel) pushes it (the oil) up." (between the barrel and tubing T, as the barrel goes down).

The misconception of counsel for Appellees concerning the operation of the pumping action of the Bray and Hofco pumps is due to the fact that counsel has not considered the space vacated by the lower portion of the traveling barrel, on its *upward* stroke, into which space an amount of oil flows from the tubing T, equal to the amount of oil displaced by the upper moving portion of the traveling barrel. Such displacement of oil results in nothing more than a change of position of the oil in the tubing T without adding a *drop of oil* to the sum total of oil in the tubing to cause overflow at the upper end of the tubing of any additional amount of oil discharged by the pump into the tubing full of oil. No oil is discharged by the pump into the tubing T on the up-stroke of the pump and consequently there is no pumping of the oil out of the tubing on the upstroke of the pump. Oil is discharged by the pump into the tubing *only* on the *down* stroke, when the traveling valve V opens, and the force of the discharged oil can act on the oil in the tubing only in an upward direction, which results in overflow of the oil at the upper end of the tubing.

In so far as the validity of the patent in suit is concerned it is immaterial whether the pump operates as the patentee Bray and counsel for Appellants have stated it, or as counsel for Appellees have stated it. The pump actually works, because the witness Rennebaum [Tr. p. 357] testified that he has operated both Appellants' and Appellees' pumps, and Bray has testified that he has installed 500 of his pumps which have operated satisfactorily for four years.   [Tr. pp. 43 and 270.]

"It is not essential that an inventor should either understand or set forth the scientific principle upon which his invention works."

Walker on Patents, 6th Ed., Sec. 160, p. 189.

On pages 118-123 of Appellees' brief, counsel for Appellees attempts to show, from the fancied theory of his expert Adams, that Bray's pump has *no jetting action*, because his pump does not "sand up", when the plunger packing wears and there is upward leakage or slippage of oil past the plunger.   This fantastic theory of inoperativeness of Bray's jet action is purely a figment of the wild imagination of one purely theoretically professional expert, against the granting of the Bray patent after careful examination by experts in the Patent Office, at least equally as capable as expert Adams; against the findings of the lower Court to the effect that claims 2 and 3 of Bray's patent are *valid,* covering an *operative and useful invention;* against the testimony of Bray, a pump manufacturer and practical oil pump expert; against the testimony of witness Rennebaum, who has operated the Bray pump; and against the successful operation of 500 of Bray's pumps, over a period of four years, without sanding up. [Tr. p. 43.] We do not have to remind this learned Court that letters patent in infringement suits are presumed to be operative and useful unless the contrary is proved beyond a reasonable doubt.   No such proof has been established by Defendants-Appellees.

The finding of the lower Court as to the *operativeness and utility* of Bray's jet, as covered by claims 2 and 3 of Bray's patent, is *stare decisis,* so far as this case is concerned, since the Defendants-Appellees, have taken no appeal from this finding of the lower Court, and Appellants have assigned no error of the lower Court on this finding.

The presumption of counsel for Appellees in attempting seriously to argue this defense of inoperativeness and lack of utility of Bray's invention before this Honorable Court, after failing to plead the same and without taking an appeal from the decree of the lower Court decreeing the invention *operative and useful,* is truly overwhelming. To say the least it does not come with good grace.

However, to prevent misleading of the Court by this alleged inoperativeness and lack of utility of the Bray pump, because of the effect of wearing of the plunger packing and leakage past the plunger, we desire to clarify the confusion caused by its injection into the evidence of the case, like the ink of a cuttlefish injected into a body of clear water.

Even if the plunger packing should wear sufficiently to reduce materially the pumping efficiency of Bray's pump, practically all of the liquid would still be drawn from the bottom of the well up through the standing column into the pump chamber, because of the free and unobstructed passage through the standing column of infinitely greater cross-sectional area than that of any alleged leakage space, of a few thousandths of an inch, between the oil-sealed alloy plunger guides 13 (Bray patent) and the traveling barrel; and any suction of liquid from the jet-pressure chamber EE (Bk. Ex. p. 178) in the lower end of the traveling barrel through any such leakage space would be nothing like as great as expert Adams would have the

Court believe, and in view of the vacuum created directly under the jet apertures at the lower end of the traveling barrel, as the barrel moved upward, practically all the liquid in the jet chamber EE would still be positively jetted downwardly through the jet apertures into the vacuum below, toward the lower clutch member and anchor, and the jetting action of the pump would continue for a considerable time, as Bray says it does, while the pump operated at reduced efficiency due to any assumed leakage, and such jet action would still function sufficiently to prevent the pump from sanding up. Conceding, for the sake of argument, an appreciable leakage, one significant fact, that expert Adams and counsel for Appellees overlook, is that if the pumping efficiency of the pump is reduced, the amount of *sand* pumped into the well tubing is correspondingly reduced, and the *jetting efficiency* of the pump could also be correspondingly reduced and still function sufficiently to wash away the correspondingly reduced amount of sand, and prevent sanding up.

The evidence referred to in counsel for Appellees' brief, p. 118 [Tr. p. 289], is misleading. The falling off of 20 to 50 barrels of oil from a well due to slippage or leakage past the plunger of the pump would not be a falling off of 20 to 50 per cent pumping efficiency or effective action unless the capacity of the well were 100 barrels, and the evidence does not show what the capacity of the well is, whether a 100 barrel or a 450 barrel well. Moreover, the witness Bray is not a learned mathematician or university engineer and could easily be confused on cross-examination, concerning pumping percentages.

The alleged leakage past the plunger due to failure of the packings 16 of the Bray patent, which expert Adams says would prevent any jetting action in the Bray pump, is grossly exaggerated by said expert. Bray's alloy guides 13, on the standing column, fit the interior of the traveling barrel very closely and are provided with annular peripheral grooves 14 which hold oil and form an oil seal against leakage or slippage of oil upwardly from the jet chamber into the pump chamber on the up-stroke of the traveling plunger. (See p. 2, lines 35-44, Bray patent.) Even if the last packing 16 wore out, the oil-sealed guides 13 would still function fairly well against leakage past the plunger, and enable the pump to continue its suction to pump oil. Bray's testimony [Tr. p. 289] to this effect is as follows:

> "I say, when that packing is gone I get a certain slippage there, enough to lose that much fluid. (*20 to 50 barrels,* not 20 to 50 per cent efficiency.) Now, I have oil seal grooves in this that do—here they are here (indicating)—each one of them is an *oil seal groove and they still leave that oil get suction.*"

The alleged inconsistency, set forth in counsel for Appellees' brief (p. 147), of our statement of the effect of the jetting action of Bray's pump producing a turbulence of the liquid directly between the lower end of the traveling barrel and the clutch and anchor, is without foundation. A jet of liquid into a body of quiescent liquid necessarily produces a turbulence of the body of liquid. The Court may take judicial notice of the turbulence of water in a pail when a jet from a garden hose is injected into the water in the pail. The turbulence of the body of liquid is the necessary physical reaction to the jet, and the projection of Bray's jet directly downward from the bottom of his traveling barrel towards the anchor and clutch, is the only form of jet that would produce such turbulence directly below the bottom of the traveling barrel that would be sufficiently effective to wash away the sand from the anchor and clutch to prevent sanding up of the pump. However, whether or not we have stated the principle of the invention correctly is of no moment, since the evidence abundantly supports the operation of Bray's pump and its effectiveness in preventing sanding up, and this is sufficient to meet the requirements of the patent law.

Walker on Patents, 6th Ed., Sec. 160, p. 189.

ISSUE 2.

Issue 2 deals with the error of the lower Court in not giving claims 2 and 3 of the patent in suit a liberal and proper construction to support the charge of infringement of said claims by the Defendants-Appellees.

Claim 2 includes the "series of venting ducts which *open directly toward* the anchor coupler face so that during the strokes of the barrel, ejected liquid is impelled toward the anchor top face to aid in prevention of sand accumulation thereon."

Claim 3 covers the same combination of elements as claim 2 in broader terms in that it specifies the "guide being *apertured* for the ejection of liquid from the barrel *directly toward* the effective lower face of the lower standing coupler part to aid in preventing sand settlement thereon."

The above elements of said claims produce the novel *jet action* of the Bray pump for washing the lower clutch coupler 2 and the anchor P (Bray patent) to carry away the sand up through the oil in the tubing T to prevent sanding up of the pump. This jet action and its *function* of washing the anchor and the lower clutch coupler for preventing "sanding up" of the pump, was unknown in the art prior to the advent of the Bray invention, and makes its first record appearance in the Bray patent in suit. There is not the slightest suggestion of this jet action and its function in the prior art, and the statement in our opening brief that Bray was the first in the art to solve the sand problem together with the gas problem in the oil fields of California, is justified despite the assertions to the contrary in Appellees' brief, page 46.

The Bradford Admore pump, Defendants' Exhibits 13 and 14, Exhibits 6 and 7 (Bk. Ex. pp. 52 and 53), which Appellees' counsel evidently considers an anticipation of claims 2 and 3 of Bray's patent (Appellees' Brief, p. 111), does not teach the jet action or its function, as embodied in said claims. It is not mentioned in the Admore catalogue (Bk. Ex. p. 53), and all of the deposition witnesses at Brandford, Penn., who were connected with the manufacture of said pump, are mute concerning any jet action of the Admore pump. It is also significant that no questions were asked these witnesses concerning such action. If the Admore pump ever had a jet action, or functioned like Bray's pump, those witnesses should have known it and their silence on this vital point speaks louder than words against any such jet action. It is only the fertile imagination of the professional expert Adams, with the Bray patent before him, that for the first time discovers a jet action in the Admore pump. From Adams' indefinite testimony [Tr. p. 211] counsel for Appellees in his brief, pages 111 and 112, assumes that there is a jet action in said pump because the area of the annular clearance in the guide is 3/7 of the total area available for the escape of fluid through the guide, the remaining 4/7 of the available area for escape of fluid being the holes in the side of the barrel above the guide. It is obvious that the greater amount of fluid (4/7) goes out of said holes through the side of the barrel directly against the immediately adjacent wall of the well tubing where said fluid will be dispersed and dissipated and no downward jetting of the fluid will be produced thereby. The greater amount of fluid escaping through the holes in the side of the barrel would materially reduce and weaken the force and action of any so-called jet through the annular clearance, so that any escaping fluid through said annular clearance would not, to use the colorful expression of expert Adams, "rise to the dignity of a jet". The meaning of said expression is elucidated by Adams' testimony [Tr. p. 197] as follows:

> "That would mean that there is practically no change of velocity in the case of the *Admore*, that is, at the middle of the stroke the *velocity of down flow would be very slight but the average would practically come to zero.*"

Adams speaks of a "down-flow" and not a jet, and if there were any such down-flow it would be through the four ⅜ of an inch holes in the side of the barrel above the guide, and its downward velocity would be zero and could not produce a jet.

A check of the figures of Appellees' counsel concerning the relative cross-sectional areas of the guide clearance and the holes in the side of the traveling barrel of the Admore pump is interesting. The clearance of the Admore guide around the one inch standing column measures *less* than 3/32 of an inch in width and this width is greater than its normal width, due to wear, because this Admore pump

(Defts. Ex. 13) is an old and worn pump. [Tr. p. 191.] The four holes in the side of the barrel above the guide measure ⅜ of an inch each in diameter. The cross-sectional area of a 3/32 of an inch clearance around a one-inch column equals .3221 square inches. The cross-sectional area of the four ⅜ of inch holes equals .44180 square inches. The cross-sectional area of the clearance is less than 3/7, and the cross-sectional area of the four holes in the barrel is greater than 4/7 of the total area of the openings at the lower end of the traveling barrel for the escape of oil from said end of the barrel. We submit that counsel for Appellees' statement in his brief [Tr. p. 111], that the annular guide clearance of the Admore pump is 3/7 of the area available for the escape of fluid, is exaggerated somewhat in favor of Appellees. The patent in suit, however is entitled to the benefit of any doubt in this matter.

The annular clearance in the Admore guide running on a standing column is nothing more than the clearance required in all pumps of the Admore class to produce the loose fit between the guide and the column, which is described by Bray [Tr. p. 37] as follows:

"You have to run it pretty loose anyway. You cannot afford to have it too tight, if you do it is not going to work very good."

The position of counsel for Appellees concerning the jetting effect through the annular clearance or "loose fit" in the "guide" of the Admore pump, which is required in all pumps of the Admore and the Bray class, as contrasted to his position that the "guide" on the Bray pump is restricted to a "tight fitting guide" (Appellees' Brief, p. 100), whereby the Hofco pump avoids infringement of Bray's claims 2 and 3, because it has a substantial clear--

ance between its guide and standing column, is a classic example of the inconsistencies of Appellees' defenses. Because the term "guide" is used in claims 2 and 3 of the Bray patent, said guide must necessarily mean a "tight fitting guide", according to Appellees' counsel, but when the same term is applied to the Admore pump, counsel for Appellees reverses his position and maintains that the term "guide" means a guide with a clearance sufficiently large to permit a jetting action of the liquid therethrough, which assumed action in the Admore pump anticipates the jetting action through the jet aperture of the guide in the Bray pump. Both positions are wrong, because Bray's guide as illustrated in his patent actually has a clearance of 3/32 of an inch to provide a *loose* fit between the guide and the column [Tr. p. 37] but said clearance is not Bray's entire jet aperture, and the Admore guide has a similar clearance for the same and no other purpose, which is proved by Adams' testimony to the effect that the velocity of the down flow through the orifice of the Admore guide would practically come to zero. [Tr. p. 197.] Ths use of the term "guide" in Bray's claims 2 and 3 is not restricted to a "tight fitting guide", because said term has a definite meaning in the art applicable to a loose-fitting guide, as employed in the Admore pump, and as used in the Admore catalogue (Bk. Ex. p. 53). In said catalogue appears the term "6-Tube Guide". The use of the term "guide" in claims 2 and 3 of Bray's patent, which means a guide with a clearance in the Admore pump (Defts. Exhibit 13) and the Admore Catalogue (Bk. Ex. p. 53), is no justification, as counsel for Appellees contends [Tr. p. 100], for restricting Bray's guide to a "tight fitting guide". The lower Court, however, did not find or decree that the Bray guide was limited to a "tight fitting guide" but found that Bray contemplated "a close or ordinary mechanical fit with the standing column" [Tr. p. 383, XI], and decreed that Claims 2 and 3 are limited to a guide with "a running fit" on the standing column. [Tr. p. 384, II.] It is submitted that in view of the clearance in the Admore guide (Defts. Ex. 13) and the Admore Catalogue (Bk. Ex. p. 53) that a close or ordinary mehanical or running fit comprises a fit of an appreciable clearance between the guide and standing column, such as provided in both the Admore and Bray guides, because there must be some play in the operation of the pump and the guide must run pretty loose to have a practical oil pump [Tr. p. 37] and a clearance is required to provide such play and loose running of the guide on the standing column.

The mere use of a "term", however, which might be construed one way or another, by counsel for opposing parties, should not, however, unduly limit the claims of a patent, to deprive a patentee of the just rewards of his patented invention.

> "Technical use of words is not necessary; an inventor his the right to use such words as seem to him to best describe his invention, and they will be so construed as to effectuate that result. A patentee is at liberty to select his own dictionary."

Walker on Patents, 6th Ed., Sec. 160, p. 189.

The only pump in the prior art, of the class specified in the claims of the Bray patent, which shows and describes substantial vertical openings through the guide, through which oil passes from the lower end of the barrel on the up-stroke of the barrel, is the pump disclosed in the Ellis patent No. 1,513,699 (Bk. Ex. pp. 133-135), but no jetting action through the passages 22 (Fig. 5) is described or suggested in said patent, and consequently the Ellis patent does not teach the jetting action of the Bray pump for washing the anchor or clutch. The Ellis specification describes (p. 1, lines 92-98) the function of its passages 22 as follows:

"The interior surface of this nut is formed with a series of oil passages 22 intermediate of the bearing surfaces 23 so as to permit the *free* discharge of oil from the dead chamber 24 which exists between the nut and the valve at the upper end of the standard."

And on page 2, lines 37-42, the specification states the following:

"Under some conditions the valve ring may be omitted as the passages in the cap nut permit *free flow* of liquid and prevent the formation of a dead chamber containing trapped liquid which will resist the free operation of the pump."

Obviously a *free flow* of liquid is not a *jetting action* of the liquid. The function of the free flow of the liquid through the passages 22 is to eliminate the trapped oil from the *dead* chamber 24 in the lower end of the barrel. A dead chamber is obviously not a pressure chamber for jetting oil therefrom, but is a chamber which would fill with incompressible oil and form an hydraulic lock against upward movement of the barrel, which would prevent operation of the pump. The Ellis patent does not mention a sand problem other than that if oil enters the dead chamber with any *velocity* it carries sand into said chamber and the function of the passages 22 is for no other purpose than to allow a free flow of the oil from the otherwise dead chamber 24 to prevent an hydraulic lock and to *avoid any jetting velocity* of the oil through said passages, which would carry sand into said chamber.

The passages 22 of Ellis were never intended to produce a jet, because if said passages must be large enough to prevent oil passing therethrough with any *velocity* into the dead chamber and jetting sand with the oil sand into said chamber, said passages would, *vice versa,* be too large to produce sufficient velocity to jet the oil out through said passages to wash the anchor on the up-stroke of the barrel. There can be no quesion that the "free flow" of the oil through the Ellis passages 22 was intended to be the very antithesis of a "jet". (Bk. Ex. p. 134, lines 22-26.) Another thing which negatives the idea of a jet in the Ellis patent is the position of the lower end of the traveling barrel and guide, considerably above the anchor, at the end of the down-stroke of the barrel, which position is so far above the anchor as to prevent the downward "free flow" of the oil from the dead chamber from ever reaching the anchor to wash sand from it. The guide in the Bray pump comes down to within six inches of the anchor on the down-stroke of the barrel, for effective jetting action. [Tr. pp. 36 and 37, also p. 34.] Fig. 1 of the Ellis patent shows the pump broken between the guide 21 and the upper end of the anchor 13, which indicates that a section of the pump is removed from between the break and the barrel and guide brought down nearer to the anchor than it comes when the removed section of the pump is in place in the actual construction of the pump. The pump is shown broken in order to foreshorten the actual length of the pump for the purpose of showing the pump on a larger scale than otherwise on the small-size patent drawing.

Moreover, one of the vital elements of the claims of the Bray patent is missing in the Ellis patent, namely, the *clutch* for locking the barrel against rotation to enable the joints of the sucker rods to be screwed up. Obviously the Ellis patent does not meet the terms of the claims of the patent in suit or suggest the Bray invention.

In the Deis patent (Bk. Ex. pp. 96-99) no oil goes out of the openings t on the up-stroke of the plunger because the valve u closes said openings. This is admitted by Defendants' expert Adams [Tr. pp. 94 and 109]. The Deis patent consequently has no jetting action.

The Downing patent (Bk. Ex. pp. 108-109) neither shows or describes any jet aperture through the puller nut H. The same is true of the O'Bannon patent (Bk. Ex. pp. 11-113), and the Thurston patent (Bk. Ex. pp. 115-117).

In the Thompson patent (Bk. Ex. pp. 119-123) the sleeve k' has a loose fit on the column f, but no jetting action is described in this patent, and even if there were a jetting action through said sleeve into the tubing r it could not remove sand from the tubing, because the plunger 1 fits tightly in the upper part of the tubing, and no sand could go past said plunger. Consequently this pump has no means of getting rid of the sand and the pump would sand up. The Thompson patent does not teach the Bray invention.

The aperture e in the Wright patent (Bk. Ex. pp. 125-127) is "to admit *free entrance or exit* of the fluid held in the space F, *the lower end e' of the working barrel operating as a piston.*" The aperture e is therefore only for the purpose of permitting *free* displacement from one side to the other of the lower end of the barrel so that it may act as a piston to force liquid from the space F through the passages b³ back down into the well to wash the rock in the bottom of the well. There is no jetting action through the aperture e and no means in this pump for washing out sand upwardly through the well tubing. This pump does not deal with any sand problem and is entirely foreign to the Bray invention, and it is not a pump of the class specified in the claims of the Bray patent, because the working barrel D, in the last analysis is the plunger which lifts the oil on the up-stroke and not on the down-stroke like the Bray pump.

The Northrup patent (Bk. Ex. pp. 129-131) in Fig. 1 of the drawing shows a narrow white space between the guide 24 and the column 1, but it is not described in the patent specification and its *raison d'etre* is left entirely to conjecture. Without a description of said white space the only reasonable interpretation of the same is that it is the usual clearance to provide a loose fit between the guide and the column which is required in all pumps of the class disclosed. [Tr. p. 37.] Since said clearance is not described in the patent specification there is nothing to indicate that there is a jet action through said white space in the Northrup patent. The patent drawing on the other hand describes the apertures 32 in the side of the barrel to permit sand, etc., to escape from the barrel. If there are any jets through said side apertures 32 said jets would strike against the immediately adjacent wall of the tubing C and be dissipated, which would destroy any downward projection of the oil to wash the anchor. The oil passing out of said apertures in the barrel above the guide might agitate the oil in the tubing but such agitation is too far above the anchor to positively force the sand therefrom and prevent the pump from sanding up. Moreover, the Northrup pump has no clutch and consequently has no jetting action washing the clutch as specified in claims 2 and 3 of the Bray patent. The clutch is a vital missing link in the Northrup pump as an anticipation of the Bray invention.

The Neilsen, *et al.*, patent (Bk. Ex. pp. 142-145) has an opening extending downwardly through the guide, but said opening is too large to effect a jet action therethrough, because the patent specification, page 1, lines 106-18, states that "on the *up-stroke* of the plunger the *spaces 15, 17 are filled* with liquid simultaneous with the filling of the pump". If the space 15 is filled with liquid on the upstroke of the plunger the liquid would pass *upwardly* through the annular space in the guide, *instead of downwardly through said space in the form of jets as in the Bray pump*. Consequently, the Neilsen patent has no jetting action *downwardly* through said annular space toward the anchor, as specified in claims 2 and 3 of the Bray patent, and the Neilsen patent therefore does not teach the Bray invention. Moreover, the Neilsen patent has no clutch and no jetting action to wash a clutch. This Neilsen pump, while it has a standing plunger and a traveling plunger, it is not a pump of the class embodied in claims 2 and 3 of the Bray patent, because its traveling barrel fits closely in a sleeve and lifts the oil on its upstroke and not on its down-stroke, under the influence of the weight of the sucker rods, like the Bray pump.

In all of the prior art pumps showing openings extending *downwardly through the guide* on the lower end of the traveling barrel, said openings fall in one or the other of the following categories:

1. The annular clearance between the guide and the standing column to provide a loose running fit of the guide on the column, which loose fit is necessary to provide a practical operating pump of the class disclosed in the Bray patent. The Admore pump (Defts. Exhibits 6 and 13) and the Northrup pump (Bk. Ex. pp. 129-131) are included in this category.

2. Openings for the purpose of permitting a *free flow* of the oil into and out of the dead chamber in the lower end of the traveling barrel, which oil, trapped in said dead chamber, would form an hydraulic lock against the up-stroke of the barrel and prevent functioning of the pump, as was the case in the inverted Bramo pump [Tr. pp. 72, 73, 84 and 85] (Bk. Ex. pp. 87 and 88) and (Dfts. Ex. 17-A). The Ellis pump (Bk. Ex. pp. 133-135) is included in this category. The openings in this type of pump which enable a *free flow* of oil therethrough are too large to produce sufficient velocity of the oil to constitute a jet.

3. Openings for the purpose of permitting a *free flow* of oil into and out of the traveling barrel to prevent a dead chamber and hydraulic lock, and to enable the admission of sufficient oil into a chamber below the traveling barrel on the *up-stroke* of the plunger, so that the *lower end of the traveling barrel* may act as a *piston* on its *down-stroke* to push said oil downwardly back into the bottom of the well to wash the rock therein, as in the Wright pump (Bk. Ex. pp. 125-127), or to wash the standing valve ball,

as in the Neilsen pump (Bk. Ex. pp. 142-145). The said openings in the guides in these pumps are too large to form a jet.

The said vertical openings in the guides of the pumps described in the above categories not forming jets have nothing to do with the washing of sand from the anchor or a clutch and discharging the sand upwardly through the well tubing.

There is not even a suggestion in the prior art of the employment of a straight downwardly projected jet from the bottom of the traveling barrel to wash sand from an anchor and a clutch and to discharge said sand upwardly through the well tubing. Even if there were any such suggestions in the prior art they would not be sufficient to anticipate the Bray invention as embodied in claims 2 and 3 of his patent.

> "But prophetical suggestions and surmises in the prior patents or publications, of what results can be achieved in a particular art, are not enough to negative the novelty of any patent on an invention which can accomplish that result."
>
> *Walker on Patents,* 6th Ed., Sec. 96, p. 122.

> "Novelty is not negatived by any prior patent or printed publication, unless the information contained therein is *full enough* and *precise enough* to enable any person skilled in the art to which it relates, to perform the process or make the thing covered by the pantent sought to be anticipated. * * * And *expert testimony,* though admissible on the question of the meaning of a prior patent or publication, is *not weighty,* unless supported by reasoning; and is *not conclusive* in court, even if so supported."
>
> *Walker on Patents,* 6th Ed., Sec. 96.

The hypothetical reversed pump structures, such as the Cummins (Bk. Ex. p. 175) and Whitling (Bk. Ex. p. 176) fail miserably to meet the terms of the claims of the patent in suit. In both of said pumps there are valves spaced so far apart that a gas lock would form therebetween. [Tr. pp. 254-255.] The patent claims specify an anchor with a complementary coupler. Cummins reversed shows this anchor on the clutch coupler, but expert Adams did not produce this element from the Cummins patent, because in the patent the lower clutch coupler 22 is on the upper end of the standing barrel (Bk. Ex. p. 174) and not on the anchor part 8. Adams got the idea of the lower clutch coupler 21 on the anchor in the Cummins reversed pump and lower clutch coupler 25 on the anchor of the reversed Whitling pump from the Bray patent. This is knowledge from the light of the patent in suit. In the Whitling patent there is no anchor, the lower end of the standing barrel being screw seated in the packer 3 (Fig. 1, Bk. Ex. 176), and the lower clutch coupler 25^a is on the upper end of the standing barrel. In neither the Cummins or the Whitling hypothetical reversed pump structures is there a downwardly projected jet aperture, nor is there any suggestion of any such jet aperture in the Cummins or Whitling patents. In the reversed Whitling structure the packing 31 on the lower end of the traveling barrel makes a close fit with the standing column which would form an oil seal and prevent any jet through the lower end of the traveling barrel. On page 155 of Appellees' brief counsel for Appellees states that the Whitling packing would wear and provide a jet opening, but this alleged jet opening did not develop in the old inverted Bramo

[Tr. pp. 72 and 73] (Defts. Ex. 17-A), and said Bramo
pump with a stuffing box at the lower end of the barrel
was abandoned. The assumption by counsel that the packing in the reversed Whitling pump structure would wear
and provide a jet aperture and jetting action is based
neither on sound reasoning or facts. The performance of
the abandoned Bramo pump negatives the false conclusion
drawn by counsel regarding his hypothetical jet action of
the Whitling pump. The theory of operation advanced
by Adams and counsel for Appellees regarding said hypothetical reversed pump structure, is truly fantastic.

The best that can be said for the Whitling reversed
structure is that it would make an ideal sand trap, if the
trapping of sand were of any commercial or sporting
utility, because its sand chambers 1 and 2 (Whitling patent, p. 2, lines 5 and 11, Fig 2) would trap sand as effectively as molasses catches flies.

Assignment of Error XVIII is as follows:

"In failing to find that, having illustrated and described one practicable form in which the invention
claimed may be embodied, that the patent in suit, in
contemplation of law, covers all equivalent forms
whether described in the specification or not, and that
the forms charged to infringe are such equivalent
forms."

This assignment of error states a long-established principle of patent law which has been recognized by this
Court in *Kings County Raisin & Fruit Co. v. U. S. Consol. S. R. Co.*, 182 Fed. 59 (C. C. A. 9th Cir.), and by the
Supreme Court in the following language:

"The patentee having described his invention and
shown its principles, and claimed it in that form which

most perfectly embodies it, is in contemplation of law, deemed to claim every form in which his invention may be copied, unless he manifests an intention to disclaim some of those forms."

Western Electric Co. v. La Rue, 139 U. S. 601, 606; 11 S. Ct. 670; 35 L. Ed. 294.

The patentee Bray of the patent in suit has manifested no intention to disclaim any of the forms in which his invention may be embodied, but on the contrary has clearly manifested his intention to include all of such forms, in the following language of his patent specification, page 1, lines 41-49, to-wit:

"Other objects, advantages and features of construction, and details of means and of operation will be made manifest in the ensuing description of the herewith illustrative embodiment; it being understood that modifications, variations and adaptations may be resorted to within the spirit, scope and principles of the invention as it is hereinafter claimed."

From such language it is obvious that the patentee Bray intended his patent to cover all forms of his invention, whether known or unknown to him at the time he filed his application for patent, and one form of his pump, which he manufactures, an *annular space* between his guide 3 and his standing column, was known to Bray at the time he filed his application for patent. [Tr. pp. 38, 41 and 42.] This form of Bray's pump is the form used by Appellees, the only difference being that the Hofco guide has a little larger annular space around the column, which space is ¼ of an inch wide. Counsel for Appellees, on pages 42 and 43 of Appellees' Brief, takes exception to a statement made by counsel for Appellants in Appellants' Opening Brief, p. 18, to the effect that Bray manufactured and sold the form of his guide with the annular space some time before Appellees' infringing pump was placed on the market. This statement of counsel for Appellants was based on the testimony of Bray [Tr. p. 40], and particularly to his testimony concerning his guide with an annular jet aperture [Tr. p. 43], as follows:

"According to my practical experience for ten years —I think I have had this kind of pump since 1926, and I was one of the first fellows that worked this kind of a pump in the field."

Counsel now representing Appellants before this Court did not try the case in the lower Court, of which fact Appellees' counsel reminds the Court in his brief, page 39, and counsel for Appellants may be mistaken in his statement about Bray manufacturing and selling his pump, with the guide having the annular space, some time before Hofco's pump was put on the market. However, no innuendo was intended by counsel for Appellants by the statement, other than what is actually supported by the evidence, and rather than quibble over this immaterial matter, counsel for Appellants withdraws the statement. The important fact that counsel for Appellants had in mind was that the form of guide with an annular jet aperture, such as employed in the guide of the Hofco pump, is one form of Bray's invention, since it was built by Bray and was Bray's invention prior to the time he filed his application for patent, and this important fact is amply supported by the evidence. [Tr. pp. 41 and 42.]

Counsel for Appellees on the other hand, in his brief, page 41, has made a confusing statement of the evidence concerning the ratio of the annular jet aperture in one form of Bray's guide to the jet aperture in the Hofco guide, but even if the ratio stated by counsel were correct, the Hofco jet aperture would be approximately only 2/10 of a square inch larger in cross-sectional area than the Bray jet aperture. This is no substantial difference. This 3/32 of an inch is only the clearance between Bray's guide and his standing column primarily to provide a loose fit of the guide running on the column which is only slightly larger than the minimum clearance, to provide such loose fit, which minimum clearance is required in guides in all pumps of the class disclosed in Bray's patent [Tr. p. 37], and as provided in the guide of the Admore pump (Defts. Ex. 13). While Bray's testimony [Tr. p. 37] is not as clear as it might be, and it was no fault of his own in the confusion of tongues in the lower Court, a careful reading of the testimony will show that Bray was not referring to the form of his pump with the annular jet aperture, but was referring to the particular form of his guide disclosed in his patent, which in addition to the 3/32 of an inch annular clearance around the standing column, to provide a little larger clearance than required by the conventional loose fit, has eight holes or jet apertures, each of which is 3/16 of an inch in diameter. [Tr. p. 188.] There is a considerable difference in cross-sectional area between the 3/32 of an inch clearance alone, and the cross-sectional area of said clearance plus the cross-sectional area of the eight 3/16 of inch holes, the sum of which areas equals the total cross-sectional area of the jet aperture of Bray's guide as disclosed in his patent.

Bray's testimony [Tr. p. 37] concerning the clearance between the guide and the column in his working pump is as follows:

"The Court: In your working pump. A. Well, I believe I will have to call for my blue print.

Q. By Mr. Sobiesky: There is a clearance of 3/32 of an inch? A. Yes.

The Court: You don't plan to have oil go through that clearance? A. Yes, a little bit, not much though. You have to run it pretty loose, anyway. You cannot afford to have it too tight, if you do it is not going to work very good.

The Court: It is more, therefore, a guide? A. Yes, the clearance is all right for making the hole.

Q. By Mr. Sobiesky: Before you applied for your patent you made a pump? A. I made it like that."

The blue print, unfortunately, was not introduced in evidence, but the Court evidently did not consider the 3/32 of an inch clearance as a jet opening from its question: "You don't plan to have oil go through in that clearance?" and Bray's answer, "Yes a little bit, not much though. You have to run it pretty loose anyway. You cannot run it too tight, if you do it is not going to work very good," and the testimony shows that Bray did not consider the 3/32 of an inch clearance as the full jet opening, but only a smaller proportion of the opening through which only a little bit of the jetted oil passed, and that the 3/32 of an inch clearance is primarily to allow the guide to *run pretty loose* on the standing column, particularly in a long stroke pump like Bray's. Bray's answer that he made the guide "like that" (with 3/32" clearance) before he applied for his patent, also shows that it must have been his guide with the eight holes (5) as shown in his patent, because that is the way, as counsel for Appellees points out in his brief (pp. 42-43), that Bray made his pump before he applied for his patent.

On cross-examination [Tr. p. 41] Bray further testified that he had made no other type of guide before he applied for his patent. Bray's testimony here is as follows:

"Q. By Mr. Herron: Now, then, the next type of guide you made was one with holes of that sort? A. Yes, I had not made any other type before I filed my application."

This is the interpretation of the evidence by counsel for Appellees on pages 42 and 43 of his brief—that Bray manufactured and sold pumps only with his guide having the holes before he applied for his patent.

Bray built only one guide with an annular jet opening like Mr. Hoferers', and tried it out successfully prior to his application. [Tr. pp. 41 and 42.]

From the evidence as above analyzed it will be obvious that the 3/32 of an inch clearance applies to Bray's guide with the additional jetting area of eight 3/16 of an inch holes and that said clearance is only slightly larger than the conventional clearance to provide the necessary loose running fit of the guide on the column. The statement of Appellees' Brief, page 41, to the effect that the jet aperature in Appellees' guide is 5/32nds of an inch greater than the 3/32nds of an inch clearance to which Bray referred [Tr. p. 37] and consequently that the cross-sectional area of the jet aperture of the Hofco pump is proportionately greater than the cross-sectional area of the jet aperture in the guide of the Bray pump as comprehended in Bray's patent, is therefore a glaringly false conclusion.

Bray testified [Tr. p. 42] on cross-examination concerning his guide with a single annular jet aperture which he built before he applied for his patent, as follows:

"I said that prior to my application for patent I had built a pump with clearance like Mr. Hoferer's. That clearance in the pump I built is on the blue print there, *you have got it.* That guide is the same as the one I built that time. The dimensions are on there. That is not the blueprint I made before I made my application, but it is the same figures. In other words, *the dimensions there are the same as the device I made. I never changed the dimensions any."*

The blue print referred to in the above testimony is obviously not the same blue print with the 3/32 of an inch clearance in the guide around the standing column [Tr. p. 37] to provide primarily the conventional loose fit of the guide, because the blue print referred to in the above testimony had a clearance in the guide like Mr. Hoferer's, which is ¼ of an inch in width. (Defts. Exhibit N. N. Bk. Ex. p. 179.) Counsel for Appellees on page 114 of his brief makes the most of the fact that the particular guide that Bray built, before Feb. 4, 1929, when he filed his patent application, which had an annular jet aperture like Hoefer's was not produced, and the fact that the blue print showing such aperture was not introduced in evidence, but counsel had the blue print in court, and with Bray on cross-examination he had an excellent chance himself to introduce said blue print in evidence, like he introduced Bray's old discarded rosette guide [Tr. p. 45, Dfts. Ex. A), but counsel muffed that chance and is estopped now from maintaining that because the blue print was not introduced in evidence, by counsel for Plaintiffs before the lower Court, that Bray never made a guide with an annular jet aperture like Hoefer's before he filed his patent application.

As for Bray's rosette guide (Defts. Ex. A) which counsel for Appellees implies had its internal ribs in close fitting contact with the standing column, which caused said ribs to wear out too quickly and form too great an opening to produce a jet, we point out to the Court that there is no evidence that the inner edges of said ribs fit the standing column with any less clearance than the conventional clearance between all guides and the standing column, in the class of pumps embodied in the patent in suit, and that with such clearance the ribs would nevertheless wear out quickly, because in crooked wells the standing column at times tends to move out of axial alignment with the traveling barrel [Tr. p. 273], and at such times, only, would the inner edges of the ribs, like the sides of the hole through any conventional guide with the conventional clearance, engage the standing column, whereby the guide performs its conventional function of preventing the standing column from moving materially out of axial alignment with the traveling barrel. There is therefore no point in counsel for Appellees' argument that Bray "deliberately sought and found a structure designed to overcome the automatic production of the very structure he now contends his claim to cover." The clearance between the hole through the guide and the standing column provides the loose fit and the necessary play between said elements to allow the pump to function freely in deep wells where the tubing necessarily cannot always be exactly perpendicular, and that there can be no such thing as a tight fitting guide on the standing column, to which Appellees' counsel would limit the claims of the patent in suit, without justification, either from the prior art or the findings of the lower court, because in the nature of the apparatus under consideration, it necessarily is not a precision instrument like a watch or an apothecary's scales.

The terms "close or ordinary mechanical fit" and "running fit" specified in the findings and decree of the lower Court, evidently do not mean a "tight-fitting guide", as asserted by counsel for Appellees, but mean the practical conventional close running mechanical fit with a clearance between the guide and the standing column, in view of the lower Court's questions as to whether oil would go through such a small clearance, and as to the fact that in view of such clearance and loose fit the jetting device 3 of Bray's patent was more, therefore, a guide. [Tr. p. 37.] If the lower Court intended to find a "tight fitting guide" it could not have found claims 2 and 3 of the patent in suit valid, because a "tight fitting guide" would be inoperative in a pump of the class disclosed in the Bray patent, in view of the required loose fit or clearance to allow for play between the standing column and traveling barrel and free movement of the barrel over the standing column to compensate for axial misalignment of the standing column and traveling barrel due to variations of the pump tubing from the perpendicular in deep wells. The Court found claims 2 and 3 valid and validity presupposes operativeness and utility, and operativeness and utility in a pump of the class included in claims 2 and 3 requires a clearance and *loose* fit between the standing column and traveling barrel. A "tight fitting guide" is purely a figment of the imagination of expert Adams and counsel for Appellees.

The only issue before this Honorable Court concerning the construction of claims 2 and 3 of the patent in suit is whether the guide is *limited to the form* particularly described in the patent in suit. The lower Court has found that said claims are so limited, but, in view of our analysis of the evidence, we maintain that this finding was gross error.

On pages 102 and 103 of Appelles' Brief a false construction is placed on the cancellation of Bray's original claim 6 and substitution therefore of the claims of the patent in suit as granted by the Patent Office, by reason of the inclusion in the granted claims of the term "bottom guide, * * * running on said column", appearing in claims 1 and 2, and the term, "bottom guide * * * *operating* on said column," appearing in claim 3. We have shown that there is necessarily a clearance between the guide and the column to enable the guide on the barrel to perform its guiding function on the column which tends to vary from the perpendicular in a deep well pump to axial misalignment with relation to the traveling barrel. There is nothing in the prior art, patent or file wrapper of the patent in suit that would restrict the patent to a tight fitting guide, nor any suggestion of such restriction, nor is there any evidence in the record that the patentee Bray ever built a pump with a "tight-fitting guide" or ever contemplated doing so. The evidence is to the contrary, because Bray testified that the guide had to run pretty loose on the standing column. [Tr. p. 37.] The granted claims, upon cancellation of original claim 6, merely express the invention in more appropriate language and limit the claims only by the inclusion of a pump having a standing guide column with a working valve and a valve traveling barrel, which contemplates a pump with the standing valve S on the upper end of the standing column in close relationship to the traveling valve V, to prevent "gas lock", according to the teaching of the specification and drawing of the patent in suit. This limitation is of moment in this litigation, because the Defendants-Appellees' pump construction embodies the same relationship of the standing valve and the traveling valve.

The positive downward jetting action for washing the clutch and the anchor and keeping the sand moving upwardly through the tubing T, to accomplish the *novel and useful result of preventing sanding up of the pump,* is a new principle of operation of the Bray invention which stands out in bold relief against the prior art, and this principle is practiced by the strong positive jets projected directly downwardly through space in the guide of suitable, cross-sectional area to provide the required downward jetting velocity of the oil out of the jet pressure chamber in the lower end of the traveling barrel. The space or opening in the guide to provide this novel jetting action, obviously lies generally within a zone of cross-sectional area dimensions between the conventional loose fit "clearance" between the guide and the standing column, to provide sufficient play in an operative pump, as exemplified in the Admore pump [Defts. Ex. 6 and 13] and the sufficiently large passages 22, Fig. 3, to allow a "free flow" of liquid there through, to prevent a dead chamber and hydraulic lock, as exemplified in the Ellis patent [Bk. Ex. pps. 133 – 135]. That zone of jet-space dimensions is comprehended within the spirit of the invention and the scope of claims 2 and 3 of the patent in suit, and there should be no poaching by infringers on the preserves of those dimensions fenced off from the generality within said zone, by said claims, for the term of the Letters Patent in suit.

e V, to prevent "gas lock",
the specification and drawing
s limitation is of moment in this
tendants-Appellees' pump con-
me relationship of the standing

jetting action for washing the
d keeping the sand moving up-
g T. to accomplish the *novel and
ng sanding up of the pump*, is a
n of the Bray invention which
against the prior art, and this
the strong positive jets projected
gh space in the guide of suit-
a to provide the required down-
the oil out of the jet pressure
nd of the traveling barrel. The
guide to provide this novel jetting
erally within a zone of cross
between the conventional loose
the guide and the standing column,
in an operative pump, as exemp-
ump [Deits. Ex. 6 and 13] and
ssages 22, Fig. 3, to allow a "free
hrough, to prevent a dead chamber
s exemplified in the Ellis patent
j. That zone of jet-space dimen-
within the spirit of the invention
2 and 3 of the patent in suit, and
aching by infringers on the pre-
as fenced off from the generality
claims, for the term of the

The right to protection of the sizes of apertures
sary to produce sufficient velocity to form the jet of
pump, is fully supported by the doctrine laid down
famous Edison incandescent light case.

*Edison Electric Light Co. v. U. S. Electric
ing Co.,* 52 F. R. 300, 1892.

"The carbon filament which constitutes th
new part of the combination of the second cl
that patent, differs from the earlier carbon b
of Sawyer and Man, only in having a diameter
sixty-fourth of an inch or less, whereas the b
of Sawyer and Man had a diameter of one
second of an inch or more."

Walker on Patents, 6th Ed., Sec. 69;

See, also

*Toledo Computing Scale Co. v. Computing
Co.,* 208 F. R. 410, 1913.

The above analysis of the. prior art covers all
pumps having openings downwardly through the
and shows conclusively that the prior art fails mi
to meet the acid test of anticipation and that the je
ture and jetting action of the Bray pump is nov
not anticipated. The Bray pump is therefore the
the art to produce an oil well pump that will no
up, and is an invention of considerable novelty and
The jetting action, which washes the anchor a
clutch, is a *new function,* and invests the Bray inv
with the character of a *primary invention,* even
it be in a limited field as considered by the lower

The Bray patent is therefore entitled to a liberal construction, sufficient to read on equivalent infringing pumps.

> *Walker on Patents,* 6th Ed. Sec. 416, p. 508
> and 417.

> *A. D. Howe Mach. Co. v. Coffield Motor Washer
> Co.,* 197 F. R. 541 and 548 (C. C. A. 4th Cir.).

See other cases cited in our Opening Brief (pages 67-70).

ISSUE III.

Our argument concerning this vital issue of *infringement* in our Opening Brief, pages 85-93, stands unshaken by any argument advanced by counsel in Appellee's Brief, and in this brief we will confine our argument to the most flagrant false conclusions drawn by counsel from the evidence on this issue. However, we respectfully request the Court to read again our short argument in our Opening Brief on this issue (pps. 85-93).

In our argument in our Opening Brief on this issue we have shown that the Defendants-Appellees have appropriated, element by element, the entire combination of elements of each and all of the claims of the patent in suit, and that the only difference between the pump of Appellees and the pump covered by Appellants' patent is merely a slight difference of *form and degree* in the *jet opening* in the guide at the lower end of the traveling barrel of the pump, whereby the jetting action of the pump is produced to wash sand from the clutch and the anchor to prevent sanding up the pump.

The difference in form between the jet apertures of the two pumps lies in the fact that in the particular form of the Bray pump disclosed in his patent, the guide 3 has

primarily the conventional loose fit, (3/32 of an inch clearance) around the standing column section 11, and is provided with eight 3/16 of an inch ducts or jet apertures 5
extending vertically through the bottom guide, while in
the Hofco pump the guide, indicated 4, D and P, and
FF, in the Book of Exhibits, pages 7, 177 and 179, respectively, has an annular jet aperture extending vertically
through the bottom guide and surrounding the standing
column A and 3, as designated in the Book of Exhibits
on pages 177 and 179, respectively, there being no reference characters applied to said jet aperture in said
exhibits, but said jet aperture appearing as a white space
between the guide D, colored red, and the standing column
A, colored blue [Bk. Ex. p. 177] and appearing as a
white space between the guide FF, colored red, and the
standing column 3, colored blue [Bk. Ex. p. 179]. The
3/32 of an inch clearing forming the loose fit, in the
Bray guide is not shown or described in his patent because
it is primarily the conventional loose fit of the guide
around the column of his practical long-stroke pump for
deep wells and is necessarily implied, and also because
the Bray patent drawing is on such a small scale that the
3/32 of an inch clearance is too small to appear
in the reduced scale of the drawing. The annular form
of jet aperture in the Hofco guide is Bray's invention
and is comprehended by the Bray patent, although not
particularly disclosed in the patent, because Bray built
this form of guide before he applied for his patent [Tr.
pps. 41 and 42]. It was not necessary for Bray to show

the Hofco form of annular jet opening in his patent, because a patentee is required only to show one form of his invention in his patent.

> *Kings County Raisin & Fruit Co. v. U .S. Consolidated S. R. Co.,* 182 Fed. 59 (C. C. A. 9th Cir.);
>
> *Western Electric Co. v. La Rue,* 139 U. S. 601, 606: 11 S. Ct. 670; 35 L. Ed. 294.

Moreover, the Bray patent expressly declares that the invention is not limited to the particular form disclosed in the patent. On page 1, lines 41 to 49, inclusive, of the patent specification, this intention is declared in the following statement:

> "Other objects, advantages and features of construction and details of means and operation will be made manifest in the ensuing description of the herewith illustrative embodiment: it being understood that *modifications, variations* and *adaptations* may be resorted to within the spirit, scope and principle of the invention as it is hereinafter claimed."

The difference in degree or size of the jet aperture between the Bray pump and the Hofco pump is so slight that there is no difference in operation or result between the two pumps.

A comparison of the sizes or cross-sectional areas of the jet apertures of the Bray pump and Hofco pump is interesting.

In the Bray patent the cross-sectional area of the clearance of 3/32 of an square inch in the guide around the one inch standing column [Tr. p. 37] is .3221 square inches, and the cross-sectional area of the 3/16 of an inch jet apertures [Tr. p. 188] is .22088 square inches. The cross-sectional areas of the clearance and the holes equals a total cross-sectional area of jetting aperture space in the Bray guide of .54298 square inches.

In the Hofco pump the cross-sectional area of the ¼ of an inch annular jetting space, or jet aperture in the guide around the 1½ inch standing column equals .6382 square inches and is so indicated in Defendants' Exhibit N.N. (Bk. Ex. p. 179) as .638.

The difference between the Bray pump and the Hofco pump in cross-sectional area of the jet-aperture space extending vertically through the guide is .09522 square inches or approximately 95/1000 of a square inch. This slight difference in cross-sectional area between the jet apertures of the two pumps is infinitesimal and makes no appreciable difference in the amount of liquid jetted or the jetting action of the two pumps. According to Defendants' expert Adams [Tr. p. 191] the amount of oil jetted through the Bray jetting space is 52.7 cubic inches and the amount of oil jetted through the Hofco jetting space is 49.5 cubic inches, which shows that there is substantially no practical difference in the amount of oil (3.2 cubic inches) jetted by the two pumps. The operation of the two pumps and results accomplished should therefore be practically the same. The testimony of Plaintiffs' witness, Rennebaum [Tr. p. 358], an oil production man of 15 years' experience, who has operated both the Bray and the Hofco pumps, concerning the operation and results of the two pumps is interesting:

"Q. You have operated both the Hofco and Bray pumps; have you noticed any difference in whether those pumps sand up or the way they operate with respect to sand accumulation below the traveling barrel?

A.   No; there is practically no difference.

Q.   You have known practically no difference?

A.   No; both give the same.

Q.   In your experience, you find they give the same results?

A.   Both work free and give the same results."

A comparison of our calculations with those set forth on page 124 of counsel for Appellees' brief will now be made.

The calculations in Appellees' brief are as follows:

	Velocity	Capacity for doing work
Bray of patent	11.76	138.
Bray as now built	4.46	19.9
Hofco as built	1.36	1.85

The above calculations of Appellees are monstrously inaccurate and entirely unsupported by the evidence, as we will show.

With only a difference of 95/1000 of a square inch in cross-sectional area between the jet aperture of the guide in Bray's patent and the jet aperture in Hofco's guide, and with only a difference of 3.2 cubic inches between the amounts of oil jetted through the apertures of the Bray guide and the Hofco guide there could not possibly be a ratio of jet velocity of 11.76 to 1.36 between the jet aperture in the guide of Bray's patent and the jet aperture in Hofco's guide, and there could be no such ratio of capacity for doing work of 138. to 1.85 between the "Bray of patent" jet and the jet of "Hofco as built". As for the Bray guide, or "Bray as now built" with the annular jet aperture, there could be no such ratio of jet velocity as 4.46 to 1.36 between said guide, and the "Hofco as built," nor any such ratio of capacity for doing work, as 19.9 to 1.85, because this Bray guide is like Hoferer's, which Bray first built before he applied for his patent, and the ratio between "Bray as now built" and "Hofco as built" should be approximately 1 to 1. Counsel for Appellees has objected to this deduction because Bray's blue print showing his annular jet aperture, or "clearance" as Bray called it, is not in evidence, but counsel for Appellees evidently did not want it in evidence, because, when he had every opportunity to introduce said blue print in evidence, he dropped it as he would drop a red hot poker. [Tr. p. 42.]

The grotesque miscalculations of Appellees' counsel only emphasizes the correctness of our conclusions that for all practical purposes the amount of oil jetted through the apertures in the Bray guide and the Hofco guide is the same; that the velocity of the jets from both pumps is the same; that the capacity for doing work is the same; that the jet action is the same; and that the results accomplished by both pumps in preventing sanding up of the pumps is the same; and our conclusions are fully corroborated by the above-quoted testimony of witness Rennebaum [Tr. p. 358] to the effect that he has operated both the Bray and the Hofco pumps and found that they operated the same and produced same results.

A closer imitation of a patented invention or a clearer case of infringement of Letters Patent could hardly be found in the annals of the patent law.

The full size drawing after the last page of this brief should aid the Court in verifying the calculations set forth in this brief, and is submitted only as part of our argument.

The most glaring inconsistency of counsel in Appellees' brief consists in the two diametrically opposed defenses, to wit: the defense of inoperativeness and lack of utility of Bray's pump, because, theoretically, it cannot have an effective jetting action to wash away sand and prevent sanding up of the pump, and the defense of no infringement because the Bray pump has a most effective jetting action which compared to the mere "surge" of the Hofco pump (Hofco Catalogue, Working Actions, Bk. Ex. p. 8), is in the ratio of 11.76 to 1.36, as to velocity, and in the ratio of 138. to 1.85, as to capacity for doing work. (Appellees' Brief, p. 124.) In one defense counsel for Appellees *denies* an effective jet action in the Bray pump, and in the other defense *admits* a most effective jet which distinguishes the Bray pump from the Hofco pump, because the "surge" of the Hofco pump is much less efficient than the Bray jet, for which reason the Hofco pump does not infringe claims 2 and 3 of the Bray patent. Counsel tries to ride two horses at the same time going in opposite directions.

Defendant Dan W. Hoferer [Tr. pp. 364-365] tries to distinguish the operation of his pump from the operation of the Bray pump, by stating that his pump *does* sand up *sometimes,* and in order to remove the sand he lengthens his sucker rods and lowers his barrel until the bottom thereof strikes the sand, whereupon he operates his pump in the usual manner of reciprocating the barrel, so that the lower end of his barrel pounds against the sand and washes it away—*the washing evidently being done by the jetting of the oil downwardly through the annular space in the bottom guide against the sand.* Hoferer is not corroborated and is flatly contradicted on this statement of operation of his pump by the witness Rennebaum, who is not a party to this suit, and who testified [Tr. p. 362], positively, despite the misleading questions of counsel on cross-examination, referring to both the Bray and the Hofco pumps, as follows:

"Q. You would expect a settling of the sand, and even though you would pump up and down your barrel in that event, the jet would not be strong enough to dig out the sand, would it?

A. *Yes, it would. I have done that.*

Q. What you have actually done is to set down and hammer?

A. *Well, we call it drilling out.*

Q. That is it exactly. In other words, you come down on that packed sand with the barrel and drill it out just as you would drill?

A. *You don't come down quite to the sand. As long as there is a little fluid coming in there the motion of the pump in the oil coming through these holes here works as a jet and forces it up.*

Q. In other words, it is coming down and as long as you come down close to that sand you in effect, drill it out.

A *Yes; Hofco.*"

The Hofco catalogue (Bk. Ex. p. 8), which Hoferer said he withdrew from circulation, but nevertheless continued to make and sell his infringing pump, after he received notice of infringement of Bray's patent, contains the statement, under "Working Actions", as follows:

> "The nearness of the two valves (balls and seats), permit pumping under extreme gas conditions. This is especially valuable when the pumps are used in gaseous wells. Likewise, *the surge of the fluids in and out of the lower end of the reciprocating barrel keeps all sand and foreign sediment stirred and permits pulling of the pump* at any time without undue trouble."

Of course, it could only be expected that Hoferer would try to minimize the jetting action of his pump by such testimony as he has given and by calling the "jet" from his guide a "surge", but even if his pump is constructed not to work as efficiently as Bray's pump, so that it does sand up *"sometime"*, as counsel for Appellees states in his brief (p. 119), this would not avoid infringement of the Bray patent, because this would amount to nothing more than impairment of function, since the means for producing the result of preventing an oil well from sanding up are substantially the same in both the Bray and the Hofco pumps, and are equivalents.

> "Infringement is not avoided by the impairment of the functions of an element of a patented device."
>
> *Walker on Patents*, 6th Ed., Sec. 414, p. 504.
>
> *Gen. Electric Co. v. Sundh Electric Co.*, 198 F. R. 116;
>
> *Murray v. Detroit Wire Spring Co.*, 206 F. R. 465 (C. C. A., 6th Cir.).

While claim 2 specifies a series of venting ducts which are shown extending around the standing column close to the column and the Hofco guide has an annular venting duct extending around the column, this difference is only a difference in *form* without a difference in *operation function* or *result* and the Hofco pump is a clear equivalent of the novel combination of elements specified in claim 2 and an infringement of said claim.

> "Upon well established principles of patent law a change of *form* and not of substance constitutes no defense to a bill for an infringement."
>
> *Morey v. Lockwood,* 75 U. S. 230, 19 L. Ed. 339.

> "It is therefore safe to define an equivalent as a thing which performs the same function and performs that function in substantially the same manner of the thing of which it is alleged to be an equivalent."
>
> *Walker on Patents,* 6th Ed., Sec. 415, at p. 506.

> "No substitution of an equivalent for an ingredient of a combination covered by any claim of a patent can avert the charge of infringement of that claim, *whether or not the equivalent is mentioned in the patent."*
>
> *Walker on Patents,* 6th Ed., Sec. 412, at p. 501;
>
> *Machine Co. v. Murphey,* 97 U. S. 120.

> "To infringe a patent it is not necessary that the thing patented should be adopted in every particular. If the patent is adopted *substantially* by the defendants they are guilty of infringement."
>
> *Sewall v. Jones,* 91 U. S. 173, 183, 23 L. Ed. 275.

Claim 3 specifies the guide being *apertured* without restriction as to the *form* of the aperture, and the language of this claim reads letter perfect on the Hofco pump and is infringed by the Hofco pump.

Conclusion.

It is submitted that Appellees have infringed the claims of the patent in suit, and the decree of the Lower Court should be reversed in favor of Appellants, as prayed in their bill of complaint.

Respectfully submitted,

ALAN FRANKLIN,
Attorney for Appellants.

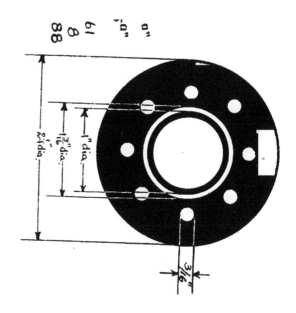

a"
a"
61
8
88

1¾" dia.
1³⁄₁₆" dia.
1" dia.
⅞" dia.

3/16"

Net Areas in Jet Apertures

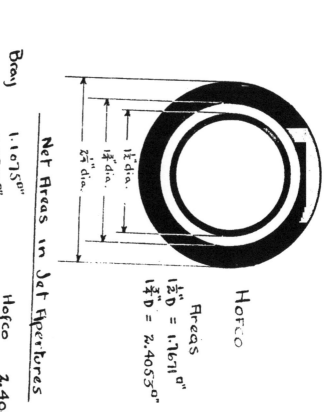

HOFCO

2¼" dia.
1¾" dia.
1½" dia.

Areas
1½"D = 1.7671 □"
1¾"D = 2.4053 □"

Bray
$\quad$ 1.1075□"
$\quad$ −0.7854□"
$\quad$ 0.3221□"
$\quad$ +0.22088□"
$\quad$ 0.54298□"

Hofco
$\quad$ 2.4053□"
$\quad$ −1.7671□"
$\quad$ Net

Total Net · 0.54298□"

Net $\quad$ 0.6382□"

No. 8171

In the United States
Circuit Court of Appeals
For the Ninth Circuit.

///

ATRICK H. BRAY and CATHERINE PATRICIA
MARQUETTE,

Appellants,

vs.

OFCO PUMP, LTD., and D. W. HOFERER,

Appellees.

PETITION FOR REHEARING.

ALAN FRANKLIN,
114 West Third Street, Los Angeles,
Attorney for Appellants.

RANK BRYANT,
Of Counsel.

FEB 15 1

PAUL P. O'BRI

Parker & Baird Company, Law Printers, Los Angeles.

TOPICAL INDEX.

TABLE OF AUTHORITIES CITED.

No. 8171

In the United States
Circuit Court of Appeals
For the Ninth Circuit.

PATRICK H. BRAY and CATHERINE PATRICIA
MARQUETTE,

Appellants,

vs.

HOFCO PUMP, LTD., and D. W. HOFERER,

Appellees.

PETITION FOR REHEARING.

*To the Honorable, the Judges of the United States Circuit
Court of Appeals for the Ninth Circuit:*

Come now the appellants, and petition this Honorable
Court for a rehearing of this case, upon the grounds hereinafter set forth.

This court in its decision of January 10, 1938, states
that the record of this case on appeal is voluminous. Appellants are fully aware of this fact and regret that they
could not control the record. Counsel for appellees made
the record voluminous by setting up the entire prior art
in the Patent Office, which had been overcome in the prosecution and granting of the patent in suit, and by taking
lengthy depositions in Pennsylvania concerning pumps
which could not possibly function to avoid sanding up in

the oil fields of California, like the pump covered by the patent in suit. It is not altogether surprising that the real issue of this case was lost sight of by the court in the confusion of the mass of irrelevant evidence with which appellees have cluttered the record.

Claim 3.

In this petition, counsel for appellants will simplify the issue of the case and reduce further work of the court to a minimum, by waiving further consideration of claims 1 and 2 and considering only claim 3 of the patent in suit, which claim, in view of its broad terms and of the decided advance in the art made by the invention in suit, clearly comprehends the pump of the appellees. The vital elements of claim 3, which are entirely new and entitled to a liberal interpretation, are as follows:

> "A bottom guide, on said barrel operating on said column and having a coupler part on its bottom end, and an anchor for said column having a top coupler part complimentary to the coupler part of said guide, whereby the engaged parts may be meshed in non-rotative interlock, and said guide being *apertured* for the ejection of liquid from the barrel *directly toward the effective face of the lower, standing coupler part* to aid in preventing sand settlement thereon.."

Prior Art.

The court has cited the Admore pump Defendants' Exhibits 13 and 14, Exhibits 6 and 7 [Book of Exhibits pp. 52 and 53] and the patent to T. A. Northrup, No. 1,378,-268, May 17, 1921. Defendants' Exhibits V and V[1] [Book of Exhibits page 129] as a basis for unduly narrowing the scope of claim 3 as above quoted in part.

The Admore pump, as pointed out in appellant's opening brief, page 39, has four holes in the side of the pump barrel above the guide on the lower end of the barrel through which oil in the lower end of the barrel escapes laterally, as the barrel moves upwardly, and strikes against the tubing in the well and is dispersed upwardly and horizontally as well as downwardly, which prevents said oil from being projected downwardly with sufficient force to stir up the sand below the bottom of the barrel to wash the sand away from the anchor and prevent "sanding up" of the pump.

The Northrup patent covers a pump like the Admore pump, with holes 32 in the sides of the barrel 22 above the guide 24, through which holes oil is projected laterally against the tubing C and dispersed as the barrel move upwardly.

The holes in the sides of the Admore and Northrup barrels are above the guides of these pumps and quite removed from the space directly beneath the lower end of the guides, where a strong positive stirring action of the oil in the bottom of the pump tubing is needed to remove the sand therefrom and prevent sanding up of the pump. Pumps with holes in the side of the barrel have not been successful in actual practice in preventing sanding up, and it was the failure of pumps of this construction that prompted appellant Bray to invent his pump covered by the patent in suit. The old established Neilsen pump with holes in the side of the barrel sanded up, and Mr. Bray had to pull the tubing (a wet job) to remove the pump in order to replace it with one of his own as covered by the patent in suit.

Mr. Bray testified that the ejection of the fluid through the side of the barrel would not give sufficient stirring effect of the fluid in the bottom of the well tubing to prevent sanding up.

"No; it hasn't got the same effect at all * * * Now, it has got a baffle to hit all the way up. When it hits that baffle—the tubing is the baffle. When it goes out of these holes it hits the tubing, it cracks; in other words, it spreads; it don't go down." [Tr. 269.]

"And that is the case a great deal that I find out with the holes in the sides. You haven't got a chance to move this sand after it gets down there. It keeps building up. The consequence is it sands up." [Tr. 271.]

"I found other pumps under the same conditions did sand up and mine did not. * * *

"Q Did you find that the Neilsen and O'Bannon pumps sanded up under conditions that yours did not?

A Yes, under the same conditions.

* * * * * * * *

Q How do you know that was true in the case you mentioned?

A Had to pull the well." [Tr. 279.]

Mr. Bray is corroborated concerning the sanding up of pumps with holes in the side of the barrel by the witness Rennebaum, a practical oil field man.

"Q Have you also operated pumps which allow for the ejection of the liquid through the *side* of the barrel of the pump?

A I have. The *Neilsen* pump is one of the old established pumps with *holes in the side*. I found it wasn't very good in some wells. (wells with sand)

Q In your opinion, as an oil man, is the operation of the pump with the oil going in and out of the bottom nut on the traveling barrel more efficient in the aid of the prevention of sand and emulsion over the barrel than where the oil goes in and out through the holes in the *side* of the barrel?

A And for this reason: When you put holes on the side here you are so much farther away from the sand. It hits up against your tubing here. You haven't the same force that you have if the oil comes straight down here." [Tr. 357-358.]

The testimony of the above two witnesses that pumps with holes in the side of the barrel "sand up" stands uncontradicted by any fact witness of appellees. Furthermore, there is no evidence that the Admore pump ever operated to prevent sanding up, nor any evidence that the Northrup pump ever operated at all. From the testimony of the above two witnesses for the appellants, the only conclusion that can be drawn concerning the Admore pump and the Northrup pump, which have holes in the side of the barrel, is that both of these Pennsylvania pumps would sand up, if used in the oil fields of California, like the Neilsen pump, which has holes in the side of the barrel.

A significant point in this case is that *there is no evidence that any pump was ever built before Bray's that would not sand up.* Bray's invention is accordingly an invention of considerable merit and is entitled to a liberal interpretation.

Walker on Patents, 6th Ed., Sec. 232;

Aeolin v. Schubert Piano Co., 261 Fed. 178.

Bray's Pump.

Bray's pump eliminates all holes in the side of the barrel and discharges all of the oil directly downward out through the guide at the bottom end of the barrel, and this principle is the only one that can be used successfully in an oil pump to prevent sanding up. The Admore and Northrup pumps with *holes in the side* of the barrel, consequently do not embody the principle of the Bray invention. The Bray patent in suit states on page 2, lines 32 and 33, that the *pump barrel is imperforate.* The difference between putting holes in the side of the pump barrel, as in the Neilsen, Ardmore and Northrup pumps, and putting the opening in the bottom guide, as in Bray's pump, to effectively stir up the sand beneath the lower end of the barrel and prevent sanding up, is the difference between failure and success. Bray has not pulled any of the 500 wells in which his pumps are installed, in four years. [Tr. 43, 45 and 270.]

"It may seem simple now. The placing the point in the heavens to catch lightning seems simple now. It is easy to confine and utilize steam now—after it is done. * * * If there was patentable invention in the case of Washburn and Another v. Barbed Wire Company, 143 U. S. 275, 12 Sup. Ct. 443, 36 L. Ed. 154, there is invention here."

Bradley v. Eccles, 138 Fed. 916, 919.

"Knowledge after the event is always easy, and problems once solved present no difficulties, indeed, may be represented as never having had any, and expert witnesses may be brought forward to show that the new thing which seemed to have eluded the search of the world was always ready at hand and easy to be seen by a merely skillful attention. But the law has other tests of invention than subtle conjectures of what might have been seen and yet was not."

Diamond Rubber Co. v. Consolidated Rubber Co.,
220 U. S. 428, 55 L. Ed. 527, 532;

Loom Company v. Higgins, 105 U. S. 580, 26 L.
Ed. 1177, 1181.

Hofco Pump.

The Court has been misled by the drawing of defendants' expert Adams of the Hofco Pump, Defendants' Exhibit NN [Book of Exhibits p. 179] as to the size of the annular jet opening in the guide of the Hofco pump. Said drawing is nearly twice the actual size of the Hofco pump and the jet opening in the guide appears to be "½" more in diameter than the pump stem," as the Court states on page 1 of its decision, but the diameter of the jet opening in the Hofco guide is only ¼" greater in diameter than the pump stem, which gives a *clearance* of only ⅛" around the stem. The diameter of the stem is 1½" and the diameter of the opening in the guide is 1¾", which gives a difference in diameter of only ¼".

Guide Clearance.

The annular jet opening or clearance of ⅛" in the Hofco guide is only *1/32" larger than the conventional clearance required in the guide of every conventional or standard pump, which clearance is 3/32"*, as testified to by Mr. Bray, according to his blue print. [Tr. 37.] Mr. Bray is fully corroborated by defendants' expert Adams as to the 3/32" clearance in standard oil pumps. Adams' Testimony [Tr. 136-137] which is newly discovered evi-

Left margin fragments:
side of the
downward
the barrel,
e used suc-
up. The
side of the
iple of the

imperforate.
side of the
d Northrup
guide, as in
beneath the
up, is the
ay has not
mps are in-

simple now.
w—after it
le invention
v. Barbed

s easy, and
ties, indeed.
id any, and
ird to show
e eluded the
at hand and
ention. But

dence, since it has been heretofore overlooked, as to the clearance in standard oil pumps is as follows:

"If the well is crooked you have to have a certain clearance. I think in most of the pumps we have been discussing we are talking very largely here of a *standard type* of what is called two and *on*-half inch pump, and the *total clearance in there* is something of the order of *three-thirty seconds of an inch on a side,* that is, three-sixteenths total, is practically that, so the clearance is small and *there is a fixed amount of fluid in that annular space—when you lift the plunger you squeeze that fluid out of there.* When it is squeezed out of there, where can it go? It has got a column of fluid 5,000 feet above it, no holes in it anywhere, just a solid fluid. *But immediately below it as soon as the tube has moved up there is a space made vacant by the upper (upward) motion of the bottom part of the plunger, into which that fluid flows. Therefore, there is a strong stirring action by the flow outside."*

Both Bray and defendants' expert Adams agree that a clearance of 3/32″ is necessary in a guide of a standard practical pump to give the pump lateral flexibility so that it may operate in *crooked wells. There are practically no straight wells and a standard pump must therefore be made to operate in crooked wells.* Bray, a practical oil man, necessarily built a practical standard pump which requires 3/32″ clearance between the guide opening and the pump stem; otherwise he could not have sold and installed 500 of his pumps. [Tr. 270.]

Concerning the crookedness of oil wells Bray testified [Tr. 273] as follows:

"Some of these pumps are actually in the well at a slight angle. a little crooked bend. *The well is not perfectly straight. Very few straight wells, very few.* Now, in my experience this bottom nut down here would tend to guide the barrel in its downward motion into the direction it was going. It won't let it come either one side or the other strong enough to hit the tubing. This guide—now, *if you get a very crooked hole you will find out that your guide will wear out* in a crooked hole, *shows that it is guiding.* Where your barrel is not worn at all your guide is worn."

In view of Bray's experience in building oil well pumps it is absurd to assume that Bray would build a pump for the trade without the conventional 3/32" clearance in the guide, and there is no justification for construing the patent in suit as covering a tight-fitting guide on the pump stem, simply because this conventional 3/32" clearance is not shown or described in the patent in suit. Patent drawings are not required to show anything but the invention. Concerning patent drawings defendants' expert Adams testified [Tr. 142] as follows:

"I made patent drawings for part of my living for a number of years, and in making those drawings *things that were not material to the drawing itself were made simply in an illustrative manner.*"

It is true that the 3/32" clearance in the guide is not claimed as part of the invention in suit. *It is not a part of the invention* because it is in every standard oil well pump, as above pointed out in the testimony of both

Bray and Adams, to enable the pump to operate in crooked oil wells, and practically all wells are somewhat crooked. The 3/32" clearance was in all oil well pumps when Bray came into the field. *But this 3/32" clearance is never large enough to enable sufficient oil to be projected through the bottom guide to stir up the sand with force enough to prevent sanding up, when there are also holes in the side of the pump barrel, as in the Neilsen, Admore and Northrup pumps, through which holes the major portion of the oil is projected laterally out of the barrel against the tubing and dissipated, whereby the force of the oil projected downwardly through said 3/32" clearance is reduced to impotency. The inventive concept of Bray's invention was in eliminating the holes in the side of the barrel and placing the entire opening of the proper size through the guide. When Bray did this his invention was a success, and the evidence fails to show a single pump constructed in any other manner that actually prevented sanding up.* Hofco did the same thing that Bray did, by eliminating holes in the side of the barrel and adding an opening through the guide to the 3/32" standard clearance found in all standard or conventional pumps, as shown in Fig. 2 of the drawings in the back of this petition. The additional opening that Hofco placed in the conventional guide consists in an enlargement of the standard clearance of 3/32" to 4/32" or 1/8" as indicated by the dotted line circle outside of the clearance in said Fig. 2. In Fig. 4, which shows the Hofco pump, the additional opening added by Hofco to the standard 3/32"

operate in
: somewhat
well pumps
"" clearance
to be pro-
: sand with
re are also
he Neilsen,
'; holes the
out of the
'y the force
'/32" clear-
concept of
in the side
the proper
his inven-
ow a single
ctually pre-
that Bray
barrel and
/32" stand-
onal pumps,
ack of this
) placed in
nent of the
as indicated

pump, the
dard 3/32"

clearance is shown between the dotted line circle, measuring 1 11/16", and the full line circle outside of said dotted line circle, which measures 1 3/4".

The standard 3/32" clearance in the guide opening, while no part of the Bray invention, is nevertheless an opening in all standard pumps, and Bray and Hofco both have it in their guides and it must be added to the additional opening placed in the guide by both Bray and Hofco. When so added we have shown that the opening in the Hofco guide, which appellants claim to be so much larger than the opening in the Bray guide, is only 95/1000 square inches larger, assuming Bray uses a one-inch pump stem, and it is mere quibbling to argue that this infinitesimal difference changes the whole principle of operation of the Hofco pump from that of the Bray pump, so that the Hofco pump produces a *surge* of the oil instead of a *jet,* as produced by Bray. It seems to have been overlooked that whether the pumps produce a jet or a surge, both the Bray and Hofco pumps project the oil directly downward out of the bottom of the barrel with the necessary force to stir up the oil beneath the lower end of the barrel to prevent sanding up of the pump, and no pump ever operated in this manner before the advent of the Bray invention. In view of this most significant fact the Bray patent, referring to claim 3, should not be limited to any particular size or form of guide aperture contrary to the express terms of the claim, which specifies the guide being *apertured* for the ejection of liquid *without any restriction or limitation as to the size or form of the aperture.*

Comparison of Guide Openings.

The Court on page 5 of its decision states that the comparison of the Bray and Hofco guide openings should be Bray .22, Hofco .64 of a square inch. This comparison is wrong because it leaves out the standard clearance of 3/32" in Bray's guide and includes the clearance in the Hofco guide. The area of the 3/32" conventional clearance around a one-inch pump stem (which size of pipe was only arbitrarily selected for illustration of the Bray pump), is .3221 square inches. To this area Bray added an additional opening in the guide amounting to .2208 square inches in area, the same being the sum of the areas of eight 3/16" openings. The area of the 3/32" conventional clearance around the 1 1/2" pump stem which Hofco uses is .4694 square inches. To this area Hofco added an additional opening in the guide amounting to .1688 square inches in area, the same being the addition of 1/32" to the conventional 3/32" clearance, as indicated by the dotted line circles in Figs. 2 and 4 of the drawings in the back of this petition. Comparing the area added by Bray to the conventional clearance in the guide, to the area added by Hofco to the conventional clearance in the guide, it will be seen that the ratio of added opening is Bray .22, Hofco .16. Bray has added a greater opening to the conventional clearance opening in the guide than Hofco. The area of the Bray guide opening is computed arbitrarily from a one-inch pump stem, but the Bray patent is certainly not limited to a pump stem of one-inch or any particular diameter. The evidence shows that both Bray and Hofco build pumps with pump stems of different diameters. The patent in suit is not restricted to any particular dimensions and this fact was conceded by defendants' expert Adams, and stipulated to by defendants' counsel. [Tr. 208-209.] This fact was brought out on page 88 of appellants' opening brief. The size and shape of the guide aperture is determined according to practical experience in operating the pump, as testified to by Bray who built and installed 500 of his pumps which had not sanded up in four years. [Tr. 279, 270, 43 and 45.]

Bray has built pumps with guides having openings of different sizes and shapes [Tr. 38, 41-42 and 273] according to the sizes of the pipe used in building the pump.

There is one of Bray's guides in evidence, which was not given an exhibit number. It was introduced in evidence by stipulation of counsel. [Tr. 207-208.] This guide has the eight 3/16″ openings and a central guide opening of 1 9/32″ diameter, as illustrated in Fig. 1 in the drawings in the back of this petition. The standard 3/32″ clearance of this guide opening around a pump stem of 1 3/32″ diameter is. 34972 square inches, which is larger than the clearance area (.32210) around a one-inch pump stem, as shown in the drawing of the Bray guide in appellants' reply brief. This larger clearance of .34972, in Bray's stipulated guide, added to the eight 3/16″ openings, gives a total guide opening of .5706 square inches. A comparison of this Bray guide opening with the Hofco guide opening of .6382 square inches shows that Hofco's guide opening is only 67/1000 of a square inch larger than the opening in this particular Bray guide.

Another pump built by Bray had a pump stem 1 1/4″ in diameter. [Tr. 272.] This pump is illustrated in Fig. 3 of the drawings in the back of this petition. The area of the standard 3/32″ clearance in the guide around this 1 1/4″ stem is .3958 square inches. which when added to the eight 3/16″ openings gives a total guide opening of .61668. A comparison of the area of the guide opening of this Bray pump with the area .6382 square inches of the guide opening of the Hofco pump shows that Hofco's guide opening is only 2/100 of a square inch larger than the Bray guide opening.

Scope of Claim 3.

Since claim 3 of the patent in suit specifies an *apertured* guide without restriction or limitation as to the size or form of the guide aperture, and since the evidence fails absolutely to show any pump that *actually operated* in the oil fields of California to prevent sanding up, prior to the advent of the Bray invention, it is submitted that there is nothing in the prior art to justify a restriction of claim 3 to the exact size or form of guide opening shown in the drawing of the patent in suit. The term "apertured" was used in claim 3 for the very purpose of covering guides of the form employed by Hofco, because Bray had made a guide like Hofco's at the time he filed his application for patent. [Tr. 41.] If claim 3 were intended to limit the invention to a guide with a series of venting ducts, there would have been no need of claim 2 of the patent, which covers specifically this particular construction of guide. The Patent Office Examiner clearly understood the scope of the two claims and was justified in allowing *claim 3 covering any suitable size and form of aperture to permit discharge of liquid from the bottom of the barrel with sufficient force to stir up the sand to prevent sanding up, because there was nothing in the prior*

around thi;
:hen addec
opening of
ide opening
e inches of
lat Hofco's
larger thar.

n apertured
the size or
idence fail;
·ated in the
p, prior to

restriction
:ning show:
"apertured"
of covering
e Bray hac
d his appli·
intended to
of venting
m 2 of the

·arly under·
justified in
nd form of
e bottom of
sand to pre·
in the prior

pump art which anticipated claim 3. The Northrup patent No. 1,378,268 was cited by the Patent Office against certain claims of Bray's patent application, but was withdrawn and never cited against claim 3, which was allowed as presented. There is therefore nothing in the prior art or in the file wrapper of the Bray patent to restrict the scope of claim 3. The granting of a patent is an adjudication by experts on the merits of the patentability of the patented invention, and should not be lightly set aside by the courts.

All of the patents set up by defendants against the patent in suit [Tr. 85-86] were before the Patent Office when Bray's patent application was pending, and all of said patents except the first three, Dies, Cummins and Whitling, Defendants' Exhibits O, P and Q, were cited by the Patent Office against Bray's patent application, and were overcome in the prosecution of Bray's patent application.

The only literature in the prior art that suggests agitation of the oil to prevent accumulation of sand, is the Northrup patent, cited by this Court, but we have pointed out in our opening brief, pages 58-59, and in this petition that the discharge of the oil in the Northrup pump is through the openings 32 in the *side of* the traveling barrel *against the well tubing* and not through openings through the bottom guide *against the anchor and the clutch,* and that pumps with holes in the side of the barrel, like the Northrup pump, sand up in actual practice, when used in the oil fields of California. This was true of the Neilsen pump, as testified to by both Bray and Rennebaum. [Tr. 271, 282-283, 357-358.] The white space shown between the guide 24 and the pump stem 1 of Fig. 1 of the drawing of the Northrup patent is nothing but the conventional clearance between the guide and pump stem to enable the pump to operate in crooked holes. It is not described in the patent and not even shown in Fig. 2 of the drawing, which shows the same pump as Fig. 1. The reason it is not described in the patent or shown in Fig. 2 of the drawing is that it is conventional and no part of the Northrup invention, and for the same reason the clearance between the guide and pump stem is not shown nor described in the patent in suit; but a 3/32″ clearance as required in standard pumps and as shown in Bray's blue print of his actual standard pump [Tr. 37], does let out some oil from the barrel on its upward stroke, and in calculating the exact amount of oil that goes out of the barrel the amount which goes out through the clearance must necessarily be included. Counsel for appellees certainly include the clearance in Admore pump in calculating the so-called jetting action of the Admore pump. (Appellees' Br. pp. 111-112.) However, with the larger openings in the side of the barrel above the guide in the Northrup and Admore pumps the liquid going down through the clearance is not of sufficient force to stir up the oil beneath the barrel to prevent sanding up, and this has been proved by the sanding up of the Neilsen pump which has holes in the side of the barrel like the Northrup and Admore pumps.

The contention of appellees that Bray's idea of washing the anchor clear of sand is not new is without foundation. Where in all of the voluminous prior art is there any suggestion of an *actual washing* of the anchor and clutch?

The Northrup patent, which is the nearest theoretical approach to the Bray pump, but a far cry from Bray's actual invention, does not state that the oil ejected through the barrel openings 32 washes the anchor. Northrup states on page 2 that oil ejected horizontally through barrel openings 32 agitate the oil in the tubing (above the guide 24) to prevent accumulation of sand upon the extension 8, and that the sand is thus held in suspension in the oil. This operation is not a positive *downward projection* of the oil directly toward the anchor, to deflect the oil and sand upwardly from the anchor between the pump and the tubing C, as produced by Bray's guide opening and imperforate barrel. The ejection of the oil into the tubing above the guide as disclosed by Northrup is to agitate the oil in the annular space between the pump and the tubing to prevent the oil settling down through said space into the bottom of the tubing upon the anchor. If the Northrup pump could prevent the sand from settling in the pump during its operation, what is there to prevent settling of the sand in the tubing upon the anchor when the pump stops running, and *when the sand settles what is there in the Northrup pump that could possibly function to wash out the settled sand?* Any downward force of the ejected liquid from the holes 32 in the barrel into the annular space between the pump and the tubing C would prevent upward movement of the sand through said annular space, the only means of escape of the sand, and consequently any such downward force of the ejected oil would only pack the sand down tighter in the bottom of the tubing and cause the pump to sand up, as illustrated in Fig. 6 in the drawings in the back of this petition. This is what happened with the Neilsen pump, which has holes in the side of the barrel like the Northrup pump.

"Drilling Out."

The projection of the oil straight downwardly out of
the guide at the bottom of the barrel of the Bray pump
causes the projected oil to impinge against the anchor P,
or any sand upon the anchor, and be reflected by the
anchor outwardly and upwardly, whereby any sand on the
anchor is washed away and carried by the reflected jetted
oil upwardly through the annular space between the pump
and the tubing T as shown in Fig. 7 in the drawings at
the end of this brief. The location of the discharge opening entirely in the guide at the bottom of the pump barrel
and the concentration of all the force of the discharged
oil from the barrel in a downward direction makes it
possible for the Bray pump to "drill out" the sand when
sand has settled in the bottom of the tubing after the
pump has stopped running, and this is a very vital advantage over the Neilsen, Northrup and Admore pumps
with holes in the side of the barrel, which pumps cannot
"drill out" the sand when it once settles in the bottom of
the tubing. This "drilling out" is testified to by the witness Rennebaum [Tr. 361-362] as follows:

"Q I take it they both would stop where you
really would stop the pump and let her set for some
days, when this sand would settle down and pack
around those parts; you would expect that, wouldn't
you?

A It all depends upon the condition of the well,
how much sand.

Q You would expect a settling of the sand, and
even though you would pump up and down your
barrel in that event, the jet would not be strong
enough to dig out the sand, would it?

A *Yes; it would. I have done that.*

Q What you have actually done is to set down and hammer?

A Well, we call it *drilling out.*

Q That is it exactly. In other words you come down on that packed sand with the barrel and drill it out just as you would drill?

A *You don't come down quite to the sand. As long as there is a little fluid coming in there the motion of the pump in the oil coming through these holes here works as a jet and forces it up.*

Q In other words, it is coming down and as long as you can come down close to that sand you, in effect, drill it out?

A Yes; Hofco.

Q As a matter of fact, or any other pump, that is true of any pump, isn't it?

A *No; I wouldn't say that.*

Q Of what pump isn't it true? Let us put it the other way: Of what pumps is it true?

A *Well, you take these two pumps of Bray and the Hofco. I have used it as an upside down pump. There was the Neilsen pump; I never had such good success with it."*

From the above testimony it will be seen that the Bray pump will "drill out" the sand after a shut down of the pump and a settling of the sand in the bottom of the tubing. This is a utility of the Bray invention not disclosed in the prior art. The Neilsen pump was not successful in "drilling out" and consequently the Northrup and Admore pumps, with holes in the sides of the barrel like the Neilsen pump, could not be operated to "drill out" of a settlement of sand. The Hofco pump with its discharge opening only in the bottom guide has been used like the Bray pump to "drill out." Consequently, here is an *identity of operation* between the Bray and Hofco pumps, and a *lack of identity of operation* between the Bray pump and the Neilsen, Northrup and Admore pumps.

Admore Pump.

What has been said of the Neilson and Northrup pumps applies with even greater force to the Admore pump. In the first place the Admore barrel has four 3/8″ openings in the side of the barrel in addition to the 3/32″ standard clearance and the total discharge area of the openings from the Admore barrel is as follows:

Four 3/8″ holes   = .44180 square inches
3/32″ clearance   = .32210
                      ————
Total jet opening   .76390 square inches

The Admore pump, which is a *small* Pennsylvania pump, has a larger jetting opening from its small barrel than the jetting opening of the larger barrel of the larger Hofco pump, which is .6382 square inches. Yet counsel for appellees contend that because the Hofco jet opening is *slightly larger* than the Bray jet opening, the Hofco pump operates on an entirely different principle from that of the Bray pump, for which reason the Hofco pump does not infringe the Bray patent, but that the Admore jet opening in a smaller barrel, which is .13 square inches larger even than the Hofco jet opening, produces exactly the same jetting action as the Bray pump, and operates on exactly the same principle as the Bray pump. This contention is obviously untenable and clearly exposes the crude fallacy of appellee's defense of no infringement. If a larger opening does not infringe, a still larger opening could not anticipate the Bray patent. *That which, if later, would not infringe, would, if earlier, not anticipate.* The Admore pump would sand up worse than the Neilsen or Northrup pumps, because the annular space between the standing valve and the tubing would form a perfect sand trap from which no jet of any kind from the barrel could remove the sand. The standing valve of the Admore pump is in the wrong place because it forms a sand trap beneath the barrel and would cause the pump to sand up, as shown in Fig. 5 in the drawings at the back of this petition. The large opening in the small Admore barrel would prevent any jetting force of the oil therethrough sufficient to lift the sand, and even if there was any downward force of the jet through the annular space between the barrel and the tubing this force would pack down the sand, because the sand could not travel *upwardly* through said annular space, the only outlet for escape of the sand, against any *downward* jetting force in said space. Holes in the side of the barrel, as in the Neilsen, Northrup and Admore pumps, if they produce a *downward* jet through the annular space between the barrel and the tubing of the well as illustrated at J, in Fig. 5 and Fig. 6 of the drawings in the back of this petition, such jet is directly in the only path of escape of the sand from the anchor below the barrel, and consequently such jet, instead of lifting the sand from the anchor and forcing the sand upwardly through said annular space and out of the well tubing, would only act to pack the sand down harder on the anchor and cause the pump to sand up sooner than otherwise. This is what has happened to this type of pump in the sanding up of the Neilsen pump. [Tr. 279, 282 and 283.]

New Principle of Bray Pump.

In the Bray pump, as illustrated in Fig. 7, in the drawings in the back of this petition, the jet J is projected directly downward through the guide opening 5 against the anchor P, and is reflected upwardly as indicated at J' by the anchor through the annular space between the barrel B and the tubing T, washing the sand from the anchor and forcing the sand upwardly through said annular space and out of the tubing.

The vital *difference in principle* between the Bray pump and the Northrup, Admore and Neilsen pumps is that the Bray pump produces an *upwardly* moving force in the oil through the annular space between the barrel B and tubing T, which carries away the sand upwardly through said space, while the Neilsen, Northrup and Admore pumps produce a *downwardly* moving force in the oil through said annular space which packs down the sand on the anchor and causes the pump to sand up.

The fact that the *idea* of producing a pump that would not sand up might not have been new, when Bray entered the field, should in no way limit the scope of claim 3 of the Bray patent, in view of the fact that the evidence fails to show that any pump ever actually functioned to prevent sanding up before the advent of the Bray pump. The *idea* of making a machine that would fly was not new when the Wright brothers actually built and operated the first flying machine. The Wright brothers were nevertheless acclaimed the inventors of the flying machine, despite the fact that many prior machines had been built and attempts made to fly them.

In view of the fact that the evidence fails to show that either the Northrup pump or the Admore pump ever functioned to prevent sanding up, said pumps may be compared to the embryo flying machines prior to the flying machine of the Wright brothers, and figuratively, it may be said of the Northrup and Admore pumps, that *these attempts to fly are not actually flying machines.*

In view of the *new principle* of operation, and the *success* of the Bray pump where other pumps had failed, there is no justification for giving claim 3 of the patent in suit a narrower interpretation than the plain import of its terms would comprehend. By eliminating holes in the side of the barrel and directing the discharge of the oil entirely through the bottom guide directly downward toward the anchor of the pump, the Bray pump was the first successful pump in the art to operate without sanding up, and it is submitted that it is immaterial what the *shape or size* of the guide opening might be, or what the *velocity of the oil* through the guide opening might be, or whether the guide was *tight-fitting or loose-fitting,* or whether the guide opening produced a "jet" or a "surge," any pump with an imperforate barrel and a guide with any opening through which the oil would be discharged from the barrel on its up-stroke with sufficient force to stir up the oil beneath the barrel to wash the sand from the anchor and direct the sand upwardly between the barrel and the tubing of the well, and thereby prevent sanding up of the pump, would come within the spirit and scope of the patent in suit and infringe claim 3 thereof.

Removal of Sand.

There is evidently some misconception about the amount of sand in the average oil well of California and the force of the "jet" or "surge" necessary to remove the sand. It is a bad well if it pumps 5 per cent sand, but if the well pumps 1 per cent sand, and there is no *effective* means for removing it, the sand builds up in the tubing and the pump sooner or later sands up. [Tr. 272.] The only effective and practical means so far produced for removing this sand is the Bray invention for the projection of the oil downwardly through an opening in the guide directly against the anchor, which reflects the projected oil outwardly and upwardly through the annular space between the pump barrel and the tubing of the well, whereby the sand is washed from the anchor and carried by the reflected stream of oil upwardly through said tubular space. The sand cannot be removed by the churning of the oil due to the reciprocation of the barrel, because there is no *upward* force produced by such churning motion to give the sand *upward direction* through the tubular space between the barrel and the tubing. If the sand could be removed in this manner there would never have been any sand problem in pumping oil wells in California, because every inverted oil pump produces a churning motion of the oil as the pump barrel reciprocates, but no oil pump has ever avoided sanding up by its mere churning action, and there is no evidence before the Court that any pump has *actually* avoided sanding up by such action.

Substantial Velocity of Jet or Surge.

The assertion of appellee's expert Adams that the velocity of the oil flowing through the guide is comparatively slow, reaching maximum velocity only when the guide has traveled a considerable distance upwardly away from the anchor, is grossly incorrect. The barrel moves upward only 5 1/8″ during the first 1/4 of its upward stroke, in a 3 foot stroke pump, to reach a substantial vertical speed component, as shown in Fig. 8 of the drawings in the back of this petition. In the diagram shown in Fig. 8, which is drawn to a scale of 1″ to 1′, it will be seen that when the crank pin swings clockwise 45 degrees from its lowermost position A to the point B, the crank pin moves vertically upward only a distance of 5 1/8″ during 1/4 of its upward stroke, to reach the point B, where the upward vertical speed component of its movement in a 20-stroke per minute pump is 135′ per minute and 5/7 of the maximum vertical speed component of its movement of 188′ per minute at the point C. At the point B where the crank pin and pump barrel have picked up 5/7 of the maximum speed of their up-stroke the lower end of the barrel guide is only 11 1/8″ above the anchor. At this short distance above the anchor the barrel is ejecting its oil with 5/7 of its maximum force directly downward through the guide opening upon the anchor, and the force of the ejected oil progressively increases with the progressive increase of speed of the crankpin as it continues its upward movement to the point C.

No Tight-Fitting Guide.

The finding that the Bray patent drawing shows a tight mechanical fit between the bushing and the stem and that the teaching of the patent that relatively strong jets of oil are forced through the apertures by the upward movement of the bushing emphasizes the importance of a tight mechanical fit around the stem, is a most unreasonably narrow construction to apply to a patent that discloses an oil well pump that operates on a *new principle* and was the *first* pump that ever functioned *successfully* to prevent sanding up; there being no evidence that the Admore, Northrup or any other prior pump ever avoided sanding up, and the evidence showing that the Neilsen pump, which was of the same type as the Admore and Northrup pumps with holes in the side of the barrel, sanded up so badly that it required pulling of the tubing, a "wet job."

It has been pointed out conclusively in the evidence that there is a 3/32″ clearance between the stem and the guide of standard oil well pumps to enable the pump to operate in crooked well holes. This clearance was not Bray's invention and Bray was not required to show it at all, or other than in a conventional manner. Bray's pump is from 12 to 15 feet long [Tr. 29] and, when the patent drawing of the pump is reduced to approximately 1/12 or 1/15 of the actual size of the pump, the proportionate reduction of the 3/32″ clearance would contract the clearance into a single line as shown in the patent in suit, but any oil pump man, one skilled in the art, would know that in building the pump from the patent drawing the conventional clearance of 3/32″ would be provided between the stem and the guide. A patent drawing meets the requirements of the patent law if others skilled in the art could make and use the invention from the drawing. A patent is addressed to those "skilled in the art." Moreover, the Bray patent (p. 3, lines 69-70) states that "The part 3 is in the form of a *bushing.*" Appellee's expert Adams states that "a bushing is something that reduces the diameter in the tube and changes the size of the tube" [Tr. 143] and that "Bushing" in mechanics is very clearly "a reduction of area" [Tr. 41], and when applied to a part in the lower end of a pump barrel the term "bushing" would apply to the Bray guide which reduces the open end of the barrel to the width of the 3/32" clearance around the pump stem. The language of the patent, and the testimony of Bray and Adams [Tr. 37-38 and 137], concerning the clearance, are a complete negation of a tight-fitting guide in the Bray pump, which is a standard pump, and the Court certainly is not justified in reading anything into the patent which it positively does not contain. Nowhere in the specification of the patent in suit is there the slightest suggestion of a "tight-fitting" guide. In fact there is no such thing in a standard or practical oil well pump, and no practical pump man, like Bray, would ever think of producing such a guide. If a guide should be made with a tight fit on the pump stem it would soon wear loose with a substantial clearance and we would have the clearance anyway, which would let out part of the oil and form part of the jet opening.

While there is ample basis in the Bray patent for including therein the standard 3/32" clearance in the guide, it is immaterial whether or not the patent discloses the clearance, because the clearance is old in the art and the patent law necessarily implies in a patent anything of the prior art that may be utilized in connection with the patented invention. Bray's invention, comprising an opening of suitable size in the guide for ejection of the liquid straight downwardly out of the bottom of the barrel, began with the 3/32″ clearance already in the guide, and it was only necessary for Bray to illustrate and describe his invention from the point where it began.

"Sec. 97. Reliance on Prior Art. While the description of the invention must be full, it must be borne in mind that the description is addressed to those skilled in the art, and those skilled in the art are presumed to know everything which has gone before in the art; therefore, it is not necessary to describe that which is old or already known in the art when describing an invention; it is sufficient to start with that which is old and proceed to describe the novelty. Lynch v. Headley, 307 O. G. 737, 1923 C. D. 150; 52 App. D. C. 269; 285 Fed. 1003."

Patent Office Practice, McCrady, p. 73.

"He may begin at the point where his invention begins, and describe what he has made that is new, and what it replaces of the old. *That which is common and well known is as if it were written in the patent and delineated in the drawing.*"

Carnegie v. Cambria, 185 U. S. 403.

The fact that relatively strong jets of oil are forced through the guide apertures on the upward movement of the bushing is no more reason for emphasizing the importance of a tight mechanical fit of the guide on the stem in the Bray pump, than it would be in the Northrup pump. If a tight fit of the guide on the stem is required in the Bray pump to force strong jets through its guide aperture, the same tight fit of the guide 24 on the stem 1 of the

Northrup pump would be required to enable this pump to "forcibly" eject the oil out through the barrel openings 32 as the Northrup patent states on page 2, lines 59-60. Yet the Northrup patent drawing, Fig. 1, shows the conventional clearance between the guide 24 and the stem 1 and this is the pump cited by the Court as a pump which would produce the jetting action of a so-called tight-fitting guide.

The evidence shows that Bray's guide with the standard 3/32" clearance and the holes in the *guide* operated successfully. [Tr. 37-38.] There is consequently no justification for limiting claim 3 of the patent in suit to a tight-fitting guide.

Conclusion.

From the foregoing analysis of the evidence the following conclusions are submitted:

1. That Bray produced the first pump that would not sand up and this fact entitles the Bray patent to a liberal interpretation and especially in view of the new principle of the Bray pump in which the discharged oil from the barrel is reflected upwardly through the annular space between the barrel and the tubing, whereby the sand is lifted out of the bottom of the tubing and carried upwardly through the tubing. (See Fig. 7 of drawings at the end of this petition.)

Walker on Patents, 6th Ed., Sec. 232.

2. That it is immaterial what the size, degree or shape of the guide opening may be, as covered by claim 3 of the patent in suit interpreted in the light of the prior art, if

the opening will allow the oil to be discharged with sufficient force to prevent sanding up.

> *Amdur Patent Law and Practice,* Chapter XIII, Infringement.
>
> Section 14, Change of Degree;
>
> Section 15, Change of Form.

3. Bray was only required to show one form of his invention in his patent to cover all forms, and particularly the form of appellee's pump, which Bray had worked out before he applied for his patent. [Tr. 41.]

> *Western Electric Co. v. La Rue,* 139 U. S. 601, 606; 11 Sup. Ct. 670; 35 L. Ed. 294;
>
> *Diamond Rubber Co. v. Consol. Tire Co.,* 220 U. S. 428.

4. It is a grave error to narrow the scope of claim 3 of the patent in suit in view of the Northrup pump and the Admore pump which are pumps of the same type as the Neilsen pump, with holes in the side of the barrel, which was a failure because it sanded up very badly, according to the uncontradicted testimony of Bray and Rennebaum. [Tr. 279, 282, 357, 358.]

5. The fact that Bray's pump is a *very successful* pump, as proved by the fact that over 500 of Bray's pumps have been installed and have not required pulling of the well tubing in four years, is *a complete answer* to appellee's expert Adams and *a complete refutation* of all the *false theoretical speculation* of said witness to the

effect that Bray's pump will not accomplish what it *actually accomplishes,* as proved by the witness Rennebaum [Tr. 358], and by Bray [Tr. 43, 45, 270.] The court should base its decision upon the **proven facts** of the case, and not upon the uncertain theoretical speculations of a professional patent expert.

6. A very important novelty and utility of the Bray pump, with its discharge opening in the guide, only, is that the oil projected straight downward out of the bottom of the barrel directly into the sand enables the pump to "drill out" the sand, which settles in the bottom of the well tubing when the pump is shut down for a day or more. Appellees' pump will "drill out" like Bray's pump and this identity of function and result of the Bray and Hofco pumps is clear evidence of infringement of the Bray patent in suit by appellees' pump and a negation of anticipation of the principle of the Bray pump by either the Northrup or the Admore pumps with holes in the side of the barrel like the Neilsen pump, which could not be used for "drilling out." [Tr. 362.]

7. Since appellees' pump is constructed with a guide of no practical difference in the size or form of its opening from that of appellants' guide, and since appellees' guide performs the same function in substantially the same manner, and accomplishes the same results as appellants' guide [Tr. 358], a clear case of infringement of claim 3 of the patent in suit by appellees is clearly made out.

Sanitary Refrigerator v. Winters, 280 U. S. 30, 42, 74 L. Ed. 147, 50 S. C. 9.

Appellants respectfully pray that a rehearing be granted them to the end that the decree of the lower court be modified, in that claim 3 of the patent in suit be given its proper scope, according to the plain import of its terms, and in view of the decided advance made in the art by the invention in suit, and that claim 3 be decreed infringed by the pump manufactured and sold by the appellees in accordance with Defendants' Exhibit NN. [Book of Exhibits, p. 179.]

Respectfully submitted,

ALAN FRANKLIN,
Attorney for Appellants.

FRANK BRYANT,
Of Counsel.

Certificate of Counsel.

I hereby certify that I am counsel for appellants and petitioner, and in my judgment the foregoing petition for rehearing is well founded and is not interposed for delay.

ALAN FRANKLIN.

No. 8200

/ 2

United States
Circuit Court of Appeals
For the Ninth Circuit.

———

CHARLES O. LONG,

Appellant,

vs.

UNITED STATES OF AMERICA,

Appellee.

———

Transcript of Record

———

Upon Appeal from the District Court of the United
States for the District of Arizona.

JAN 18 1937

PAUL P. O'BRIEN,

K

Parker Printing Company, 545 Sansome Street, San Francisco

No. 8200

United States
Circuit Court of Appeals

For the Ninth Circuit.

CHARLES O. LONG,

Appellant,

vs.

UNITED STATES OF AMERICA,

Appellee.

Transcript of Record

Upon Appeal from the District Court of the United States for the District of Arizona.

Parker Printing Company, 545 Sansome Street, San Francisco

INDEX

[Clerk's Note: When deemed likely to be of an important nature, errors or doubtful matters appearing in the original certified record are printed literally in italic; and, likewise, cancelled matter appearing in the original certified record is printed and cancelled herein accordingly. When possible, an omission from the text is indicated by printing in italic the two words between which the omission seems to occur.]

GEORGE F. MacDONALD,
   Luhrs Tower,
   Phoenix, Arizona,
      Attorney for Appellant.

F. E. FLYNN,
   United States Attorney,
   Federal Building,
   Phoenix, Arizona,
      Attorney for Appellee. [3*]

In the District Court for the United States
in and for the District of Arizona

C-5200-Phx.

Viol: 26 U. S. C. 696 and 692
(Harrison Narcotic Act)

THE UNITED STATES OF AMERICA,
                       Plaintiff,

vs.

CHARLES O. LONG,

                       Defendant.

United States of America,
District of Arizona.—ss.

INDICTMENT

In the District Court of the United States, in and for the District aforesaid, at the March Term thereof, A. D. 1935.

The Grand Jurors of the United States, impaneled, sworn, and charged at the Term aforesaid, of the Court aforesaid, on their oath present, that Charles O. Long, on or about the 7th day of April, in the year of our Lord nineteen hundred thirty-five, in the said district and within the jurisdiction of said Court, in the City of Phoenix, County of Maricopa, did then and there wilfully, unlawfully, knowingly and falsely sell, barter, exchange and give away, certain derivatives and salts of opium, to-wit, approximately 245 ¼-grain morphine tablets, to one, Elmer Ethridge, not in pursuance of a written order from the said Elmer Ethridge, on a form issued in blank for that purpose by the Commissioner of Internal Revenue, under the provisions of Section 2 of the Act of Congress, dated December 17, 1914, called the Harrison Narcotic Act, the said Elmer Ethridge not being then and there a patient of the said Charles O. Long, and the said morphine was then and there dispensed and distributed by the said Charles O. Long not in the course of his professional practice only, contrary to the form of the statute in such case made and provided, and against the peace and dignity of the United States of America.

SECOND COUNT: And the Grand Jurors aforesaid, on their oath aforesaid, do further present and show that Charles O. Long, on or about the 7th day of April, 1935, and within the District, City and County aforesaid, did then and there wilfully, unlawfully, knowingly and falsely sell, dispense and distribute to one, Elmer Ethridge, a certain derivative of opium, to-wit, approximately 245 ¼ grains of morphine, which said morphine was not then and there in nor from the original stamped package containing said morphine; contrary to the form of the statute in such case made and provided, and against the peace and dignity of the United States.

F. E. FLYNN

United States Attorney

A True Bill.

MARK BRADLEY

Foreman

[Endorsed]: Filed Jul. 6, 1935. [4]

[Title of Cause.]

VERDICT

WE, THE JURY, duly empaneled and sworn in the above-entitled action, upon our oaths, do find the defendant, Charles O. Long, on the first count Guilty; on the second count Not Guilty.

ANTHON E. JACOBSON,

Foreman.

[Endorsed]: Filed Feb. 29, 1936. [5]

[Title of Court and Cause.]

MOTION IN ARREST OF JUDGMENT

COMES NOW the Defendant, Charles O. Long, by his attorneys, Lewkowitz & Zaversack, Herman Lewkowitz of Counsel, appearing this 4th day of March, 1936, and moves the Court that the Judgment in the above entitled cause be arrested as to him, the said Charles O. Long, Defendant, upon the following grounds and for the following reason:

That the defendant having been found guilty of Count 1, of the indictment, the evidence of both Count 2 and Count 1, being the same, the verdict is contrary to law.

WHEREFORE, defendant prays that this Motion be granted.

Dated: This 4th day of March, 1936.

LEWKOWITZ & ZAVERSACK
By HERMAN LEWKOWITZ
Attorneys for Defendant

[Endorsed]: Filed Mar. 4, 1936. [6]

————

In the United States District Court
for the District of Arizona

October 1935 Term   At Phoenix

Minute Entry of March 4, 1936
(Phoenix General Minutes)

Honorable JAMES H. BALDWIN, United States District Judge, specially assigned, Presiding.

C-5200

UNITED STATES OF AMERICA,
                                        Plaintiff,
        vs.

CHARLES O. LONG,
                                        Defendant.

F. E. Flynn, Esquire, United States Attorney, and G. E. Wood, Esquire, Assistant United States

Attorney, appear for the Government. The defendant, Charles O. Long, is present in person, with his counsel, Herman Lewkowitz, Esquire, this being the time heretofore fixed for judgment herein.

Herman Lewkowitz, Esquire, now presents and files Motions in Arrest of Judgment and for New Trial, and

IT IS ORDERED that said Motions be continued and set for hearing at the hour of 1:30 o'clock, P. M., this date.

Subsequently, at the hour of 1:30 o'clock, P. M., Defendant's Motion for New Trial, is duly argued by Herman Lewkowitz, Esquire, and F. E. Flynn, Esquire, United States Attorney, and

IT IS ORDERED that said Motion be denied; that an exception be entered on behalf of Defendant, and that Defendant be allowed to and including Tuesday, March 10, 1936, within which to file Bill of Exceptions.

Thereupon, IT IS ORDERED that said Motion in Arrest of Judgment be denied, and that an exception be entered on behalf of Defendant.

Said defendant is now duly informed by the Court of the nature of the crime charged in Count One of the Indictment herein, to-wit: unlawfully, wilfully, fraudulently, knowingly and feloniously selling, bartering, dispensing, administering, exchanging and giving away not in the course of professional practice only, certain derivatives and salts of opium, to-wit: morphine, not in pursuance of a written order of the person to whom such article was sold, bartered, dispensed, administered, exchanged or given, said person being then and there not a patient of said defendant, on a form issued in blank for that purpose by the Commissioner of Internal [7] Revenue as required by the Act of December 17, 1914, committed within the County of Maricopa, State and District of Arizona, on or about April 7, 1935, in violation of Title 26, United States Code Annotated, Sections 1043-1044; of his arraignment on said charge, and of his plea of Not Guilty thereto, and of his trial and conviction thereof by jury, and no legal cause appearing why judgment should not now be imposed, the Court renders judgment as follows:

That the said defendant having been duly convicted of said crime, the Court now finds him Guilty thereof, and as a punishment therefor, does now

ORDER, ADJUDGE AND DECREE that said defendant, Charles O. Long, be committed to the custody of the Attorney General of the United States or his authorized representative for imprisonment in a Penitentiary or other penal institution, for a term of eighteen (18) months, said term of imprisonment to date from March 4, 1936. [8]

[Title of Court and Cause.]

ASSIGNMENT OF ERRORS

COMES NOW the defendant, O. Charles Long, by his attorney, George F. Macdonald, and in connection with his appeal herewith filed, makes it known that in the record, proceedings, judgment and sentence appealed from, manifest error has intervened to the prejudice of O. Charles Long in these things, to-wit:

1. The Court erred in charging the jury, as follows:

"Now I charge you, if you believe that narcotic inspector Moore is entitled to full credit and believe that his story of this transaction is true, it is your duty to convict. On the other hand we have him supported by the testimony of two police officers here."

2. The Court erred in charging the jury, as follows:

"Now the defendant in this case has seen fit to produce testimony here of people by depositions whom you have not seen, that he is a man of good repute in the community in which he has lived. He has a right to do this, and because the defendant has introduced testimony tending to show his previous good character as to being a law-abiding citizen and has conformed with the ethics of his profession and the standards of practice, the court instructs you that it is competent for a person accused of crime to prove as a circumstance in his defense that his previous [9] character as a law-abiding citizen was good and that he conformed with the ethics and standards in the practice of his profession. Previous good character is not of itself a defense but it is a circumstance to be considered by the jury in connection with all of the other evidence in the case, in determining the guilt or innocence

of the accused. It is a circumstance which may be
shown for the purpose of rebutting the presumption of guilt arising from circumstantial evidence.
It may be used as tending to show that men of such
character would not be likely to commit the crime
charged. It should be taken and used in consideration as to whether other evidence is conclusive or
inconclusive, and it is for the jury to determine under all of the facts and circumstances in the case
what weight should be given to such evidence, when
it is considered with the other evidence, and if a
reasonable doubt is created as to the defendant's
guilt, he is entitled to an acquittal.

"In connection with that instruction I want to
say to you what George M. Borquin, my predecessor
in office, often said: 'That character is one thing
and reputation is what others believe you to be.
Character would prevent the commission of a crime
of this kind, but reputation is merely a circumstance tending to show that it is not probable that
a man with a reputation of that kind could do the
things that are charged to have been done here. But
reputation is merely built upon the fact that you
have not been caught, and when you are caught it
shows the character and destroys the reputation.' "

3. The Court erred in refusing to give to the
jury defendant's requested instruction No. 4, which
is as follows:

"The Court instructs the Jury that before you
can convict the defendant, the evidence must be so
strong as to convince every juror of the defendant's
guilt beyond a reasonable doubt; and if, [10] after

considering all the evidence, a single juror has a reasonable doubt of the defendant's guilt arising out of any part of the evidence, then you cannot convict the defendant.

> Blashfield on Instructions, paragraph 1949.
> Mitchell vs. State, 129 Ala. 23.''

4. The Court erred in denying defendant's Motion in Arrest of Judgment based on the ground that the verdict was contrary to law in that the verdict of guilty on the first count of the indictment and not guilty on the second count were wholly inconsistent, each of the alleged offenses clearly appearing from the indictment and the instructions of the Court to be based upon an alleged single transaction, namely, one sale of one bottle of morphine to one person at one time and one place.

5. The judgment pronounced and entered in this cause is contrary to law in this, to-wit, that said judgment was pronounced upon a verdict of guilty in the first count of the indictment, which verdict was wholly inconsistent with the verdict of not guilty on the second count of the indictment, in that, as appears from the indictment and the instructions of the Court, each of the alleged offenses in counts one and two was based on an alleged single transaction, namely, one sale of one bottle of morphine to one person at the same time and place.

And upon the foregoing Assignment of Errors and upon the record in said cause, said defendant prays that the said judgment and sentence against and upon him, the said O. Charles Long, under the indictment herein, may be reversed and held for naught.

GEO. F. MACDONALD

Attorney for Defendant

Received a copy of the within Assignment of Errors this 27th day of October, 1936.

F. E. FLYNN,

United States Attorney,

By GEORGE E. WOOD,

Asst. U. S. Atty.

[Endorsed]: Filed Oct. 27, 1936. [11]

[Title of Court and Cause.]

BILL OF EXCEPTIONS

BE IT REMEMBERED, that the trial of the above entitled cause came on regularly to be heard before the Hon. James N. Baldwin, Judge of the District Court of the United States, and a jury, at the court room of said Court in the Federal Building, City of Phoenix, State and District of Arizona, on the 28th day of February, 1936, at the hour of 10:00 o'clock A. M.; plaintiff, The United States of America, being represented by Frank E. Flynn, United States Attorney for the District of Arizona, and George Wood, Assistant United States Attorney; and the defendant, O. Charles Long, being present in person and being represented by his attorney, Herman Lewkowitz; and the parties having announced ready for trial, L. O. Tucker was

·duly sworn as shorthand reporter. The United States Attorney read the indictment to the jury; counsel for the defendant stated that he reserved his statement until the conclusion of the Government's case.

Whereupon, witnesses were sworn and evidence both oral and documentary was offered and admitted upon behalf of the plaintiff, and the plaintiff rested.

Whereupon, witnesses were sworn and evidence both oral and documentary was offered and admitted upon behalf of the defendant, and the defendant rested. There was no rebuttal [12] testimony offered on behalf of the plaintiff.

Whereupon, the case was argued to the jury by counsel for the plaintiff and defendant.

Whereupon, the Court instructed the jury, as follows:

"Gentlemen of the Jury, it now becomes my duty to charge you as to the law of the case. You know what I say as to the law of the case you take that from me, and you decide the facts. The oath you took required it, and I have no doubt that in this case, as in any other case, you obey that oath. However, I deem it wise, especially in a case as we have here, when we have a man who is in a position that may affect your sympathy and where we have an attack upon another man because of an unfortunate addiction. There has never yet been a drug addict who did not receive the drug, and the fact is he gets it from some physician in practice who gives it to him.

Your oath is, gentlemen, that you will well and⸱ truly try this cause at issue and a true verdict render therein by the law given you by the court and the evidence, not according to your sympathy in the case, not according to your feeling which is naturally against a drug addict who has testified in the case, not according to your sympathy for the young wife of a defendant in a case, but your duty as declared to your God, society, to the court and yourselves is to try this case on the law and the facts, not on sympathy and not on feeling. Your oath is to take the law from me. My duty is to take the facts from you, for the law is that all of the facts are for you to determine.

As requested by the defendant in this case, you are instructed that you are made by law, gentlemen, the sole judges of the facts in this case and as to the credibility of each of the witnesses who have testified in the case, and of the weight [13] that will be given to their testimony. And in determining their credibility and the weight you will give to their testimony, you have a right to take into consideration their manner and appearance while testifying. .

At that point, there has been considerable criticism of Etheridge. Did you notice anything in his appearance or manner on the witness stand that would cause you to disbelieve him? Ask yourselves if you do not know by his own confession that he is a drug addict, that he is the least credible of the witnesses. The law is that the testimony of a drug addict, where uncorroborated, is to be distrusted. And also the law is that you have a right to take into consideration the fact that a man is the defendant in the case, but it does not necessarily mean that either one or the other is lying, and you have no right to say merely because he is a drug addict or that he is the defendant in this case that he is lying. They are both witnesses and their testimony is weighed as the testimony of other witnesses, as will be pointed out to you later.

You have the right to take into consideration their manner and appearance while testifying, their means of knowledge——

Is not Etheridge in just as good a position to know as the defendant here?

Any interest or motive which they may have——

Can you gentlemen see any more reason to show that Etheridge has a motive than the Doctor might have one? These are all questions that you have a right to ask, not only that you have the right to ask but you have a right to answer them in determining the credit between these two witnesses. These rules apply to the other witnesses.

And the probability of the truth of the statements when taken into consideration with the other evidence in the case. [14]

Now, as I say, you are the sole judges of the weight and the effect to be given to the testimony, but you are not arbitrary judges, and you are bound to do your judging in the light of the rules of evidence as fixed by law, not for your guidance in this case but for your guidance and my guidance in all cases where a man is on trial charged with a crime.

Now, as I have said, you have no right to say that the defendant is necessarily falsifying merely because he is the defendant in this case. The law permits a defendant charged with a crime to testify in his own behalf. He does not have to; the government could not make him, but the law permits him to testify in his own behalf and the jury must fairly and impartially consider his testimony together with all of the other evidence in the case. The jury does not have the right to disregard the testimony of the defendant merely because of the fact that he is the defendant, but it is the duty of the jury to view the testimony of the defendant to test it by the same rules under which the jury must determine the credibility of the other witnesses; and in doing so, that is, in considering the credit of the witness, the defendant here, you may consider the fact that he is the defendant, and the nature and enormity of the crime with which he is charged. In other words, the defendant starts as other witnesses start, with the presumption that he is telling the truth. He is weighed in the same balance that other witnesses are weighed, and it is a question whether he is interested, whether any witness has an interest or motive for what they have said on the witness stand; and I charge you that in determining whether the defendant has an interest you have a right to and you should consider the fact that he is the defendant, and the nature and enormity of the crime with which he is charged in this case. [15]

You start, as I say, with the presumption that he is innocent, with the presumption that he is telling the truth, but you also start with the knowledge of the fact that he is the defendant, that he has a direct and personal interest in the result of this case, and that your verdict is of vital consequence to him.

So you ask yourselves, "Is that a sufficient motive for him to color the facts in this case?"

Now as I have said he and Etheridge start out together. The crime charged is not the things that took place in California. This court would have no jurisdiction of anything that happened across the California line. Our jurisdiction extends to that line and not an inch beyond.

You can not convict a man of things he did in California if he did the things that Etheridge said he did, but the question here presented is, "Did he do the things that are said to have been done on Washington street between Third and Fourth Avenues in Phoenix?"

However, since counsel has seen fit to tell you that you can not believe a drug addict merely because he is a drug addict, which is not the law—the law is that you have a right to consider that fact, and the law is too that a drug addict is not entitled to full credit because of the fact that he is an addict, and you can not convict upon his testimony alone, not that you can not believe him.

But let us see just how far the witness Etheridge is corroborated by the defendant himself. What did the drug addict say? That they met in California on the seventh day of January of last year.

What did the Doctor say? That they met on the seventh day of January in California, last year.

[16]

What did Etheridge say? That he went to the Doctor and said he was suffering pain and that he wanted morphine.

What did the Doctor say? That he came to his office and suffering with pain and wanting treatment. A difference in wording.

What did Etheridge say? That he told the Doctor that he had been an addict but was off the stuff.

What did the Doctor say? That in making this examination of this man he rolled up his sleeve and saw a line down his forearm four inches long. That indicated what? That he had been a drug addict. No contradiction there.

Now you can not convict on the uncorroborated testimony of a drug addict. That is the law. But ask yourselves if there was any contradiction. Etheridge says that the doctor refused to give him morphine until he could get money. What does the Doctor say? That for a period of ten days he refused to treat or that he refused to treat Etheridge until he got the money, after having given him some treatment, and Etheridge did not come back for ten days.

And when did Etheridge come back? What does the doctor say to Etheridge when he got a five dollar bill?—in other words, when he got the money, that he and the doctor both say that it was necessary to continue the treatment. Then what happened?

Now I am saying that you can not—and I am instructing you along this line because the law is that you can not convict upon the testimony of Etheridge unless it is corroborated by the circumstances in the case or the testimony of a witness testifying here.

What does the doctor say? He tells you, gentlemen of the jury—or what does Etheridge say? He says he went to the doctor's office. When? Noon and evening. For what purpose? [17] To get a shot in the arm. A shot of what? A shot of morphine. And what did he tell you first? That the doctor gave to him a one-quarter grain of morphine in the evening to carry him through the morning and to be self administered.

Does the doctor contradict it? The doctor's testimony, if I recall it correctly, though that is a question for you, was that on February fifteenth of one nine three five, until March seventeenth, the day that Etheridge left, and the doctor says—there they corroborate again—for Phoenix, that Etheridge came to his place and was treated twice a day.

Now is there any contradiction there? Etheridge says that the doctor charged him what? Two dollars. And the doctor says the same thing. Is there any contradiction there?

The only contradiction between the doctor's story and Etheridge's story is that the doctor says he gave one drug and Etheridge says he gave another, but the fact remains, gentlemen of the jury, that Etheridge testified when he was getting pay checks for the work he was doing he gave his checks to the doctor and the doctor says he got them. Etheridge says that the doctor got the entire amount of these checks excepting such as was nedeed for room and food.

What does the doctor say? He says that he got the check, and he gives you the figure, which is clearly not more than enough, in my opinion—however, that is a question of fact for you to determine—to pay the room and board of a man for two weeks, or whatever the period was.

Now, gentlemen of the jury, if it was so simple that the man just went in and got a shot in the arm of some acid compound that was not narcotic in its effect, why was it necessary for him to get a supply from California when he was leaving or in Phoenix where there is an abundance of good doctors? Ask yourselves those questions. [18]

You have the right to reason to a conclusion in this case; you must decide it upon your reason and experience as men, and, gentlemen of the jury, you again have the right to ask yourselves, and you should ask yourselves, whether Etheridge was trying to act the honest part here. What is the fact? He was leaving California, Brawley, I believe, and coming to Phoenix. And what did he do? He went down to his employer and said, "Give the last pay check I earned to the doctor who has been giving me what I want." Is there any contradiction upon that?

The doctor shows and his records show that these checks with his endorsement upon them, the last dime that this addict earned across the California line, went to the doctor.

Now, do you believe, gentlemen of the jury, in the light of reason that it is reasonable to suppose that he would be writing to this doctor in California and telephoning him? But Etheridge and the doctor says that he did write and that he did telephone. And the doctor says that in response to one of those telephone calls a clerk in his office sent this exhibit B, I believe, for the defendant. If it was such a simple treatment and not forbidden, why did he not direct him to go to some Phoenix doctor? Ask yourselves those questions.

You do not test a case entirely upon the spoken words of witnesses. You test it in the light of reason, in the light of your experience as men, and you test it in the light of the occurrences as they would ordinarily happen under the same conditions. The checks are here; there is no dispute about them. The doctor says that he got them. He says he gave Etheridge enough to live on.

And that is the story all down the line from the fifteenth day of February. There seems to be no contradiction and there is not any contradiction that the doctor did give to [19] Etheridge narcotic drugs and that those narcotic drugs were morphine, because the doctor himself says that in the period commencing on February fifteenth down to March seventeenth he gave Etheridge one prescription for four tablets of morphine. That is one. He says that he himself gave to Etheridge four of these little tablets on four or five occasions. In other words, the doctor by his own oath has told you that on at least six occasions he has given to the witness Etheridge the narcotic drug that brought him to the state where they say he can not be believed.

I tell you again that he can not be believed so far as considering his proof alone as sufficient to justify a verdict of guilty in this case unless he is corroborated by somebody else. But search your memory and see if he was not corroborated in every statement that he made by the doctor testifying as a witness for himself, except that the doctor said that he did not give him or prescribe for him or allow him to use any morphine until some time after February fifteenth of last year. There is a difference of days.

Etheridge said he got the morphine, got a shot in the arm, every time he went in, which was twice a day. The doctor said he was there but he gave him some acid compound until he saw there was no money in it, then he told him he would not continue the treatment, and Etheridge stayed away for ten days. Now, draw your own conclusions upon that.

But the fact is that is not the question in this case. The charge is that at Phoenix the seventh day of April one nine three five the defendant sold to Etheridge two hundred and forty five one-quarter grain morphine tablets and not on a prescription and not in the ordinary course of his practice as a physician treating a patient. [20]

In the second count it is charged that the doctor at the same time and place sold to the witness Etheridge two hundred and forty five one-quarter grains of morphine which were not then and there in the original stamped package nor taken therefrom.

Now there are lots of words in this indictment, gentlemen, and the burden is upon the government to prove the truth of each and every allegation therein except the negative allegations.

Before you can convict upon the second count of the indictment you must find that the container that is in evidence here, I think as government's exhibit number one, a little glass bottle, which nobody contradicts contained two hundred and forty-five one-quarter grain tablets of morphine, was sold by this defendant to Etheridge, or given to him, or exchanged with him so that he got it in Phoenix, and some time on or about the seventh day of April. So that is all there is in the second count. No one disputes the contention that the bottle is without stamp. No one suggests that it is a bottle that was taken from a bottle properly stamped. The presumption is that it was not, and the sole question in the second count is did the defendant sell or give or dispose or dispense or distribute to Etheridge the bottle in evidence here with its contents, morphine.

On the first count which is identical so far as the substance in the charge is contained—it is charged that at the same time and place the defendant did sell, barter, exchange, and give away certain derivatives and the salts of opium, to-wit: morphine—and the court charges you that morphine is a salt and derivative of opium—to Etheridge not in pursuance of an order form—there is not suggestion of an order form—and not in the course of his practice as a physician.

The doctor by his plea of not guilty has placed upon the government the burden of proving these things but the doctor himself goes upon the witness stand and declares flatly, and at [21] the outset of the trial his attorney stated distinctly to you and me that he was going to fix his defense upon one point, that is, the fact that he did not do—that the defendant did not give this bottle of morphine to Etheridge the witness. And that is the theory upon which the case has been tried and argued.

When the defendant puts a plea of not guilty to the charges contained in the indictment in this case upon the record he places upon the government the burden of proving his guilt beyond a reasonable doubt. Now, you will note "beyond a reasonable doubt;" not beyond all doubt, not beyond a possibility of error, but beyond a reasonable doubt, which should convey a definite and fixed meaning to you. However, the law requires that I should define a reasonable doubt, and it is defined in law as a doubt based on reason. You ask yourself in this case, after considering all of the facts in the record and all of the circumstances concerning the transactions as they are shown here, "Is there a doubt which can be based upon the reason of an intelligent man and which is reasonable in view of all of the evidence," and if after an impartial comparison and consideration of all of the evidence you can candidly say that you are not satisfied of the defendant's guilt, you have a reasonable doubt. In other words, if after fairly and impartially considering all of the evidence in the case, not what you would like to find, not what your sympathy would lead you to find, but what your reason dictates your finding should be, can you say that you are not satisfied of the defendant's guilt. If you can, you have a reasonable doubt and should not convict. On the other hand, if after such an impartial consideration of all of the evidence you can truthfully say that you have an abiding conviction of the defendant's guilt such as you would be willing to act upon in the more weighty and important matters relating to your own affairs, you have no reasonable doubt. In other words, if you would be willing [22] to act in affairs of importance to yourselves upon the conclusion that the charge here is true, then you have no reasonable doubt.

And in determining whether you have a doubt in this case such as would justify an acquittal or whether the government has proven its case beyond a reasonable doubt, that is, to the point which will justify a conviction, you must have in mind the fact that the law presumes the innocence and not the guilt of the defendant. That is the starting point in

every criminal case, a presumption of innocence.
This presumption of innocence continues throughout
the trial and entitles the defendant to an acquittal
unless the evidence in the case taken as a whole
satisfies you of the defendant's guilt beyond a reasonable doubt as defined in these instructions, not
beyond all doubt, not beyond the possibility of
error, but beyond a reasonable doubt, that is, with
that degree of certainty that would cause you to be
willing to act upon the truth of the charge here
made if it were a matter of importance to yourselves.

Now, the law again is that if you believe any witness has wilfully testified falsely as to any material fact in the case, you are at liberty to disregard the entire testimony of that witness except
in so far as it may be corroborated by the testimony of other credible witnesses or supported by
other evidence in the case. That applies to the doctor; it applies to Etheridge; it applies to everyone
else in this case. If you feel that anyone in this
case has wilfully testified falsely as to a material
matter, that is, to anything that would have a reasonable bearing upon the decision by you in this
case, then you are at liberty to disregard his entire
testimony except in so far as it may be corroborated
by the testimony of other credible witnesses or supported by the evidence in the case. [23]

It is for you to determine whether there is or is
not any direct conflict. The only conflict I recall in

the case is the conflict beginning between the doctor and the witness Etheridge, the doctor saying that he gave an acid compound by injecting it into the blood stream of the witness Etheridge, and Etheridge saying that he put a narcotic drug, morphine, into that blood stream from about the time he met him. The doctor said he did that on four or five occasions between February fifteenth and March seventeenth and it is for you to determine who is telling the truth there.

As I say, that is not the important thing, although it is an element you have a right to consider in determining the credit of these witnesses.

Now, in proving these crimes the general rule is, and the rule in this case is, that in order to convict the defendant of the crime charged it is incumbent upon the prosecution to prove the truth of every material allegation of the indictment.

In this case the law is that where there are no stamps upon the container of the narcotic drug the presumptions are that there never were any there. That prsumption continues and should control your decision in this case because there is no proof to the contrary.

Nothing is to be presumed or taken by implication against the defendant. The law presumes him to be innocent of the crime with which he is charged until he is proven guilty beyond a reasonable doubt by competent evidence, and this presumption continues throughout the progress of the trial and until such time as you are satisfied from the evidence beyond a reasonable doubt of his guilt. This is

merely a repetition of the old instruction we have
given you in every criminal case for years, that the
burden is upon the government to prove its case,
and it is for you to say whether that case has been
proven or not. [24]

Now the defendant in this case has seen fit to
produce testimony here of people by depositions
whom you have not seen, that he is a man of good
repute in the community in which he has lived. He
has a right to do this, and because the defendant
has introduced testimony tending to show his previous good character as to being a law-abiding citizen and has conformed with the ethics of his profession and the standards of practice, the court
instructs you that it is competent for a person
accused of crime to prove as a circumstance in his
defence that his previous character as a law-abiding
citizen was good and that he conformed with the
ethics and standards in the practice of his profession. Previous good character is not of itself a
defence but it is a circumstance to be considered by
the jury in connection with all of the other evidence
in the case, in determining the guilt or innocence
of the accused. It is a circumstance which may be
shown for the purpose of rebutting the presumption
of guilt arising from circumstantial evidence. It
may be used as tending to show that men of such
character would not be likely to commit the crime
charged. It should be taken and used in consideration as to whether other evidence is conclusive or
inconclusive, and it is for the jury to determine
under all of the facts and circumstances in the case

what weight should be given to such evidence, when it is considered with the other evidence, and if a reasonable doubt is created as to the defendant's guilt, he is entitled to an acquittal.

In connection with that instruction I want to say to you what George M. Bourquin, my predecessor in office, often said: "That character is one thing and reputation is another. Character is what you are; reputation is what others believe you to be. Character would prevent the commission of a crime of this kind, but reputation is merely a circumstance tending to show that it is not probable that a man with a reputation of that kind could do the things that are charged to have been done here. But reputation is merely built upon the fact that you have not been [25] caught, and when you are caught it shows the character and destroys the reputation."

It is not a defense. It is merely a circumstance which you may or may not give weight in considering this case. But I say to you, gentlemen of the jury, that in this case you have the direct and positive testimony of three men that everything that is charged in this indictment was a natural occurrence within a few blocks of where we are now sitting, and there is no contradiction to the statements of those three men except the statement of the defendant, which is, "I did not do it; I got the money; it is marked money. I met Etheridge at the time and on the day specified and at the place set out in this indictment, but I came here to collect money that he owed me because he said he was gambling and getting a lot of money and he could pay the entire bill if I would come and get it."

Now ask yourself is it reasonable for a man to drive two hundred and seventy-five miles to collect money from a man who says he has it and who seems to be willing to give it to him, and is it according to the experience of men and does it ring true to you to believe that anyone under those conditions would make the trip the doctor made unless he had some other motive.

On the other hand we have the testimony of a man who appears to have no interest in the case, Mr. Moore, the narcotic agent. He is employed by the government to perform his duties, not to catch men who are not guilty of crime, but men who are guilty of crime. What does he say? He says that he saw that bottle in the hand of the doctor. He says that he saw it held up, and that coroborates the testimony of the witness Etheridge who said that the doctor held it in his hand. The doctor says, "Where is the money?" And he said, "Where is the dope?" [26]

The doctor took the bottle out and held it where he could not reach it, over toward his wife's side and toward the windshield, and when he commenced counting the money the doctor slid it over and he got it when the counting was finished.

Now I charge you if you believe that narcotic inspector Moore is entitled to full credit and believe that his story of this transaction is true, it is your duty to convict. On the other hand we have him supported by the testimony of two police officers here.

The question involved in this case is, did Doctor Long deliver that bottle with narcotics in it to the witness Etheridge. Etheridge says he did. The doctor says he got the money but he did not deliver the drugs. The narcotic agent, Moore, says that the doctor put the drug in his hand, that he got the money and passed the drug to Etheridge. We have in addition to that the testimony of two police officers, one of them Mr. Moore, I believe, who testified that he saw the entire transaction, that he saw the money passed, that he saw the bottle passed, and that everything that Etheridge said was true. And we have the testimony of one other police officer who says he did not see the bottle passed though he saw the rest of it. And we have heard much talk about the possibility that Etheridge may have gone somewhere and gotten that bottle, that he may have had it on him. Now that would be interesting it seems to me in the light of reason, although that is a question for you, it is one of fact to consider that problem if it were not for the testimony of Etheridge, if it were not for the testimony of the police officer Moore, and if it were not for the testimony of narcotic agent Moore, that they saw the bottle actually in the hand of Doctor Long, that they saw it passed from that hand into the hands of Etheridge; it was not taken from Etheridge pocket, it was not taken from a concealed place about [27] his person, but if these men, all officers, are telling the truth, that bottle was in their sight from the time it appeared in the doctor's hand at the time the money counting commenced until it passed from that hand when the counting had been completed and into the hands of Etheridge who did what? They say he held it up that way (the court illustrating to the jury). In other words, it was not concealed from the time it appeared in the hands of the defendant here, if we believe the testimony of those three men, which is corroborated by the second police officer, although he does not say that he saw the article passed from the hands of the defendant in this case to the hands of the man who counted the money and paid for it.

And did the doctor's wife say that it was not done? She said Etheridge and her husband did things that she did not see, because she was doing what? Watching others. In other words, she put it very definitely that she was not watching to see everything that happened. And it appears to me, although that again is for you, that it might reasonably be said that everything that is said to have occurred here, by the officers, according to the officers' statements, may have occurred without her seeing it. She testified that she was sitting there for half an hour before Etheridge showed up, and during the time that her husband and Etheridge were talking to each other, until she got sunburned.

Did her husband corroborate that statement? He says that after Etheridge went and put his suitcase in the lobby of the Patrick Hotel they talked for many minutes more and she demanded that the talk stop so she could get something to eat.

Does her story that her husband had given this fur to her as a present carry out this story that it belonged to someone else? Is that the ordinary course of business for a man to give to his wife a present something that he does not own? [28]

Those are all circumstances. The weight to be given to them is for you to decide. But, as I say, in this case as it appears to me, you have to say either that the doctor, because of his interest, has forgotten some of the important circumstances, or you have to say that the two police officers, a narcotic agent of the federal government, and a narcotic user, all testified falsely for the purpose of convicting an innocent man. It is for you to say what theory you will adopt.

And in conclusion, at the outset the defendant is presumed to be innocent. That presumption continues throughout the trial of the case and he is entitled to an acquittal until and unless you are satisfied beyond a reasonable doubt of the charge laid in this indictment.

Has the government any exception to the charge made?

Mr. FLYNN: None, your Honor.

The COURT: Has the defendant?

Mr. LEWKOWITZ: If I caught one statement —I did not interrupt your Honor—I have an exception to enter to the statement, if I am correct upon it, when your Honor says, "I charge you if you believe narcotic agent Moore you should convict."

The COURT: I said if they believed he was entitled to full credit and that the statements made by him were true, which is according to the law.

In other words, gentlemen, the testimony of one witness who is entitled to full credit is sufficient proof of any fact in this case, and if you find that the witness Moore, the narcotic agent, or that police officer Moore and that Etheridge were telling the truth, at the time when he says that he saw this container of narcotics pass from the hand of the defendant to the hand of Etheridge, that is sufficient to justify a conviction. In other words, the testimony of one witness who in your opinion is telling [29] the truth and who is entitled to full credit is sufficient for the proof of any fact in this case.

Now so that the record may be complete, at the request of the defendant, the court has given defendant's requested instructions one, two, three, five, six, and seven, and has refused to give defendant's requested instruction number four. The instructions requested and the request that they be given will be filed.

The defendant will be granted an exception to the ruling of the court in refusing to give his requested instruction number four."

Which said requested instruction number four is as follows:

"The Court instructs the Jury that before you can convict the defendant, the evidence must be so strong as to convince every juror of the defendant's guilt beyond a reasonable doubt; and if, after considering all the evidence, a single juror has a reasonable doubt of the defendant's guilt arising out of any part of the evidence, then you cannot convict the defendant.

Blashfield on Instructions, paragraph 1949.
Mitchell vs. State, 129 Ala. 23."

This cause was thereupon submitted to the jury for its deliberation and verdict.

And now in furtherance of justice, and that right may be done, the appellant, O. Charles Long, tenders and presents the foregoing as his Bill of Exceptions in this case, and prays that the same may be settled and allowed and signed and sealed by the Court and made a part of the record.

Dated: October 26th, 1936.

GEO. F. MACDONALD
Attorney for Defendant. [30]

JUDGE'S CERTIFICATE TO
BILL OF EXCEPTIONS

The foregoing Bill of Exceptions contains:

1. The charge of the Court to the Jury and the exceptions thereto by the defendant.

2. Defendant's requested instruction No. 4, and defendant's exceptions to the Court's refusal to give such requested instruction.

The said Bill of Exceptions correctly shows the proceedings had and is correct in all respects and is a true and complete Bill of Exceptions. The said Bill of Exceptions was duly proposed and duly and regularly filed with the Clerk of said Court and thereafter duly and regularly served within the time authorized by law; and that no amendments were proposed to said Bill of Exceptions except such as are embodied therein; that due and regular notice of time for settlement and certifying the said Bill of Exceptions was given, and the same is hereby approved, settled and allowed, and is hereby made a part of the record in this cause.

Dated this 9th day of November, 1936.

DAVE W. LING
United States District Judge

Received a copy of the within Bill of Exceptions this 27th day of October, 1936.

F. E. FLYNN,
United States District Attorney.
By GEORGE E. WOOD,
Assistant.

Proposed Bill of Exceptions lodged Oct. 27, 1936.

[Endorsed]: Filed Nov. 9, 1936.

EDWARD W. SCRUGGS,
Clerk,
United States District Court for
the District of Arizona,
By W. T. CHOISSER,
Deputy Clerk. [31]

_____

[Title of Court.]

October 1936 Term   At Phoenix

Minute Entry of November 9, 1936
(Phoenix General Minutes)

Honorable DAVE W. LING, United States District Judge, Presiding.

[Title of Cause.]

ORDER ALLOWING, SETTLING AND
APPROVING BILL OF EXCEPTIONS.

No appearance is made on behalf of the Government. George F. Macdonald, Esquire, appears on behalf of the Defendant, and now presents to the Court for settlement Defendant's Proposed Bill of Exceptions, and it appearing that service has been regularly had upon the same and that no Proposed Amendments thereto have been filed or submitted,

IT IS ORDERED that said Proposed Bill of Exceptions be, and the same is hereby allowed, settled and approved as the Bill of Exceptions herein and made a part of the record in this cause. [32]

[Title of Court and Cause.]

PRAECIPE FOR TRANSCRIPT OF RECORD

To the Clerk of the United States Court for the District of Arizona :

COMES NOW O. Charles Long, defendant and appellant, and files this Praecipe and indicates the portions of the record to be incorporated in the Transcript, as follows:

1. The Indictment;
2. The Verdict;
3. The Motion in Arrest of Judgment;
4. Assignment of Errors;
5. Bill of Exceptions, together with Certificate of the United States District Judge to the Bill of Exceptions, approving, settling and allowing the same;
6. Minute entry dated March 4, 1936, denying motion for new trial and motion in Arrest of Judgment, and Judgment;

7. Minute entry of order allowing, approving and settling Bill of Exceptions;

8. This Praecipe.

Dated this 9th day of November, 1936.

GEORGE F. MACDONALD

Attorney for Defendant and Appellant

Received a copy of the within foregoing Praecipe this 13th day of November, 1936.

F. E. FLYNN

United States Attorney

[Endorsed]: Filed Nov. 13, 1936. [33]

[Title of Court.]

United States of America,
District of Arizona.—ss.

I, EDWARD W. SCRUGGS, Clerk of the United States District Court for the District of Arizona, do hereby certify that I am the custodian of the records, papers and files of the said Court, including the records, papers and files in the case of United States of America, Plaintiff, vs. Charles O. Long, Defendant, numbered C-5200 Phoenix, on the docket of said Court.

I further certify that the attached pages, numbered 1 to 33, inclusive, contain a full, true and correct transcript of the proceedings of said cause and all the papers filed therein, together with the endorsements of filing thereon, called for and designated in the praecipe filed in said cause and made a part of the transcript attached hereto, as the same appear from the originals of record and on file in my office as such Clerk, in the City of Phoenix, State and District aforesaid.

I further certify that the Clerk's fee for preparing and certifying to this said transcript of record amounts to the sum of $5.00 and that said sum has been paid to me by counsel for the appellant.

WITNESS my hand and the seal of the said Court this 25th day of November, 1936.

[Seal]   EDWARD W. SCRUGGS,

Clerk. [34]

[Title of Court and Cause.]

NOTICE OF APPEAL

Name and address of Appellant:

CHARLES O. LONG,
Brawley, California.

Name and address of Appellant's Attorney:

. HERMAN LEWKOWITZ,
of Lewkowitz & Zaversack,
First National Bank of Arizona Bldg.,
Phoenix, Arizona.

Offense:

Vio. of 26 U. S. C. 696 (Harrison Narcotic Act) being Par. 1044, Tit. 26 U. S. C. A. Defendant being charged with selling to Elmer Etheridge 245 ¼ grain morphine tablets not pursuant to a written order and not in the course of his professional practice.

Date of judgment:

March 4, 1936.

Brief description of sentence:

Defendant was sentenced to serve a term of eighteen (18) months in a penitentiary to be designated by the Attorney General of the United States.

Name of prison where now confined:

In the City Jail, Phoenix, Maricopa County, Arizona.

I, the above named appellant, hereby appeals to the United States Circuit Court of Appeals for the 9th Circuit from the judgment above mentioned on the grounds set forth below.

O. CHARLES LONG, M. D.,

Appellant

Dated: March 6, 1936.

LEWKOWITZ & ZAVERSACK

By HERMAN LEWKOWITZ

Attorneys for Appellant

Grounds of Appeal:

As Grounds of Appeal, the Appellant relies on the following:

(a) That the Verdict is contrary to the Law in that, the evidence adduced at the trial upon the first count of the Indictment was to the same effect as that introduced on the second count of the Indictment and the Jury having found the Defendant, Appellant, NOT GUILTY on the second count of the Indictment, they should have rendered a similar verdict as to the first count of the Indictment. The Appellant contends that the verdict of NOT GUILTY on the second count precludes a verdict of GUILTY on the first count.

(b) That the statements of the Trial Judge as hereinafter set forth were argumentative and prejudicial to the Appellant. Said statements being:

1. "Now I charge you if you believe that Narcotic Inspector Moore is entitled to full credit and believe that his story of this transaction is true, it is your duty to convict. On *the hand,* we have him supported by the testimony of two police officers here."

2. "On the other hand, we have the testimony of a man who appears to have no interest in the case, Mr. Moore, the Narcotic Agent. He is employed by the Government to perform his duties, not to catch men who are not guilty of crime, but men who are guilty of crime. * * *"

3. "and in doing so—that is, considering the credit of the witness, the defendant here, you may consider the fact that he is the defendant, and the nature and enormity of the crime with which he is charged. * * *

You start, as I say, with the presumption that he is innocent, with the presumption that he is telling the truth, but you also start with the knowledge of the fact that he is the defendant, that he has a direct and personal interest in the result of this case, and that your verdict is of vital consequence to him. So you ask yourselves, 'is that a sufficient motive for him to color the facts in this case?''

[Endorsed]: Filed May 13, 1936. Paul P. O'Brien, Clerk of the United States Circuit Court of Appeals for the Ninth Circuit.

———

[Endorsed]: No. 8200. United States Circuit Court of Appeals for the Ninth Circuit. Charles O. Long, Appellant, vs. United States of America, Appellee. Transcript of Record. Upon Appeal from the District Court of the United States for the District of Arizona.

Filed November 27, 1936.

PAUL P. O'BRIEN,

Clerk of the United States Circuit Court of Appeals for the Ninth Circuit.

No. 8200

United States

Circuit Court of Appeals

For the Ninth Circuit.

———

CHARLES O. LONG,

Appellant,

vs.

UNITED STATES OF AMERICA,

Appellee.

———

Appellant's Opening Brief

———

Upon Appeal from the District Court of the United
States for the District of Arizona

———

GEORGE F. MACDONALD,
Attorney for Appellant.

No. 8200

United States
Circuit Court of Appeals
For the Ninth Circuit.

CHARLES O. LONG,

Appellant,

vs.

UNITED STATES OF AMERICA,

Appellee.

Appellant's Opening Brief

Upon Appeal from the District Court of the United
States for the District of Arizona

GEORGE F. MACDONALD,
Attorney for Appellant.

INDEX

TABLE OF CASES

STATEMENT C
DISCLC

Appellant w...
United States ...
the first chargin...
the United St...
April, 1935, t...
five ¼ grain ...
written order i...
in blank for th...
nal Revenue, ...
a patient of th...
then and there ...
Long, not in th...
(Tr. 1, 2) The...
having on the ...
sold, dispensed ...
phine tablets, ...
the original ...
Title 26, of th...

A plea of n...
fore a jury (Tr. ?
the first count a...

A Motion i...
March 4, 1935 ...
ground of its...
guilty on the ...
count, and b...
(Tr. 5) On t...
pronounced a...
tice of Appe...
sistency of the...
structions of the C...

STATEMENT OF PLEADINGS AND FACTS
DISCLOSING JURISDICTION

Appellant was indicted in the District Court of the United States for the District of Arizona on two counts, the first charging a violation of Section 696, Title 26, of the United States Code, in the sale on the 7th day of April, 1935, to one Etheridge, of two hundred and forty-five ¼ grain morphine tablets, not in pursuance of a written order from the said Etheridge, on a form issued in blank for that purpose by the Commissioner of Internal Revenue, the said Etheridge not being then and there a patient of the said Long; and that the morphine was then and there dispensed and distributed by the said Long, not in the course of his professional practice only. (Tr. 1, 2) The second count charges the defendant with having on the same date and place to the same person sold, dispensed and distributed the same number of morphine tablets, which were not then and there in nor from the original stamped package, in violation of Section 692, Title 26, of the United States Code. (Tr. 2, 3)

A plea of not guilty was entered and upon trial before a jury (Tr. 2) the defendant was found guilty on the first count and not guilty on the second count. (Tr. 3)

A Motion in Arrest of Judgment (Tr. 3, 4) was, on March 4, 1935, duly filed, attacking the verdict on the ground of its inconsistency in finding the defendant guilty on the first count and not guilty on the second count, and being overruled, exception was duly entered. (Tr. 5) On the same date judgment and sentence was pronounced and entered (Tr. 4-6) and thereafter Notice of Appeal, raising the same question of the inconsistency of the verdict and assigning as error several instructions of the Court, was duly filed. (Tr. 37 et seq.)

STATEMENT OF THE CASE

From the foregoing it will be noted that there is involved a question of the utter inconsistency of the verdict, the jury having said, as to the first count, that the defendant was guilty of having made the sale; whereas, under identical facts and circumstances, covering the identical transaction, the jury with equal conviction said, as to the second count, that the defendant was not guilty. The question of the irreconcilability of such a verdict was raised by Motion in Arrest of Judgment (Tr. 3, 4) which was denied. (Tr. 5) Confirmation of this is found in the Court's instruction to the jury. It is conceded that if a sale was made, no proper written order was received from Etheridge and that in the particular transaction there did not exist the relationship of doctor and patient between appellant and Etheridge. It is further conceded that if the sale of morphine was made, it was not in nor from the original stamped package containing it. The Court, therefore, was correct in instructing the jury in the following particulars, after rehearsing the evidence of former relations between appellant and Etheridge in California. (Tr. 11-15) The Court said:

> "You can not convict a man of things he did in California if he did the things that Etheridge said he did, but the question here presented is; 'Did he do the things that are said to have been done on Washington Street between Third and Fourth Avenues in Phoenix?'" (Tr. 15)

Again reviewing the relations between appellant and Etheridge in California, the Court said:

> "But the fact is that is not the question in this case. The charge is that at Phoenix the seventh

day of April one nine three five the defendant sold to Etheridge two hundred and forty-five one-quarter grain morphine tablets and not on a prescription and not in the ordinary course of his practice as a physician treating a patient." (Tr. 20, 21)

Referring to the second count of the indictment (on which appellant was found not guilty), we find this instruction: (Tr. 21)

"In the second count it is charged that the doctor at the same time and place sold to the witness Etheridge two hundred and forty-five one-quarter grains of morphine which were not then and there in the original stamped package nor taken therefrom."

And, same page:

"Before you can convict upon the second count of the indictment you must find that the container that is in evidence here, I think as government's exhibit number one, a little glass bottle, which nobody contradicts contained two hundred and forty-five one-quarter grain tablets of morphine, was sold by this defendant to Etheridge, or given to him, or exchanged with him so that he got it in Phoenix, and some time on or about the seventh day of April. So that is all there is in the second count. No one disputes the contention that the bottle is without stamp. No one suggests that it is a bottle that was taken from a bottle properly stamped. The presumption is that it was not, and the sole question in the second count is did the defendant sell or give or dispose or dispense or distribute to Etheridge the bottle in evidence here with its contents, morphine."

"And again, immediately following the above instruction: (Tr. 21, 22)

"On the first count which is identical so far as the substance in the charge is contained—it is charged that at the same time and place the defendant did sell, barter, exchange, and give away certain derivatives and the salts of opium, to-wit: morphine—and the court charges you that morphine is a salt and derivative of opium—to Etheridge not in pursuance of an order form—there is not suggestion of an order form—and not in the course of his practice as a physician."

Accentuating all of the foregoing, the Court put the restricted issues squarely before the jury, as follows: (Tr. 22)

"The doctor by his plea of not guilty has placed upon the government the burden of proving these things but the doctor himself goes upon the witness stand and declares flatly, and at the outset of the trial his attorney stated distinctly to you and me that he was going to fix the defense upon one point, that is, the fact that he did not do— that the defendant did not give this bottle of morphine to Etheridge the witness. And that is the theory upon which the case has been tried and argued."

There is also raised the question of the correctness of the Court's following instruction to the jury:

"Now I charge you if you believe that narcotic inspector Moore is entitled to full credit and believe that his story of this transaction is true, it is your duty to convict. On the other hand we have him supported by the testimony of two police officers here." (Tr. 28) to which instruction an exception was duly taken. (Tr. 31)

While not raised by any exception in the Court below, appellant questions the correctness of and has assigned as error (Tr. 7) that part of the instruction given by the Court as to the legal effect of evidence of good character offered by the appellant. (Tr. 26) No quarrel is made with the introductory part of the charge beginning on Transcript page 26,

> "Now the defendant in this case has seen fit to produce testimony here of people by depositions whom you have not seen, that he is a man of good repute in the community in which he has lived."

and ending with the fourth line of page 27 of the Transcript,

> ". . . and if a reasonable doubt is created as to the defendant's guilt, he is entitled to an acquittal."

The question of error is based upon the adoption by the Court in his instructions of a charge which the presiding Judge stated was often said by Judge George M. Bourquin, his predecessor. The objectionable language of the charge is found on page 27 of the Transcript and is as follows:

> "But reputation is merely built upon the fact that you have not been caught, and when you are caught it shows the character and destroys the reputation."

Appellant tendered his Requested Instruction No. 4 (Tr. 32) in effect, that the evidence must convince every juror of the defendant's guilt beyond reasonable doubt and that if it does not, the jury cannot convict the defendant. Instruction was refused and exception to the ruling duly taken. (Tr. 32)

SPECIFICATION OF ASSIGNED ERRORS RELIED UPON

There being but five Assignments of Error (Tr. 6-9), the last two of which (Tr. 9) raise but one question of law, to-wit: the inconsistency of the verdict, we here specify that the appellant will rely upon each of them.

ARGUMENT

Assignments of Error Nos. 4 and 5

Assignment No. 4:

"The Court erred in denying defendant's Motion in Arrest of Judgment based on the ground that the verdict was contrary to law in that the verdict of guilty on the first count of the indictment and not guilty on the second count were wholly inconsistent, each of the alleged offenses clearly appearing from the indictment and the instructions of the Court to be based upon an alleged single transaction, namely, one sale of one bottle of morphine to one person at one time and one place."

Assignment No. 5:

"The judgment pronounced and entered in this cause is contrary to law in this, to-wit, that said judgment was pronounced upon a verdict of guilty in the first count of the indictment, which verdict was wholly inconsistent with the verdict of not guilty on the second count of the indictment, in that, as appears from the indictment and the instructions of the Court, each of the alleged offenses in counts one and two was based on an alleged single transaction, namely, one sale of one bottle of morphine to one person at the same time and place."

As will be observed, each of these assignments raises the question of the inconsistency of the verdict. We fully recognize the correctness of the doctrine laid down in many cases that where different or additional evidence is required in one count than that required in another, a verdict of guilty on the one and not guilty on the other may be easily reconciled. That, however, does not exist in the present case. From what was disclosed to the jury from the Court's various statements, heretofore in this brief set out and to be found in Tr. 15, and which we accept as a correct statement of what the evidence disclosed as to these matters, it is most clear that the sole and only issue given to the jury and upon which the jury found its verdict was, to quote the exact language of the Court, "Did he (referring to appellant) do the things that are said to have been done on Washington Street between Third and Fourth Avenues in Phoenix?" This question was answered by the jury as to the first count, Yes, and as to the second count, No.

Again the Court, speaking of the second count, after stating the elements of the offense and showing the absence of any dispute as to whether the bottle of morphine was without stamp and telling the jury that the presumption is that it was not stamped and that no one suggested that it was a bottle taken from a bottle properly stamped, instructed them that "the sole question in the second count is, did the defendant sell or give or dispose or dispense or distribute to Etheridge the bottle in evidence here with its contents, morphine." (Tr. 21) To this question the jury by its verdict said No. Immediately thereafter, reverting to the first count, the Court

(Tr. 21, 22) told the jury that this first count was identical, so far as the substance in the charge is "contained" (no doubt the Court meant "concerned" instead of "contained")—and gave to the jury the additional element that in this first count the sale was "not in pursuance of an order form—and not in the course of his practice as a physician."

Again the Court (Tr. 22) correctly advised the jury of the one issue and one theory upon which the case was tried and argued, and when he used the language, "And that is the theory upon which the case has been tried and argued," he unquestionably meant the case as including not only the first but also the second count. And that theory was, upon the part of the government, that the defendant did make the sale to Etheridge, and upon the part of appellant, that he did not sell, give or dispose or dispense or distribute to Etheridge the bottle in evidence with its contents, morphine. Upon that theory, to reiterate, the jury by its verdict in the first breath said he did sell, barter, exchange or give away the morphine, and in the second and final breath said he didn't do any such thing.

This Court, in Rosenthal v. United States, 276 Fed. 714, declared the rule, and we think correctly, where the separate offenses contained in each of the two counts of the indictment arose out of the same transaction. In that case the defendant was acquitted of the charge of buying and receiving certain goods stolen from an interstate commerce shipment, with knowledge that they had been so stolen. He was convicted on the second count, charging possession of the same goods, with the same knowledge. The Court rightly held that if he did not buy or receive the goods, which the jury by its verdict as to the first count said was true, the defendant could not have been guilty of possessing the same goods where there was but one transaction. Keeping in mind that the facts in the instant case covered, as Assignment No. 5 said, "An alleged single transaction, namely, one sale of one bottle of morphine to one person at the same time and place," (Tr.9) the doctrine laid down in that case has more forcible application to the instant case,

The verdicts were even more contradictory than those in the Rosenthal case. There the jury said said that while the defendant did not buy and did not "receive," he nevertheless "possessed." True, one can hardly receive anything and not at the same time, at least momentarily, possess it. Nevertheless, there is a distinction, however nebulous, between the two words. In the present case the issue was narrowed by the Court, to the exclusion of all others, to the one question, Did the defendant sell the morphine to Etheridge, or did he not?

The jury gave two verdicts that absolutely forbid any reconciliation, either in law or common sense. We can appreciate these verdicts under a different condition of the record as being reconcilable. The jury might well have found him guilty by reason of the absence of the written order, and the jury might equally well have acquitted him on the second count where the controlling element, in addition to the sale, was, not the absence of a written order, but the fact that the morphine sold "was not then and there in nor from the original stamped package." But if there was a sale, each of these elements, distinguishing the second from the first count, was present, as is readily seen from the Court's positive instructions to the jury that they were, and from the appellant's sole defense that there was, in fact, no sale made by him to Etheridge on that date under any circumstances.

In the case of Pankratz Lumber Co. v. United States, 50 Fed. (2d) 174, the defendant company was indicted separately from its officers. In each indictment there were two counts, charging falsification of income tax returns for 1927-28. The acts and transactions charged in each indictment are substantially the same. The indictments, by agreement, were consolidated and tried together. The company was convicted on count two—falsification of the 1928 return, and the individuals, George and John Pankratz, who were named in the other indictment, were acquitted on both counts. The appellant raised the question of inconsistent verdicts. District Judge Neterer, who wrote the decision, upheld the verdicts, holding that the company could not complain because the individuals were acquitted, even though the transactions were the same, but he said this:

"The theory opposing inconsistent verdicts is no doubt former acquittal. If A is charged in several indictments with the same transactions and acts, and these indictments are consolidated and he is acquitted on one count and convicted on another for the same transactions and acts, to permit the inconsistent verdict to stand would punish him for acts and transactions of which the jury had acquitted him. The finding of not guilty becomes res adjudicata as to A; but B, on the same charge, may not on conviction invoke a not guilty verdict as to A as res adjudicata. In the instant case, irrespective of any view the Court may have as to inconsistent verdicts, the appellant may not claim that the verdict of not guilty as to the officers is inconsistent as to it and therefore, res adjudicata. There is no intimation that the evidence is insufficient to convict appellant; ample evidence is disclosed by the record." (P174)

In Baldini v. United States, 286 Fed. 133, Judge Ross (who also wrote the opinion in the Rosenthal case, supra) again declared the same rule in these words:

> "Counsel for the government rightly concede that, if the two counts related to the same transaction, the position taken on behalf of the plaintiff in error is valid. It was so held by this court in the recent case of Rosenthal v. United States, 276 Fed. 714, and is the *well-established law.*"
>
> (underscoring ours) P 134

The only distinction between the two cases was one of fact. The Rosenthal case was reversed because the Court found that there was but one transaction. It affirmed the Baldini conviction because, as Judge Ross further said, that while it was true that in each count the offense was alleged to have been committed at Reno, the second count charged that the offense therein described was committed at a specified place in the City of Reno, to-wit: the Tuscano Hotel, which hotel was not mentioned in the first count; that the record in the case, according to the opinion, showed a conviction of appellant's co-defendant Gemignani on the possession count and the appellant's acquittal on that count. The record also affirmatively showed that no mention whatever was made of Gemignani as to the second count, charging the common nuisance upon which appellant was convicted; further, that the record failed to disclose that the two alleged offenses did relate to the same transaction and the Court would not, in the absence of a showing, presume that they did so.

Again, in Lambert v. United States, 26 Fed. (2d)

773, this Court, through Judge Rudkin, held that a verdict of guilty on a liquor nuisance count was not inconsistent with a verdict of not guilty on possession or sale counts of earlier dates.

Again, in Hesse v. United States, 28 Fed. (2d) 770, this Court, speaking through Judge Rudkin, said that it was apparent from the record that the jury was satisfied beyond a reasonable doubt that the appellant sold the liquor, as charged, but was not satisfied as to who possessed the liquor before the sale; and that "one may obviously possess without selling and one may sell and cause to be delivered a thing of which he has never had possession; or one may have possession and later sell, as appears to have been done in this case." We have no quarrel with that doctrine. The verdict could well be reconciled on the non-concurrence of the sale and the possession. No reconciliation could be had in the instant case.

Again, in United States v. Anderson, 31 Fed. (2d) 436, 437, this Court, speaking through District Judge Bean, held there was no inconsistency in a verdict of guilty of the defendant on a conspiracy count to violate the National Prohibition Act and the acquittal of the defendant on the remaining two counts charging, respectively, carrying on the business of a distiller without giving bond, and the making and fermenting of mash fit for distillation of spirits not in a distillery authorized by law. That case and the instant case are readily distinguishable. None of the three overt acts charged, any one of which, if proven, would complete the crime of conspiracy, is necessarily conclusive of the direct violations charged in counts two or three of the indictment, for neither the possession of mash fit for distillation, nor the operation of a still, nor the possession of intoxicating liquors, is in itself an offense, however proper each of them might be as evidence tending, with other necessary acts, to prove that the defendant was carrying on the business of a distiller without giving bond, or that he was making and fermenting mash fit for distillation, not in an authorized distillery. The direct offenses charged could very well have been committed after the crime of conspiracy. Such a situation is, beyond any question, absent from the instant case. There was but one transaction; either Etheridge received the morphine from appellant and immediately paid him the money, or he did not. The whole transaction must have been accomplished within a few minutes.

All of the foregoing cases were referred to by this Court in Dunn v. United States, 50 Fed. (2d) 779, and the most recent of them, Pankratz Lumber Co. v. United States, supra, is cited. As stated supra, the quoted language of Judge Neterer fits the instant case beyond cavil. The charge in the two counts of the instant indictment covers the same transactions and acts. If, as happened, the defendant

". . . is acquitted on one count and convicted on another for the same transactions and acts, to permit the inconsistent verdict to stand would punish him for acts and transactions of which the jury had acquitted him." (50 Fed. (2d) 174)

As Judge Neterer continued, "The finding of not guilty becomes res adjudicata as to A." There was no defendant B in the instant case.

Assignment of Error No. 1

"The Court erred in charging the jury, as follows:

'Now I charge you if you believe that narcotic inspector Moore is entitled to full credit and be-

lieve that his story of this transaction is true,
it is your duty to convict. On the other hand
we have him supported by the testimony of two
police officers here.'" (Tr. 7, 28)

The vice of this charge is apparent in the Court's singling out and accentuating the credibility of the testimony
of Narcotic Inspector Moore and the two police officers
and over-looking the application of the law to the testimony of the appellant and his supporting witnesses. We
submit they were entitled to the same charge but did not
get it. The Court (Tr. 27, 28) gave scant consideration
to what the appellant, in part, said in his own behalf.
Proceeding, (Tr. 28) the jury was told who Moore was
and while it is true he was "employed by the government
to perform his duties, not to catch men who are not guilty
of crime, but men who are guilty of crime," which is but
stating a truism, that fact puts him on no greater pedestal
for veracity than the appellant and his witnesses, and for
that reason the test should not vary. The exclusive use
of it as to Inspector Moore obviously resulted in prejudice to the defendant. It was argumentative in building
up the character of Inspector Moore, without any proof
of his good character, and assumed that such a man
would hardly tell an untruth to the jury; whereas, it, impliedly as to this assignment and expressly in an assignment to follow, left the jury to understand that as to appellant, the least said the better.

Upon exception being taken to this instruction (Tr.
31) the Court, being given full opportunity to apply that
instruction equally to appellant and his witnesses, not
only declined to do so, but repeated and elaborated upon
the instruction and reiterated its application to inspector
Moore and officer Moore and informer Etheridge. And
the Court's concluding language (Tr. 32):

> "In other words, the testimony of one witness
> who in your opinion is telling the truth and who
> is entitled to full credit is sufficient for the proof
> of any fact in this case,"

did not cure the error, but undoubtedly influenced the
jury in the belief it had application only to the named
witnesses and no others.

In Meadows v. United States, C.C.A. District of Columbia, 82 Fed. (2d) 881 et. seq., the well recognized
principle is again laid down to the effect that an instruction which singles out and declares the effect of certain
facts, without consideration of other modifying facts,
constitutes prejudicial error. The charge (p. 884) that
assault with bare fists only does not of itself justify
one, in repelling such assault, to restore to the use of a
dangerous or deadly weapon in a manner calculated to
produce death or serious bodily harm was followed by
this: (p. 885)

> "There must be something more than such an
> assault with fists. There must be the honest be-
> lief, on the part of the defendant, that death or
> serious bodily harm was imminent at the hands
> of the decedent."

The Circuit Court of Appeals held (p. 885) that while
this contained the implication that under such circumstances, despite the use of bare fists only by Jordan, there
might be lawful resort to a dangerous weapon by the
defendant, it was "unaided by any substantial mention
or enumeration of the various items of testimony which,
if believed by the jury, supported the defendant's
theory . . ."

The singling out in that case, which required reversal,
seems to us less objectionable than the singling out of
which we complain. The appellant and his witnesses

had as much right to an instruction as did inspector Moore, whom the Court gave the jury to understand was "employed by the government to perform his duties, not to catch men who are not guilty of crime, but men who are guilty of crime." And in view of the appellant's previous good repute prior to the transaction involved, as will be shown in the next assignment, we submit he was more entitled to such an instruction than inspector Moore, whose employment gave him no such sacrosanct character merely because he was a governmental employee.

The above case cites Weddel v. United States (C.C.A. 8th) 213 Fed. 208, wherein a requested instruction, similar in character, was properly refused.

Assignment of Error No. 2:

"The Court erred in charging the jury, as follows:

'Now the defendant in this case has been fit to produce testimony here of people by depositions whom you have not seen, that he is a man of good repute in the community in which he has lived. He has a right to do this, and because the defendant has introduced testimony tending to show his previous good character as to being a law-abiding citizen and has conformed with the ethics of his profession and the standards of practice, the court instructs you that it is competent for a person accused of crime to prove as a circumstance in his defense that his previous character as a law-abiding citizen was good and that he conformed with the ethics and standards in the practice of his profession. Previous good character is not of itself a defense but it is a circumstance to be considered by the jury in connection with all of the other evidence in the case, in determining the guilt or innocence of the accused. It is a circumstance which may be shown for the purpose of rebutting the presumption of guilt arising from circumstantial evidence. It may be used as tending to show that men of such character would not be likely to commit the crime charged. It should be taken and used in consideration as to whether other evidence is conclusive or inconclusive, and it is for the jury to determine under all of the facts and circumstances in the case what weight should be given to such evidence, when it is considered with the other evidence, and if a reasonable doubt is created as to the defendant's guilt, he is entitled to an acquittal.

'In connection with that instruction I want to say to you what George M. Borquin, my predecessor in office, often said: "That character is one thing and reputation is what others believe you to be. Character would prevent the commission of a crime of this kind, but reputation is merely a circumstance tending to show that it is not probable that a man with a reputation of that kind could do the things that are charged to have been done here. But reputation is merely built upon the fact that you have not been caught, and when you are caught it shows the character and destroys the reputation.' "

While, under the ruling, we quote the whole instruction, we single out the concluding language, which the Court attributed to and adopted from "George M. Borquin, my predecessor in office." It is:

"But reputation is merely built upon the fact

that you have not been caught and when you are caught it shows the character and destroys the reputation." (Tr. 8)

Why no exception was taken to that instruction by he appellant's trial lawyer, we do not know. We in-voke the doctrine enunciated by Courts and thus ex-pressed by the Supreme Court of the United States:

"In exceptional circumstances, especially in criminal cases, appellate courts, in the public interest, may, of their own motion, notice errors to which no exception has been taken, if the errors are obvious, or if they otherwise seriously affect the fairness, integrity or public reputation of judicial proceedings. See New York C. R. Co. v. Johnson, 279 U. S. 310, 318, 73 L. ed. 706, 710, 49 S. Ct. 300; Brasfield v. United States, 272 U. S. 448, 450, 71 L. ed. 345, 346, 47 S. Ct. 135." United States v. Atkinson, 80th Law ed. p. 555 at 557.

297 U. S. 157-160

From whomsoever such an instruction was taken or adopted, and from what source his predecessor in turn got it, we submit that it is not only not a correct state-ment of the law on the effect of good reputation, as es-blished by character witnesses, but utterly annihilates he true doctrine which the Court had theretofore in the instruction fairly stated. Admitting the distinction made between character and reputation, we submit that repu-ation is not, as the Court said it was, "merely built upon he fact that you have not been caught." It is built pon the daily life of a man and his contact with his ellow citizens in the community in which he has lived ufficiently long to have that reputation established, and ts kind, as to whether good, bad, or indifferent, is to be,

as it was in this case, determined by the sworn and uncontradicted testimony of those qualified, by reason of
their frequent contact with him in the affairs of life. The
instruction shows crystal clear that the Court was ursur-
ping in advance the functions of the jury to say whether
or not the appellant had been caught and was saying
that in fact he had been caught and that, therefore,
there was nothing for the jury to do but to disregard all
of the testimony, including that of the character witnesses given in his behalf and to re-echo by its verdict
what the Court had told it—that he had been caught.
The error is so gross and prejudicial that we ask this
Court, even though no exception was taken, to notice it,
because, we submit, it does affect the fairness of judicial
proceedings.

Assignment of Error No. 3:

> "The Court erred in refusing to give to the
> jury defendant's requested instruction No. 4,
> which is as follows:
>
> 'The Court instructs the Jury that before you
> can convict the defendant, the evidence must be
> so strong as to convince every juror of the de-
> fendant's guilt beyond a reasonable doubt; and
> 'if, after considering all the evidence, a single
> juror has a reasonable doubt of the defendant's
> guilt arising out of any part of the evidence, then
> you cannot convict the defendant."

We submit that this requested instruction was proper
and its refusal constituted reversible error. It was not
covered by any other instruction. Had the Court given
an instruction that it took concurrence of all twelve of
the jurors to render a verdict, we could not well argue
the correctness of this assignment, but he did not do
so, and in the absence of such instruction we were en-

titled to the one submitted. State v. Swain, 156 So. 162. (180 La. 20)

In State v. Magnuson, 202 N.W. 638, (48 S.D. 112) a refusal to give an instruction, substantially as the one here requested, was held not prejudicial in view of all instructions that jurors were not mislead as to their duties to acquit, if any of them entertained a reasonable doubt. No such instruction was given in this case.

In State v. Edgell, 118 S.E. 144 (94 W. Va. 198) an instruction on the presumption of innocence, charging the jury that it was its duty to give defendant the full benefit of the presumption and to acquit him unless they felt compelled to find him guilty by the law and the evidence, but convincing them of his guilt beyond a reasonable doubt, was held not to cover a requested instruction that defendant must be convicted beyond a reasonable doubt, and if any juror, after due consideration of the evidence and consultation with his fellows, has such doubt, he should acquit the defendant.

In People v. Dole, 55 P. 581 at 585, the Court said that the defendant was entitled to an instruction that if after consideration of the whole case, any juror should entertain a reasonable doubt of the guilt of the defendant, it is the duty of such juror * * * not to vote for a verdict of guilty, nor to be influenced in so voting, for the single reason that a majority of the jury should be in favor of a verdict for guilty. And the refusal to give such a verdict was one of the grounds for reversal.

In Blashfield's Instructions Juries, Second Edition, Vol. 2, p. 1947, Sec. 2428, many approved instructions are given, showing the propriety of many instructions similar to the one requested and rejected here, that "each juror must be separtely and segregately so satisfied to support a conviction." (p. 1948)

Again, ". . . and if after considering all the evidence a single juror has a reasonable doubt of the defendant's guilt, arising out of any part of the evidence, then they cannot convict him." Again, the Court instructs "the jury that the defendant cannot be convicted in this case unless each and every juror is . . . satisfied from the evidence, and the evidence alone, beyond all reasonable doubt, and to a moral certainty, of his guilt." (p. 1949)

Similar instructions are found (p. 1950), ". . . the law contemplates the concurrence of the twelve minds of the jury in the proposition of guilt before conviction can be had." And again, (p. 1951) "Each juror must for himself find that the defendant is guilty beyond a reasonable doubt." And, same page, ". . . and so long as you, or any one of you, have a reasonable doubt . . ."

As the Court, in its general instructions, failed to instruct the twelve jurors they must agree, or that there must be unanimity before a verdict of guilty could be rendered, we respectfully submit that he should have given our defendant's Requested Instruction No. 4, and its refusal was reversible error.

CONCLUSION

For all of the foregoing reasons, we respectfully submit that the judgment of conviction on the first count of the indictment should be reversed.

Respectfully submitted,

George F. Macdonald

Attorneys for Appellant.

Circu

3

CHARLES O. L

UNITED STATI

Upon Appeal from

for

APPELLEE'S
ASSIG

The United ;
by F. E. Flynn,
trict of Arizona.
record the assig
grounds and fo:

───
*Unless otherwise in
of the printed tra

No. 8200

IN THE

United States
Circuit Court of Appeals
For the Ninth Circuit

CHARLES O. LONG,
> *Appellant,*

vs.

UNITED STATES OF AMERICA,
> *Appellee.*

**Upon Appeal from the District Court of the United States
for the District of Arizona**

APPELLEE'S MOTION TO STRIKE OUT
ASSIGNMENTS OF ERROR

The United States of America, appellee herein, by F. E. Flynn, United States Attorney for the District of Arizona, moves the Court to strike from the record the assignments of error (6-10)*, upon the grounds and for the reasons following, to-wit:

---

*Unless otherwise indicated, figures in parentheses refer to pages of the printed transcript of record.

I.

That no assignments of error were filed at the time of the taking of the appeal, nor within thirty days thereafter.

II.

That the assignments of error were not filed during the same term in which judgment was rendered.

The record in this case shows that the jury returned a verdict of guilty on February 29, 1936 (3). Defendant was sentenced March 4, 1936 (5-6). Notice of appeal was dated March 6, 1936 (37-38). Assignments of error were filed October 27, 1936 (10).

Respectfully submitted,

F. E. FLYNN,
    United States Attorney.

POINTS AND AUTHORITIES IN SUPPORT OF APPELLEE'S MOTION TO STRIKE OUT ASSIGNMENTS OF ERROR

This motion should be granted because appellant failed to comply with the provisions of Rule IX of the Supreme Court governing criminal appeals.

Bradford J. St. Charles v. United States,
    86 F. (2d) 463 (C.C.A.9).

F. T. Carey and E. H. Williams v. United States,
    86 F. (2d) 461 (C.C.A.9).

Slade v. United States, 85 F. (2d) 786.

The verdict was returned (3) and judgment passed (4-6) in the October, 1935, term of the District Court. The assignments of error were not filed until the October, 1936, term of the District Court (10).

28 *U.S.C.A.* 143.

Respectfully submitted,

F. E. FLYNN,
            United States Attorney.

BRIEF FOR APPELLEE

We shall discuss appellant's assignments of error in the numerical order in which they appear in the transcript of record.

ASSIGNMENT NO. 1

This assignment is based upon the following charge to the jury:

"Now I charge you, if you believe that narcotic inspector Moore is entitled to full credit and believe that his story of this transaction is true, it is your duty to convict. On the other hand, we have him supported by the testimony of two police officers here."

The exception to this instruction at the trial is not sufficient. We quote the exception:

"MR. LEWKOWITZ: If I caught one statement —I did not interrupt your Honor—I have an exception to enter to the statement, if I am correct upon it, when your Honor says, "I charge you if you believe narcotic agent Moore, you should convict". (31).

This Court has said:

"Merely identifying the statement by saying "the remarks of the court with reference to Insull as touching the proposition of character and good reputation", without stating the ground of the objection to the instruction, is insufficient to raise the question of error in giving the instruction, if any error was committed."

Baldwin v. United States, 72 F. (2d) 810-815.

The exception taken at the trial is not set out

in the assignment. Therefore, the assignment does not comply with Rule XI of this Court, and should not be considered.

> *Baldwin v. United States*, 72 F. (2d) 810-815, and authorities cited therein.

The instruction in question correctly states the law.

In the case of *Meadows v. United States*, 82 F. (2d) 881, cited by appellant (p. 15, appellant's brief), the instruction complained of singled out one item in the evidence which, in itself, was not sufficient to determine the guilt or innocence of the defendant. In the present case, the guilt or innocence of defendant depends on whether or not he delivered the narcotics to the witness Ethridge, as testified to by Ethridge and Government witness Moore. This is the transaction referred to in the Court's instruction, where the Court said:

"Now I charge you, if you believe that narcotic inspector Moore is entitled to full credit and believe that his story of this transaction is true, it is your duty to convict." (28).

The Supreme Court of the United States has said that it is not error for the Court to comment on the consideration to be given certain parts of the evidence.

> *Hyde v. United States*, 225 U.S. 347-381.

After the attempted exception to the instruction,

the Court explained the instruction further. After this explanation, there was no exception taken (32).

ASSIGNMENT NO. II

This assignment is based upon the Court's charge to the jury covering the evidence given at the trial to the effect that the defendant was a man of good reputation and had conformed with the ethics of his profession and the standards of practice (27). No exception was taken to this instruction at the time of the trial. Therefore, the assignment should not be considered.

Baldwin v. United States, 72 F. (2d) 810-815.

In the *Baldwin case*, supra, an instruction in substantially the same words as the one in question was sustained by this Court.

ASSIGNMENT NO. III.

This assignment is based upon the Court's refusal to give to the jury defendant's requested instruction No. 4, which was as follows:

"The Court instructs the jury that before you can convict the defendant, the evidence must be so strong as to convince every juror of the defendant's guilt beyond a reasonable doubt; and if, after considering all the evidence, a single juror has a reasonable doubt of the defendant's guilt arising out of any part of the evidence, then you cannot convict the defendant."

The substance of this instruction was contained in the instruction given by the Court. We quote from the instruction:

"When the defendant puts a plea of not guilty to the charges contained in the indictment in this case upon the record he places upon the government the burden of proving his guilt beyond a reasonable doubt. (22)

" * * if after impartial comparison and consideration of all of the evidence you can candidly say that you are not satisfied of the defendant's guilt, you have a reasonable doubt. In other words, if after fairly and impartially considering all of the evidence in the case, not what you would like to find, not what your sympathy would lead you to find, but what your reason dictates your finding should be, can you say that you are not satisfied of the defendant's guilt. If you can, you have a reasonable doubt and should not convict. (23)

"And in determining whether you have a doubt in this case such as would justify an acquittal or whether the government has proven its case beyond a reasonable doubt, that is, to the point which will justify a conviction, you must have in mind the fact that the law presumes the innocence and not the guilt of the defendant. That is the starting point in every criminal case, a presumption of innocence. This presumption of innocence continues throughout the trial and entitles the defendant to an acquittal unless the evidence in the case taken as a whole satisfies you of the defendant's guilt beyond a reasonable doubt as defined in these instructions." (23-24).

From the above excerpts, it clearly appears that the Court fully and fairly gave instructions on the question of reasonable doubt and the presumption of innocence. This being so, the trial court was not required to employ the language used in the requested instruction.

Dean v. United States, 51 F. (2d) 481-485 (C. C.A.9).

ASSIGNMENTS NOS. IV AND V
Both of these assignments are based on the contention that the verdict of guilty on the first count of the indictment was inconsistent with the verdict of not guilty on the second count and that, therefore, the judgment pronounced and entered was contrary to law and that the motion in arrest of judgment should have been granted.

The motion in arrest of judgment was not filed within three days after the verdict, as required by Rule II, subdivision 2, of Rules of Practice. The verdict was returned on February 29, 1936 (3), and the motion in arrest of judgment was filed March 4, 1936 (4). The time limit fixed by the Rules is jurisdictional.

Fewox v. United States, 77 F. (2d) 699.

Consistency in the verdict is not necessary.

Dunn v. United States, 284 U.S. 390-393.

Lee Choy v. United States, 293 Fed. 582 (C.C. A. 9).

Yep v. United States, 81 F. (2d) 637.

The first count upon which defendant was found guilty charges defendant with having sold morphine to one, Elmer Ethridge, *not in pursuance of a written order * * the said Elmer Ethridge not being then and there a patient of the said Charles O. Long and the said morphine was then and there dispensed and distributed by the said Charles O. Long not in the course of his professional practice only.*

In the second count, upon which the defendant was found not guilty, the defendant was charged with selling morphine which was *not then and there in nor from the original stamped package.*

The elements of the two crimes are entirely different, and Judge Rudkin, of this Court, said in *Lee Choy v. United States*, supra:

"Under such circumstances, the verdict of not guilty as to one count is not inconsistent with the verdict of guilty as to the other."

In proving the charge contained in the second count, it would be necessary to prove that the morphine sold was not in nor from the original stamped package. Proof of this fact is not necessary as to the first count.

This Court does not have the evidence before it, but even if we were to concede that at the trial it was undisputed that the morphine was not in nor from the original stamped package, there would be no error in the verdict of guilty.

In the case of *Dunn v. United States*, supra, in which the Supreme Court affirmed the decision of this Court, the Supreme Court quoted with approval from the case of *Steckler v. United States*, 7 F. (2d) 59-60:

"The most that can be said in such cases is that the verdict shows that either in the acquittal or the conviction the jury did not speak their real conclusions, but that does not show that they were not convinced of the defendant's guilt. We interpret the acquittal as no more than their assumption of a power which they had no right to exercise, but to which they were disposed through lenity."

CONCLUSION

We respectfully submit:

1. That the assignments of error are not properly before this Court.

2. That this Court, not having the evidence before it, cannot say there was prejudicial error, of which it can take notice, that would warrant reversal.

3.   That the instructions correctly state the law.

Therefore, the judgment should be affirmed.

Respectfully submitted,

F. E. FLYNN,
United States Attorney.

I N D E X

15

IN THE

United States
Circuit Court of Appeals
For the Ninth Circuit.

CHARLES O. LONG,

Appellant,

vs.

UNITED STATES OF AMERICA,

Appellee.

ANSWER TO APPELLEE'S MOTION TO STRIKE OUT ASSIGNMENTS OF ERROR AND REPLY BRIEF OF APPELLANT

Upon Appeal from the District Court of the United States for the District of Arizona

FILED

MAR 6 - 1937

GEORGE F. MACDONALD,
Attorney for Appellant.

PAUL P. O'BRIEN,
CLERK

BOWER COMPANY, INC., PHOENIX, ARIZONA

IN THE

United States
Circuit Court of Appeals
For the Ninth Circuit.

CHARLES O. LONG,

Appellant,

vs.

UNITED STATES OF AMERICA,

Appellee.

ANSWER TO APPELLEE'S MOTION TO STRIKE OUT ASSIGNMENTS OF ERROR AND REPLY BRIEF OF APPELLANT

Upon Appeal from the District Court of the United States for the District of Arizona

GEORGE F. MACDONALD,
Attorney for Appellant.

BOWER COMPANY, INC., PHOENIX, ARIZONA

INDEX

TABLE OF CASES

ANSWER TO APPELLEE'S MOTION TO STRIKE

There is no merit in the first ground (Appellee's Brief, p. 2) that no Assignments of Error were filed at the time of the taking of the appeal, nor within thirty days thereafter, because,

1. There were assignments filed at the time of the taking of the appeal and they may be found in the very instrument which constitutes the taking of the appeal, to-wit, the Notice of Appeal (Tr. 38), filed in the District Court two days after the date of judgment, pursuant to Rule III of the Supreme Court. These Grounds of Appeal are clearly recognized by the Rules of the Supreme Court as Assignments of Error. Rule III requires that they shall be contained in the Notice of Appeal and designates them as "a succinct statement of the grounds of appeal." If further assignments are to be asserted, Rule VIII says they "may amplify or add to the grounds stated in the notice of appeal." And where further assignments are based upon claimed errors contained in the Bill of Exceptions, Rule IX says, "Within the same time (referring to the time of filing the Bill of Exceptions) the appellant shall file with the clerk of the trial court an assignment of errors of which appellant complains." Certainly an assignment once stated in the grounds attached to the Notice of Appeal need not, for its preservation, be reiterated in future assignments. And these Grounds of Appeal raise the same questions as were later raised in the additional or amplified assignments (Tr. 6-10), particularly Assignments No. 4 and 5 (Tr. 9) and Assignment No. 1 (Tr. 7).

2. There is no law or rule requiring the Assignments of Error to be filed within thirty days after the

taking of the appeal, and Rule IX of the Supreme Court, which is invoked by appellee (Brief, p. 3), puts no such arbitrary limit upon the appellant. The Assignments of Error were filed within the same time as the Bill of Exceptions and they merely reiterate and amplify the Assignments of Error heretofore set out as the Grounds of Appeal and attached to the Notice of Appeal; and, as to them, that is all Rule IX requires. (Tr. 10 and Tr. 34).

The cases of Cary, et al, v. United States, 86 Fed. (2d) 461, and St. Charles v. United States, idem p. 463, cited by appellee, have been carefully examined. We confine the argument to the requirements of time and the filing of Assignments of Errors. In the first of these cases the Court said (p. 462), "Cary did not obtain an order extending the time within which to file his assignments of error until May 9, 1935," and (same page) "With respect to appellant Cary, the time within which he might file his assignments of error expired at the end of 30 days after the appeal was taken, for there was no order extending that time. The assignments of error were not filed within that time."

In the St. Charles case (p. 463) the Court said, "The assignments of error were not mentioned in this order," (speaking of the first order of extension respecting the conviction on the first count) and speaking of the sentence on the second count and the first order extending the time to file Bill of Exceptions, the Court said, "No mention was made of assignments of error in this order."

We read the above quoted language of the Court in the Cary case to mean that, aside from whether Cary did get a proper order within thirty days extend-

ing the time to file his Bill of Exceptions, he did not
(having failed to file his Assignment of Errors within
the thirty days granted by the rule) in the same, or a
separate order, also get an extension of time within
which to file his Assignment of Errors, a motion to
strike the Assignment of Errors must prevail. In the
St. Charles case the Court, interpreting criminal rule
IX, said (p. 464), "It means that an order is ineffectual to extend the time within which a bill of
exceptions and *assignment of errors* (italics ours) may
be filed, unless such order is entered within 30 days
after the appeal was taken," and cited the Cary case
in support of that interpretation.

With all due respect to this Court, we beg leave
to submit, in so far as that ruling has exclusive reference to Assignment of Errors, the Court's interpretation is foreign to both the letter and the spirit of Rule
IX, and adds a harshness not intended by that rule.
The rule that the Bill of Exceptions must be filed
within thirty days, unless within that period an extension is granted, is obviously clear. Thus far nothing
in the rule is said that Assignment of Errors fall within
the same limitation, and it is equally clear that two
periods of time are thus far mentioned, the original
thirty day period of time and, if the order is secured in
due time, the additional time which the Court may grant,
each having exclusive reference to the Bill of Exceptions. Obviously also, if no extension is granted, not
only the Bill, but also the Assignment must be filed
within the original thirty day period of time. But when
additional time is given to file the approved Bill of
Exceptions by an order made within the original time,
then the subsequent language of the rule, "Within the

same time (italics ours) the appellant shall file with the clerk of the trial court an assignment of the errors of which appellant complains," can not in reason find a limited ascription to the first mentioned time to the resultant arbitrary exclusion of the second equally mentioned time.

The language "within the same time" must mean that the Assignment of Errors follow the Bill of Exceptions and must be filed within the "same time" as the former, be that time the original thirty day period or the time allowable under the order of extension as to the Bill. Admitting that these rules are intended to accelerate the procedure on appeal in criminal cases, what good purpose could be served by requiring the Assignment of Errors to be filed within the Procrustean fixity of thirty days, there to peacefully slumber until a future day, be it long or short, when the Bill, having been settled and approved, would share its company, both to be simultaneously transmitted by the clerk to this Court "together with such matters of record as are pertinent to the appeal." (same rule).

If such be the rule, appellant would have to forego assigning any of the errors discovered after the Assignment was filed, arising in the Bill of Exceptions, for the Assignment may be filed long in advance of the settling and filing of his Bill of Exceptions. In fairness he ought to be permitted to add such meritorious assignments as an inspection of his Bill justifies; and to do that the work in the preparation of each of them should be concurrent so that both could be filed at the same time.

So with no express requirement, under Rule IX, that the Assignment must be filed within the original thirty day period, unless an extension of time is granted, and it being quite clear that the *time* within which the Bill must be filed may be either the original period of thirty days or an extended period, it is, we submit, very unreasonable to insist that the words "within the same time" (having reference to the Assignment) must be narrowed to the original thirty day period, event though an extension of time has been granted for the filing of the Bill, and equally unreasonable to insist that these same words cannot be interpreted to include the additional period of time.

Applying the well known principle that where the reason for the rule ceases, the rule itself ceases and having shown that there is no reason for a construction that obviously cannot make for haste where the record on appeal contains assignments necessarily predicated upon errors only to be found in the Bill of Exceptions, we submit that the view herein expressed is the correct one and in the interest of justice and fairness to appellant ought to prevail.

There is the same lack of merit in appellee's second ground, implying that the assignments must be filed during the same term in which judgment was rendered. Counsel for appellee cites 28 U. S. C. A. 143, providing for the commencement of terms in various places within the District of Arizona. New terms beginning for the transaction of matters arising since the last term are not conclusive of the termination of the former term in specific cases, and for specific purposes in those cases, where the appeal, as it was in this case, has been perfected in the judgment term. Assume that the date of the verdict, February 29, 1936, was the last day of the October, 1935, term; Rule II, subdivision 2, gives the defendant three days to file motions in arrest of judgment, or for a new trial; necessarily, they would be filed at a new term, but certainly, under this Rule II, the former term is automatically extended for the purposes stated, otherwise the defendant would be cut off from the right thus given, and—which would compensate this loss—the court would have lost jurisdiction to impose sentence. Aside from all this, the Supreme Court, by its rules applying to criminal appeals, has said nothing about terms, but fixes its time when things are to be done in days.

In Slade v. United States, 85 Fed. (2d) 786, cited in Appellee's Brief, p. 3, the order of extension covered only Bill of Exceptions, was made in due time, January 4, 1936, and gave defendant until February 1, 1936. On the latter date he obtained another order extending his time to March 1, 1936. The Court held that this order was ineffectual. No doubt anticipating such a ruling, the appellant on February 8 abandoned the second order and secured a third, which attempted to amend the first by making the period of extension March 1 instead of February 1. The Court held this order equally ineffectual, in that this was not an order correcting a former actual order of the Court, which had been erroneously entered, but an order inserting the date which counsel for appellant had merely intended when the original order was made. In passing, it calls to mind the masterly (?) objection of a certain learned counsel, "Having failed to 'nunc' it then, he can't 'pro tunc' it now." There is nothing in the case to indicate other than that the Assignment of Errors, presumably filed at the same time as the Bill of Exceptions, went down with the latter, which was filed too late.

REPLY BRIEF

Issue was taken by appellee (Brief, p. 4) that the exception taken to the instruction at the trial (same page) is unsufficient. No quarrel is made there that the instruction is not properly assigned. The Assignment sets out the instruction totidem verbis and Rule XI, in effect at the time these assignments were filed, did not require, as the new Rule XI does, that the Assignment must include "the grounds urged at the trial for the objection and the exception taken." Baldwin v. United States, 72 Fed. (2d) 810-815, held that the instruction in that case was not properly assigned as error. As the Assignment is not given, a compromise between it and the Assignment in the instant case cannot be had. Nor, as we have said, is it necessary, as Appellee's Brief insists, that the exception taken at the trial must be set out in the Assignment, because old Rule XI, applicable here, did not so require, and the Baldwin case does not so hold.

The doctrine in the Meadows case (Appellant's Brief, p. 15), and discussed in Appellee's Brief, p. 5, still stands in its application to the instruction assigned as error. The instruction did single out one item in the evidence, namely, the testimony of Inspector Moore, and did tell the jury that if they believed him it was their duty to convict. (Tr. 7) This is more than a mere comment. It utterly disregarded all the other evidence in the case. We do not care what the transaction was (Appellee's Brief, p. 5) referred to in the Court's instruction. We are not complaining of transactions; we are complaining of what, we submit, was a vicious instruction in telling the jury that they could convict this appellant upon the testimony of one government witness, thus excluding from their minds the consideration of the testimony of appellant and his witnesses and, in effect, telling them that they did not have to consider it and that they must not believe it.

Reverse the language of the charge. Insert for "Narcotic Inspector Moore", "Defendant Long" and "it is your duty to convict" to "it is your duty to acquit" and we have a fair sample to the government's prejudice of an equally vicious charge, which the government, of course, could have no means of redressing, and while the result might be an acquittal of the defendant, he would no more be acquitted under the law than he was convicted under it by virtue of the erroneous instruction.

Counsel for appellee (Brief, p. 5) invoked the doctrine of estoppel against appellant because, forsooth, he did not also except to the further explanation of the Court after his exception was duly entered to the instruction complained of. The further explanation may be seen, Transcript pages 31, 32, and we have treated it in Appellant's Brief, pages 14, 15. The explanation (?) was no cure for the fatal error already committed and nowhere advised the jury that the same rule applied to the appellant and his witnesses and that in case they believed them it was their duty to acquit.

Assignment No. II

True, as we admitted, no exception was taken to this instruction, but under the authorities there cited, (Appellant's Brief, p. 18) we submitted, that it was such gross error that the Court ought to take notice of it. The only answer appellee suggests (Brief, p. 6) was another invocation of the Baldwin case, 72 Fed. (2d) 815, and counsel's conclusion—without a word from the Baldwin case showing wherein any instruction in that case was substantially the same as the one in question here. Where, in the Baldwin case, we ask them, are words substantially as follows: "But reputation is merely built upon the fact that you have not been caught . . ." To the contrary, we have the Baldwin case telling us that, "Now reputation is a valuable possession. It is what seems to the public at large," and is a positive declaration that reputation is something of substance and not, as the lower court substantially said, that is merely built upon the sands of ability, knowing that you are an utter rascal, to raise upon these sands an edifice of evasion and deception, so that your friends and acquaintances will never know that you are such a rascal. "Character," again says the Baldwin case, p. 815, "is what a man really is and reputation is supposed to reflect his character." Of course, as this Court said in the Baldwin case, the trial judge "has a right to inform the jury that a man may have a good reputation and yet commit crime," but that is a far cry from telling the jury that reputation, which this Court said is a valuable possession, has no value whatever because it is merely built upon the fact that you have not been caught.

Assignment No. III

Counsel for appellee impliedly admits that the refusal to give appellants requestd instruction No. 4 (Tr. 32) is reversible error unless the substance of the refused instruction was contained in the instructions given, and seeks to avoid the consequences by attempting to show that in the instructions given the substance of the refused instruction was incorporated. The instruction given upon which appellee relies are found in its Brief, p. 7, 8. From these three instructions it will

be plainly noted that the Court was dwelling solely upon
what constituted reasonable doubt, presumption of innocence and how that presumption must be overcome.
We re-assert and counsel for appellee has failed to point
out either in the three instructions noted in its Brief, or
in any other instructions, that the failure to instruct upon
this question of unanimity of the verdict was not cured
by any other instruction. Surely the repeated use of
the words "you" and "your" (in the instructions cited in
Appellee's Brief) cannot be construed as an instruction
that unanimity must be had, or that it took the concurrence of all twelve of the jurors to render a verdict,
or that every juror, as the requested instruction had it,
must be convinced of the defendant's guilt beyond a
reasonable doubt. And if this be so, then it was error
to refuse to instruct the jury as requested by the appellant, and the cases cited by us (Brief, p. 20-21) sustain
us in that contention.

The Dean case, 51 Fed. (2d) 481-485, cited by appellee (Brief, p. 8) supports the contention of appellee
on its erroneous hypothesis that appellant is quarreling
with the correctness of the Court's definition of reasonable doubt and presumption of innocence, but palpably
appellant is doing no such thing, and so the Dean case
has no application whatever to the question here, which
is solely one of required unanimity in the juries verdict.

Assignments No. VI and V

Counsel for appellee, again seemingly hard pressed
for a real answer to our contentions, attacks the Motion
in Arrest of Judgment as not filed within three days
after the verdict and again invokes the criminal rules.
From the record and from that part of Rule II, Subdivision 2, which he cites, he has built up an apparently

invincible situation. His dates are correct and with only Rule II, Subdivision 2 as a guide, we should confess utter failure, but counsel has simply failed to note other rules than II and XI, which he frequently invokes. If "the last shall be first" or as equally important, we note that the concluding paragraph of Rule XIII (O'Brien's 1936 Cumulative Supplement, p, 34), being the Supreme Court's tribute to hard working attorneys, and following the Biblical injunction, "Six days shalt thou labor but on the seventh thou shalt rest" and being mindful also of the Fourth Commandment, "Remember the Sabbath Day to keep it holy", excludes Sundays and legal holidays in computing the time "specified in the foregoing rules." Our 1936 calendar tells us that February 29, 1936, the date of the verdict, was Saturday, so, omitting March 1, which was Sunday, we had three working days left, March 2, 3, and 4, and the Motion in Arrest was filed on March 4, 1936. It was an oversight, of course, a mere Homeric nod.

Again as to these two assignments, appellant's contention that the verdicts were inconsisent is seemingly admitted the appellee's contention (Brief, p. 9) being that consistency in verdicts is not necessary.

We will briefly touch upon the cases cited by apappellee (Brief, p. 9). Judge Rudkin is quoted in Lee Choy v. United States, 293 Fed. 582, thus:

> "Under such circumstances, the verdict of of not guilty as to one count is not inconsistent with the verdict as to the other."

We take no issue with that, for under the circumstances of that case, it may well be true, but Judge Rudkin did

not say that if the verdict was inconsistent, it nevertheless should stand.

In the Yep case, cited by appellee, 81 Fed. (2d) 637,
the Tenth Circuit, while not conceding that the verdicts
of conviction and acquittal were inconsistent, did say
that if it were true, it would be no ground for reversal,
and cited Dunn v. United States, 284 U. S. 390 76 L.
Ed. 356 in support thereof.

This brings us to the Dunn case, which went from
this Court, 50 Fed. (2d) 779, to the above cited case, in
the United States Supreme Court. In that .case this
Court cited many of its former cases, which we fully
discussed in our Opening Brief, p. 8-13, the Court saying (p. 780) that "it is pointed out in the briefs and it
has been in a number of our decisions that there is a
conflict in the decisions of our Circuit Courts of Appeals upon the question of inconsistent verdicts," and
speaking for itself the Court said (same page) that "this
court has found it unnecessary to determine which line
of authority it would adopt". It (p. 781) stated that by
its subsequent decisions it modified its own decision in
Rosenthal v. United States, 276 Fed. 714, as that case
was explained in its later case of Baldini v. United States,
286 Fed. 133. It did not modify its more recent opinion,
Pankratz Lumber Company v. United States, 50 Fed.
(2d) 174, decided May 25, 1931, and discussed in our
Brief p. 13. And in the Dunn case itself, it held that
a verdict of guilty of maintaining a common nuisance
and keeping liquor was not, under the evidence, inconsistent with acquittal on counts of selling and unlawful
possession. It did not hold, as the Second Circuit held
in the Yep case, that where the verdict was absolutely

inconsistent as between counts, it would nevertheless uphold such a verdict.

On certiorari to the Supreme Court, that Court agreed with the contention of the government and this Court's decision that the verdict of guilty on the common nuisance count was not inconsistent with the verdicts of not guilty on the sale and possession counts. That was sufficient for an affirmance of this Court's decision. True, the Supreme Court did, for what reason we cannot assign, proceed with what we submit is pure dictum to hold that "consistency in the verdicts is not necessary", and on that score, we further submit that the learned, vigorous and lengthy dissenting opinion of Justice Butler is entitled to greater consideration than the brief dictum of the majority. Furthermore, the illustrations given by the Supreme Court in its dictum are not applicable to the case at bar. The correctness of the illustrations upon the theory of separate indictments, which the Court indulged in, may be conceded as sound, but only upon the same theory as the Court in its decision, aside from its dictum, stated. If there was no inconsistency in the verdicts of guilty on the common nuisance count and not guilty on the sale and possession counts, then it makes absolutely no difference whether the same jury tried it on the three counts, or whether three separate juries passed upon the several questions involved in each count. The jury trying the sales count may have acquitted the defendant on the same reasoning used by the Court in its decision. The jury trying the possession count may have reached the same verdict for the same reasons as the decision states, but the verdict of a third jury, convicting the defendant on the common nuisance count, in no sense decided the issues in the two previous trials and most certainly the defendant could not plead those acquittals as a bar either as once jeopardy or a former acquittal; and that situation cannot and did not arise in the instant case. Not to repeat our argument of the Opening Brief, the jury was not called upon to say separately whether the defendant sold the morphine, not in pursuance of a written order from Etheridge on the proper form, or whether the same sale was made to Etherdige, the morphine not being then and there in nor from the original stamped package. If the jury was so called upon to decide and it convicted because the sale was made without a proper order blank and acquitted because defendant had established that the morphine was in or from the original stamped package, we could have no quarrel with the verdict of conviction. But, as we have conclusively shown in our Opening Brief, no such separate elements as would justify such a result were submitted by the Court to the jury and the statements of the Court, also found in our Opening Brief, show positively that he submitted but one issue to the jury and that was, to quote, "Did he (referring to appellant) do the things that are said to have been done on Washington Street between Third and Fourth Avenues in Phoenix?" Tr. 15. And again, referring to the second count, "The sole question in the second conut is, Did the defendant sell or give or dispose or dispense or distribute to Etheridge the bottle in evidence here with its contents, morphine." Tr. 21.

The Steckler case, C. C. A. 2nd, 7 Fed. (2d) 59, 60, cited by the Supreme Court in support of its dictum, to our mind bears its own condemnation. It admits that in the acquittal or the conviction the jury did not speak its real conclusions and still,—remarkable to say,—one part of the verdict, that of conviction, was held to support a judgment and sentence. The Steckler opinion further says, "But that does not show that they were not convinced of the defendant's guilt." The answer to that is that it as much shows that they were convniced of his innocence as they were of his guilt. And the Court in the Steckler case had no more right to interpret the acquittal "as no more than their (the jury's) assumption of a power which they had no right to exercise" than the Court had the right to interpret the conviction as an equal assumption of a power which they had no right to exercise.

Language of Courts seems very impressive on its surface, but it is just as much subject to critical analysis as that of the decisions of administrative or executive bodies, and the reasoning of the Steckler case does not and can not stand under a fair analysis.

CONCLUSION

For all the above reasons, the appellant submits that the judgment of conviction in this case should be reversed.

Respectfully submitted,

GEORGE F. MACDONALD,
Attorney for Appellant.

Received copy of the within
Reply Brief this 4th day of
March, 1937.

F. E. FLYNN
United States Attorney.

No. 8200

1 6

IN THE

United States

Circuit Court of Appeals

For the Ninth Circuit.

CHARLES O. LONG,

Appellant,

vs.

UNITED STATES OF AMERICA,

Appellee.

APPELLANT'S PETITION FOR A REHEARING

and

APPLICATION FOR STAY OF ISSUANCE
OF MANDATE

GEORGE F. MACDONALD,

Attorney for Appellant.

No. 8200

IN THE

United States
Circuit Court of Appeals
For the Ninth Circuit.

CHARLES O. LONG,

Appellant,

vs.

UNITED STATES OF AMERICA,

Appellee.

APPELLANT'S PETITION FOR A REHEARING

and

APPLICATION FOR STAY OF ISSUANCE
OF MANDATE

GEORGE F. MACDONALD,

Attorney for Appellant.

TO THE HONORABLE CURTIS D. WILBUR, PRE-
SIDING JUDGE, AND TO THE ASSOCIATE
JUDGES OF THE UNITED STATES CIRCUIT
COURT OF APPEALS FOR THE NINTH CIR-
CUIT:

The appellant in the above entitled action hereby respectfully petitions for a rehearing of this cause on the following grounds, to-wit:

The Court states in its opinion that "appellee has moved this court to strike out the bill of exceptions and the assignment of errors." It is true that prior to the time of oral argument the appellee had filed its motion to strike the assignment of errors and that motion was fully answered in appellant's reply brief, and undoubtedly considered by the Court when the case was taken under advisement. Our position as to that motion was, and is, that nothing in the Supreme Court Rules requires the appellant to get any extension of time to file his assignment of errors, if he does secure an order extending the time to file his bill of exceptions, and the time for the filing of the assignments necessarily follows the time within which the bill may be properly filed.

The motion to strike the bill of exceptions was not filed for several weeks after the case was argued and submitted. Appellant filed no reply to the latter motion for he was never noticed for any hearing on this motion, although Rule IV of the Supreme Court Rules requires that five days notice be given him, and as appears from the Opinion this motion was entertained by the Court in the absence of any such notice. Upon the face of the record this motion needed no answer, for it mis-stated Rule IX of the Supreme Court Rules in arguing that said rule requires the settlement of a bill of exceptions within thirty days after the taking of an appeal. Rule IX contains no such requirement. Nothing in the record, upon which appellee then relied, supports his contention. For the first time we find in the Opinion that this Court, without any motion or notice to appellant invoking this Court's powers, "called upon the clerk of the District Court to certify the orders of extension."

Counsel for appellee were duly and properly served with appellant's tendered bill of exceptions. They had ample time to have formulated any amendments or additional matters which they deemed necessary to complete the bill of exceptions so as to properly make it a part of the record on appeal. For ten days they sat silent, failed to file any suggested amendments or additions, and permitted the appellant's tendered bill of exceptions to go before the Court as complete and sufficient, and it was so certified. This Court assumes that, notwithstanding this, and notwithstanding that it had formerly held that the sufficiency of the bill as to terms and time fixed by law may be shown in the certificate of the trial judge approving and settling the bill, under the new Criminal Appeals Rules, giving this Court supervisory power over the preparation of the record on appeal, it may without motion or without request disregard the certificate of the trial judge and supplement the record upon which the case was briefed, argued and submitted in this Court, and upon such new record base its decision.

We take serious issue with that position and respectfully submit that Rule IV was never intended by the Supreme Court to confer any such power upon this Court. And as strongly indicative of the correctness of our position, we call the Court's attention to the concluding paragraph of said Rule IV, to the effect that the Court's powers to so act must be predicated upon a motion made by the interested party "to dismiss the appeal, or for directions to the trial court, or to vacate or modify any order made by the trial court, or by any judge in relation to the prosecution of the appeal, including any order for the granting of bail." We are advised by the clerk of the Arizona District Court that he received advice from this Court on May 17, 1937, to forward the orders of extension, and counsel for appellant were not served with appellee's motion to strike the bill of exceptions until May 20, 1937. Clearly, from this, at the time this Court acted in sending for the record there was no motion to strike the bill of exceptions filed and when it was filed, on or about May 21, 1937, the motion, as we have shown, was based solely on the false premise that Rule IX requires the settlement of the bill of exceptions within thirty days after taking the appeal. The motion made no suggestion as to diminution of the record or request for an order supplementing it. The motion to strike the Assignments and the later motion to strike out the bill of exceptions, being based, as they clearly show, upon the record existing at the time each was filed, have no support either in the law applicable to or in the facts shown by that record.

Indeed, the Opinion of this Court, while reciting the fact that appellee has moved the Court to strike out the Bill and the Assignments, does not, as most clearly it could not from anything said in either motion, grant the one or the other.

The remaining point is the Court's brief dismissal (based on the authority of Dunn v. United States, 284 U. S. 390; 76 L. Ed. 356, and Yep v. United States, 81 Fed. (2d) 637) of appellant's contention under his Assignment of Error No. 5 and based on the inconsistency of the verdict.

that the decision in the Dunn case was, as was demonstrated in both the opening and closing brief of appellant, dictum of the purest ray serene. As such, from its acceptance by this Court as controlling, we can only hope that upon a reconsideration this Court will not put the stamp of finality upon such a dictum and interpret it to mean that in any case, however glaring the facts—and they can never be clearer or more glaring than in the instant case—sufficient foundation for a solemn judgment depriving a man of his liberty may be founded upon a verdict so devoid of meaning as that involved here.

We do not hesitate to predict that for whatever inexplicable reason the Supreme Court gave vent to such dictum in the Dunn case, absolutely unnecessary for an affirmance of this Court's decision in that same case, it will never, in a proper case with the simple issue as in this case made, be heard to say that its language there is to be forever understood and accepted as meaning that a jury may say "yes" and "no" to a simple issue of fact given it to determine, may say "guilty" and "not guilty" under the same simple issue, and that from such meaningless and stupid answers the Court may assume that the jury, when it said "no", meant "yes"; and when it said "not guilty", meant "guilty", and that from such a solemn exhibition of, in effect, saying nothing, the Court may proceed with the imposition of judgment and sentence based, as it must be, not on what the jury said, for it said nothing, but upon the Court's interpretation and selection of what it thought the jury ought to have said.

In closing, we cannot subscribe to the limited declaration of the Court that "this question arises on the face of the indictment and verdict." That view would destroy every argument we have made showing the inconsistency of the verdict. And the grounds for our Fifth Assignment, which the Court quotes (Opinion p. 2) immediately preceeding the Court's language as above quoted, show that the question arises, and must arise, on more than the face of the indictment and verdict. If it arose solely on the face of the indictment and verdict, as the Court says, reconcilability is readily apparent, for, as we have labored to show, if we had only the indictment and the verdict to compare, consistency may well be predicated. Something more must be shown to assert inconsistency, for with only the indictment and the verdict as our guide, it was neither repugnant nor inconsistent for the jury to say that appellant did sell the morphine to Etheridge, not in pursuance of a written order, and Etheridge not being then and there his patient; and to say also that he did not sell Etheridge morphine which was not in nor from the original stamped package because the jury may have found the evidence was clear that it was from the original stamped package, or at least there was a reasonable doubt on the question of whether it was or was not.

Irreconcilability follows only by the confession by appellant that there was no written order, that Etheridge was not his patient, that the morphine was not in nor from the original stamped package, and his insistence only that no sale was made by him to Etheridge. And the Court accepted that confession and put the bare, simple issue of the sale to the jury; and it is in its "yes" and "no" and the "guilty" and "not guilty" verdict, based upon that simple issue, that we assert utter inconsistency. And without that we confess no right to be heard.

CONCLUSION

We respectfully submit that in fairness and in justice to the appellant further consideration should be given to this case and we sincerely and respectfully urge this Honorable Court to grant a rehearing and to correct the errors in its printed decision.

Dated at Phoenix, Arizona, this 15th day of July, 1937.

Respectfully submitted,

GEORGE F. MACDONALD,
Attorney for Appellant
and Petitioner.

The undersigned hereby certifies that in his judgment, the foregoing petition for rehearing is well founded and meritorious and that it is not interposed for delay.

Dated at Phoenix, Arizona, this 15th day of July, 1937.

Attorney for Appellant
and Petitioner.

No. 8200

IN THE

United States
Circuit Court of Appeals
For the Ninth Circuit.

CHARLES O. LONG,

Appellant,

vs.

UNITED STATES OF AMERICA,

Appellee.

APPLICATION FOR STAY OF ISSUANCE OF MANDATE

TO THE HONORABLE CURTIS D. WILBUR, PRESIDING JUDGE, AND TO THE ASSOCIATE JUDGES OF THE UNITED STATES CIRCUIT COURT OF APPEALS FOR THE NINTH CIRCUIT:

In the event that this petition for rehearing should be denied, it is the purpose and desire of appellant to apply to the Supreme Court of the United States for the issuance of a writ of certiorari and for that reason application is hereby formally made for a stay of the issuance of mandate of this Honorable Court pending the presentation and determination of such petition for writ of certiorari.

Dated at Phoenix, Arizona, this 15th day of July, 1937.

Respectfully submitted,

GEORGE F. MACDONALD, ₵ℓ-℥

Attorney for Appellant ₵ℓ-9

and Petitioner. ₵ ℓ- ı